SOLAR AND STELLAR MAGNETIC FIELDS: ORIGINS AND MANIFESTATIONS

IAU SYMPOSIUM 354

T0271174

IMAGE CREDIT:

The solar eclipse image was obtained on 2 July 2019 near Tres Cruses, Chile, and processed to visualize coronal structures by Miloslav Druckmüller and Peter Aniol.

IAU SYMPOSIUM PROCEEDINGS SERIES

Chief Editor
MARIA TERESA LAGO, IAU General Secretariat
IAU-UAI Secretariat
98-bis Blvd Arago
F-75014 Paris
France
mtlago@astro.up.pt

Editor
JOSÉ MIGUEL RODRÍGUEZ ESPINOSA, IAU Assistant General Secretary
IAU-UAI Secretariat
98-bis Blvd Arago
F-75014 Paris
France
IAU_AGS@iap.fr

INTERNATIONAL ASTRONOMICAL UNION

UNION ASTRONOMIQUE INTERNATIONALE

SOLAR AND STELLAR MAGNETIC FIELDS: ORIGINS AND MANIFESTATIONS

PROCEEDINGS OF THE 354th SYMPOSIUM OF THE INTERNATIONAL ASTRONOMICAL UNION HELD IN COPIAPO, CHILE
30 June–6 July, 2019

Edited by

ALEXANDER KOSOVICHEV
New Jersey Institute of Technology, USA

KLAUS STRASSMEIER
Leibniz-Institute for Astrophysics Potsdam, Germany

and

MOIRA JARDINE
University of St Andrews, U.K.

CAMBRIDGE
UNIVERSITY PRESS

CAMBRIDGE
UNIVERSITY PRESS

University Printing House, Cambridge CB2 8BS, United Kingdom

One Liberty Plaza, 20th Floor, New York, NY 10006, USA

477 Williamstown Road, Port Melbourne, VIC 3207, Australia

314-321, 3rd Floor, Plot 3, Splendor Forum, Jasola District Centre, New Delhi - 110025, India

79 Anson Road, #06-04/06, Singapore 079906

Cambridge University Press is part of the University of Cambridge.

It furthers the University's mission by disseminating knowledge in the pursuit of education, learning and research at the highest international levels of excellence.

www.cambridge.org
Information on this title: www.cambridge.org/9781108482493

First published 2020

A catalogue record for this publication is available from the British Library

This journal issue has been printed on FSC™-certified paper and cover board. FSC is an independent, non-governmental, not-for-profit organization established to promote the responsible management of the world's forests. Please see www.fsc.org for information.

ISBN 978-1-108-48249-3 Hardback
ISSN 1743-9213

Table of Contents

Preface   xi

Editor   xiii

Conference Photograph   xiv

Participants   xv

Chapter 1. Total Solar Eclipse of 2019

Early results from the solar-minimum 2019 total solar eclipse   3
    Jay M. Pasachoff, Christian A. Lockwood, John L. Inoue,
    Erin N. Meadors, Aristeidis Voulgaris, David Sliski, Alan Sliski,
    Kevin P. Reardon, Daniel B. Seaton, Ronald M. Caplan, Cooper Downs,
    Jon A. Linker, Glenn Schneider, Patricio Rojo and Alphonse C. Sterling

Chapter 2. New observational diagnostics of solar, stellar and interstellar magnetic fields

Diagnosing coronal magnetic fields with radio imaging-spectroscopy
technique   17
    Yihua Yan, Baolin Tan, V. Melnikov, Xingyao Chen, Wei Wang,
    Linjie Chen, Fei Liu and MUSER Team

Observing the Sun with the Atacama Large Millimeter/submillimeter
Array – from continuum to magnetic fields   24
    Sven Wedemeyer, Mikolaj Szydlarski, Jaime de la Cruz Rodriguez and
    Shahin Jafarzadeh

Revisiting the building blocks of solar magnetic fields by GREGOR   38
    Dominik Utz, Christoph Kuckein, Jose Iván Campos Rozo,
    Sergio Javier González Manrique, Horst Balthasar, Peter Gömöry,
    Judith Palacios Hernández, Carsten Denker, Meetu Verma,
    Ioannis Kontogiannis, Kilian Krikova, Stefan Hofmeister
    and Andrea Diercke

Ca II 854.2 nm spectropolarimetry compared with ALMA and with scattering
polarization theory   42
    J. W. Harvey and SOLIS Team

Diagnosing chromospheric magnetic field through simultaneous spectropolarimetry
in Hα and Ca II 854.2 nm   46
    K. Nagaraju, K. Sankarasubramanian and K. E. Rangarajan

The magnetic structure and dynamics of a decaying active region   53
    Ioannis Kontogiannis, Christoph Kuckein,
    Sergio Javier González Manrique, Tobias Felipe, Meetu Verma,
    Horst Balthasar and Carsten Denker

Coordinated observations between China and Europe to follow active
region 12709   58
    S. J. González Manrique, C. Kuckein, P. Gömöry, S. Yuan, Z. Xu,
    J. Rybák, H. Balthasar and P. Schwartz

Chapter 3. Progress in understanding the solar/stellar interior dynamics and dynamos

Global simulations of stellar dynamos   65
    G. Guerrero

3D Modeling of the Structure and Dynamics of a Main-Sequence
F-type Star   86
    Irina N. Kitiashvili and Alan A. Wray

Helioseismic insights into the generation and evolution of the Sun's internal
magnetic field   94
    Anne-Marie Broomhall and René Kiefer

Resolving Power of Asteroseismic Inversion of the Kepler Legacy Sample   107
    Alexander G. Kosovichev and Irina N. Kitiashvili

Cycle times of early M dwarf stars: mean field models versus observations   116
    Manfred Küker, Günther Rüdiger, Katalin Oláh and Klaus Strassmeier

Searching for the cycle period in chromospherically active stars   120
    F. Villegas, R. E. Mennickent and J. Garcés

Are there local dynamo in solar polar region?   123
    Chunlan Jin

A Clock in the Sun?   127
    C. T. Russell, J. G. Luhmann and L. K. Jian

Various scenarios for the equatorward migration of sunspots   134
    Detlef Elstner, Yori Fournier and Rainer Arlt

A solar cycle 25 prediction based on 4D-var data assimilation approach   138
    Allan Sacha Brun, Ching Pui Hung, Alexandre Fournier, Laurène Jouve,
    Olivier Talagrand, Antoine Strugarek and Soumitra Hazra

Global Evolution of Solar Magnetic Fields and Prediction of Activity Cycles . . .   147
    Irina N. Kitiashvili

Solar Open Magnetic Flux Migration Pattern over Solar Cycles   157
    Chia-Hsien Lin, Guan-Han Huang and Lou-Chuang Lee

Probing solar-cycle variations of magnetic fields in the convection zone using
meridional flows   160
    Chia-Hsien Lin and Dean-Yi Chou

Chapter 4. Stellar rotation and magnetism

Magnetic field evolution in solar-type stars   169
    Axel Brandenburg

Magnetic field and prominences of the young, solar-like, ultra-rapid
rotator AP 149 181
 Tianqi Cang, Pascal Petit, Colin Folsom and Jean-Francois Donati

Dipolar stability in spherical simulations: The impact of an inner
stable zone 185
 Bonnie Zaire and Laurène Jouve

A large rotating structure around AB Doradus A at VLBI scale 189
 J. B. Climent, J. C. Guirado, R. Azulay and J. M. Marcaide

The impact of magnetism on tidal dynamics in the convective envelope of
low-mass stars 195
 A. Astoul, S. Mathis, C. Baruteau, F. Gallet, A. Strugarek,
 K. C. Augustson, A. S. Brun and E. Bolmont

The rotation of low mass stars at 30 Myr in the cluster NGC 3766 200
 Julia Roquette, Jerome Bouvier, Estelle Moraux, Herve Bouy,
 Jonathan Irwin, Suzanne Aigrain and Régis Lachaume

Chapter 5. Role of magnetic fields in solar and stellar variability

Possible evidence for a magnetic dynamo in hot Algols 207
 R. E. Mennickent, J. Garcés, G. Djurašević, G. Rojas, D. Schleicher
 and S. Otero

New Candidates for Chromospherically Young, Kinematically Old Stars 211
 Eduardo Machado Pereira and Helio J. Rocha Pinto

The dynamo-wind feedback loop : Assessing their non-linear interplay 215
 Barbara Perri, Allan Sacha Brun, Antoine Strugarek and Victor Réville

Statistical analysis of geomagnetic storms and their relation with the
solar cycle 224
 Paula Reyes, Victor A. Pinto and Pablo S. Moya

Examining the optical intensity and magnetic field expansion factor in the open
magnetic field regions associated with coronal holes 228
 Chia-Hsien Lin, Guan-Han Huang and Lou-Chuang Lee

Solar oblateness & asphericities temporal variations: Outstanding some unsolved
issues 232
 Jean P. Rozelot, Alexander G. Kosovichev and Ali Kilcik

Chapter 6. Star-planet relations

Solar activity influences on planetary atmosphere evolution: Lessons from
observations at Venus, Earth, and Mars 241
 J. G. Luhmann

Different types of star-planet interactions 259
 A. A. Vidotto

Influence of the magnetic field of stellar wind on hot jupiter's envelopes 268

Dmitry V. Bisikalo and Andrey G. Zhilkin

Star-planet interaction through spectral lines 280

C. Villarreal D'Angelo, A. A. Vidotto, A. Esquivel, M. A. Sgró, T. Koskinen and L. Fossati

From the Sun to solar-type stars: radial velocity, photometry, astrometry and $\log R'_{HK}$ time series for late-F to early-K old stars 286

Nadège Meunier and Anne-Marie Lagrange

Could star-planet magnetic interactions lead to planet migration and influence stellar rotation ? 295

Jérémy Ahuir, Antoine Strugarek, Allan-Sacha Brun, Stéphane Mathis, Emeline Bolmont, Mansour Benbakoura, Victor Réville and Christophe Le Poncin-Lafitte

TESS light curves of low-mass detached eclipsing binaries 300

Krzysztof G. Hełminiak, Andrés Jordán, Nestor Espinoza and Rafael Brahm

Tuning in to the radio environment of HD189733b 305

R. D. Kavanagh, A. A. Vidotto, D. Ó Fionnagáin, V. Bourrier, R. Fares, M. Jardine, Ch. Helling, C. Moutou, J. Llama and P. J. Wheatley

Chapter 7. Formation, structure and dynamics of solar and stellar coronae and winds

Observational constraints for solar-type stellar winds 313

Manuel Güdel

Semi-empirical 2D model of the solar corona and solar wind using solar eclipse images: Progress report 333

Edward C. Sittler Jr. and Linda M. Sittler

Realistic 3D MHD modeling of self-organized magnetic structuring of the solar corona 346

Irina N. Kitiashvili, Alan A. Wray, Viacheslav Sadykov, Alexander G. Kosovichev and Nagi N. Mansour

Coherent structures and magnetic reconnection in photospheric and interplanetary magnetic field turbulence 351

Rodrigo A. Miranda, Abraham C.-L. Chian, Erico L. Rempel and Suzana S. A. Silva

Analysis of the chromosphere and corona of low-activity early-M dwarfs. 355

Gaetano Scandariato, E. González Álvarez, J. Maldonado, A. Suárez Mascareño, M. Perger and the HADES collaboration

Reversibility of Turbulent and Non-Collisional Plasmas: Solar Wind 363

Belén Acosta, Denisse Pastén and Pablo S. Moya

Temporal evolution of the velocity distribution in systems described by
the Vlasov equation; Radiation Belts: Analytical and computational results . . . 367
> *Abiam Tamburrini C, Iván Gallo-Méndez, Sergio Davis and Pablo S. Moya*

On the multifractality of plasma turbulence in the solar wind 371
> *Sebastián Echeverría, Pablo S. Moya and Denisse Pastén*

Chapter 8. Mechanisms of flaring and CME activity on the Sun and stars

The UV/X-ray radiation fields and particle (CME) flows of M dwarf
exoplanet host stars 377
> *Alexander Brown*

Exploring Flaring Behaviour on Low Mass Stars, Solar-type Stars and
the Sun 384
> *L. Doyle, G. Ramsay, J.G. Doyle, P. F. Wyper, E. Scullion, K. Wu and*
> *J. A. McLaughlin*

Trigger mechanisms of the major solar flares 392
> *Shuhong Yang*

(Simulating) Coronal Mass Ejections in Active Stars 407
> *Julián D. Alvarado-Gómez, Jeremy J. Drake, Cecilia Garraffo,*
> *Sofia P. Moschou, Ofer Cohen, Rakesh K. Yadav and*
> *Federico Fraschetti*

Diagnostics of non-thermal-distributions from solar flare EUV line spectra 414
> *Elena Dzifčáková, Alena Zemanová, Jaroslav Dudík,*
> *and Juraj Lörinčík*

Linking radio flares with spots on the active binary UX Arietis 418
> *Christian A. Hummel and Anthony Beasley*

CME deflections due to magnetic forces from the Sun and Kepler-63 421
> *F. Menezes, Y. Netto, C. Kay, M. Opher and A. Valio*

Coronal dimming as a proxy for stellar coronal mass ejections 426
> *M. Jin, M. C. M. Cheung, M. L. DeRosa, N. V. Nitta,*
> *C. J. Schrijver, K. France, A. Kowalski, J. P. Mason and R. Osten*

Chapter 9. Surface magnetic fields of the Sun and stars

On the properties of the magnetic Chemically Peculiar B, A, and
F-type stars 435
> *Kutluay Yüce, Saul J. Adelman, Diane M. Pyper*
> *and Robert J. Dukes*

Impact of small-scale emerging flux from the photosphere to the corona: a case
study from IRIS 439
> *Salvo L. Guglielmino, Peter R. Young, Francesca Zuccarello,*
> *Paolo Romano and Mariarita Murabito*

Multi-flux-rope system in solar active regions 443

    Yijun Hou, Jun Zhang, Ting Li and Shuhong Yang

The 3D structure of the penumbra at high resolution from the bottom of the
photosphere to the middle chromosphere 448

    Mariarita Murabito, Ilaria Ermolli, Fabrizio Giorgi,
    Marco Stangalini, Salvo L. Guglielmino, Shahin Jafarzadeh,
    Hector Socas-Navarro, Paolo Romano and Francesca Zuccarello

On the Role of Magnetic Fields in an Erupting Solar Filament 452

    Qiao Song, Shuhong Yang and Jing-Song Wang

Fast downflows in a chromospheric filament 454

    K. Sowmya, A. Lagg, S. K. Solanki and J. S. Castellanos Durán

Chapter 10. Observations of solar eclipses and exoplanetary transits

Characterization of stellar activity using transits and its impact on
habitability 461

    Raissa Estrela, Adriana Valio and Sourav Palit

Discovering the atmospheres of hot Jupiters 467

    P. Wilson Cauley

Sun-as-a-star observations of the 2017 August 21 solar eclipse 473

    Ekaterina Dineva, Carsten Denker, Meetu Verma, Klaus G. Strassmeier,
    Ilya Ilyin and Ivan Milic

Solar astrometry with planetary transits 481

    Marcelo Emilio, Rock Bush, Jeff Kuhn and Isabelle Scholl

Author Index 495

Preface

Recent observational results from space and ground-based telescopes have convincingly demonstrated that the progress in our understanding of how magnetic fields are generated, how they emerge from the interior, organize in active regions, and cause powerful eruptions can be achieved only by developing a unified approach from relationships between solar and stellar magnetism. Developing a general synergy of solar and stellar astronomy is essential for solving grand-challenge problems like the primary mechanisms of magnetic activity and its impact on planetary atmospheres. An important key issue is that the same or similar phenomena occur on the Sun and other stars under different conditions (different mass, age, metallicity, rotation rate, etc.). Studying these similarities and differences helps to uncover the underlying physical mechanisms of magnetic activity, its evolution in time, and its impacts on planetary environments.

The Proceedings presents recent results and discussions of new emerging topics that include magnetic field diagnostics using high-resolution observation; initial data from ALMA, Chinese Radio Spectroheliograph and other instruments; detection of stellar magnetospheres; detailed mapping of the magnetic fields on the surface of stars using new unique instrumentation, such as the PEPSI spectrograph that provided first high-resolution spectropolarimetry with a 12m telescope. The new observations stimulate comparisons of solar and stellar results, and advance our understanding of how surface magnetic structures and their evolution are related to the generation of magnetic fields by dynamos in solar and stellar interiors.

In this respect, tremendous progress has been achieved from helioseismology and asteroseismology with data from SDO, Kepler, and TESS, as well as from synoptic observations of solar and stellar variability. Discussions of the current long-term trend of declining solar activity and its initial results on the prediction of the next solar cycles are among the hot topics. The new picture of stellar cycles that is emerging from analysis of the Kepler and supporting ground-based spectroscopic data reveals scaling laws and relations that need to be taken into account in solar magnetism studies. Recent theoretical studies based on advanced supercomputer simulations have demonstrated the key role of magnetism for establishing solar and stellar differential rotation laws, and the importance of observational tests to validate theoretical predictions.

One of the puzzles of solar and stellar magnetism is related to the origin of extreme flare events. During the last weak magnetic cycle, the Sun produced some of the strongest flares in the history of observations. This raises questions on how the flare energetics are related to the magnetism of other stars that produce giant superflares, and what physical mechanism drives such extreme events. Another important topic of joint solar-stellar studies is the influence of solar and stellar variability on planetary space environments which become more and more important.

The interest in understanding the role of stellar magnetism in star-planet relations is driven by the need to determine conditions for habitability. In this aspect, the discussion is focused on properties of solar and stellar coronae and winds, and their interactions with planetary magnetospheres. Compared to the solar system, in many recently discovered planetary systems stellar winds are substantially stronger, and planets are much closer to their parent stars. This creates extreme conditions for magnetic interactions and radiation environments, which depend on the state of stellar magnetic activity. The discussion of this renewed old problem, that is beyond traditional studies, raises interest in understanding the broader impacts of magnetic activity on planetary space weather and habitability.

These Proceedings present recent advances and key problems of solar and stellar magnetic fields and their impact on planetary atmospheres, discussed at the IAU Symposium 354 "Solar and Stellar Magnetic Fields: Origins and Manifestations", from June 30–July 6, 2019. The Symposium was organized in conjunction with the Total Solar Eclipse of July 2, 2019. The opening paper in Chapter 1 presents the initial observational results of this eclipse. Chapter 2 is focused on new observational diagnostics of solar magnetic fields. The progress in understanding the solar and stellar interior dynamics and dynamos is discussed in Chapter 3. Chapter 4 is devoted to investigations of relationships between stellar rotation and magnetism. The role of magnetic fields in solar and stellar variability is discussed in Chapter 5. Star-planet relations are discussed in Chapter 6. Chapter 7 is focused on the problem of the formation of solar and stellar coronae and winds. The progress in the understanding of solar and stellar flares and coronal mass ejections is presented in Chapter 8. Some key aspects of magnetic field structures and dynamics on the surface of the Sun and other stars are described in Chapter 9. The final Chapter 10 discusses the role of observations of solar eclipses and exoplanetary transits for characterization of solar and stellar activity and its impacts on the habitability of exoplanets.

The Symposium was organized in close cooperation and support of the University of Atacama, other Chilean universities, as well as of local authorities of the city of Copiapo. In particular, we thank Luis Campusano, Natalie Huerta, Pablo Moya, and Giovanni Leone for their support, enthusiasm, dedication, and hard work that made the IAUS 354 such as success.

Editor

Alexander Kosovichev
New Jersey Institute of Technology, USA

Klaus Strassmeier
Leibniz-Institute for Astrophysics Potsdam, Germany

Moira Jardine
University of St Andrews, U.K.

Organizing Committee
Scientific Organizing Committee

Alexander Kosovichev (co-chair)	New Jersey Institute of Technology, USA
Klaus Strassmeier (co-chair)	Leibniz-Institute for Astrophysics Potsdam, Germany
Moira Jardine (co-chair)	University of St Andrews, U.K.
Pablo Moya (co-chair)	Physics Department, U of Chile
Mauro Barbieri	Universidad de Atacama, Chile
Alfio Bonanno	Catania Astrophysical Observatory, Italy
Dmitry Bisikalo	Institute of Astronomy, Russia
Elisabete de Gouveia Dal Pino	University of São Paulo, Brazil
Margit Haberreiter	PMOD/WRC Davos, Switzerland
Ahmed Abdel Hady	Cairo University, Egypt
Elena Khomenko	Instituto de Astrofisica de Canarias, Spain
Tetsuya Magara	Kyung Hee University, South Korea
Nadege Meunier	University of Grenoble, France
Marina Stepanova	Physics Department, U of Santiago, Chile
Sergio Flores Tullian	Universidad de Atacama, Chile
Santiago Vargas Dominguez	National University of Colombia, Colombia
Michael Wheatland	University of Sydney, Australia
Jun Zhang	National Astronomical Observatory, China

Local Organizing Committee

Luis Campusano (co-chair)	Astronomy Department of Universidad de Chile
Mauro Barbieri (co-chair)	Universidad de Atacama
Zhong Wang (co-chair)	NAOC Astronomy Office in Santiago
Jaime Araneda	Physics Department, U of Concepción
Cristobal Espinoza	Physics Department, U of Santiago
Giovanni Leone	Universidad de Atacama
Ronald Mennickent	Astronomy Department – University of Concepción
Lorenzo Morelli	Universidad de Atacama
Pablo Moya	Physics Department, U of Chile
Roberto Navarro	Physics Department, U of Concepción
Andreas Reisenegger	Astronomy Department, U Católica
Dominik Schleicher	Astronomy Department, U of Concepción
Mario Soto	Universidad de Atacama
Sergio Flores Tullian	Universidad de Atacama
Alejandro Valdivia	Physics Department, U of Chile
Wei Wang	NAOC Astronomy Office in Santiago

CONFERENCE PHOTOGRAPH

PARTICIPANTS

First Name	Last Name	Country
Alicia	Aarnio	USA
Belén	Acosta	Chile
Jérémy	Ahuir	France
Julián David	Alvarado Gómez	USA
Louis	Amard	UK
Eliana	Amazo-Gomez	Germany
Aurélie	Astoul	France
Graham	Barnes	USA
Luis	Bellot Rubio	Spain
Christopher	Bert	USA
Lionel	Bigot	France
Dmitry	Bisikalo	Russia
Veronique	Bommier	France
Axel	Brandenburg	Sweden
Anne-Marie	Broomhall	UK
Alexander	Brown	USA
Allan Sacha	Brun	France
Paul	Cally	Australia
Luis	Campusano	Chile
Tianqi	Cang	France
Thorsten	Carroll	Germany
Paul	Cauley	USA
Juan Bautista	Climent Oliver	Spain
Manuel	Collados	Spain
Simon	Daley-Yates	France
Jaime	de la Cruz Rodriguez	Sweden
Marc	DeRosa	USA
Andrea	Diercke	Germany
Ekaterina	Dineva	Germany
Lauren	Doyle	UK
Elena	Dzifčáková	Czech Republic
Sebastian	Echeverrua	Chile
Henrik	Eklund	Norway
Detlef	Elstner	Germany
Marcelo	Emilio	Brazil
Raissa	Estrela	USA
Adam	Finley	UK
Iván	Gallo	Chile
Marta	Garca Rivas	Czech Republic
Manuel	Güdel	Austria
Gustavo	Guerrero	Brazil
Salvo	Guglielmino	Italy
John	Harvey	USA
Krzysztof	Helminiak	Poland
Aleida	Higginson	USA
Yijun	Hou	China
Christian	Hummel	Germany
Patryk	Iwanek	Poland
Moira	Jardine	UK
Meng	Jin	USA
Chunlan	Jin	China
Christoffer	Karoff	Denmark
Robert	Kavanagh	Ireland
Elena	Khomenko	Spain
Irina	Kitiashvili	USA

First Name	Last Name	Country
Lucia	Kleint	Germany
Ioannis	Kontogiannis	Germany
Kelly	Korreck	USA
Alexander	Kosovichev	USA
Matthieu	Kretzschmar	France
Sowmya	Krishnamurthy	Germany
Nagaraju	Krishnappa	India
Christoph	Kuckein	Germany
Manfred	Küker	Germany
Błażej	Kuźma	Poland
Zhi-Chao	Liang	Germany
Chia-Hsien	Lin	Taiwan
Joe	Llama	USA
Janet	Luhmann	USA
Eduardo	Machado-Pereira	Brazil
Fernando	Marques	Brazil
Ralph	McNutt	USA
Amber	Medina	USA
Jorge	Melendez	Brazil
Fabian	Menezes	Brazil
Ronald	Mennickent	Chile
Nadege	Meunier	France
Andrea	Miglio	UK
Rodrigo	Miranda	Brazil
David	Montes	Spain
Alberto Jose	Morales-Rodriguez	Costa Rica
Roberta	Morosin	Sweden
Pablo	Moya	Chile
Mariarita	Murabito	Italy
Aimee	Norton	USA
Dualta	O Fionnagain	Ireland
Carlos	Orquera-Rojas	Chile
Mayukh	Panja	Germany
Jay	Pasachoff	USA
Barbara	Perri	France
Pascal	Petit	France
Alexander	Pietrow	Sweden
Geisa	Ponte	Brazil
Renzo	Ramelli	Switzerland
Paula	Reyes	Chile
Alexis Eder	Rodríguez Quiros	Peru
Bonnie	Romano Zaire	France
Julia	Roquette	UK
Jean Pierre	Rozelot	France
Christopher	Russell	USA
Aoife Maria	Ryan	Ireland
Antonia	Savcheva	USA
Gaetano	Scandariato	Italy
Caius	Selhorst	Brazil
Douglas	Silva	Brazil
Javier	Silva	Chile
Edward	Sittler	USA
Sami	Solanki	Germany
Qiao	Song	China
Klaus	Strassmeier	Germany
Antoine	Strugarek	France
Alejandro	SuárezMascareño	Spain

First Name	Last Name	Country
Ambiam	Tamburrini	Chile
Javier	Trujillo Bueno	Spain
Adriana	Valio	Brazil
Aline	Vidotto	Ireland
Carolina	Villarreal D'Angelo	Ireland
Fabricio	Villegas	Chile
Mariangela	Viviani	Germany
Maria	Weber	USA
Sven	Wedemeyer	Norway
Thomas	Wilson	UK
Yihua	Yan	China
Shuhong	Yang	China
Kutluay	Yuce	Turkey
Tomas	Zurita	Chile

Chapter 1. Total Solar Eclipse of 2019

Observing the solar eclipse

Solar and Stellar Magnetic Fields: Origins and Manifestations
Proceedings IAU Symposium No. 354, 2019
A. Kosovichev, K. Strassmeier & M. Jardine, ed.
doi:10.1017/S1743921320001453

Early results from the solar-minimum 2019 total solar eclipse

Jay M. Pasachoff[1], Christian A. Lockwood[1], John L. Inoue[1],
Erin N. Meadors[1], Aristeidis Voulgaris[2], David Sliski[3], Alan Sliski[4],
Kevin P. Reardon[5,6], Daniel B. Seaton[7], Ronald M. Caplan[8],
Cooper Downs[8], Jon A. Linker[8], Glenn Schneider[9], Patricio Rojo[10]
and Alphonse C. Sterling[11]

[1]Williams College–Astronomy Department, Williamstown, MA 01267-2565, USA

[2]Icarus Optomechanics, Thessaloniki, Greece

[3]Astronomy Department, U. Pennsylvania, Philadelphia, PA 19104, USA

[4]273 Concord Road, Lincoln, MA 01773, USA

[5]National Solar Observatory, Boulder, CO 80303, USA

[6]Astrophysics and Planetary Sciences Dept., University of Colorado, Boulder, CO 80303, USA

[7]Cooperative Institute for Research in Environmental Sciences, Univ. of Colorado, and NOAA
National Centers for Environmental Information, Boulder, Colorado, USA

[8]Predictive Science Inc., San Diego, CA 92121, USA

[9]Steward Observatory, U. Arizona, Tucson, AZ 85721, USA

[10]Astronomy Department, U. Chile, Santiago, Chile

[11]NASA Marshall Space Flight Center, Huntsville, AL 35812

Abstract. We observed the 2 July 2019 total solar eclipse with a variety of imaging and spectro-scopic instruments recording from three sites in mainland Chile: on the centerline at La Higuera, from the Cerro Tololo Inter-American Observatory, and from La Serena, as well as from a chartered flight at peak totality in mid-Pacific. Our spectroscopy monitored Fe X, Fe XIV, and Ar X lines, and we imaged Ar X with a Lyot filter adjusted from its original H-alpha bandpass. Our composite imaging has been compared with predictions based on modeling using magnetic-field measurements from the pre-eclipse month. Our time-differenced sites will be used to measure motions in coronal streamers.

Keywords. Sun: corona, eclipses, instrumentation: spectrographs

---

1. Introduction

We tackled the observations of the 2 July 2019 total solar eclipse, which occurred at extreme solar minimum, with a variety of imaging and spectroscopy tools, following surveys of recent coronal research (Pasachoff 2017a; Pasachoff & Fraknoi 2017).

General background for eclipse studies has been available over a span of years (Pasachoff 1973, 2017b; Golub & Pasachoff 2014; Golub & Pasachoff 2017), with a more technical treatment in Golub & Pasachoff (2009). Observational techniques were discussed in Pasachoff (2019).

Maps showing the path of totality across the Earth's surface have been computed since the work of Edmond Halley for the eclipse of 1715 (Olson & Pasachoff 2019). In spite of worries about the prospective cloudiness or the marine layer†, especially

† http://eclipsophile.com

Figure 1. Totality in clear sky from our site above La Higuera on totality's centerline.

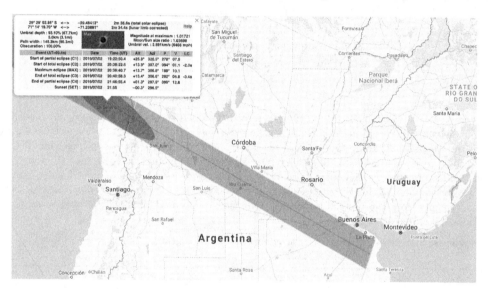

Figure 2. Orientation map showing details for our centerline site (Courtesy of Xavier Jubier and Google maps).

given that totality occurred with the Sun only 13° above the western horizon†, we have observations in clear skies from our three ground-based observing sites: (1) The Cerro Tololo Inter-American Observatory, 7,240-foot altitude, 2 min 6 sec of totality; (2) La Higuera, centerline, 2,500-foot altitude, 2 min 35 sec of totality (Figures 1 and 2); (3) La Serena, sea level, 2 min 25 sec of totality. Prominences on the limb provided orientation and coordination with spacecraft observations from NOAA's GOES-16 Solar Ultraviolet Imager (SUVI) and the Atmospheric Imaging Assembly (AIA) on NASA's Solar Dynamics Observatory (SDO). We also have imaging and spectroscopy from a chartered Boeing 787 along the centerline, with nearly 8 min 30 sec of totality!

† http://xjubier.free.fr/en/site_pages/solar_eclipses/xSE_GoogleMap3.php?Ecl=+20190702

Figure 3. Sunset view on 1 July 2019 from the La Higuera site (courtesy Ian Kezsbom).

Figure 4. Drone views of the La Higuera site.

From our sites for the 2019 eclipse, we planned to expand on our observations from the 2017 American eclipse, though without the coronal-oscillation observations in the Fe XIV and Fe X coronal lines (the "green line" and the "red line," respectively), because we worried that oscillations in the terrestrial atmosphere would overwhelm the slight effect in coronal intensity that we are monitoring. For our 2017 observations, see Pasachoff (2018); Pasachoff *et al.* (2018).

2. Observations from the centerline at La Higuera

For about two years prior to totality, we had planned our observations from the centerline, and our travel agent, Mark Sood, reconnoitered at La Higuera and chose a site on a high ridge overlooking the town.

Our scientific team consisted of JMP and three Williams College students, using equipment sent to La Serena with the assistance of the Kitt Peak National Observatory and the Cerro Tololo Inter-American Observatory. With the clear weather and absent marine layer the day before the eclipse (Figure 3), we had confidence that we would be able to observe totality (Figure 4).

Geosynchronous weather satellite GOES-16 showed that the umbra reached a cloud-free region of Chile (Figure 5). The new series of Geostationary Operational

Figure 5. A view of the Earth showing the umbra approaching the Chilean Pacific coast, from NOAA GOES-16 satellite.

Figure 6. A 2 July 2019 eclipse-day view of 50,000-kelvin chromospheric gas (He II, 304 Å) from SUVI on GOES-16, eclipsed with prominences showing, centered in an early composite of our white-light eclipse images (courtesy Daniel B. Seaton and NOAA/U Colorado CIRES).

Environmental Satellites, including GOES-16, includes a Solar Ultraviolet Imager (SUVI) on each, carrying a sun-pointing set of telescopes that take ultraviolet images of continuum and corona (Figure 6). We also continue our interest in the effect of the extreme eclipse darkening and cooling on the terrestrial atmosphere, continuing our joint work with Marcos Peñaloza-Murillo of Universidad de Los Andes, Mérida, Venezuela (Peñaloza-Murillo & Pasachoff 2018).

The solar cycle was in extreme minimum phase, with 111 days (61%) of 2019 prior to the eclipse showing no sunspots (Figure 7).

Our imaging includes series with an Astro-Physics 630 mm refracting telephoto, courtesy of Dan Schechter, Long Beach, CA, with a Nikon D850, courtesy of Nikon Professional Services (Figure 8).

On our site, we also had a team from Yunnan Observatory, China, headed by Zhongquan Qu, and Alphonse Sterling of NASA's Marshall Space Flight Center,

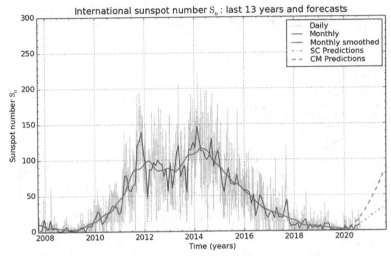

Figure 7. The recent sunspot cycle, from the Solar Indices Data Center (http://sidc.oma.be/silso) at the Royal Observatory, Belgium, 2020.

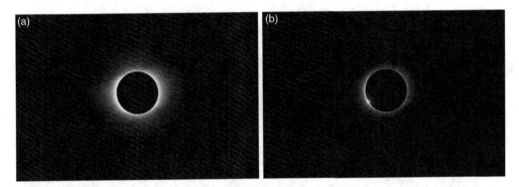

Figure 8. The eclipse seen from La Higuera during (a) totality; (b) the second diamond ring.

Huntsville, AL, in addition to a tour group of about 100 people, arrangements for whom helped in the logistics. The morning drive from La Serena took about 1 hour, while the evening return to La Serena, because of the large number of tourists who had come from many parts of Chile, took over 5 hours.

3. Observations from the Cerro Tololo Inter-American Observatory

Nine months before totality, five teams of up to four scientists each were awarded the opportunity to observe the eclipse from the Cerro Tololo Inter-American Observatory†. Four of the teams carried out coronal experiments, while a fifth was studying the effect of the eclipse on the Earth's atmosphere and ionosphere. Our team was headed by Williams College alumnus Kevin Reardon from the National Solar Observatory, and also included instrumentation specialist Aristeidis Voulgaris from Greece and father-and-son instrumentationalists Alan Sliski from Lincoln, MA, USA, and David Sliski from U. Pennsylvania. The site (2,207 m = 7,241 ft) sacrificed a half minute of totality (duration there was 2 min 6 sec, see Figure 9) in a trade-off for facilities and altitude (Figure 10). In the event, the sky was exceptionally clear for totality (Figure 11). With a view out

† http://www.ctio.noao.edu/noao/node/14748

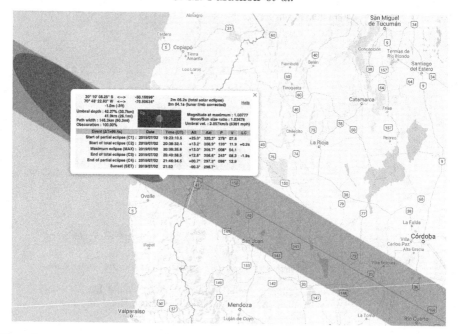

Figure 9. Map showing details for our Cerro Tololo site (courtesy of Xavier Jubier and Google maps).

Figure 10. The sky with finger-tip occulting of the Sun at eclipse time one day before totality. Conditions on the day of the eclipse were very similar.

over the ocean, the shadow could be seen to move over the city of La Serena, over 50 km away, several tens of seconds before it arrived at Cerro Tololo.

The double-diamond ring that appeared at second contact (Figure 12) will extend our determination at the 2017 American total solar eclipse of a new IAU-recommended value of the solar diameter through comparison with simulations (Pasachoff *et al.* 2017). The details of the lunar limb used for the simulations are now available from observations by the Japanese Kaguya spacecraft and the American Lunar Reconnaissance Orbiter.

Figure 11. Totality from Cerro Tololo was observed in an extremely dark and clear sky.

Figure 12. Second contact showed a double diamond ring.

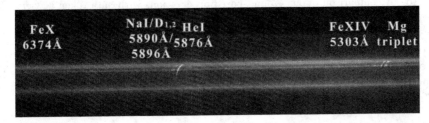

Figure 13. A slitless spectrum from the Cerro Tololo Inter-American Observatory.

Our coronal spectra from slitless spectrographs (Figure 13), from CTIO, showed the Fe XIV 530.3 nm green line substantially weaker than the Fe X 637.4 nm red line, corresponding to the relatively low coronal temperature at this phase of the solar-activity cycle.

On the spectra we also detected the weak coronal emission line of Ar X at 553.3 nm, as we also detected at the previous total solar eclipse of August 21, 2017, in the USA. We also obtained on-band and off-band images of the corona in the Ar X line (Figure 14), using a Lyot filter transformed by Voulgaris from an H-alpha filter borrowed from the New Jersey Institute of Technology's Big Bear Solar Observatory.

We again worked with a theoretical team from Predictive Science Inc. (PSI) in San Diego to compare our observations with their prediction released days before totality, as we published for the 2017 American eclipse (Mikić *et al.*. 2018). The 2019 prediction of the structure of the corona from an MHD model (Figure 15) carried out by PSI compares well with a composite of our images. This was presented within two days after

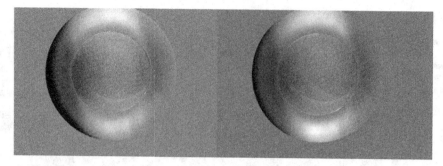

Figure 14. On-band and off-band images in Ar X, data still to be reduced.

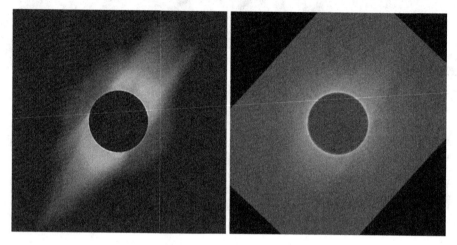

Figure 15. A comparison of our ground-based observations with data taken at Cerro Tololo (*right*), with pre-eclipse prediction produced by Predictive Sciences (*left*).

the eclipse in a NASA on-line display that allowed users to move a sliding vertical bar to transition between the prediction and the observed image to aid in the comparison†. We also reported the comparison to an American Astronomical Society meeting (Pasachoff *et al.* 2020; Lockwood *et al.* 2020).

4. Observations from La Serena

The city of La Serena, headquarters for several international observatories, was well within the path of totality (Figure 16).

In spite of months of worrying about the potential marine layer, the sky was clear (Figure 17). The corona and diamond rings were perfectly visible (Figure 18).

5. Observations from the e-flight on a Boeing 787-9

In a collaboration with Glenn Schneider, Voulgaris sent a spectrograph and Pasachoff sent a telephoto aloft for 8 minutes and 27 seconds of totality from mid-Pacific (Figure 19). The over 4 minutes of totality available at the intercept point were extended beyond 8 minutes by the aircraft keeping partial pace with the lunar umbral shadow‡.

† https://www.nasa.gov/feature/how-scientists-used-nasa-data-to-predict-appearance-of-july-2-eclipse

‡ Glenn Schneider, http://nicmosis.as.arizona.edu:8000/ECLIPSE_WEB/TSE2019/TSE2019_EFLIGHTMAX.html

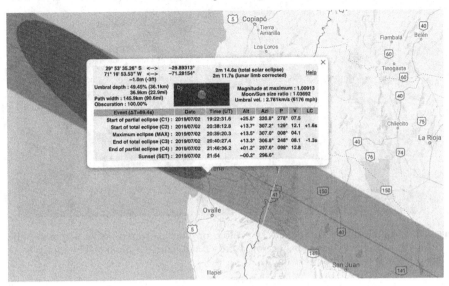

Figure 16. Orientation map showing details for La Serena (courtesy of Xavier Jubier and Google maps).

Figure 17. Westward totality view from La Serena. Credit: Ian Kezsbom

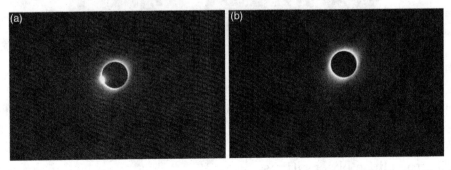

Figure 18. (a) The first diamond ring. (b) A still image of the corona. Credit: Ian Kezsbom and Sam Glaisyer.

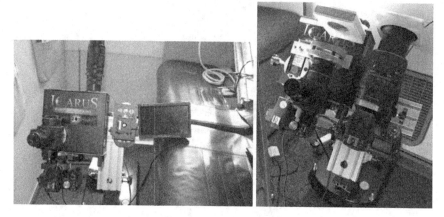

Figure 19. Collaborative equipment sent aloft on the chartered 787-9 airplane. The NIR spectrograph and the telephoto lens are on a motorized azimuth mount. The image to the left was taken during a pre-eclipse test flight.

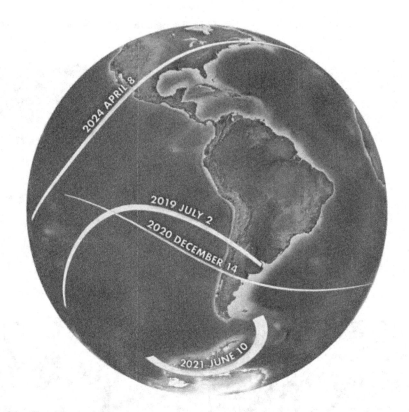

Figure 20. Totality paths in the western hemisphere of Earth through the American eclipse of 2024. Credit: Michael Zeiler, https://www.GreatAmericanEclipse.com.

6. Future eclipse observations

Totality in both 2019 and 2020 hits land in Chile and Argentina. Attempts to observe the 2019 total solar eclipse from Oeno Island in the Pitcairn group failed because of clouds. One cruise ship out of Tahiti did succeed in observing totality.

The peak of the 2020 totality will be over Argentina. The 2021 eclipse will be visible on ocean or land only in regions with poor cloudiness statistics†. The 2023 totality, not shown on this hemispherical map, will clip the westernmost protrusion of Australia and go over East Timor. The 2024 totality will hit land at Mazatlán, Mexico, and proceed over Mexico, the central and northeastern United States, and eastern Canada (Figure 20).

Our IAU Working Group on Solar Eclipses has colleagues from all over the world, and includes colleagues who make maps and predictions, as well as consult on safe observing. Our website at http://eclipses.info, an easy-to-remember URL, has useful links. Members are: scientists: Jay Pasachoff (USA, Chair), Iraida Kim (Russia), Hiroki Kurokawa (Japan), Jagdev Singh (India), Vojtech Rusin (Slovakia), Yoichiro Hanaoka (Japan), Zhongquan Qu (China), Beatriz Garcia (Argentina), Patricio Rojo (Chile); technical contributors to eclipse efforts: Xavier Jubier (France), web mapping; Fred Espenak (US), mapping and http://EclipseWise.com website, updated from "NASA website"; Jay Anderson (Canada), eclipse meteorology; Glenn Schneider (US), airborne planning; Michael Gill (UK), Solar Eclipse Mailing List, now SEML@groups.io; Michael Zeiler (USA), eclipse maps; Bill Kramer (USA), eclipse statistics; Michael Kentrianakis (USA), USA 2017 American Astronomical Society Project Manager; and Ralph Chou (Canada), eye safety.

Acknowledgments

Our expedition to Chile and subsequent data reduction received major support from grant AGS-903500 from the Solar Terrestrial Program, Atmospheric and Geospace Sciences Division, U.S. National Science Foundation. The Cerro Tololo Inter-American Observatory site was courtesy of Associated Universities for Research in Astronomy (AURA). We had additional student support from the Massachusetts NASA Space Grant Consortium; Sigma Xi; the Global Initiatives Fund at Williams College; the National Solar Observatory; and the University of Pennsylvania. Predictive Science Inc. was supported by AFOSR, NASA, and NSF. ACS was supported by a NASA Heliophysics HGI grant, and by the MSFC Hinode Project. KPR was supported by the National Solar Observatory. AV thanks the mathematician Christophoros Mouratidis for his help with the data reduction of the spectra. We thank Aegean Airlines company for kindly providing the opportunity of a test flight for the airborne NIR spectrograph. SUVI was described in Tadikonda *et al.* (2019). The Associated Universities for Research in Astronomy (AURA) also hosted scientific and legislative dignitaries at CTIO to view the eclipse.

References

Golub, L. & Pasachoff, J. 2017, *The Sun, Kosmos (Reaktion Books)*
Golub, L. & Pasachoff, J. M. 2009, *The Solar Corona, 2nd ed. (Cambridge U. Press)*
Golub, L. & Pasachoff, J. M. 2014, *Nearest Star: The Surprising Science of Our Sun, 2nd ed.* (Cambridge U. Press)
Lockwood, C. A., Pasachoff, J. M., Seaton, D. B., Sliski, D. H., & Lefaudeux, N. 2020, *Research Notes of the AAS*, 4, 133
Mikić, Z., Downs, C., et al. 2018, *Nature Astronomy*, 2, 913
Olson, R. & Pasachoff, J. 2019, *Cosmos: The Art and Science of the Universe (Reaktion Books)*
Pasachoff, J. M. 2019, *A Field Guide to the Stars and Planets, 4th Edition (Houghton Mifflin)*
Pasachoff, J. M. 1973, *Scientific American*, 229, 68
Pasachoff, J. M. 2017a, *Nature Astronomy*, 1, 0190
Pasachoff, J. M. 2017b, *Scientific American*, 317, 54
Pasachoff, J. M. 2018, Astronomy and Geophysics, 59, 4.19
Pasachoff, J. M. & Fraknoi, A. 2017, *American Journal of Physics*, 85, 485

† Jay Anderson, 2019: http://eclipsophile.com

Pasachoff, J. M., Jubier, X., & Wright, E. 2017, in AAS/Division for Planetary Sciences Meeting Abstracts #49, AAS/Division for Planetary Sciences Meeting Abstracts, 417.17

Pasachoff, J. M., Lockwood, C., Meadors, E., *et al.* 2018, *Frontiers in Astronomy and Space Sciences*, 5, 37

Pasachoff, J. M., Lockwood, C. A., Inoue, J. L., *et al.* 2020, in American Astronomical Society Meeting Abstracts, Vol. 52, American Astronomical Society Meeting Abstracts, 359.03

Peñaloza-Murillo, M. A. & Pasachoff, J. M. 2018, *Journal of Geophysical Research: Atmospheres*, 123, 13, 443

Tadikonda, S. K., Freesland, D. C., Minor, R. R., Seaton, D. B., *et al.* 2019, *Solar Phys.*, 294, 14

Chapter 2. New observational diagnostics of solar, stellar and interstellar magnetic fields

Solar and Stellar Magnetic Fields: Origins and Manifestations
Proceedings IAU Symposium No. 354, 2019
A. Kosovichev, K. Strassmeier & M. Jardine, ed.
doi:10.1017/S1743921320000629

Diagnosing coronal magnetic fields with radio imaging-spectroscopy technique

Yihua Yan[1,2] ⓘ, **Baolin Tan**[1,2], **V. Melnikov**[1,3], **Xingyao Chen**[1], **Wei Wang**[1], **Linjie Chen**[1], **Fei Liu**[1] **and MUSER Team**[1]

[1]CAS Key Lab of Solar Activity, National Astronomical Observatories,
Chinese Academy of Sciences, Beijing 100101, China
email: yyh@nao.cas.cn

[2]School of Astronomy & Space Science, University of Chinese Academy of Science,
Beijing, China

[3]Pulkovo Observatory of Russian Academy of Sciences, Saint Petersburg, Russia

Abstract. Mingantu Spectral Radioheliograph (MUSER) is an aperture-synthesis imaging telescope, dedicated to observe the Sun, operating on multiple frequencies in dm to cm range. The ability of MUSER to get images and measure Stokes I and V parameters simultaneously at many frequencies in a wide band is of fundamental importance. It allows one to approach/solve such important problems as measuring the strength, geometry and dynamics of magnetic field at coronal heights. Here we consider some of the recently developed radio physics methods to be used for solving the problems. These methods allow us to obtain information that is unattainable in other areas of the electromagnetic spectrum.

Keywords. Sun: magnetic fields, Sun: corona, Sun: radio radiation, Sun: flares, Sun: coronal mass ejections (CMEs)

1. Introduction

Radio dynamic imaging spectroscopy is important to understand the primary energy release process, particle accelerations and propagations. The radio FS locations are mostly not clear. The emission mechanism corresponding to different regions in the solar corona is in general different: bremsstrahlung emission in the quiet corona; cyclotron (gyro-resonance) and gyro-synchrotron emission in the active regions; and synchrotron, plasma emission, and/or electron cyclotron maser in the flaring regions. Therefore, the coronal magnetic field diagnostics is a most complicated work including determining exactly physical partitions in source regions, identifying emission mechanism, applying diagnosing functions, etc. The free parameters need to be specified in order to diagnose coronal magnetic fields(Casini *et al.* (2017)).

Either solar or non-solar dedicated new facilities, e.g., JVLA, ALMA, LOFAR, MWA, MUSER, SRH, GRH, etc., have been developed to make progress in addressing these solar radio problems (Ramesh *et al.* (1998); Yan *et al.* (2009); Lesovoi *et al.* (2012); Chen *et al.* (2013, 2015); Shimojo *et al.* (2017); Reid & Kontar (2017); Kontar *et al.* (2017); Mohan & Oberoi (2017); Chen *et al.* (2019)).

In the next section, we will present the diagnosing methods for coronal magnetic fields with radio observations. Then we introduce the initial observations of Mingantu Spectral Radioheliograph (MUSER), which is expecting to make progress for diagnosing coronal magnetic fields.

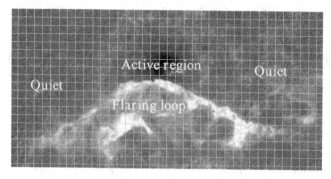

Figure 1. Different regions correspond to different emission mechanisms.

2. Diagnosing Coronal Magnetic Fields by Radio Techniques

The imaging-spectroscopy ability of MUSER (Yan *et al.* (2009)) allows to approach/ solve such important problems such as: measuring the strength, geometry and dynamics of magnetic field at coronal heights; identifying triggers of solar flares and CMEs; and selecting the most appropriate mechanism/model of electron acceleration in solar flares, as shown in in Fig. 1. Factors to diagnose coronal magnetic fields are discussed in a recent review of coronal magnetic field diagnostics by Casini *et al.* (2017).

The observed radio parameters include: frequency, intensity, polarization, duration, and drift rate, etc. The diagnosing functions correspond to the different emission mechanism. The Corona magnetic fields will be obtained in terms of the strength, the direction and the location.

Different region may correspond to different mechanism. Exactly physical partition, and the corresponding diagnosing functions are the key factors to obtain coronal magnetic fields.

Quiet solar corona: The bremsstrahlung emission may be applied to weak field. The coronal B_l can be measured from the difference of free-free absorption coefficients between O-mode and X-mode emission by measuring the polarization degree (P) and spectral index (δ) (Gelfreikh *et al.* (1987); Gelfreikh (2004)):

$$B_l \approx \frac{10714}{\lambda \delta}\, P, \quad (\text{unit}: \text{Gs}), \quad P = \frac{T_b^R - T_b^L}{T_b^R + T_b^L}, \delta = -\frac{\partial \log T_b}{\partial \log f} = \frac{\partial \log T_b}{\partial \log \lambda}.$$

When the source region is optical thin, $\delta \sim 2$, then

$$B_l \approx \frac{5357}{\lambda} P,$$

where the unit of the magnetic field B_l is Gs, and that of the wavelength λ is cm.

The sensitivity mainly depends on the accuracy of polarization measurements. The applications can be found in Zhang *et al.* (2002) and Iwai *et al.* (2014).

Above active region: The cyclotron (gyro-resonance) & gyro-synchrotron emission can be applied to the active regions or the network magnetic regions with strong field.

The edges where the optically thick layer drops below the transition region can be identified by the rapid drop in the brightness temperature. From the ratio of the edge frequencies in the two modes, the harmonic numbers of the corresponding gyroresonance layers can be identified, and the magnetic field B can also be identified unambiguously (White *et al.* (1991)).

Zhou & Karlicky (1994) obtained a set of modified expressions of coronal magnetic fields after Dulk & Marsh (1982) with

$$B \approx \left[\frac{c^2}{kT_{br}A_1} f_{pk}^{1.3+0.98\delta} f^{-0.78-0.9\delta} A_2^{-2.52-0.08\delta} \right]^{\frac{1}{0.52+0.08\delta}},$$

$$A_1 \approx 4.24 \times 10^{14+0.3\delta}(\sin\theta)^{0.34+0.07\delta}, \quad A_2 \approx 2.8 \times 10^6,$$

where c is the speed of light, f_{pk} is the peak frequency, δ is the spectral index, T_b is the brightness temperature, and $2 < \delta < 7$, $10 <$ harmonics $(s) < 100$, $E_0 > 10$ keV for simple source. The magnetograms derived from 17 GHz of NoRH was presented in Huang (2006).

Recently Fleishman & Kuznetsov (2010) developed fast GS codes with sufficient accuracy, and applicable to both isotropic and anisotropic electron distributions. Computation time for the exact formulae grows exponentially with the harmonic number, while for the fast algorithm this is nearly constant. The codes are freely available for use, and are incorporated into SSW.

Flaring region: The gyrosynchrotron, plasma emission, or electron cyclotron maser mechanism (ECME) can be applied to the flaring source regions with strong field and fast changing processes. For the flaring source regions, it is highly dynamic, including plasma instabilities, particle accelerations, and fast energy releases.

Radio spectral fine structures, e.g., microwave quasi-periodic pulsations (QPP), zebra patterns, Type III pairs, fiber bursts, etc., are the indicators of flaring source regions and the coronal magnetic field can be diagnosed.

For the observed microwave QPP structures (Tan *et al.* (2010)), if they are from fast sausage oscillation modes, one can have the period P related to the magnetic field B as follows:

$$P_{sausage}^{fast} \approx 2.02 \times 10^{-16} \frac{a\sqrt{n_e}}{B}$$

and if they are from due to the fast kink oscillation modes, one obtains the relation between the period P and the magnetic field B as follows:

$$P_{kink}^{fast} \approx 6.48 \times 10^{-17} \frac{L\sqrt{n_e}}{B}$$

where a is the section radius, L is the loop length, and n_e is the plasma density.

For the observed fiber burst structures, the magnetic field can be derived as follows (Wang & Zhong (2006)),

$$B = \frac{4\pi H_n \sqrt{m_i m_e}}{ec} \frac{df}{dt} \approx 10.15 \times 10^{-14} H_n \frac{df}{dt}, \quad \text{(Gs)}$$

where m_i, m_e are ion and electron mass, respectively, e is the electron charge, c is the speed of light, H_n is the density scale height, and f is the frequency.

According to the physical classification (Tan *et al.* (2014)), different types of zebra pattern (ZP) structures may be due to different coupling processes. Therefore, we should adopt different mechanism to derive the magnetic fields in source regions:

(1) Bernstein model (Zheleznyakov & Zlotnik (1975)):

$$B \approx \frac{2\pi m_e}{e} \Delta f;$$

(2) Whistle Wave model (Chernov (1996); Chernov *et al.* ((2005)):

$$B \approx 71.1 \times 10^{-8} \Delta f;$$

(3) Double plasma layer resonance (DPR) model (Zheleznyakov & Zlotnik (1975)):

$$B \approx 35.6 \times 10^{-8} \frac{2H_n - H_b}{H_b} \Delta f.$$

In the above, Δf is the frequency difference between the adjacent zebra strips and H_n, H_b are scale heights of the density and the magnetic field, respectively.

For the ZPs observed in an X2.2 flare event on 15 Februry 2011, the magnetic field decreases obviously from the flare rising phase to its decay phase (Tan *et al.* (2012)). The relaxation of magnetic field relative to plasma density was also found as revealed from microwave ZPs (Yan *et al.* (2007), Yu *et al.* (2012))

For the microwave Type III pairs (Tan *et al.* (2016)), the plasma density around source region is obtained from the separation regime of the opposite frequency drifting directions as $n_x = \frac{f_x^2}{81s^2}$, m^{-3}, where s is the harmonic number, The temperature T can be derived from the ratio of SXR emission fluxes at two energy bands. Then the magnetic field B is obtained as:

$$B \approx (B_L + B_H)/2,$$

$$B_L > 3.402 \times 10^{-19} (n_x T \bar{D} R_c)^{1/2}, \quad B_H < 3.293 \times 10^{-16} \left[\frac{n_x T \bar{D} R_c}{(n_x \tau)^{1/3}} \right]^{1/2},$$

with B_L and B_H as the magnetic field corresponding to the normal and RS type III branches, respectively. The beam velocity can be obtained as

$$v_b \approx \frac{2\mu_0 n_x k_B T}{B^2} \bar{D} R_c.$$

The length of the acceleration regions can be estimated as

$$L_c = \frac{\mu_0 n_x k_B T}{B^2} R_c \frac{\Delta f}{f_x}.$$

In the above, R_c is the curvature radius of the magnetic field lines, τ is the burst lifetime defined as the time difference between the start and end of each individual type III burst, μ is the magnetic permeability, k_B is the Boltzmann constant, $\bar{D}$ is the relative frequency drift rate and Δf is the the observed frequency gap between the normal and RS type III branches.

For the Type-III pairs observed in a flare event on 26 September 2011 as shown in Fig. 2(a) (Tan *et al.* (2016)), the observed parameters are as follows. For normal branches, the frequency drift rate is 2.12–7.38 GHz/s. The relative frequency drift rate is 1.52–5.06 s^{-1}. The separate frequency is in 1.22–1.49 GHz and the frequency gap is in 172–442 MHz. For RS branches, the frequency drift rate is 281–647 MHz/s. The relative frequency drift rate is 0.23-0.50 s^{-1}. We can derive the diagnosed results as shown in Fig. 2(b). It can be seen that the plasma $\beta > 1$, which means that it is highly dynamic and unstable near the source region. Although RS branches drift about 10 times slower than the normal branches, the energies of upward electrons are still close to the downward electrons. Both of the upward and downward electrons are possibly accelerated by similar mechanism (Tan *et al.* (2016)).

3. MUSER Progress and Initial Observations

MUSER is a solar-dedicated imaging-spectroscopy facility with its brief specification as shown in Table 1 (Yan *et al.* (2009)). MUSER data is processed using CASA and some own developed programs-Direct FT. They have made all code available at https://github.com/astroitlab/museros (Mei *et al.* (2018)). During 2014-2019, a total number of 83 solar radio burst events have been registered by MUSER, as shown in Table 2.

Table 1. Brief Specifications of MUSER.

Array	MUSER-I	MUSER-II
Frequency range	0.4 - 2 GHz	2 - 15 GHz
Antennas	40(Φ4.5 m)	60 (Φ2 m)
Frequency channel	64 (25 MHz)	520 (25 MHz)
Space resolution	$51.6'' - 10.3''$	$10.3'' - 1.4''$
Time resolution	$\sim$25 ms	$\sim$200 ms
Polarizations	LCP, RCP	LCP, RCP
Maximum baseline	$\sim$3 km	$\sim$3 km

Table 2. Event list of MUSER in 2014-2019.

Flare class	Number of radio burst events
X	2
M	15
C	38
B	19
A	4
below A	5
Total number	83

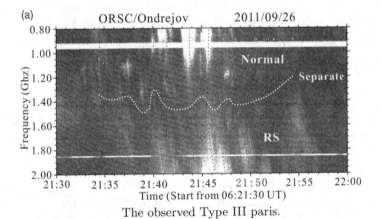

The observed Type III paris.

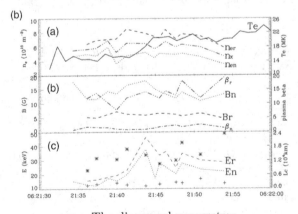

The diagnosed parameters.

Figure 2. The Type-III pairs observed in a flare event on 26 September 2011 and the diagnosed results (Tan *et al.* (2016)).

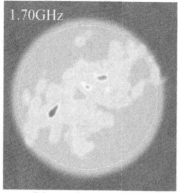 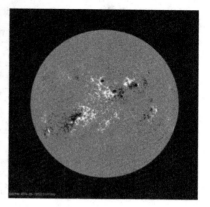

30 ms integral time

Figure 3. Radio intensity image (*left panel*) of the quiet Sun observations on 12 May 2014 by
MUSER and the full disk HMI/SDO magnetogram (*right panel*).

Fig. 3 shows the radio intensity image of the quiet Sun observations on 12 May 2014
by MUSER and the full disk magnetogram of the line of sight component of the photospheric magnetic field by HMI/SDO. It can be seen that the radio features agree with
the magnetic field properly. Further data processing is needed for future studies to obtain
coronal magnetic fields from observed data based on the above methods.

4. Summary

In summary, solar radio approaches for diagnosing coronal magnetic fields have
been developed for quiet Sun, active regions and flaring processes. The imaging-
spectroscopy observations are needed for radio fine structure analyses. MUSER progress
and observations are promising and further data processing is needed for future studies.

Acknowledgements

This work is supported by NSFC grants (11790301, 11433006, 11973057, 11573039).
The MUSER data processing is partially supported by MOST grant (2018YFA0404602).
The visit of VM at NAOC was supported by the CAS PIFI program.

References

Casini, R., White, S. M., & Judge, P. G. 2017, *Space Sci. Rev.*, 210, 145
Chen, B., Bastian, T. S., Shen, Chengcai, Gary, D. E., Krucker, S., & Glesener, L. 2015, *Science*,
    350, 1238
Chen, B., Bastian, T. S., White, S. M., Gary, D. E., Perley, R., Rupen, M., & Carlson, B. 2013,
    ApJ, 763, 21
Chen, B., Shen, C., Reeves, K. K., Guo, F., & Yu, S. 2019, *ApJ*, 884, 63
Chernov, G. P. 1996, *Astronomy Reports*, 40, 561
Chernov, G. P., Yan, Y. H., Fu, Q. J., & Tan, Ch. M. 2005, *A&A*, 437, 1047
Dulk, G. A. & Marsh, K. A. 1982, *ApJ*, 259, 350
Fleishman, G. D. & Kuznetsov, A. A. 2010, *ApJ*, 721, 1127
Gelfreikh, G. B. 2004, in: Gary D.E., Keller C.U. (eds) *Solar and Space Weather Radiophysics.*
    Astrophysics and Space Science Library(Dordrecht: Springer), vol. 314, p.115
Gelfreikh, G. B., Peterova, N. G., & Riabov, B. I. 1987, *Sol. Phys.* 108, 89
Huang, G. 2006, *Sol. Phys.*, 237, 173
Iwai, K., Shibasaki, K., Nozawa, S., Takahashi, T., Sawada, S., Kitagawa, J., Miyawaki, S., &
    Kashiwagi, H. 2014, *Earth Planets & Space*, 66, 149

Kontar, E. P., Yu, S., Kuznetsov, A. A., Emslie, A. G., Alcock, B., Jeffrey, N. L. S., Melnik, V. N., Bian, N. H., & Subramanian, P. 2017, *Nature Comm.*, 8, 1515

Lesovoi, S. V., Altyntsev, A. T., Ivanov, E. F., & Gubin, A. V. 2012, *Sol. Phys.*, 292, 168

Mei, Y., Wang, F., Wang, W., Chen, L. J., Liu, Y. B., Deng, H., Dai, W., Liu, C. Y., & Yan, Y. H. 2018, *PASP*, 130, 14503

Mohan, A. & Oberoi, D. 2017, *Sol. Phys.*, 292, 168

Ramesh, R., Subramanian, K. R., Sundararajan, M. S., & Sastry, Ch. V. 1998, *Sol. Phys.*, 181, 439

Reid, H. A. S. & Kontar, E. P. 2017, *A&A*, 606, 141

Shimojo, M., Bastian, T. S., Hales, A. S., *et al.* 2017, *Sol. Phys.*, 292, 87

Tan, B. L., Karlický, M., Mészárosová, H., Kashapova, L., Huang, J., Yan, Y., & Kontar, E. P. 2016, *Sol. Phys.*, 291, 2407

Tan, B. L., Tan, C. M., Zhang, Y., Mészárosová, H., & Karlický, M. 2014, *ApJ*, 780, 129

Tan, B. L., Yan, Y., Tan, C. M., Sych, R., & Gao G. N. 2012, *ApJ*, 744, 166

Tan, B. L., Zhang, Y., Tan, C. M., & Liu Y. Y. 2010, *ApJ*, 723, 25

Wang, S. J. & Zhong, X. C. 2006, *Sol. Phys.*, 236,155

White, S. M., Kundu, M. R., & Gopalswamy, N. 1991, *ApJ*, 366, 43

Yan, Y. H., Huang, J., Chen, B., & Sakurai, T. 2007, *PASJ*, 59, 815

Yan, Y. H., Zhang, J., Wang, W., Liu, F., Chen, Z. J., & Ji, G. 2009, *Earth Moon Planets*, 104, 97

Yu, S. J., Yan, Y. H., & Tan, B. L. 2012, *ApJ*, 761, 136

Zhang, C.-X., Gelfreikh, G. B., & Wang, J.-X. 2002, *Chin. J. Astron. Astrophys.*, 2, 266

Zheleznіакоv, V. V. & Zlotnik, E. Ia. 1975a, *Sol. Phys.*, 43, 431

Zhelezniakov, V. V. & Zlotnik, E. Ia. 1975b, *Sol. Phys.*, 44, 461

Zhou, A. H. & Karlicky, M. 1994, *Sol. Phys.*, 153, 441

Discussion

IRINA KITIASHVILI: You detected high plasma-β in the solar corona. Could you, please, describe how for this region located above photosphere and how it evolves?

YIHUA YAN: The region corresponds to the frequency regime of 1.22 GHz – 1.49 GHz, which correspond to the plasma density of about 6.0×10^{15} m^{-3} (with harmonics s= 2). Therefore this region located about 8000 km – 40000 km above the photosphere. It evolves dynamically during about 20 second period, e.g., as shown in Fig. 2(b) (for details, please refer to Tan *et al.* 2016).

Solar and Stellar Magnetic Fields: Origins and Manifestations
Proceedings IAU Symposium No. 354, 2019
A. Kosovichev, K. Strassmeier & M. Jardine, ed.
doi:10.1017/S1743921319009906

Observing the Sun with the Atacama Large Millimeter/submillimeter Array – from continuum to magnetic fields

Sven Wedemeyer[1,2]🅾, Mikolaj Szydlarski[1,2], Jaime de la Cruz Rodriguez[3] and Shahin Jafarzadeh[1,2]

[1]Rosseland Centre for Solar Physics, University of Oslo, Postboks 1029 Blindern, N-0315 Oslo, Norway

[2]Institute of Theoretical Astrophysics, University of Oslo, Postboks 1029 Blindern, N-0315 Oslo, Norway

[3]Institute for Solar Physics, Dept. of Astronomy, Stockholm University, Albanova University Center, SE-10691 Stockholm, Sweden

Abstract. The Atacama Large Millimeter/submillimeter Array offers regular observations of our Sun since 2016. After an extended period of further developing and optimizing the post-processing procedures, first scientific results are now produced. While the first observing cycles mostly provided mosaics and time series of continuum brightness temperature maps with a cadence of 1-2s, additional receiver bands and polarization capabilities will be offered in the future. Currently, polarization capabilities are offered for selected receiver bands but not yet for solar observing. An overview of the recent development, first scientific results and potential of solar magnetic field measurements with ALMA will be presented.

Keywords. Sun: atmosphere, Sun: chromosphere, Sun: radio radiation, radiation mechanisms: thermal, magnetic fields, polarization, techniques: interferometric

1. Introduction

The magnetic field plays an essential role in the structuring, dynamics, and energy balance in the atmospheres of solar-like stars. Measurements of the magnetic field in the upper atmospheric layers, however, are still challenging even for the Sun because of technical limitations in connection with the few currently available diagnostics. Common measurement techniques employ magnetically sensitive spectral lines that are formed in the chromosphere, such as the lines of singly ionized calcium (CaII). These lines typically have only weak polarisation signals, which requires an instrument with high sensitivity. In addition, the meaningful interpretation of chromospheric spectral lines requires to properly take into account deviations from equilibrium conditions, in particular regarding non-local thermodynamic equilibrium (NLTE) and the ionization state of the chromospheric gas.

The continuum radiation at millimeter wavelengths is an alternative diagnostic tool for measuring magnetic fields in the atmosphere of the Sun and other stars. The Atacama Large Millimeter/Submillimeter Array (ALMA) has the potential of facilitating such measurements in the future. Regular continuum observations of the Sun with ALMA began in 2016. Full polarization capabilities are not yet available for solar observations but for non-solar targets for a few receiver bands. It should be emphasized that solar observations with ALMA are very different from observations of other targets because the Sun evolves on very short time scales and because the extent of the Sun is much wider than ALMA's field-of-view so that the primary antenna beam is filled with complex emission. Consequently, developing robust procedures for calibration, imaging and further post-processing of regular (and also Commissioning and Science Verification) solar ALMA data was time-demanding. Now (in 2019), some first results are already produced and published (see, e.g. Bastian *et al.* 2017; Shimojo *et al.* 2017b; Yokoyama *et al.* 2018; Brajša *et al.* 2018; Jafarzadeh *et al.* 2019; Rodger *et al.* 2019; Selhorst *et al.* 2019; Nindos *et al.* 2018; Loukitcheva *et al.* 2019; Molnar *et al.* 2019; Wedemeyer *et al.* 2020. See also Wedemeyer *et al.* (2016)) for an overview of potential science cases with ALMA (see also Wedemeyer 2016; Bastian *et al.* 2018).

We focus here on ALMA's potential for measuring the magnetic field in the solar chromosphere since it is a challenging but important first step before the same can be attempted for other (and thus spatially unresolved) stars. In the following, we give a concise overview of ALMA's current observing capabilities relevant in this context (Sect. 2) and discuss ALMA's potential for future magnetic field measurements based on observations and numerical simulations (Sect. 3). A summary and outlook is provided in Sect. 4.

2. Observations so far

2.1. *The Atacama Large Millimeter/Submillimeter Array (ALMA)*

ALMA† is located on the Chajnantor plateau in the Chilean Andes close to San Pedro de Atacama at an altitude of 5100 m. The array consists of several components: The 12-m Array with 50 movable 12-m antennas and the Atacama Compact Array (ACA) with 12 antennas with 7 m diameter and 4 Total Power (TP) antennas with 12 m diameter for single dish observations. The ACA remains in a fixed compact configuration whereas the 12-m Array changes through different predefined configurations throughout the observing cycle. That way, ALMA can provide maximum baselines from a few hundred meters in a compact configuration to a maximum of 16 km in the most extended configuration.

ALMA is an aperture synthesis telescope with an angular resolution that corresponds to a telescope aperture equivalent to the longest baseline, i.e. the longest distance between two individual antennae. Each antenna pair in the array is sensitive to a certain spatial scale and orientation angle in the source image corresponding to one component in the spatial Fourier space. Each unique combination of two antennae in the array thus maps one Fourier component of the source image as determined by the length and orientation of the baseline with regard to the target position on the sky. The more unique Fourier components are measured, the better the sampling of the Fourier space and the better can the source image be reconstructed from the measured Fourier components. ALMA is therefore using a comparatively large number of antennae. The smaller antennae in the ACA have the advantage of providing shorter baselines as short as 8.5 m, which are otherwise impossible to obtain with 12-m antennae, and thus maps larger spatial scales unaccessible with the 12-m Array. The TP antennae, which are used for single-dish observations, measure the zero component in Fourier space and thus provide the absolute offset for the interferometric observations.

ALMA currently offers 8 different receiver bands for observations of non-solar targets covering a range from 84 GHz (3.6 mm) to 950 GHz (0.3 mm). The range will be eventually expanded to lower frequencies (35 GHz, 8.5 mm) once Band 1 and 2 become available. The receivers have two linear polarisation feeds with perpendicular orientation (X and Y). ALMA offers both single and dual polarisation modes, providing the cross-correlation product $\langle XX^* \rangle$ (with better spectral resolution) or $\langle XX^* \rangle$ and $\langle YY^* \rangle$ (with

† The ALMA project is an international partnership between Europe, North America, and East Asia in cooperation with the Republic of Chile and minor partners.

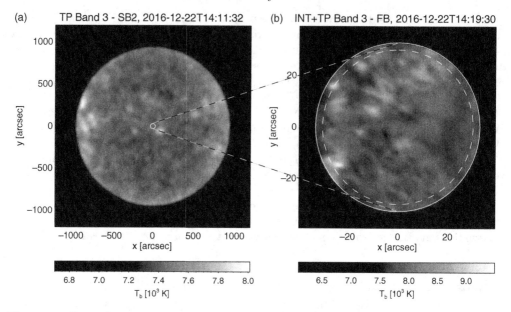

Figure 1. Examples of Band 3 observations of the Sun from the first regular solar observing campaign: **a)** Single-dish TP map and **b)** a (combined TP+) interferometric map. The dashed circle in the latter marks the single-dish beam, whereas the solid circles in both panels mark the chosen interferometric field-of-view. Please note that the whole Sun is shown in panel a whereas the map in panel b is only a small portion of the Sun.

better sensitivity), respectively. In full polarisation mode, all cross-correlation products are provided (see Sect. 2.3). The solar observing capabilities are described in Sect. 2.2.

2.2. *ALMA's current capabilities for observing the Sun*

Regular ALMA observations of the Sun were first offered in Cycle 4, which started in October 2016 with the first solar observations attempted in December 2016. Solar projects consist of interferometric observations either at a fixed target tracked for solar rotation or as a larger mosaic with up to 149 positions. In addition, the whole solar disk is scanned with single Total Power (TP) dishes in a double-circular motion in the course of a few minutes. As many of the fast TP scans are attempted at the same time (or close to) the interferometric observation. More technical details about the interferometric and the TP part are given by Shimojo *et al.* (2017a) and White *et al.* (2017), respectively. As example, a full-disk TP scan and a combined interferometric map, both obtained in Band 3, are shown in Fig. 1. The interferometric map has been combined ("feathered") with the TP map and thus shows absolute brightness temperatures. Please see Wedemeyer *et al.* (2020) for details regarding the imaging and post-processing of this data set.

In Cycle 4 and 5, only the receiver bands 3 (3 mm) and 6 (1.3 mm) were offered but Band 7 (0.9 mm) followed in Cycle 7 (from October 2019, see Table 1). The interferometric data can be split into 4 sub-bands with a band-width of 2 GHz each and each sub-band can be split into 128 spectral channels each. The provided scripts for the first cycles do not split but combine all 4×128 channels for the reconstruction of one "full-band" continuum image. This approach (multi-frequency synthesis) is chosen in order to maximize the spatial Fourier coverage and thus maximize the image fidelity. The full-disk TP scans are produced for all four sub-bands but cannot be further split into spectral channels.

The interferometric observations are typically carried out for a single pointing (following solar rotation) at a cadence of 1 s (2 s in Cycle 4) for a duration of $\sim$10 min before the

Table 1. Available solar observing modes: Receiver bands with central frequency ν_{LO1} and corresponding wavelength λ, available array configurations with maximum baseline and when these capabilities were offered for the first time.

Band	ν_{LO1} [GHz]	λ [mm]	First offered	Configuration (max. baseline)			
				1 (161 m)	2 (314 m)	3 (500 m)	4 (784 m)
3	100.0	3.00	Cycle 4	yes	yes	yes	since Cycle 7
6	239.0	1.25	Cycle 4	yes	yes	yes	no
7	346.6	0.86	Cycle 7	yes	yes	no	no

observation is interrupted for a 2-3 min intermediate calibration break. An observation often consists of 3-4 of these scans, resulting in a net observing time of up to ~40 min. The field-of-view (FOV) is set by the primary beam of a single antenna or more precisely the chosen cut-off of the primary beam response. The resulting FOV is then typically 60" for Band 3, 25" for Band 6, and 17" for Band 7. As mentioned above, larger areas on the Sun can be mapped with mosaics with up to 149 points, thus sacrificing the time domain to obtain one large map (see, e.g., Jafarzadeh *et al.* 2019, and references therein). The achievable angular resolution is determined by the observing wavelength, the longest available baseline (see Table 1), the position of the Sun on the sky, and in practice also the seeing conditions. According to the ALMA Technical Handbook (Cycle 7†), the resolution for Band 3 can be as low as 1.42" (in configuration 3) and since Cycle 7 down to 0.92" (in configuration 4), down to 0.615" for Band 6, and down to 0.666" for Band 7. For solar Band 3 observations so far, effective resolutions of down to 1.4" for the minor axis of the synthesized beam are reported whereas the major axis is larger (on the order of ~2") and changes with the Sun moving across the sky.

Solar observations are typically carried out in dual linear polarization mode ($\langle XX^* \rangle$, $\langle YY^* \rangle$) although in principle the single-polarization ($\langle XX^* \rangle$) mode is offered for solar observations, too. Using dual polarization mode has the advantage that $\langle XX^* \rangle$ and $\langle YY^* \rangle$ represent mostly independent measurements of the source, which can be exploited to suppress noise and improve the final image quality. Continuum images are therefore typically derived from the average of $\langle XX^* \rangle$ and $\langle YY^* \rangle$ under the assumption that there is no significant net linear polarisation and that averaging thus improves the image quality. ALMA offers no full polarisation for solar observations yet.

2.3. *ALMA's current polarisation capabilities for (non-solar) targets*

For non-solar targets, full polarisation capabilities are currently (Cycle 7) offered for single-pointing observations with the 12-m Array in bands 3, 4, 5, 6 and 7. All cross-correlation products are available for such observations: $\langle XX^* \rangle$, $\langle XY^* \rangle$, $\langle YX^* \rangle$, $\langle YY^* \rangle$ (or written as cross-correlation visibilities: $V_{XX}, V_{XY}, V_{YX}, V_{YY}$). Under ideal conditions, all Stokes components including linear and circular polarization states can be calculated from the visibilities:

$$I = \tfrac{1}{2}(V_{XX} + V_{YY}) \tag{2.1}$$

$$Q = \tfrac{1}{2}(V_{XX} - V_{YY}) \tag{2.2}$$

$$U = \tfrac{1}{2}(V_{XY} + V_{YX}) \tag{2.3}$$

$$V = \tfrac{1}{2i}(V_{XY} - V_{YX}) \tag{2.4}$$

† https://almascience.nrao.edu/documents-and-tools/cycle7/alma-technical-handbook

In practice, the separation into the two orthogonal polarization states (X and Y) is not perfect and effects like instrumental polarization need to be corrected for. In addition, this correction and thus the calibration depends on the position within the primary beam. Consequently, only on-axis polarization is offered for Cycle 7, restricting the full polarization capabilities to the innermost parts of the primary beam, whereas off-axis polarization (i.e., position further away from the axis, i.e. away from the center of the FOV) is still under commissioning. Please see the ALMA Technical Handbook (Cycle 7†) for more details.

According to the ALMA Cycle 7 Proposer's Guide‡, the accuracy level of linear polarization imaging of a compact source on-axis (both in continuum and with full spectral resolution) is 0.1 % (3 σ, i.e. 3 times the systematic calibration uncertainty) within the inner third of the primary beam (i.e. within 1/3 of its FWHM). The corresponding accuracy for on-axis circular polarisation is currently 1.8% (3 σ, 0.6 % at 1 σ). The field of view of the observations is limited to the inner one-third for linear polarisation observations and to the inner one-tenth of the primary beam FWHM for circular polarisation observations, respectively. Proper calibration of full polarization observations currently takes about three hours. Substantial efforts are being mode to further develop and improve the accuracy of the full polarization mode and eventually offer it as a standard mode and for solar observations.

3. Prospects for measuring magnetic fields with ALMA

3.1. *Basic considerations*

The radiation at millimeter wavelengths emitted from the Sun originates from its chromosphere. Non-thermal contributions such as gyrosynchrotron radiation are expected to be only significant during flares so that the solar millimeter radiation is of thermal nature in most situations. The solar atmosphere is permeated by magnetic field of varying strength, which is polarizing the thermal free-free radiation. A linearly polarized wave that is propagating through the chromosphere in the presence of magnetic fields is split into two components, namely the ordinary wave and extraordinary wave, also referred to as o-mode and x-mode, respectively. The absorption coefficient and thus the opacity κ is higher in the extraordinary mode than in the ordinary ($\kappa_x > \kappa_o$), which results in a difference of the heights where the two modes become optically thick and thus from where the observed emission originates from. The x- and o-mode thus probe (slightly) different atmospheric layers with the difference of the formation heights of the x- and o-modes increasing with magnetic field strength. This effect produces an observable net circular polarisation

$$P = \frac{T_b^x - T_b^o}{T_b^x + T_b^o}, \tag{3.1}$$

which is thus connected to the observable (brightness) temperature difference between these two layers ($T_b^x - T_b^o$). The average of the measurable brightness temperature for both modes gives the (continuum) brightness temperature:

$$T_{b,cont} = \frac{T_b^x + T_b^o}{2}. \tag{3.2}$$

Bogod & Gelfreikh (1980) and Grebinskij *et al.* (2000) developed a technique that exploits this effect and allows for deriving the longitudinal component B_l of the magnetic

---

† https://almascience.nrao.edu/documents-and-tools/cycle7/alma-technical-handbook
‡ https://almascience.nrao.edu/documents-and-tools/cycle7/alma-proposers-guide

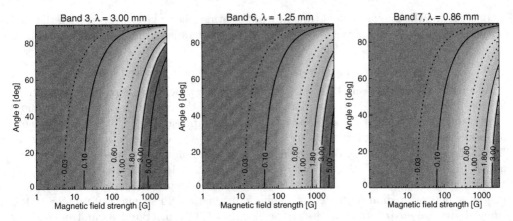

Figure 2. Estimate of the degree of circular polarisation for the central wavelengths of the ALMA bands 3, 6, and 7 as function of the magnetic field strength and the angle between the field orientation and the line of sight.

field from the observed degree of circular polarisation P and the slope $\zeta = d(\log T_{\rm b})/d(\log \nu)$ of the brightness temperature spectrum $T_{\rm b}(\nu)$. For the simple case of an optically thin, homogenous, and isothermal atmospheric gas (cf. Kundu 1965), the degree of circular polarization can then be approximated as

$$P(\%) \simeq 1.85 \times 10^{-3} \lambda \;(\mathrm{mm}) \; B \;(\mathrm{G}) \; \cos \theta \tag{3.3}$$

for an observing wavelength λ and for the longitudinal magnetic field component

$$B_{\rm l} = |\vec{B}| \cos \theta, \tag{3.4}$$

where θ is the angle between the magnetic field vector $\vec{B}$ and the line of sight. The resulting values are visualized for a wider range of field strengths and inclination angles for the central wavelengths of the ALMA bands 3, 6, and 7 in Fig. 2. According to this approximation, the degree of circular polarization is expected to be on the order of a few percent for a magnetic field strength of $1\,\mathrm{kG}$ as it occurs above sunspots (see, e.g., Wedemeyer *et al.* 2016, and references therein). More precisely, the approximation predicts values of up to $P \approx 5.5\,\%$ for Band 3, $2.3\,\%$ for Band 6, and $1.6\,\%$ for Band 7, respectively. These numbers are in line with more realistic calculations by, e.g., Fleishman *et al.*, (in Wedemeyer *et al.* 2016) who derive $|P|_{\rm max} = 3.6\,\%$ for an Active Region at a frequency of $114.8\,\mathrm{GHz}$ ($\lambda = 2.61\,\mathrm{mm}$, close to Band 3). The magnetic field strength in the Quiet Sun chromosphere is much lower and thus results in a correspondingly lower degree of circular polarization. Detailed modeling for the Quiet Sun chromosphere predicts values of only $|P| \sim 0.05\%$ at $\lambda = 1\,\mathrm{mm}$, which is clearly below the current detection limits mentioned in Sect. 2.3.

The same principle also offers a straightforward method for deriving the longitudinal component of the magnetic field from measurements of the circular polarization $|P|$ and the (local) brightness temperature slope ζ as

$$B_{\rm l} = |\vec{B}| \cos \theta = \frac{P\,\nu}{\zeta} \; (2.8 \times 10^{6})^{-1} \; \mathrm{Hz}^{-1} \; (G) \tag{3.5}$$

for the observing frequency ν. The measured field component refers to an atmospheric height corresponding to the effective formation height at the observed frequency. In principle, the magnetic field $B_{\rm l}$ can be derived for different frequencies and thus for a corresponding range of heights in the solar atmosphere. Such measurements could allow for reconstructing the magnetic field topology in the atmospheric layer probed with the chosen receiver band. See Loukitcheva *et al.* (2017) for a detailed evaluation of this method.

3.2. *Model predictions for solar continuum polarisation*

As already implied in Sect. 3.1, forward modelling can provide more realistic predictions for the polarization to be expected for observations of the Sun with ALMA. Even more importantly, such calculations can provide important input and tests for the development of more sensitive techniques that are applicable for weaker magnetic field strengths. Loukitcheva *et al.* (2017) calculate the degree of circular polarization based on a numerical model produced with the Bifrost code (Carlsson *et al.* 2016; Gudiksen *et al.* 2011) for a wavelength range covered by ALMA. The model is representative of an enhanced magnetic network region with a chromospheric magnetic field strength of up to $\sim$100 G. They derive values $|P| < 0.15\%$ for a wavelength of 3.2 mm (Band 3) and only a few 0.01% for shorter wavelengths. Their polarisation signal would thus be lower than the current 3σ-level of 1.8 % (see Sect. 2.3), making the measurement of the magnetic field for the type of modelled region in Band 3 impossible given the current technical limitations. In Figs. 3–5, we present similar results calculated with the Advanced Radiative Transfer (ART) code (de la Cruz Rodriguez *et al.*, in prep.) for Band 3, 6, and 7, respectively. The model used for these calculations is a continuation of the simulation by Carlsson *et al.* (2016) with a newer version of the Bifrost code. The brightness temperatures are shown for the x- and o-mode separately, which can be combined to the continuum brightness temperature T_b and the circular polarization P according to Eqns. 3.2 and 3.1. The central region at 10 % of the primary beam FWHM is marked for each band, illustrating the small region that will be initially be usable for circular polarization measurements. For Band 3 (Fig. 3), values of up $|P|_{\max} = 0.18\%$ are found for this particular model snapshot but most pixels exhibit much lower values as can be seen from the corresponding distribution in Fig. 6b. The same is true for the other two bands but the values decrease with wavelength. For Band 6 and 7, we find maximum values of $|P|_{\max} = 0.09\%$ and $|P|_{\max} = 0.07\%$, respectively. Much of the simulated field of view and in particular the outer parts, which can be considered representative of Quiet Sun conditions, feature circular polarization signals that would be too weak even with a 10 times lower detection limit. However, the initial measurements will be confined to a small sub-region (on-axis) that even in Band 3 is just large enough to enclose one of the simulated magnetic patches. Such measurements will therefore require very accurate pointing and tracking of the solar rotation. The magnetic field measuring technique discussed in Sect. 3.1 (cf. Loukitcheva *et al.* 2017) might be improved by exploiting the fact that the circular polarisation can in principle be measured at several frequencies across the employed receiver band and thus at slightly different atmospheric heights. The resulting increased number of data points as input for data inversion codes might allow for lowering the effective detection limit to well below the 3-σ level. More numerical studies are needed in this regard.

3.3. *Continuum polarisation from solar observations*

As mentioned above, full polarization capabilities are not yet offered for solar observations. The data can nonetheless be split into the $\langle XX^* \rangle$ and $\langle YY^* \rangle$ cross-correlation products and used to construct images from on the corresponding visibilities separately. As we will show below, this procedure is still tentative and the resulting images must be interpreted with caution. The resulting maps in Fig. 7a and b are at first glance very similar. That is expected under the assumption that the mapped radiation from the chromosphere has no substantial linear polarization. The sum of the two maps or more precisely $\frac{1}{2}(V_{XX} + V_{YY})$ (see Fig. 7c) should provide Stokes I (see Eq. 2.1) and is indeed

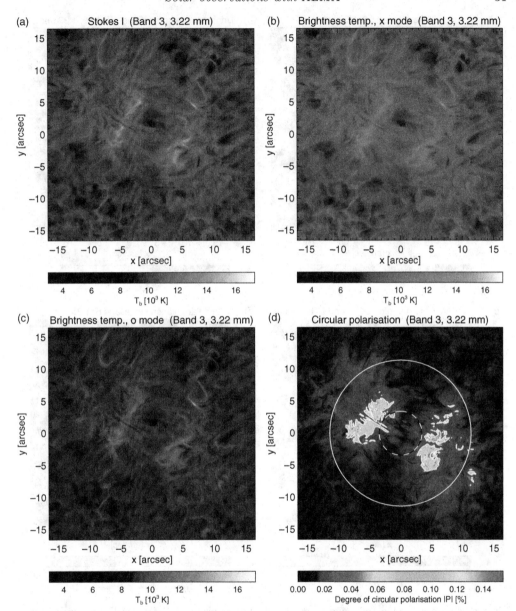

Figure 3. Results of a radiative transfer calculation with ART for a selected time step of the 3D numerical simulation. **a)** Continuum brightness temperature (Stokes I) for Band 3. **b)** Extraordinary mode (x-mode). **c)** Ordinary mode (o-mode). **d)** Absolute degree of circular polarisation $|P|$ in percent. The white contour mark the $|P| = 0.05\%$ level. The circles in panel d mark the FOV for $1/3$ (white solid) and 0.1 (black-white dashed) times the width of the primary beam in Band 3.

very close to the continuum brightness temperature map derived from all visibilities for the same time step (similar to Fig. 1b but without the TP offset). The difference of the $\langle XX^* \rangle$ and $\langle YY^* \rangle$ maps (see Fig. 7d) should be zero if no significant linear polarization is expected. The difference can under that assumption be understood as brightness temperature uncertainty. The resulting value distribution can be well approximated with a Gaussian with $\sigma = 15\,\mathrm{K}$ (FWHM= $35\,\mathrm{K}$) for the region with a diameter equivalent to the

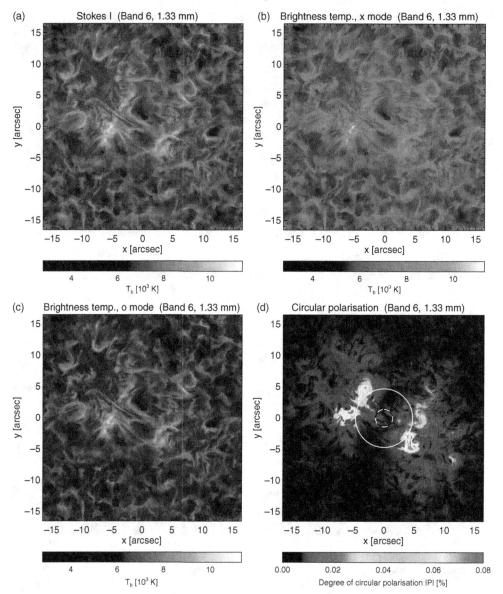

Figure 4. Results of a radiative transfer calculation with ART for a selected time step of the 3D numerical simulation. **a)** Continuum brightness temperature (Stokes I) for Band 6. **b)** Extraordinary mode (x-mode). **c)** Ordinary mode (o-mode). **d)** Absolute degree of circular polarisation $|P|$ in percent. The white contour mark the $|P| = 0.05\,\%$ level. The circles in panel d mark the FOV for 1/3 (white solid) and 0.1 (black-white dashed) times the width of the primary beam in Band 6.

primary beam FWHM (blue-white dashed circle in Fig. 7d) and $\sigma = 9\,\mathrm{K}$ (FWHM$= 20\,\mathrm{K}$) for the inner region with one third of that diameter (red-white dashed circle in Fig. 7d). Averaging over ten consecutive time steps does not change these numbers significantly.

In principle, the degree of linear polarization can be derived from the Stokes components as

$$L = \frac{\sqrt{Q^2 + U^2}}{I} \; . \tag{3.6}$$

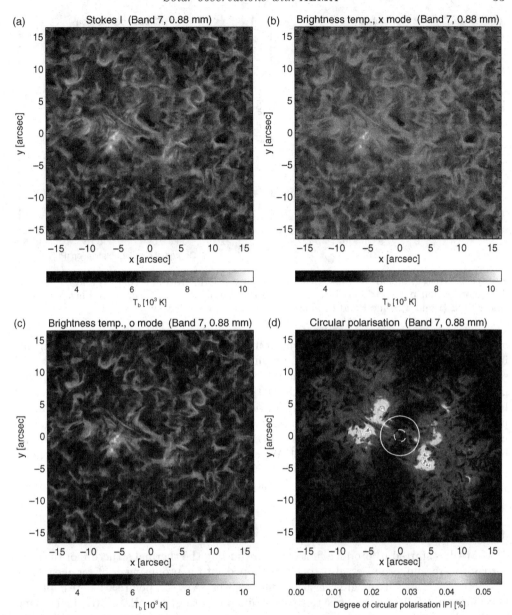

Figure 5. Results of a radiative transfer calculation with ART for a selected time step of the 3D numerical simulation. **a)** Continuum brightness temperature (Stokes I) for Band 7. **b)** Extraordinary mode (x-mode). **c)** Ordinary mode (o-mode). **d)** Absolute degree of circular polarisation $|P|$ in percent. The white contour mark the $|P| = 0.05\%$ level. The circles in panel d mark the FOV for 1/3 (white solid) and 0.1 (black-white dashed) times the width of the primary beam in Band 7.

The currently available $\langle XX^* \rangle$ and $\langle YY^* \rangle$ cross-correlation products for solar observations thus do not allow for deriving the full degree of linear polarization because the Stokes U component cannot be calculated. For now, only a lower limit of the linear polarization degree can be derived from $(\langle XX^* \rangle - \langle YY^* \rangle)/(\langle XX^* \rangle + \langle YY^* \rangle)$ (see Fig. 7e). The distribution of the resulting values is shown for the two different regions at full and one third the primary beam FWHM. The distributions have a Gaussian core with

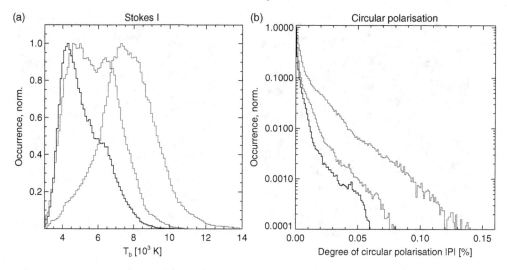

Figure 6. Histograms for **a)** the continuum brightness temperature (Stokes I) and **b)** absolute degree of circular polarisation $|P|$ in percent for the Band 3, 6, and 7 maps (see Figs. 3–5), here colored red (Band 3), blue Band 6) and black (Band 7), respectively.

$\sigma \sim 4\%$ (FWHM = 9-11 %), which is much more than expected based on measurements in other wavelength ranges (mostly lower in the atmosphere) and based on numerical simulations. For instance, Kuridze *et al.* (2018) derive a maximum degree of linear polarization of 1.5 % from Ca II 854.2 nm observations of a flaring active region, which would be rather defining the maximum that can be expected at that wavelength. The linear polarization degree for the Quiet Sun target shown in Fig. 7 should be much lower. The maps shown in Fig. 7 have therefore most likely still large uncorrected contributions that hide the true linear polarization signal. We conclude that the linear polarisation can in principle be derived from ALMA observations of the Sun but that the calibration has yet to be improved and that a scientific analysis has to wait until the solar polarization capabilities are properly commissioned.

4. Discussion and outlook

ALMA has started to produce first results regarding the thermal structure of the solar chromosphere and promises also to be developed into a promising tool for measuring magnetic fields in this enigmatic atmospheric layer. A solar polarisation test plan is developed and full solar polarisation (most likely starting with Band 3) will be commissioned and offered for regular observations in the future. Polarisation capabilities for further bands (most likely first Band 6, then Band 7) will follow. Based on the experience with ALMA observations of the Sun so far, it must be emphasized that solar observing is a non-standard mode and that consequently the development of adequate calibration and post-processing techniques for solar polarisation data and further improvements are necessary. These tasks will require significant efforts from the involved scientific community. The necessary development can and should be guided by forward modelling as presented here and already by, e.g., Loukitcheva *et al.* (2017), Loukitcheva (2020) and Nita *et al.* (2015) (see also references therein). The circular polarisation signals in Quiet Sun regions are expected to be typically smaller than 0.1 % for the currently available receiver bands, whereas a few percent are expected for regions above sunspots. It is thus most promising to start solar magnetic field measurements with ALMA with sunspot observations in Band 3. Those observations will then help to further develop the proposed method in

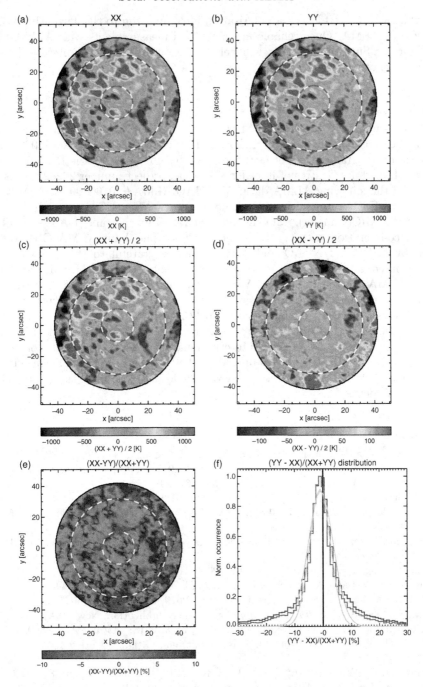

Figure 7. A solar Band 3 observation obtained on 2016-12-22 UT14:20:00 and split into the cross-correlation products **a)** $\langle XX^* \rangle$ and **b)** $\langle YY^* \rangle$. Under ideal conditions, the combinations **c)** $\frac{1}{2}(V_{XX} + V_{YY})$ and **d)** $\frac{1}{2}(V_{XX} - V_{YY})$ would correspond to Stokes I and Q, respectively. **e)** The ratio of these two maps, which would correspond to a lower limit of the linear polarization degree under ideal conditions, is also shown as histogram in panel f as blue line for the region defined by the primary beam FWHM (blue-white dashed circle in the maps) and as red line for the innermost region with a radius of 1/3 of the primary beam FWHM (red-white dashed circle in the maps). It must be emphasized that the results clearly show that a scientific analysis must wait until the polarization capabilities are commissioned.

view of the low expected levels of polarisation in comparison to the currently achievable accuracy with ALMA. Once reliable magnetic field measurements with ALMA become possible for the Sun, these results will help to interpret (spatially unresolved) ALMA observations of other stars.

Acknowledgments

This work is supported by the SolarALMA project, which has received funding from the European Research Council (ERC) under the European Union's Horizon 2020 research and innovation programme (grant agreement No. 682462), and by the Research Council of Norway through its Centres of Excellence scheme, project number 262622. This paper makes use of the following ALMA data: ADS/JAO.ALMA#2016.1.00423.S. ALMA is a partnership of ESO (representing its member states), NSF (USA) and NINS (Japan), together with NRC(Canada), MOST and ASIAA (Taiwan), and KASI (Republic of Korea), in co-operation with the Republic of Chile. The Joint ALMA Observatory is operated by ESO, AUI/NRAO and NAOJ. We are grateful to the many colleagues who contributed to developing the solar observing modes for ALMA and for support from the ALMA Regional Centres. We acknowledge support from the Nordic ARC node based at the Onsala Space Observatory Swedish national infrastructure, funded through Swedish Research Council grant No 2017–00648, and collaboration with the Solar Simulations for the Atacama Large Millimeter Observatory Network (SSALMON, http://www.ssalmon.uio.no). The ISSI international team 387 "A New View of the Solar-stellar Connection with ALMA" was funded by the International Space Science Institute (ISSI, Bern, Switzerland). We thank Vasco Manuel de Jorge Henriques and Tobia Carozzi for helpful comments.

References

Bastian, T. S., Bárta, M., Brajša, R., *et al.* 2018, The Messenger, 171, 25
Bastian, T. S., Chintzoglou, G., De Pontieu, B., *et al.* 2017, *Astrophys. J. Lett.*, 845, L19
Bogod, V. M. & Gelfreikh, G. B. 1980, *Solar Phys.*, 67, 29
Brajša, R., Sudar, D., Benz, A. O., *et al.* 2018, *A&A*, 613, A17
Carlsson, M., Hansteen, V. H., Gudiksen, B. V., Leenaarts, J., & De Pontieu, B. 2016, *A&A*, 585, A4
de la Cruz Rodriguez, J., Szydlarski, M., Wedemeyer, S., & *et al.* in prep., *The Advanced Radiative Transfer code*
Grebinskij, A., Bogod, V., Gelfreikh, G., *et al.* 2000, *A&AS*, 144, 169
Gudiksen, B. V., Carlsson, M., Hansteen, V. H., *et al.* 2011, *A&A*, 531, A154+
Jafarzadeh, S., Wedemeyer, S., Szydlarski, M., *et al.* 2019, *A&A*, 622, A150
Kundu, M. R. 1965, *Solar radio astronomy* (Interscience Publication, New York)
Kuridze, D., Henriques, V. M. J., Mathioudakis, M., *et al.* 2018, *Astrophys. J.*, 860, 10
Loukitcheva, M., White, S. M., Solanki, S. K., Fleishman, G. D., & Carlsson, M. 2017, *A&A*, 601, A43
Loukitcheva, M. A., White, S. M., & Solanki, S. K. 2019, *Astrophys. J. Lett.*, 877, L26
Loukitcheva, M. 2020, *Front. Astron. Space Sci.*, 11 August 2020 (https://doi.org/10.3389/fspas.2020.00045)
Molnar, M. E., Reardon, K. P., Chai, Y., *et al.* 2019, *Astrophys. J.*, 881, 99
Nindos, A., Alissandrakis, C. E., Bastian, T. S., *et al.* 2018, *A&A*, 619, L6
Nita, G. M., Fleishman, G. D., Kuznetsov, A. A., Kontar, E. P., & Gary, D. E. 2015, *Astrophys. J.*, 799, 236
Rodger, A. S., Labrosse, N., Wedemeyer, S., *et al.* 2019, *Astrophys. J.*, 875, 163
Selhorst, C. L., Simões, P. J. A., Brajša, R., *et al.* 2019, *Astrophys. J.*, 871, 45
Shimojo, M., Bastian, T. S., Hales, A. S., *et al.* 2017a, *Solar Phys.*, 292, #87

Shimojo, M., Hudson, H. S., White, S. M., Bastian, T. S., & Iwai, K. 2017b, *Astrophys. J. Lett.*, 841, L5

Wedemeyer, S. 2016, The Messenger, 163, 15

Wedemeyer, S., Bastian, T., Brajša, R., *et al.* 2016, *Space Sci. Rev.*, 200, 1

Wedemeyer, S., Szydlarski, M., Jafarzadeh, S., *et al.* 2020, *A&A*, 635, A71

White, S. M., Iwai, K., Phillips, N. M., *et al.* 2017, *Solar Phys.*, 292, #88

Yokoyama, T., Shimojo, M., Okamoto, T. J., & Iijima, H. 2018, *Astrophys. J.*, 863, 96

Discussion

CALLY: Does your data processing to remove jumping in the images also delete real oscillation data?

WEDEMEYER: We have been very careful with developing the processing routines and indeed investigated the oscillation properties. The results will be published in a forthcoming article. I would say that it is most important to gain confidence in the processed data first. After that, we can certainly see if the processing routines can be improved even further to make sure that the oscillation properties of the observations are preserved.

BOMMIER: My question is very naive. Is it possible to observe/measure the solar corona? If spectral lines and polarization can be measured, this wavelength range (mm) is the most favorable to measure the coronal field by Zeeman effect in spectral lines. This field remains unknown up to day.

WEDEMEYER: As I showed during my presentation, there seem to be weak imprints of the corona in the ALMA Band 3 continuum maps but it is not clear yet how much can be learned about the corona from that. Unfortunately, spectral lines have not been detected yet in the available solar ALMA observations. Only after and if we have found suitable spectral lines, we can evaluate if measuring the coronal magnetic field with ALMA is possible.

Solar and Stellar Magnetic Fields: Origins and Manifestations
Proceedings IAU Symposium No. 354, 2019
A. Kosovichev, K. Strassmeier & M. Jardine, ed.
doi:10.1017/S174392131900989X

Revisiting the building blocks of solar magnetic fields by GREGOR

Dominik Utz[1] 🄯, **Christoph Kuckein**[2] 🄯, **Jose Iván Campos Rozo**[1],
Sergio Javier González Manrique[3] 🄯, **Horst Balthasar**[2],
Peter Gömöry[3], **Judith Palacios Hernández**[4], **Carsten Denker**[2],
Meetu Verma[2], **Ioannis Kontogiannis**[2] 🄯, **Kilian Krikova**[1],
Stefan Hofmeister[1] and **Andrea Diercke**[2]

[1]IGAM/Institute of Physics, Karl-Franzens University Graz, Universitätsplatz 5/II,
Graz, Austria
email: Dominik.Utz@uni-graz.at

[2]Leibniz-Institut für Astrophysik Potsdam (AIP), An der Sternwarte 16, Potsdam, Germany
email: ckuckein@aip.de

[3]Astronomical Institute, Slovak Academy of Sciences-AISAS, Astronomický ústav SAV,
Tatranská Lomnica, Slovak Republic
email: gomory@ta3.sk

[4]Leibniz-Institut für Sonnenphysik (KIS), Schöneckstr. 6, Freiburg im Breisgau, Germany
email: jpalacios@leibniz-kis.de

Abstract. The Sun is our dynamic host star due to its magnetic fields causing plentiful of activity in its atmosphere. From high energetic flares and coronal mass ejections (CMEs) to lower energetic phenomena such as jets and fibrils. Thus, it is of crucial importance to learn about formation and evolution of solar magnetic fields. These fields cover a wide range of spatial and temporal scales, starting on the larger end with active regions harbouring complex sunspots, via isolated pores, down to the smallest yet resolved elements – so-called magnetic bright points (MBPs). Here, we revisit the various manifestations of solar magnetic fields by the largest European solar telescope in operation, the 1.5-meter GREGOR telescope. We show images from the High-resolution Fast Imager (HiFI) and spectropolarimetric data from the GREGOR Infrared Spectrograph (GRIS). Besides, we outline resolved convective features inside the larger structures – so-called light-bridges occurring on large to mid-sized scales.

Keywords. Sun: sunspots, Sun: magnetic fields, Sun: photosphere, Sun: chromosphere, instrumentation: high angular resolution

1. Sunspots – the large building blocks of solar magnetism

The most obvious and impressive solar magnetic features are sunspots. They are known to humans and observed at least since the ancient times (e.g., by ancient Chinese astronomers). Detailed observations started with Galileo Galilei in the Renaissance. Hale (1908) first discovered that sunspots are actually formed by strong magnetic fields and he identified the line splitting in sunspots caused by the Zeeman effect. In Figure 1, we show sunspots observed in 2017 during an observation campaign using the 1.5-m GREGOR telescope (Schmidt *et al.* 2012) with its powerful adaptive optics system (Berkefeld *et al.* 2012). On the left side, we see the simple active region NOAA 12681 consisting of a single sunspot showing a well-evolved penumbra around its umbral core. On the right panel of Figure 1 and in Figure 2, active region NOAA 12682 can be seen on September 29 and 28. This active region is strongly evolving and showing particularly

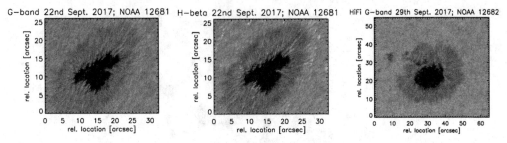

Figure 1. Sunspot in active region NOAA 12681 observed by HiFI (see, Denker *et al.* 2018) after speckle restoration. The left panel shows the G-band image, a magnetic field sensitive photospheric molecular band whereas the middle panel shows the same sunspot at the same time but observed by a broad-band filter centered at the Hβ line. This line is more sensitive to the chromospheric network and thus outlining better small-scale magnetic fields. The right panel shows a G-band image of the next appearing, more complex, sunspot in active region NOAA 12682.

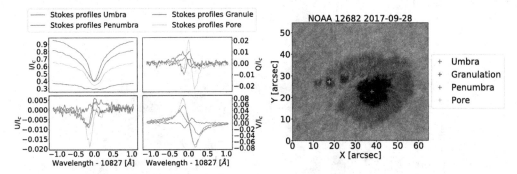

Figure 2. Active region NOAA 12682 (see, Figure 1, rightmost panel, but on the previous day). The right illustration depicts a HiFI speckle restored G-band image of the sunspot. The left panels illustrate Stokes spectra of a few selected pixels marked in the right panel observed by GRIS (Collados *et al.* 2012). Clearly, the penumbra substantially evolved by losing the pores within the penumbra while enclosing the whole sunspot within 24 hours.

interesting features like pores within the not yet closed and still evolving penumbra. The pores within the penumbra of September 28 (Figure 2, right panel) practically disappeared on September 29 (Figure 1, right panel). Besides, the penumbra starts to enclose more and more of the sunspot as it grows in the gap observed on September 28.

2. Features within sunspots

The two main features of sunspots are the central umbra and the surrounding filamentary penumbra. In addition, sunspots often display complex fine and small-scale structures. Among them are (see Figure 3 and, for example, Rimmele 2008):

– Umbral dots, i.e., small-scale convective features in the umbra where the convection is not fully suppressed by the strong magnetic fields and hot plasma can rise to the surface, and

– Light-bridges, i.e., regions of convection separating umbral cores (see Figure 3). Typically, they indicate the boundary of stronger flux elements (i.e., that the umbra of a sunspot can be formed by several strong magnetic centres and/or elements) and are often seen towards the end of the lifetime of a sunspot indicating the starting point of the terminal decline.

– Orphan penumbrae, i.e., regions of strong horizontal magnetic fields similar to normal penumbrae, but missing the umbra of a sunspot.

HiFi G−band 15 May 2016; NOAA 12544

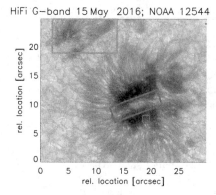

Figure 3. High-resolution HiFI speckle-restored G-band image depicting a complex sunspot (NOAA 12544) with a light-bridge (*red-box*), umbral dots (*pink-box*), and an orphan penumbra (*green-box*). Data courtesy of S. P. Rajaguru. More details of this region can be found in Felipe *et al.* (2017).

G−band 29th Sept. 2017

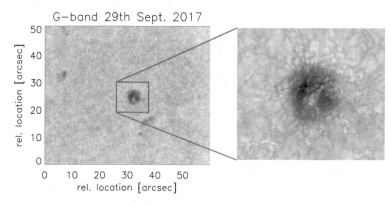

Figure 4. High resolution HiFI speckle-restored G-band image showing a strongly evolving pore. The left-hand side presents the full field-of-view within an emerging flux region. A region-of-interest is marked by a red square and a zoomed image is shown in the right panel. The pore is strongly evolving and exhibits a light-bridge on September 29 before dissolution. This light-bridge is evolving because of magneto-convection as is already evident on September 28 in Figure 5.

3. Pores and single flux fibers – the smaller building blocks

Solar magnetic fields are organised on many scales. Mid-sized magnetic features are pores, which generally consist of a strong vertical magnetic field in the absence of a penumbra (strong horizontal fields). In Figure 4, we show a pore observed on September 29. This particular pore showed an evolving light-bridge between September 28 and 29. Light-bridges are common features for sunspots often signaling their decay. Indeed, this pore disappeared on (or before) September 30. Figure 5 depicts the pore on September 28 as well as the smallest yet detectable solar magnetic features – so called MBPs (see, e.g., Utz *et al.* 2014; Kuckein 2019).

Acknowledgements

This work was supported by FWF grant: P27800. PG, SJGM, and JK acknowledge project VEGA 2/0004/16. The 1.5-meter GREGOR solar telescope was built by a German consortium under the leadership of KIS with AIP, IAG, and MPS as

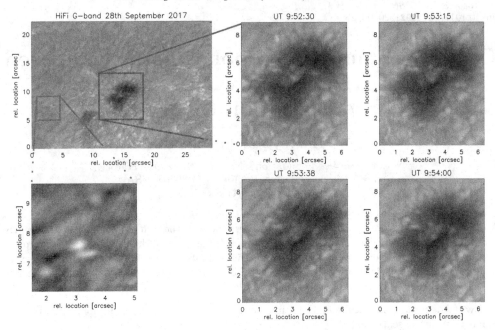

Figure 5. The same pore as in Figure 4, but one day earlier observed on September 28. Magneto-convective features can be seen in the centre of the pore. Magneto-convection is probably taking part at the boundary of stronger flux elements. This pore apparently consists of at least three monolithic magnetic flux elements. Probably the forces, which earlier brought the strong flux elements together, are weakening and magneto-convection can set in between the magnetic entities leading ultimately to the formation of the light-bridge (September 29; Figure 4) and the dissolution of the pore within the next 48 hours. The lower left close-up shows MBPs – currently the smallest detectable magnetic features.

partners, and with contributions by the IAC and ASU. This work is part of a collaboration between AISAS and AIP supported by the German Academic Exchange Service (DAAD), under project No. 57449420. CD, CK, IK, and MV acknowledge support by grant DE 787/5-1 of the Deutsche Forschungsgemeinschaft (DFG). The support by the European Commission's Horizon 2020 Program under grant agreement 824135 (SOLARNET – Integrating High Resolution Solar Physics) is highly appreciated.

References

Berkefeld, T., Schmidt, D., Soltau, D., *et al.* 2012, *AN*, 333, 863

Collados, M., López, R., Páez, E., *et al.* 2012, *AN*, 333, 872

Denker, C., Kuckein, C., Verma, M., *et al.* 2018, *ApJ Suppl. Ser.*, 236, 5

Felipe, T., Collados, M., Khomenko, E., *et al.* 2017, *A&A*, 608, 97

Hale, G. E. 1908, *ApJ*, 28, 315

Kuckein, C. 2019, *A&A*, 630, 139

Rimmele, T. 2008, *ApJ*, 672, 684

Schmidt, W., von der Lühe, O., Volkmer, R., *et al.* 2012, *AN*, 333, 796

Utz, D., del Toro Iniesta, J. C., Bellot Rubio, L. R., *et al.* 2014, *A&A*, 796, 79

Solar and Stellar Magnetic Fields: Origins and Manifestations
Proceedings IAU Symposium No. 354, 2019
A. Kosovichev, K. Strassmeier & M. Jardine, ed.
doi:10.1017/S1743921320000125

Ca II 854.2 nm spectropolarimetry compared with ALMA and with scattering polarization theory

J. W. Harvey and SOLIS Team

National Science Foundation's National Solar Observatory,
3665 Discovery Drive, Boulder, CO 80303, USA
email: jharvey@nso.edu

Abstract. Ca II 854.2 nm spectropolarimetric observations of the Sun are compared with nearly simultaneous ALMA observations. These two types of chromospheric observations show rough agreement but also several notable differences. High-sensitivity ($\simeq 0.01\%$) observations reveal ubiquitous linear polarization structures across the solar disk in the core of the 854.2 nm line that are consistent with previous theoretical studies.

Keywords. Sun: chromosphere, Sun: magnetic fields, Sun: radio radiation, polarization, radiative transfer, scattering.

1. Introduction

Since 2003, regular spectropolarimetric observations of the full solar disk have been made at the National Science Foundation's National Solar Observatory using the Synoptic Optical Long-term Investigations of the Sun (SOLIS) Vector Spectro-magnetograph (VSM) (Keller *et al.* (2003)). Starting in 2016 VSM observations include full Stokes spectra using the 854.2 nm Ca II line (Gosain (2017)). Such observations are valuable for studying the physical conditions in the chromosphere that produce spectrum line polarization; in particular, magnetic and velocity fields, and light scattering in the anisotropic atmosphere. This poster presents some examples of VSM observations compared with ALMA images and also first 854.2 nm Ca II observations of ubiquitous linear polarization structures across the solar disk. Some of these latter results were previously shown in preliminary form (Harvey *et al.* (2016)).

2. 854.2 nm Ca II Compared with ALMA

Fig. 1 compares ALMA 100 and 239 GHz full-disk solar images from 2015 December with nearly simultaneous VSM Ca II images. Intensity images from near the core of the Ca II line are roughly similar to ALMA brightness images when the former are raised to the 0.2 power. However, ALMA prominently shows network features while they are much less evident relative to plages in the Ca II line core. While ALMA shows limb brightening, Ca II shows limb darkening. Both types of observations show dark circumfacule structures around active regions and dark filament channels where horizontal magnetic fields are dominant. The poster includes Ca II images blurred to match the quoted ALMA resolutions. This does not significantly improve the comparisons. The differences between ALMA and Ca II may be due to different height ranges of the two observations or Ca emission arising from radiative transfer processes in addition to just temperature.

Fig. 2 compares Ca II with ALMA array observations of a small area near a sunspot with improved resolution. Filamentary structures emerge from the sunspot in the ALMA

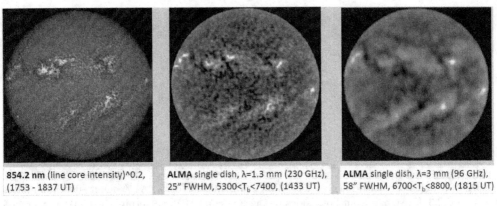

Figure 1. Full-disk SOLIS/VSM 854.2 nm and single-dish ALMA observations on 2015.12.17

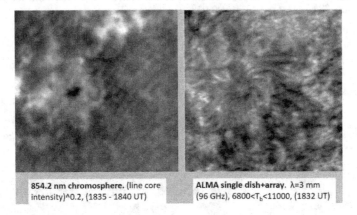

Figure 2. AR 12470 SOLIS/VSM 854.2 nm and ALMA array observations on 2015.12.16

image but are less obvious in the lower resolution Ca image. Again, network features are prominent in the ALMA image but less so in the Ca image. Curiously, the plage near the top of the Ca image is scarcely visible in ALMA observations. We note recent publication of higher resolution comparisons of small-area ALMA images with several chromospheric lines including Ca II 854.2 nm (Molnar *et al.* (2019)).

3. Scattering Linear Polarization in 854.2 nm Ca II

Ubiquitous linear polarization structures across the solar disk in the core of the Ca II line are revealed in VSM results from 2016.03.23 with spectropolarimetric sensitivity close to 0.01% (Fig. 3). The poster includes context images that show where these high-sensitivity observations were made. These observations confirm theoretical studies that predicted this scattering linear polarization and also earlier observations of linear polarization in small areas near the solar limb. Sequential measurements show that the linear polarization features change rapidly in time. Spatially, the polarized structures are closely associated with bright Ca II mottles and dark elongated fibril structures. Bright (dark) features tend to be polarized parallel (perpendicular) to the closest limb as illustrated in the poster. The transverse Zeeman effect is generally overwhelmed by the scattering polarization, except in sunspots and strong plages (see panel A in Fig. 3). This dominance of linear polarization by scattering presents a severe challenge to measuring the chromospheric vector magnetic field using the 854.2 nm line.

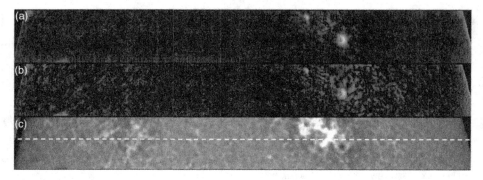

Figure 3. Linear polarization fraction in line wing and line core. (A) Zeeman effect dominates the 0.1 to 0.3 Å average from both sides of the line core. (B) Scattering polarization averaged ±0.1 Å across the line core. (C) Line core intensity with dashed line indicating the slit position of the Stokes spectra in Fig. 4. Linear polarization strength is displayed on a log intensity scale ranging from 0.001% to 5% for (A) and 0.01% to 5% for (B).

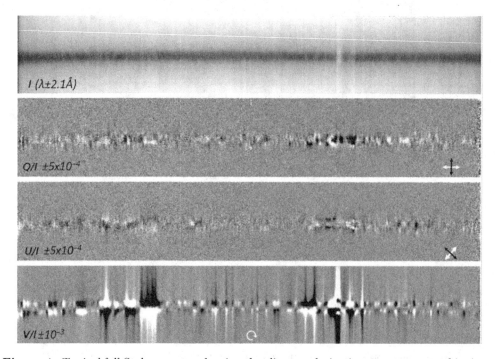

Figure 4. Typical full Stokes spectra showing that linear polarization structures are ubiquitous across the disk in the line core of 854.2 nm Ca II. The location of these spectra is the dashed line in Fig. 3. Top panel is intensity. Middle panels are Q/I and U/I linear polarization. Bottom panel is V/I circular polarization. The linear polarization noise levels are ±0.012%.

In Fig. 4 we see that transverse Zeeman splitting patterns are present only in strong magnetic regions. Elsewhere, scattering linear polarization dominates spectra near the core of the Ca line. Compared with circularly polarized spectral features arising from the line-of-sight magnetic field, the scattering linear polarization shows much more erratic wavelength variations. There is a gradual increase of linear polarization strengths moving toward the limb. Not shown here is a strong increase of a correlation between linear polarization strength and intensity toward the limb. Also noted in the spectra, and visible near the bottom left of panels A and B of Fig. 3, is a strong linear polarization signal in a disk filament. The azimuth of this linear polarization appears to be modified by the magnetic field of the filament.

Jurčák *et al.* (2018) observed with a slit near the solar disk center and found linear polarization features similar to the ubiquitous ones reported in this poster. Both sets of observations are remarkably similar to the models of Štěpán & Trujillo Bueno (2016) confirming that the models include the relevant physics. An essential finding is that the chromosphere is extremely inhomogeneous and dynamic, which is no surprise.

Acknowledgements

This work utilizes SOLIS data obtained by the NSO Integrated Synoptic Program (NISP), managed by the National Science Foundation's National Solar Observatory, which is operated by the Association of Universities for Research in Astronomy (AURA), Inc. under a cooperative agreement with the National Science Foundation.

This paper makes use of the following ALMA data: ADS/JAO.ALMA#2011.0. 00020.SV.. The Joint ALMA Observatory is operated by ESO, AUI/NRAO and NAOJ. ALMA is a partnership of ESO (representing its member states), NSF (USA) and NINS (Japan), together with NRC (Canada) and NSC and ASIAA (Taiwan), and KASI (Republic of Korea), in cooperation with the Republic of Chile. The National Radio Astronomy Observatory is a facility of the National Science Foundation (NSF) operated under cooperative agreement by Associated Universities, Inc.

References

Gosain, S. 2017, *SOLARNET IV. The Physics of the Sun from the Interior to the Outer Atmosphere*, Online at https://ui.adsabs.harvard.edu/link_gateway/2017psio.confE..50G/PUB_PDF

Harvey, J. W., Bertello, L., Branston, D., Britanik, J., Bulau, S., Cole, L., Gosain, S., Harker, B., Jones, H. P., Marble, A., Martinez Pillet, V., Pevtsov, A., Schramm, K., Streander, K., & Villegas, H. 2016, *AAS/Solar Physics Division Abstracts #47*, Online at https://ui.adsabs.harvard.edu/abs/2016SPD....4710106H/abstract

Jurčák, J., Štěpán, J., Trujillo Bueno, J., & Bianda, M. 2018, *A&A*, 619, A60

Keller, C. U., Harvey, J. W., & Solis Team 2003, *Astron. Soc. Pac. Conf. Series*, 307, 13

Molnar, M. E., Reardon, K. P., Chai, Y., Gary, D., Uitenbroek, H., Cauzzi, G., & Cranmer, S. R. 2019, *ApJ*, 881, 91

Štěpán, Jiří, & Trujillo Bueno, Javier 2016, *ApJ*, 826, L10

Solar and Stellar Magnetic Fields: Origins and Manifestations
Proceedings IAU Symposium No. 354, 2019
A. Kosovichev, K. Strassmeier & M. Jardine, ed.
doi:10.1017/S1743921320000186

Diagnosing chromospheric magnetic field through simultaneous spectropolarimetry in H α and Ca II 854.2 nm

K. Nagaraju[1]⊙, K. Sankarasubramanian[2] and K. E. Rangarajan[1]

[1]Indian Institute of Astrophysics, Sarjapur Road, Bengaluru, India
email: `nagarajuk@iiap.res.in`; rangaraj@iiap.res.in
[2]URSC & ISRO, Vimanapura, Bengaluru, India
email: `sankark@ursc.isro.gov.in`

Abstract. Measurement of magnetic field in this layer is challenging both from point of view of observations and interpretation of the data. We present in this work about spectropolarimetric observations of a pore, simultaneously in Ca II (CaIR) at 854.2 nm (CaIR) and H α (656.28 nm). The observed region includes a small scale energetic event (SSEE) taking place in the region between the pore and the region which show opposite polarity to that of pore at the photosphere. The energetic event appears to be a progressive reconnection event as shown by the time evolution of the intensity profiles. Closer examination of the intensity profiles from the downflow regions suggest that the height of formation of CaIR is higher than that of Hi α, contrary to the current understanding about their height of formation. Preliminary results on the inversion of Stokes-I and V profiles of CaIR are also presented.

Keywords. Photosphere: Magnetic field, Chromosphere: Magnetic field, Multi-line-Spectropolarimetry

1. Introduction

Accurate knowledge of magnetic field at multiple heights in the solar atmosphere is of paramount importance to understand the various dynamics such as energy transportation and dissipation, energetic events of various spatial and energy scales, wave-dynamics, just to name a few. Measurement of magnetic field at the photosphere is carried out more or less on routine basis. But at the higher layers it is much more challenging due to weak polarimetric signals in the spectral lines because of the weaker magnetic fields and the inference of magnetic field from the observed spectrum requires taking into account the NLTE effects. Inspite of this there has been a significant progress towards chromospheric magnetic field measurements in the recent past (Lagg *et al.* 2017). Some of the most widely used spectral lines for inferring chromospheric magnetic field are CaIR and He I 10830 lines (Lagg *et al.* 2017) while efforts on exploring the diagnostic capability of other spectral lines is still going on. However, each of these lines have their own strengths and limitations under certain conditions. The He I samples mostly upper chromospheric layers and it's formation depends on the coronal EUV emission (Andretta & Jones 1997). On the other hand CaIR line gets completely ionized during flares (Kerr *et al.* 2016; Kuridze *et al.* 2018; Bjørgen *et al.* 2019). In order to probe wide range of dynamics those take place in the chromosphere, multi-line spectropolarimetry is needed. In this context H α may play an important role in probing chromospheric magnetic field as it's ubiquitously present in the wide variety of chromospheric plasma conditions such as quiet sun, active regions(Rutten 2007, 2008) and flaring regions (Bjørgen *et al.* 2019).

Inspite of its ubiquitous presence H α has been rarely used for probing chromospheric magnetic field. To mention a few, Abdusamatov & Karat (1971); Balasubramaniam *et al.* (2004); Hanaoka (2005); Nagaraju *et al.* (2008) have presented chromospheric magnetic field measurements through H α spectropolarimetry. One of the reasons for rarely using this line for magnetic field measurements is interpreting the observations is notoriously difficult. For instance Socas-Navarro & Uitenbroek 2004 have shown through numerical simulations that H α Stokes-V is sensitive to photospheric magnetic field in quiet sun model while its temperature sensitivity is in chromosphere. However, Leenaarts *et al.* (2012) have shown through numerical simulation that solving 3D radiative transfer equation is essential to model H α line and shown that it has a good sensitivity to chromospheric magnetic field. Very recently Bjørgen *et al.* (2019) have shown again through simulation that flare ribbons are seen in all chromospheric lines including H α but least visible in CaIR.

Hence exploring diagnostic capability of H α is important. In this context we compare in this paper about the simultaneous spectropolarimetric observations in CaIR and H α. The comparison is done above pore, opposite polarity region and SSEE. Though the observations were of full Stokes polarimetry, we present only Stokes-*I* and *V* profiles because Stokes-*Q* and *U* signals are weak or negligible compared to the noise level.

2. Observations and Data reduction

Simultaneous spectropolarimetric observations in H α and CaIR were carried out using the Spectropolarimeter for Infrared and Optical Regions (SPINOR: Socas-Navarro *et al.* 2006) at the Dunn Solar Telescope, National Solar Observatory, Sunspot, New Mexico. The spatial sampling in scan and slit directions is 0.3″/pixel on the Sarnoff camera used for recording the H α spectrum. The spectral sampling is 22.45 mÅ/pixel and the spectral resolution is 32.18 mÅ. For CaIR region the spatial sampling is same as that of H α while the spectral sampling is 33.73 mÅand the spectral resolution is 49.17 mÅ.

The region of observations on the sun was around a small pore (Fig. 1). The cosine of the heliocentric angle μ is 0.536. Though the observations were carried out on four successive days from 4 through 7 December 2008, only a few scans obtained on 4 December are with a reasonably good seeing. Four scans of of 20 slit positions with a corresponding spatial extent of 6″ recorded at 15:35, 15:42, 15:57 and 16:08 UT are considered for analysis. Each raster map took about 3 minutes to scan. The noise level in the Stokes profiles is in the order $10^{-3}I_c$ (I_c is the continnum intensity).

The data were subjected to bias and flat corrections. The data were corrected for polarimeter response using the calibration data obtained on the same day of observation. To correct for the telescope induced cross-talks we have used the telescope calibration data obtained during May 2005. Since it is a very old calibration the cross-talk among Stokes parameters was not corrected very well. To correct for these residual cross-talks we have applied scatter plot method (Sanchez Almeida & Lites 1992; Schlichenmaier & Collados 2002). Only the cross-talk from Stokes *V* to Stokes *Q* and *U* is corrected as the amplitude of former is always greater than the amplitudes of later ones. The cross-talk from intensity to polarisation is corrected by assuming the continuum to be non-polarized. Estimation of stray-light factor was done by comparing disc center Stokes-*I* profiles with that of atlas spectrum (FTS). The estimated straylight was about 18%. Nearby quiet sun profile was used as profile function for the straylight.

The observing period covers a small scale energetic event (SSEE). The raster maps of the observed region is shown in Fig. 1. The first map in this figure corresponds to the continuum close to H α, the second, third and fourth maps correspond to the line core intensities of Fe I at 656.922 nm, H α and CaIR, respectively and the fifth, the sixth and the seventh maps correspond to the respective Stokes-*V* amplitudes. The region of SSEE

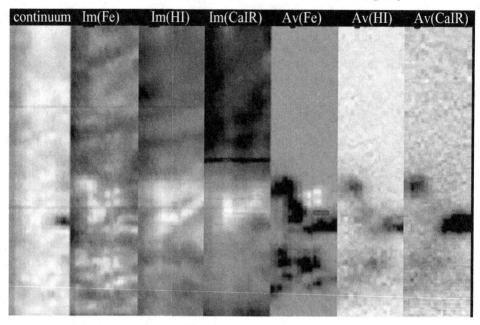

Figure 1. Raster maps of the observed region. The left most raster map corresponds to the continuum (close to H α) followed by the line core intensity images of Fe I at 656.922 nm, H α and CaIR, respectively and the last three column correspond to the respective Stokes-V amplitudes.

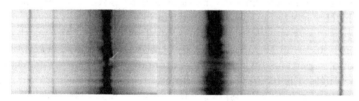

Figure 2. Stokes-I spectral images of H α and CaIR corresponding to a SSEE (as indicated by an arrow on the CaIR spectrum).

is indicated by a blue rectangular box placed on the Stokes-V amplitude map of Fe I line at 656.922 nm. Also seen in this map is the region of magnetic field which is opposite to that of pore. Though it is visible only in the photospheric line, henceforth we refer this region as the opposite polarity region.

Sample spectral images of CaIR (left panel) and H α (the right panel) corresponding to a slit position above SSEE are shown in Fig. 2. The SSEE is indicated by an arrow in this figure shows up as a mustache structure in the CaIR spectrum, typical of Ellerman bomb event (Ellerman 1917). However, there's no signature EB in the H α spectral image. The time evolution of intensity profiles as shown Fig. 3 suggest that SSEE is a progressive reconnection event smilar to that of flaring arch filament (Vissers *et al.* 2015). Initially emission in the wings of CaIR is observed indicating that the heating taking place in the lower atmospheric layers. The H α shows evelated intensity profiles. Towards the end of our observations, CaIR line core goes in to emission and H α line gets further elevated due to the heating taking place in the region. The magnetic field geometry associated with the SSEE seems to have a simple configuration with the newly emerged opposite polarity fields reconnecting with the canopy fields of the pore.

In the beginning of the SSEE the region shows strong downflows. Sample profiles are shown in Fig. 4. The left and the right panels in this figure show CaIR and H α intensity

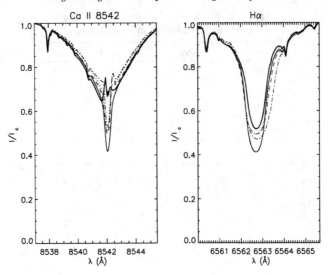

Figure 3. Stokes-*I* profiles of H α and CaIR corresponding to the SSEE. The thin solid curves correspond to a quiet sun region, The dotted, dashed, dash-dotted and thick solid curves correspond to the spectra observed at 15:35, 15:42, 15:57 and 16:08 UT, on 4 December, 2008, respectively.

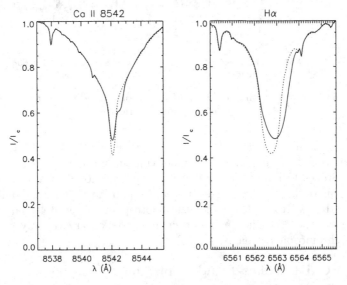

Figure 4. Stokes-*I* profiles of CaIR (the left panel) and H α (the right panel) showing downflows during the reconnection event.

profiles (the solid curves), respectively. The dotted curves correspond to the respective quiet sun profiles. While the CaIR shows redshift away from the line core, H α shows redshift including its line core. This possibly indicates that the height of formation of H α is lower than that of the CaIR. This is contrary to the current understanding that height of formation of CaIR is very close to that of H α or slighty below.

Sample Stokes-*I* (the top row) and *V* (the bottom row) profiles of CaIR (the panels on the left column) and H α (the panels on the right column) are shown in Fig. 5 above the opposite polarity region (see Stokes-*V* amplitude map of Fe I in Fig. 1). The Stokes-*V* profile of H α has the polarity opposite to that of Fe I line at 656.922 nm which forms at

Table 1. Number of nodes used for each
parameter in different inversion cycles.

Parameters	Cycle 1	Cycle 2	Cycle 3
Temperature	4	8	10
V_{los}	1	4	4
B_z	1	3	4
B_x	0	0	0
B_y	0	0	0
Microturbulence	1	2	2

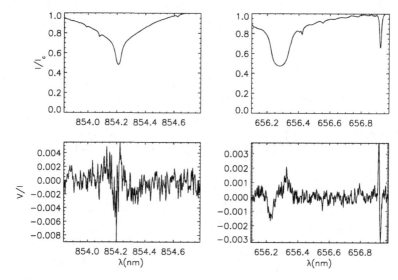

Figure 5. Stokes-I and V profiles of H α and CaIR corresponding to the region of opposite polarity (photospheric).

the photosphere. Though the signal looks very noisy, the Stokes-V of CaIR also shows the sign same as that of H α, particularly the blue-wing amplitude of the line. As seen in the raster maps of Stokes-V amplitudes as shown in Fig. 1 do not show any opposite polarity in the chromospheric lines. This suggests on the one hand that the chromospheric field that is sampled by CaIR and H α above the photospheric opposite polarity region is of canopy fields form the pore and on the other hand this observationally proves that H α indeed probes chromospheric magnetic fields.

3. Inversion CaIR Stokes-I and V profiles

Spectropolarimetric inversions of the CaIR Stokes profiles have been carried out using the NICOLE inversion code (Socas-Navarro *et al.* 2015). The inversion of quiet sun and pore profiles have been carried out in two cycles and those from SSEE region carried out in three cycles. In the first cycle lower number of nodes were used and in the next cycle the number of cycles were increased (see Table. 1). For the first cycle FALC model was used as an initial guess model. In the second cycle, the output model from the cycle-1 was used as an input guess model but with the magnetic field values replaced by those estimated through weak-field-approximation. Since, the Stokes-Q and U profiles are noise they were not considered for fitting. This was done by setting zero nodes to the horizontal components during the inversion (cf. Table. 1). Only Stokes-I and V were fitted. In the third cycle we altered the output model from cycle-2 to introduce velocity gradients. This is done only for the profiles from the region of SSEE. Signatures of velocity gradients are

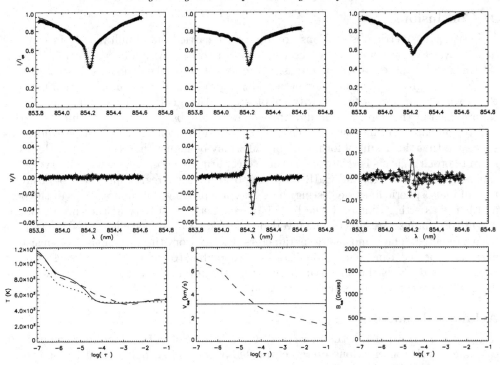

Figure 6. Plots of Stokes I (the top rows) and V (the middle rows) profiles of quiet sun (the left column), pore (the middle column) and SSEE (the right column). The observed profiles are plotted in plus symbols and the profiles fitted using NICOLE are over plotted as the solid curves. The third row shows the temperature (the left), velocity (the middle) and the LOS magnetic field (the right panel). The solid, dotted and dashed curves correspond to the quiet sun, pore and SSEE region, respectively.

clearly seen in the Stokes-I and V profiles as asymmetries. Typical observed and fitted Stokes-I (the top row) and V (the middle row) profiles are shown in Fig. 6. The panels from the left to the right in the first and the second row correspond to the regions of quiet sun, pore and SSEE. The left, the middle and the right panels in the third row show the temperature, line-of-sight (LOS) velocity and LOS magnetic field stratification (on $\log(\tau)$ scale) output from the NICOLE inversion. The solid, the dotted and the dashed curves correspond to the quiet sun, pore and SSEE regions, respectively.

The temperature derived from the inversion is lower above the pore compared to that of quiet sun except between the heihgts with $\log(\tau)$ equal to -4 and -3 where the values are comparable. The atmosphere above the pore is cooler by upto $\approx$2000 K. The temperature in the SSEE region is enhanced compared to quiet sun up to about $\approx$800 K. The maximum temperature enhancement take place close to $\log(\tau)$ equal to -4. However, the fitted profile shows the wing enhancement further away than that of observed wing enahncement (cf. the top right panel in Fig. 1) suggesting that the temperature enahcement is taking place higher up in the atmosphere than estimated by the inversion. In order to fit the profiles from SSEE regions presence velocity gradients was necessary. A typical velocity stratification from this region is shown in the middle panel of the bottom row of Fig. 6. The magnetic field output from the inversion is shown in the bottom panel of Fig. 6. Though the LOS component of the magnetic field is consistent with the observed region that field is stronger above the pore than above SSEE, there's no variation observed with height. Introduction of field gradient actually worsened the fitting of the profiles.

4. Conclusions

Spectropolarimetric observations around a pore carried out simultaneosly in H α and CaIR have been presented. The observing period covered a small scale energetic event (SSEE). The time evolution of intensity profiles of both the lines has indicated that SSEE is a progressive reconnection event. This is produced due to an interaction between the newly emerged opposite polarity (photospheric) magnetic field and the pre-existing canopy field of the pore. Strong downflows have been observed in the beginning of this event. Comparison of intensity profiles of H α and CaIR from these downflow regions has suggested that the height of formation of CaIR may be higher than that of H α. However, more rigorous analysis is required in order to confirm this. Further, it was observed that Stokes-V profiles of H α and CaIR above opposite polarity region have sign opposite to that of photospheric lines suggesting these lines are probing the canopy fields of the pore and hence asserting that both the lines have magnetic sensitivity at the chromospheric heights. Preliminary results on the inversion of Stokes-I and V profiles of CaIR also have been presented. The atmospheric parameters derived from the inversions are consistent with the expected atmospheric conditions corresponding to different regions such as quiet sun, pore and SSEE. Further improvement in the inversions are expected which may improve the accuracy in the derived atmospheric parameters.

Acknowledgement

We would like to thank the observers at the Dunn Solar Telescope for their help in carrying out the observations presented in this paper and H. Socas-Navarro for help in setting up of NICOLE code in the initial stages.

References

Abdussamatov, H. I. 1971, *Sol. Phys*, 16, 384
Andretta, V., Jones, H. P. 1997, *ApJ*, 489, 375
Balasubramaniam, K. S., Christopoulou, E. B., Uitenbroek, H. 2004,*ApJ*, 606, 1233
Bjørgen, J. P., Leenaarts, J., Rempel, M., Cheung, M. C. M., Danilovic, S., de la Cruz Rodríguez,
    J., Sukhorukov, A. V. 2019, arXiv:1906.01098
Ellerman, F., 1917, *ApJ*, 46, 298
Hanaoka, Y. 2005, *Pub. Astro. Soc. Japan*, 57, 235
Kerr, G. S., Fletcher, L., Russell, A.e.J. B., Allred, J. C., 2016, *ApJ*, 827, 101
Kuridze, D., Henriques, V. M. J., Mathioudakis, M., Rouppe van der Voort, L., de la Cruz
    Rodríguez, J., Carlsson, M., 2018, *ApJ*, 860, 10
Lagg, A., Lites, B., Harvey, J., Gosain, S., Centeno, R., 2017, *Space Science Rev.*, 210, 37
Leenaarts, J., Carlsson, M., Rouppe van der Voort, L., 2012, *ApJ*, 749, 136
Nagaraju, K., Sankarasubramanian, K., Rangarajan, K. E., 2008, *ApJ*, 678, 531
Rutten, R. J. 2007, in Heinzel, P., Dorotovič, I., Rutten, R. J. eds, The Physics of Chromospheric
    Plasmas, ASP Conference Series, Volume 368, p. 27
Rutten, R. J. 2008, in Matthews, S. A., Davis, J. M., Harra, L. K., eds, First Results From
    Hinode, ASP Conference Series,Volume 397, p. 54
Sanchez Almeida, J., Lites, B. W., 1992, *ApJ*, 398, 359. DOI. ADS.
Schlichenmaier, R., Collados, M., 2002, *Astron. & Astrophys.*, 381, 668
Socas-Navarro, H., Uitenbroek, H., 2004, *ApJ (letters)*, 603, L129
Socas-Navarro, H., Martínez Pillet, V., Elmore, D., Pietarila, A., Lites, B. W., Manso Sainz,
    R.m 2006, *Sol. Phys.*, 235, 75
Socas-Navarro, H., de la Cruz Rodríguez, J., Asensio Ramos, A., Trujillo Bueno, J., Ruiz Cobo,
    B., 2015, *Astron. & Astrophys.*, 577, 7
Vissers, G. J. M., Rouppe van der Voort, L. H. M., Rutten, R. J., Carlson, M., de Pontieu, B.,
    2015, *ApJ*, 812, 11

Solar and Stellar Magnetic Fields: Origins and Manifestations
Proceedings IAU Symposium No. 354, 2019
A. Kosovichev, K. Strassmeier & M. Jardine, ed.
doi:10.1017/S1743921319009955

The magnetic structure and dynamics of a decaying active region

Ioannis Kontogiannis[1]📖, Christoph Kuckein[1]📖, Sergio Javier González Manrique[2]📖, Tobias Felipe[3,4]📖, Meetu Verma[1]📖, Horst Balthasar[1]📖 and Carsten Denker[1]📖

[1]Leibniz-Institut für Astrophysik (AIP), An der Sternwarte 16, 14482, Potsdam, Germany
email: ikontogiannis@aip.de

[2]Astronomical Institute, Slovak Academy of Sciences, Tatranská Lomnica, Slovakia

[3]Instituto de Astrofísica de Canarias (IAC), La Laguna, Tenerife, Spain

[4]Universidad de La Laguna, Tenerife, Spain

Abstract. We study the evolution of the decaying active region NOAA 12708, from the photosphere up to the corona using high resolution, multi-wavelength GREGOR observations taken on May 9, 2018. We utilize spectropolarimetric scans of the 10830 Å spectral range by the GREGOR Infrared Spectrograph (GRIS), spectral imaging time-series in the Na I D$_2$ spectral line by the GREGOR Fabry-Pérot Interferometer (GFPI) and context imaging in the Ca II H and blue continuum by the High-resolution Fast Imager (HiFI). Context imaging in the UV/EUV from the Atmospheric Imaging Assembly (AIA) onboard the Solar Dynamics Observatory (SDO) complements our dataset. The region under study contains one pore with a light-bridge, a few micro-pores and extended clusters of magnetic bright points. We study the magnetic structure from the photosphere up to the upper chromosphere through the spectropolarimetric observations in He II and Si I and through the magnetograms provided by the Helioseismic and Magnetic Imager (HMI). The high-resolution photospheric images reveal the complex interaction between granular-scale convective motions and a range of scales of magnetic field concentrations in unprecedented detail. The pore itself shows a strong interaction with the convective motions, which eventually leads to its decay, while, under the influence of the photospheric flow field, micro-pores appear and disappear. Compressible waves are generated, which are guided towards the upper atmosphere along the magnetic field lines of the various magnetic structures within the field-of-view. Modelling of the He I absorption profiles reveals high velocity components, mostly associated with magnetic bright points at the periphery of the active region, many of which correspond to asymmetric Si I Stokes-V profiles revealing a coupling between upper photospheric and upper chromospheric dynamics. Time-series of Na I D$_2$ spectral images reveal episodic high velocity components at the same locations. State-of-the-art multi-wavelength GREGOR observations allow us to track and understand the mechanisms at work during the decay phase of the active region.

Keywords. Sun: atmosphere, Sun: chromosphere, Sun: magnetic field, Sun: photosphere, Sun: oscillations

1. Overview

Active region NOAA 12708 was decaying during the observations on May 9, 2018 (Fig. 1). It contained several pores, a large one with a light-bridge at the negative polarity part, and a few smaller ones scattered in both polarities of the region. The magnetic field lines in Fig. 1 indicate the coexistence of several magnetic loops. The longer loops are

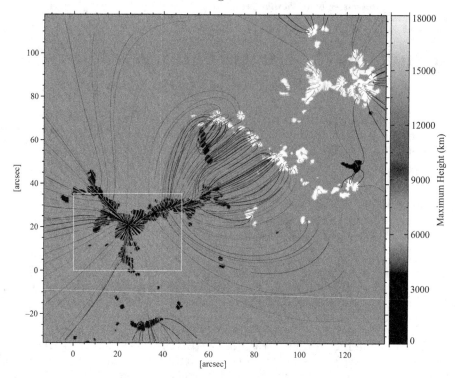

Figure 1. Active region NOAA 12708. Map of the photospheric radial component, Br, of the magnetic field provided by the HMI (Scherrer *et al.* 2012) onboard SDO (Pesnell *et al.* 2012). Overploted are the magnetic field lines which correspond to the extrapolated current-free (potential) magnetic field vector (Alissandrakis 1981). The yellow rectangle marks the region observed with GREGOR.

associated with higher emission in the hotter channels of AIA, e.g. in 171 Å (Fig. 2, bottom right). Upper chromospheric/transition region emission in He II 304 Å is more intense at regions with shorter loops, which connect the negative polarity magnetic elements with the nearby positive ones (see e.g. Kontogiannis *et al.* 2018).

The upper photosphere and lower chromosphere are observed in great detail by the GREGOR (Schmidt *et al.* (2012)) instruments (Fig. 2, top row). The broad-band filters of HiFI (Kuckein *et al.* 2017; Denker *et al.* 2018) show the granulation pattern while at the Na I D$_2$ flanks, sampled by GFPI (Puschmann *et al.* 2012), reversed granulation is more prominent, due to the upper photospheric/lower chromospheric origin of the line. The blue wing of this line can also be used to track bright points (BPs; Kuckein 2019). These observations reveal the impact of convective motions on the magnetic bright points, the micro-pores and the pore, the latter being slowly eroded and the light-bridge becoming wider.

In the upper chromosphere, in the region observed in He I thin absorption features are jutting out from the magnetic bright points and the pores (Fig. 3). These are more dense at the upper right part of the region, where the concentration of BPs and micro-pores is more prominent. They exhibit wider and deeper He I profiles and intense downflows and upflows near the footpoints.

2. Light-Bridge

The convective motions intrude into the magnetized region of the pore forming a light-bridge (Fig. 4, left panel) signifying the decay of the structure. The granule inside the

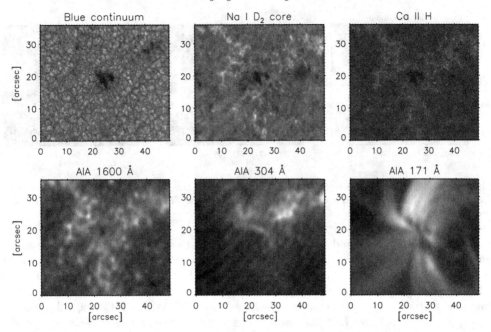

Figure 2. The region enclosed by the yellow rectangle in Fig. 1 seen with the HiFI blue continuum (top left) and the Ca II H broad band filter (top right), at the Na I D₂ spectral line flanks scanned by the GFPI (top middle), and at the three AIA channels (bottom row).

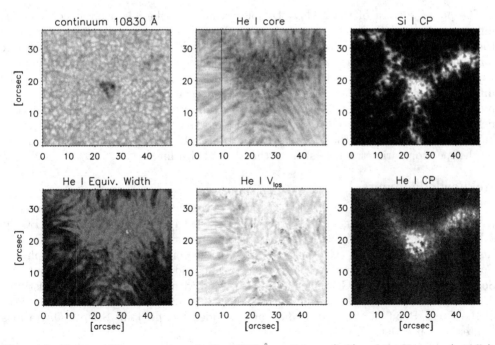

Figure 3. Top row: The pore as seen in the 10830 Å continuum (left) and the He I core (middle). Bottom row: Map of the He I equivalent width (left) and the line-of-sight velocity (middle; see e.g. González Manrique *et al.* 2018 for the spectral line fitting technique). The last column contains the maps of the circular polarization (CP) in Si I and He I (top and bottom row respectively).

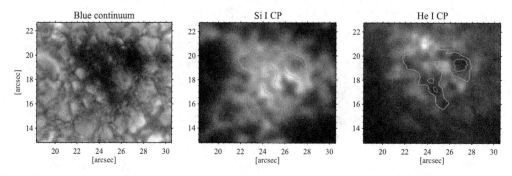

Figure 4. Close-ups of the pore and the light-bridge in the blue continuum (left) along with the corresponding maps of the circular polarization in the Si I and He I spectral lines (middle and right).

light-bridge is distorted and exhibits small-scale, fast evolving structures. The atmosphere within a light-bridge is complicated due to the interaction between the strongly magnetized plasma and the convection (e.g. Lagg *et al.* 2014). The magnetic field of a light-bridge usually forms a canopy-like structure, with field lines from the adjacent magnetic footpoints converging above the light-bridge. The polarization signal in the Si I (upper photosphere) and He I lines agree with such an interpretation. These maps also reveal that the pore is not a monolithic structure but consists of regions with different strengths and geometries of magnetic field.

3. Velocity Oscillations in Na I D_2

The power spectra of the Na I D_2 Doppler velocity oscillations exhibit considerable differences in the different regions of the pore. Power spectra from the light-bridge and the left part of the pore show a clear peak at 3.5 mHz, which is typical of the photospheric 5-minute oscillations. This peak is lower at the power spectra of other parts of the pore, which exhibit, also, a second peak at 3 min ($\sim$5 mHz), of chromospheric origin. The differences will be further investigated, taking into account the formation of the Na I D_2 line in the presence of strong magnetic field (Rutten *et al.* 2011), the detailed magnetic field and atmospheric stratification of the pore and their impact on the propagation of magnetoacoustic waves in the pore (e.g. Felipe *et al.* 2019). The latter will be determined by the spectropolarimetric inversions of the Si I and He I lines.

Acknowledgements

This work was supported by grant DE 787/5-1 of the Deutsche Forschungsgemeinschaft (DFG). SJGM acknowledges the project VEGA 2/0004/16. The 1.5-meter GREGOR solar telescope was built by a German consortium under the leadership of the Leibniz-Institute for Solar Physics (KIS) in Freiburg with the Leibniz Institute for Astrophysics Potsdam, the Institute for Astrophysics Göttingen, and the Max Planck Institute for Solar System Research in Göttingen as partners, and with contributions by the Instituto de Astrofísica de Canarias and the Astronomical Institute of the Academy of Sciences of the Czech Republic. SDO HMI data are provided by the Joint Science Operations Center - Science Data Processing.

References

Alissandrakis, C. E. 1981, *A&A*, 100, 197
Denker, C., Kuckein, C., Verma, M, *et al.*, 2018, *ApJSS*, 236, 5
Felipe, T., Kuckein, C., Khomenko, E., & Thaler, I. 2019, *A&A*, 621, A43

González Manrique, S. J., Kuckein, C., Collados, M., *et al.* 2018, *A&A*, 617, A55

Kontogiannis I., Gontikakis, C., Tsiropoula, G. & Tziotziou, K. 2018, *Sol Phys*, 293, 56

Kuckein, C. 2019, *A&A*, 630, A139

Kuckein, C., Denker, C, Verma, M, *et al.*, 2017, *IAU Symposium 327*, 20

Lagg, A., *et al.* 2014, *A&A*, 568, A60

Pesnell, W. D., Thompson, B. J., & Chamberlin, P. C., 2012, *Sol Phys*, 275, 3

Puschmann, K. G., Denker, C., Kneer, F., *et al.*, 2012, *AN*, 333, 880

Rutten, R., *et al.* 2011, *A&A*, 531, A17

Scherrer, P. H., Schou, J., Bush, R. I, *et al.*, 2012, *Sol Phys*, 275, 207

Schmidt, W., von der Lühe, O., Volkmer, R., *et al.*, 2012, *AN*, 333, 796

Solar and Stellar Magnetic Fields: Origins and Manifestations
Proceedings IAU Symposium No. 354, 2019
A. Kosovichev, K. Strassmeier & M. Jardine, ed.
doi:10.1017/S1743921320000101

Coordinated observations between China and Europe to follow active region 12709

S. J. González Manrique[1]📖, C. Kuckein[2]📖, P. Gömöry[1], S. Yuan[3],
Z. Xu[3], J. Rybák[1], H. Balthasar[2] and P. Schwartz[1]

[1]Astronomical Institute, Slovak Academy of Sciences, 05960 Tatranská Lomnica,
Slovak Republic
email: `smanrique@ta3.sk`

[2]Leibniz-Institut für Astrophysik Potsdam (AIP), An der Sternwarte 16,
14482 Potsdam, Germany

[3]Yunnan Observatories, Chinese Academy of Sciences, Kunming, 650011, China

Abstract. We present the first images of a coordinated campaign to follow active region NOAA 12709 on 2018 May 13 as part of a joint effort between three observatories (China-Europe). The active region was close to disk center and enclosed a small pore, a tight polarity inversion line and a filament in the chromosphere. The active region was observed with the 1.5-meter GREGOR solar telescope on Tenerife (Spain) with spectropolarimetry using GRIS in the He I 10830 Å spectral range and with HiFI using two broad-band filter channels. In addition, the Lomnicky Stit Observatory (LSO, Slovakia) recorded the same active region with the new Solar Chromospheric Detector (SCD) in spectroscopic mode at Hα 6562 Å. The third ground-based telescope was located at the Fuxian Solar Observatory (China), where the active region was observed with the 1-meter New Vacuum Solar Telescope (NVST), using the Multi-Channel High Resolution Imaging System at Hα 6562 Å. Overlapping images of the active region from all three telescopes will be shown as well as preliminary Doppler line-of-sight (LOS) velocities. The potential of such observations are discussed.

Keywords. Sun: activity, Sun: chromosphere, Sun: filaments, Sun: magnetic fields

1. Introduction

Solar phenomena extend over many layers of the atmosphere. Frequently magnetic fields are rooted in the photosphere and expand with height and twist, forming helical configurations such as filaments. In order to follow the field lines across the solar atmosphere, multiwavelength observations are crucial. However, current ground-based telescopes such as GREGOR only offer up to three simultaneous instruments which cover different wavelength ranges. To better understand the formation and evolution of solar phenomena we need more co-temporal observations with many wavelengths. This is also the main objective of future high-resolution telescopes such as DKIST (Tritschler *et al.* 2015) and EST (Collados *et al.* 2013). In the meantime, efforts are being done to coordinate high-resolution ground-based telescopes to study our Sun. This work shows the viability of coordinated observations between telescopes in Europe and China, as well as the potential when combining these observations.

2. Observations

Active region (AR) NOAA 12709 was observed on 2018 May 13 very close to disk center at heliographic coordinates (S6°, W3°). The AR was of interest because it enclosed several solar phenomena. In the center of the AR, a narrow polarity inversion line separated an

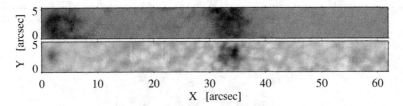

Figure 1. Continuum intensity (*bottom*) and Stokes V (*top*) slit-reconstructed image of AR NOAA 12709 recorded with GRIS between 08:53:13 UT and 08:57:02 UT on 2018 May 13.

extensive area of opposite polarities. Furthermore, a decaying pore was present at the border of the AR and an intermediate filament was seen in the chromosphere.

Four instruments, which were located at three different telescopes around the world, were recording simultaneously this active region: (1) The GREGOR Infrared Spectrograph (GRIS, Collados *et al.* 2012) and (2) the High-resolution Fast Imager (HiFI, Kuckein *et al.* 2017, Denker *et al.* 2018) attached to the 1.5-meter GREGOR solar telescope on Tenerife, Spain (Schmidt *et al.* 2012); (3) The new Solar Chromospheric Detector (SCD, Kucera *et al.* 2015) based on the Lomnicky Stit Observatory (LSO), Slovakia; and (4) the Multi-Channel High Resolution Imaging System (Xu *et al.* 2013) placed at the the 1-meter New Vacuum Solar Telescope, China (NVST, Liu *et al.* 2014).

We observed with HiFI anf GRIS around two hours and twenty minutes the same region (starting at 08:08 UT), mainly focusing on the main pore but also observing the polarity inversion line. The SCD science data were acquired in the time intervals 05:57–06:28, 06:43–07:19, 08:52–09:27, and 09:53–10:10 UT. We observed with a good afternoon seeing at the Fuxian lake around 1.5 hours with the Hα filter band on 08:48–10:22 UT while we only observed around 20 minutes with the TiO starting at 9:38 UT.

All data were dark and flat-field corrected, as well as polarimetrically calibrated following standard procedures. HiFI and NVST data were restored using the speckle code from Wöger & von der Lühe (2008) and Liu *et al.* (1998), respectively.

3. Combining different telescopes

One big challenge is the combination of data acquired with different instruments. While SCD, NVST, and HiFI are imaging instruments, GRIS is a slit spectrograph. Hence, GRIS only provides slit-reconstructed 2D images (Fig. 1). Furthermore, all instruments have a different image scale. While HiFI provides very high spatial resolution images in the blue wavelength range (about $0.025''$pixel^{-1}), a lower spatial sampling of $0.136''$pixel^{-1} or $0.340''$pixel^{-1} is recorded by NVST and SCD in Hα, respectively. Yet, having different image scales can be beneficial because they produce different sizes of the FOV. SCD provides the largest FOV with about $410''\times335''$ (Fig. 2) and therefore it is an excellent instrument to cover context information surrounding the largely spread AR. The images from the 1-meter NVST provide high resolution images of the chromosphere and still cover a large FOV of about $126''\times126''$ (left panel of Fig. 3).

4. Potential of coordinated multiwavelength observations

For the analysis of the central pore in Fig. 3, the combination of the present instruments, together with space telescopes, have the potential to provide: (1) the vector magnetic field and LOS velocities in the photosphere and chromosphere by applying spectral line inversion tools to the four Stokes parameters (GRIS data set); (2) tracking of horizontal flows using the high-resolution data of HiFI and NVST; (3) Doppler velocities of the chromosphere in the whole AR using SCD, to study the connectivity of the pore to the entire AR; (4) context data from the Solar Dynamics Observatory

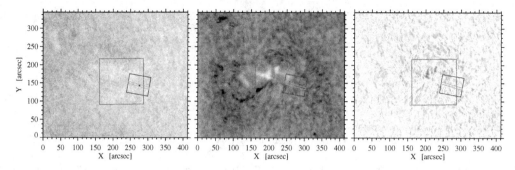

Figure 2. Overview of AR NOAA 12709 obtained with the SCD at 08:53:09 UT on 2018 May 13. Quasi-continuum intensity in the far Hα blue line wing (*left*), Hα line core intensity (*middle*), and Doppler shifts clipped between ± 2 km s^{-1} (*right*). The red and blue rectangles represent FOVs of NVST and HiFI instrument at GREGOR, respectively. The green rectangle represents the FOV covered by GRIS.

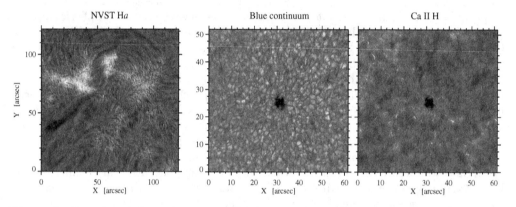

Figure 3. Overview of the active region NOAA 12709 obtained on 2018 May 13. Restored Hα 6562.8 Å image recorded (08:53:05 UT) with the Multi-Channel High Resolution Imaging System placed on the NVST (*left*), and HiFI Speckle-reconstructed (08:53:08 UT) blue continuum 4505 Å (*middle*) and Ca II H 3968 Å (*right*). NVST and HiFI are not showing the same FOV.

(SDO, Pesnell *et al.* 2012) will be used to follow the dynamics of the AR at different coronal layers of the atmosphere. Additionaly, the Helioseismic and Magnetic Imager (HMI, Scherrer *et al.* 2012; Schou *et al.* 2012) provides the possibility to investigate the evolution of the vector magnetic field of the whole AR by carrying out magnetic field extrapolations.

Acknowledgements

The 1.5-meter GREGOR solar telescope was built by a German consortium under the leadership of the Leibniz-Institut für Sonnenphysik in Freiburg (KIS) with the Leibniz-Institut für Astrophysik Potsdam (AIP), the Institut für Astrophysik Göttingen (IAG), the Max-Planck-Institut für Sonnensystemforschung in Göttingen (MPS), and the Instituto de Astrofísica de Canarias (IAC), and with contributions by the Astronomical Institute of the Academy of Sciences of the Czech Republic (ASCR). This work is part of a collaboration between the AISAS and AIP supported by the German DAAD, with funds from the German Federal Ministry of Education & Research and Slovak Academy of Science, under project No. 57449420. SJGM, PG, and PS acknowledge the support of the project VEGA 2/0004/16. SJGM also is grateful for the support of the Stefan Schwarz grant of the Slovak Academy of Science.

References

Collados, M., Bettonvil, F., Cavaller, L., Ermolli, I., Gelly, B., Pérez, A., Socas-Navarro, H., Soltau, D., Volkmer, R., and EST Team 2013 *Mem. Soc. Astron. It.*, 84, 379

Collados, M., López, R., Páez, E., Hernández, E., Reyes, M., Calcines, A., Ballesteros, E., Díaz, J. J., Denker, C., Lagg, A., Schlichenmaier, R., Schmidt, W., Solanki, S.K., Strassmeier, K. G., von der Lühe, O., Volkmer, R. 2012, *Astron. Nachr.*, 333, 872

Denker, C., Dineva, E., Balthasar, H., Verma, M., Kuckein, C., Diercke, A., González Manrique, S. J. 2018, *Sol. Phys.*, 293, 44

Kucera, A., Tomczyk, S., Rybak, J., Sewell, S., Gomory, P., Schwartz, P., Ambroz, J., Kozak, M. 2015, *in IAU General Assembly*, 29, 2246687

Kuckein, C., Denker, C., Verma, M., Balthasar, H., González Manrique, S. J., Louis, R. E., Diercke, A. 2017, in n Fine Structure and Dynamics of the Solar Atmosphere, eds. S. Vargas Domínguez, A. G. Kosovichev, P. Antolin, & L. Harra, *in IAU Symp.*, 327, 20

Lexa, J. 1963, *Bulletin of the Astronomical Institutes of Czechoslovakia*, 14, 107

Liu, Z., Xu, J., Gu, B.-Z., Wang, S., You, J.-Q., Shen, L.-X., Lu, R.-W., Jin, Z.-Y., Chen, L.-F., Lou, K., Li, Z., Liu, G.-Q., Xu, Z., Rao, C.-H., Hu, Q.-Q., Li, R.-F., Fu, H.-W., Wang, F., Bao, M.-X., Wu, M.-C., Zhang, B.-R. 2014, *RAA*, 14, 705

Liu, Z., Qiu, Y., & Lu, R. 1998, *in Proc. SPIE*, 3561, 326

Pesnell, W. D., Thompson, B. J., Chamberlin, P. C. 2012, *Sol. Phys.*, 275, 3

Scherrer, P. H., Schou, J., Bush, R. I., Kosovichev, A. G., Bogart, R. S., Hoeksema, J. T., Liu, Y., Duvall, T. L., Zhao, J., Title, A. M., Schrijver, C. J., Tarbell, T. D., Tomczyk, S., 2012, *Sol. Phys.*, 275, 207

Schmidt, W., von der Lühe, O., Volkmer, R., Denker, C., Solanki, S. K., Balthasar, H., Bello Gonzalez, N., Berkefeld, T., Collados, M., Fische, A., Halbgewachs, C., Heidecke, F., Hofmann, A., Kneer, F., Lagg, A., Nicklas, H., Popow, E., Puschmann, K. G., Schmidt, D., Sigwarth, M., Sobotka, M., Soltau, D., Staude, J., Strassmeier, K. G., Waldmann, T. A. 2012, *Astron. Nachr.*, 333, 796

Schou, J., Scherrer, P. H., Bush, R. I., Wachter, R., Couvidat, S., Rabello-Soares, M. C., Bogart, R. S., Hoeksema, J. T., Liu, Y., Duvall, T. L., Akin, D. J., Allard, B. A., Miles, J. W., Rairden, R., Shine, R. A., Tarbell, T. D., Title, A. M., Wolfson, C. J., Elmore, D. F., Norton, A. A., Tomczyk, S., 2012, *Sol. Phys.*, 275, 229

Tritschler, A., Rimmele, T. R., Berukoff, S., Casini, R., Craig, S. C., Elmore, D. F., Hubbard, R. P., Kuhn, J. R., Lin, H., McMullin, J. P., Reardon, K. P., Schmidt, W., Warner, M., Woger, F. 2015, *in 18th Cambridge Workshop on Cool Stars, Stellar Systems, and the Sun*, 18, 933

Wöger, F., von der Lühe, O., II 2008, *in Proc. SPIE*, 7019, 70191E

Xu, Z., Jin, Z., Xu, F., Liu, Z. 2013, *in Proceedings of the International Astronomical Union*, 8(S300), 117

Chapter 3. Progress in understanding the solar/stellar interior dynamics and dynamos

Lorenzo Morelli and Pablo Moya

Solar and Stellar Magnetic Fields: Origins and Manifestations
Proceedings IAU Symposium No. 354, 2019
A. Kosovichev, K. Strassmeier & M. Jardine, ed.
doi:10.1017/S1743921320000721

Global simulations of stellar dynamos

G. Guerrero ⓘ

Physics Department, Universidade Federal de Minas Gerais, Av. Antonio Carlos, 6627, Belo Horizonte, MG 31270-901, Brazil
email: guerrero@fisica.ufmg.br

Abstract. The dynamo mechanism, responsible for the solar magnetic activity, is still an open problem in astrophysics. Different theories proposed to explain such phenomena have failed in reproducing the observational properties of the solar magnetism. Thus, *ab-initio* computational modeling of the convective dynamo in a spherical shell turns out as the best alternative to tackle this problem. In this work we review the efforts performed in global simulations over the past decades. Regarding the development and sustain of mean-flows, as well as mean magnetic field, we discuss the points of agreement and divergence between the different modeling strategies. Special attention is given to the implicit large-eddy simulations performed with the EULAG-MHD code.

Keywords. Sun: rotation, Sun: magnetic fields, Stars: rotation, Stars: magnetic fields, MHD

---

1. Introduction

The Sun exhibits a large-scale magnetic field which is believed to be driven by a dynamo somewhere in its interior. As a consequence of the dynamo several processes in the upper layers of the Sun define the solar activity (i.e., the 11-years sunspot cycle, the solar wind, coronal mass ejections and flares). It establishes a close interaction between the Sun and the earth defining the space weather and, perhaps, influencing long term variations of the earth's temperature. Besides being fascinating as a physical problem, the astrophysical dynamo is a highly relevant area for our society.

The main properties of the solar cycle are presented in Fig. 1(top panel) and can be summarized as follow: (1) Sunspots appear in pairs of opposite polarity at latitudes of about 30° and migrate towards the equator, (2) Spots in the northern hemisphere have the opposite polarity than their analogs in the southern hemisphere, (3) When the number of sunspots is maximum, the poloidal field reaches its minimum and reverses polarity, (4) accordingly, when the poloidal field is maximum, sunspots of a new cycle start to appear with opposite polarity in both hemispheres.

In addition to the solar data, recent observations of magnetic fields in main sequence stars of types F, G and K have imposed further constrains to the dynamo mechanism. Observations indicate two important relations between rotational period of the stars, $P_{\rm rot}$ and the observed magnetic field. First, the field strength exhibits two regimes: for fast rotation it is independent of $P_{\rm rot}$, for slow rotation the field amplitude decays with $P_{\rm rot}$ with power law dependence (e.g., Wright *et al.* 2011; Vidotto *et al.* 2014)†. The second relation regards the stars that exhibit activity cycles. A comparison between the magnetic cycle period, $P_{\rm cyc}$, and $P_{\rm rot}$ shows two regimes of activity, the so called active and inactive branches. For both of them, the longer $P_{\rm rot}$, the longer the $P_{\rm cyc}$ (Fig 1, bottom-right).

† Fully convective stars follow also this trend.

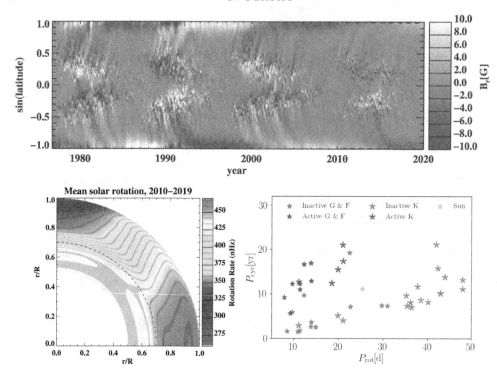

Figure 1. Top: time-latitude butterfly diagram (courtesy: A. Kosovichev); bottom-left: solar differential rotation (courtesy: A. Kosovichev); bottom-right: relation between P_{rot} and P_{cyc}. The blue and red stars correspond to the observed active and inactive branches, respectively. The data was taken from Brandenburg *et al.* (2017).

Models to understand the solar/stellar dynamo problem are based on the theoretical framework known as mean-field dynamo theory (Parker 1955; Steenbeck *et al.* 1966). It describes the generation of the mean magnetic field as the results of the inductive action of large-scale shear-flows (Ω-effect) and small scale, turbulent, helical motions and currents (α-effect). The properties of the Ω-effect, i.e., the solar differential rotation, have been accurately identified by helioseismology (Fig. 1, bottom-left). Recent asteroseismology results have inferred that similar rotational shear profiles occur in solar type stars (Benomar *et al.* 2018). On the other side, the values and profiles of the α-effect in the solar convection zone are uncertain. Some heuristic guesses for these contribution have found some success in reproducing the main characteristics of the solar cycle, nevertheless the results are ambiguous.

From the seminal work of Gilman & Miller (1981), three dimensional, first principles, global simulations were developed with the goal of unambiguously determine the location and properties of the dynamo sources. These models have been successful in reproducing and explaining the mechanisms that sustain the differential rotation (DR, see e.g., Miesch *et al.* 2006; Käpylä *et al.* 2011; Guerrero *et al.* 2013; Featherstone & Miesch 2015), yet the magnetism has still some caveats. The growing of magnetic field in spherical turbulent convection resembling the solar case was obtained by Brun *et al.* (2004). They found magnetic fields with erratic behaviour without large-scale dynamics. More recently, Ghizaru *et al.* (2010) obtained for the first time cyclic dynamo action in a simulation using implicit sub-grid scale (SGS) modeling with the EULAG-MHD code. The magnetic cycle period of their simulation was about 60 years, yet the migration of the magnetic field did not reproduce quite well the solar activity pattern. Later on, other groups have obtained oscillatory dynamos (e.g., Käpylä *et al.* 2012; Augustson *et al.* 2015) in spherical shells that consider only the convection zone. Their results are characterized by short dynamo cycle periods. Guerrero *et al.* (2016a) compared dynamo models with and without the radial shear layer at the interface between the radiative and the convective zone, the tachocline. They found important differences in the process of generating large scale magnetic fields and noticed that when the tachocline is included, P_{cyc} could be of the order of decades.

More recently the role of rotation was explored for simulations mimicking the solar interior. At odds with the observations, simulations including the convection zone only found that the cycle period decreases as the rotational period increases (Strugarek *et al.* 2017). On the other hand, Guerrero *et al.* (2019b) found proportionality between P_{cyc} and P_{rot} in simulations including also a fraction of the convection zone. The results, however, do not completely fit to the active or inactive branches of activity.

The numerical resolution used in the simulations discussed above is still far from capturing all the relevant scales of the solar (or stellar) interior. To the date, the simulations with highest resolution were run for less than 100 years (Viviani *et al.* 2018; Hotta *et al.* 2016). The simulations of Hotta *et al.* (2016) have the maximal resolution reported in the literature and were run for 50 yr. These simulations show reversals of magnetic field and, more importantly, are able to develop small scale dynamo action. Thus, the contribution of the small scale Maxwell stresses is captured in the simulations. Unfortunately, the characteristics of the magnetic field, evolution and periodicity, still diverge from the solar case. Parametric analyses at this resolution are still prohibitive. For simulations where the cycle periods are of the order of decades, the evolution time should be several hundred years for the variables to achieve statistically steady state. Therefore, large computing resources are needed.

The results described above make clear that in spite of the recent progress in observations and the development of high resolution simulations, the solar/stellar dynamo problem is not closed. There are some convergent results from which we have learn much about the behavior of stellar interiors. However, there is not yet a complete theory to satisfactory explain the details captured by the observations. In this work we review the recent advances in *ab-initio* global simulations of solar and stellar dynamos. We discuss the encouraging results as well as the caveats in the different approaches of global modeling. Even though great progress has been done in simulating fully convective stars, we focus here in solar-like stars with an inner radiative zone and a convective envelope.

This paper is organized as follows. In the following section we describe the equations of magnetohydrodynamics (MHD), the framework on which the global simulations are built, and some characteristics of the simulations setups. In §2.2 we discuss the direct numerical simulations and the large eddy simulations approaches for modeling rotating turbulent convection. In §3 and in §4 we examine the results regarding mean-flow and large-scale magnetic field development in stellar interiors. A brief discussion of the theoretical interpretation of the results is also presented. Finally, we draw some conclusions in §5.

2. Magnetohydrodynamics in stellar interiors

The mathematical formalism describing the dynamics of the plasma in stellar interiors combines the equations of Navier-Stokes for a magnetized fluid with the magnetic field induction equation. The goal is simulating the motions in convection zones of the Sun and other similar stars. With exception of the upper part of the solar convection zone, these motions are sub-sonic. Thus, the anelastic approximation, which relaxes the Courant condition imposed from sound waves on the simulations time step, has been broadly used. Under this approach the MHD equations are the following:

$$\nabla \cdot (\rho_s \boldsymbol{u}) = 0, \tag{2.1}$$

$$\frac{D\boldsymbol{u}}{Dt} + 2\boldsymbol{\Omega} \times \boldsymbol{u} = -\nabla \left(\frac{p'}{\rho_s} \right) + \mathbf{g}\frac{\Theta'}{\Theta_s} + \frac{1}{\mu_0 \rho_s}(\boldsymbol{B} \cdot \nabla)\boldsymbol{B} + \mathcal{D}_{\mathbf{u}}, \tag{2.2}$$

$$\frac{D\Theta'}{Dt} = \mathcal{D}_{\Theta}, \tag{2.3}$$

$$\frac{D\boldsymbol{B}}{Dt} = (\mathbf{B} \cdot \nabla)\boldsymbol{u} - \boldsymbol{B}(\nabla \cdot \boldsymbol{u}) + \mathcal{D}_{\mathbf{B}}, \tag{2.4}$$

$$\nabla \cdot \boldsymbol{B} = 0, \tag{2.5}$$

where $D/Dt = \partial/\partial t + \boldsymbol{u} \cdot \nabla$ is the total time derivative, $\boldsymbol{u}$ is the velocity field in a rotating frame with $\boldsymbol{\Omega} = \Omega_0(\cos\theta, -\sin\theta, 0)$, p' is a pressure perturbation variable that accounts for both the gas and magnetic pressure, $\boldsymbol{B}$ is the magnetic field, and Θ' is the potential temperature perturbation with respect to a background state. It is related to the specific entropy via $s = c_p \ln\Theta + \text{const}$; $\boldsymbol{g} = GM/r^2 \hat{e}_r$ is the gravity acceleration, G and M are the gravitational constant and the stellar mass, respectively, and μ_0 is the magnetic permeability. The $\mathcal{D}$ terms in the equations (2.2)-(2.4) are dissipative terms which diffuse momentum, heat and magnetic field. Inside stars the dissipation coefficients may be computed with the Spitzer formula (Spitzer 1962). In the upper part of the convection zone there is radiative cooling because of hydrogen ionization, nevertheless, global simulations do not include this effect because of its numerical cost.

2.1. Simulation domains

Convective dynamos are modeled in spherical coordinates. Most of the simulations cover the entire longitudinal and latitudinal extent, i.e., $0 \leq \varphi < 2\pi$, and $0 \leq \theta \leq \pi$, respectively. The radial extent spans for most of the convection zone, $0.71R_\odot \leq r \leq 0.95R_\odot$, for the case of the Sun. Simulations with the pencil-code consider wedges of different longitudinal extents, and a latitudinal extent that does not reach the poles to avoid shorter time scales due to the spherical grid (e.g., Käpylä et al. 2012). However, since it solves fully compressible equations, the radial extent in simulations with this code reaches up to $1R_\odot$. Brun et al. (2011, 2017), with the ASH code, and Guerrero et al. (2013) with the EULAG-MHD code have performed hydrodynamic (HD) simulations including a stable stratified layer. Masada et al. (2013) with a code based on the Yin-Yang grid and second order finite differences, as well as Ghizaru et al. (2010); Guerrero et al. (2016a), with the EULAG-MHD code, performed dynamo simulations which also include a fraction of the radiative zone.

2.2. Direct numerical simulations (DNS) and large eddy simulations (LES)

The dynamo phenomena is a MHD problem associated to rotating turbulent convection, i.e., the Reynolds number (Re $= u_{\text{rms}}L/\nu$, where u_{rms} is the turbulent velocity, L a characteristic length scale of the system, and ν is the kinematic viscosity of the plasma) and the magnetic Reynolds number (Re$_M = u_{\text{rms}}L/\eta$, where η is the magnetic diffusivity) have large values. In the solar convection zone both quantities exceed 10^6. Numerical simulations capturing all the relevant scales of a turbulent flow are called direct numerical simulations (DNS). Since the Reynolds numbers give a rough measurement of the range of scales, between the advective and the dissipative processes, it can be estimated that the number of grid points necessary to resolve all the contributing scales must be of the order of $N \sim \text{Re}^{9/4}$. Thus, DNS simulations reproducing the dynamics of the

Sun's interior are still prohibitive for the current supercomputers (e.g., the simulations of Hotta *et al.* (2016) reach a maximum Reynolds number of about 7000). Thus, most simulations use values of the dissipation coefficients much larger than the molecular ones and consistent with the turbulent rates of dissipation.

Other alternatives have been developed over the years to simulate turbulent flows. These theories aim to mimic the contribution of the non-resolved motions by adding a sub-grid scale (SGS) terms in the prognostic equations. This allows to model high Re flows with less expensive simulations. Since the computations in this approach resolve only the scales of relatively large structures they are called large-eddy simulations. In most of the SGS models the contribution of the non-resolved scales is proportional to the strain tensor of the large-scale flows (Smagorinsky 1963; Germano *et al.* 1991). To test the accuracy of this kind of modeling, LES are compared, when possible, with experiments or with high resolution DNS. For instance, the incompressible simulations with forced turbulence carried out at a resolution of 4096^3 mesh points by Kaneda *et al.* (2003) were compared to the Smagorinsky and the hypervisocity SGS models by Haugen *et al.* (2004); Haugen & Brandenburg (2006). The results showed a good agreement between the three cases in the turbulent inertial range.

More recently, a new class of numerical schemes have obtained results compatible with LES simulations without the need of a SGS model. These schemes are based on non-oscillatory finite volume (NFV) approximations and the strategy is called implicit large-eddy simulation (ILES, see e.g., Grinstein *et al.* 2007). There is not yet a turbulence theory justifying the basis of the ILES approach, nevertheless, Margolin & Rider (2002) presented a solid rationale for their use by studying the Burger's equation. More recently, Margolin (2019) showed that the terms resulting from the NFV formulation might indeed represent physical phenomena.

ILES modeling of the solar interior were performed during the last decade with the EULAG-MHD code. These attempts are summarized in the sections below, where we point out the success and caveats of the obtained results.

2.3. *Global solar/stellar simulations with the EULAG-MHD code*

EULAG-MHD (Smolarkiewicz & Charbonneau 2013) is an extension of the hydrodynamic model EULAG predominantly used in atmospheric and climate research (Prusa *et al.* 2008). Its solver is adapted to simulate the turbulent subsonic flows found in the majority of the solar interior. EULAG-MHD is powered by a non-oscillatory forward-in-time MPDATA method (multidimensional positive definite advection transport algorithm; see Smolarkiewicz (2006) for an overview). This is a nonlinear, second-order-accurate iterative implementation of the elementary first-order-accurate flux-form upwind scheme.

Elliott & Smolarkiewicz (2002) performed the first HD ILES simulations of the solar convection zone. They were able to create solar-like differential rotation profiles by modeling turbulent convection rotating at the solar rotation rate. MHD simulations including the radiative zone were performed by Ghizaru *et al.* (2010). They found, for the first time, cyclic reversals of magnetic field occurring at the tachocline. In their work, and subsequent simulations with EULAG-MHD, the heat and radiative transfer terms, $\mathcal{D}_\theta$, are replaced by the simple forcing and cooling parametrization,

$$\frac{D\Theta'}{Dt} = -\boldsymbol{u} \cdot \boldsymbol{\nabla}\Theta_{\mathrm{amb}} - \frac{\Theta'}{\tau}. \tag{2.6}$$

In this form, the Newtonian cooling (second term on the RHS) relaxes perturbations of potential temperature towards an axisymmetric ambient state, Θ_{amb}, in a time scale, τ (see Held & Suarez 1994; Cossette *et al.* 2017, and references therein). Under this scheme,

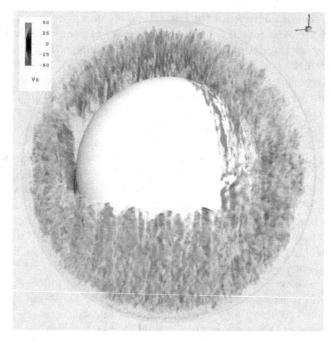

Figure 2. Radial velocity component of a characteristic EULAG-MHD simulation of rotating convection. The elongated convective rolls are the so-called banana cells, typical from rotationally constrained motions.

convection is driven by these perturbations whenever the ambient is super-adiabatic, and relaxes towards a statistically steady state on timescales much shorter than the ones needed by the transport of heat due to radiation or diffusion. In addition, it simplifies the energy boundary conditions because only the radial derivative of the radial convective flux has to be specified (Smolarkiewicz & Charbonneau 2013). As it will be discussed below, the choice of the ambient state is pivotal to define not only the strength of the convective motions, but also the frequency of inertial gravity waves in stable stratified atmospheres. Suitable ambient states may be obtained for the Sun and other stars from evolutionary codes. The viscosity and magnetic diffusivity terms are also dropped out from Eqs. (2.2 and 2.4). Thus, the only dissipation in the system is due to the ILES numerical scheme, MPDATA. Figure 2 depicts the vertical velocity of a characteristic EULAG-MHD simulation with a rotational period of 28 days. The banana cell structures, typical of rotationally constrained convection, are evident in the snapshot.

Although this approach has been instrumental to obtain mean-flows and cyclic dynamos resembling the observations, it still has some issues that must be addressed. For instance, Guerrero *et al.* (2013) used EULAG-MHD to study how mean-flows develop and sustain in simulated stars rotating at different rates. Their reference simulations have $128 \times 64 \times 64$ grid points in the φ, θ and r directions, respectively. Rotating at the solar rotation rate it develops solar-like DR. When the resolution is doubled or increased four fold, the resulting DR exhibits different patterns. Thus, understanding the interplay between a scale independent Newtonian cooling and the implicit numerical dissipation (depending on numerical resolution) is necessary.

3. Mean flows

Differential rotation and meridional circulation (MC) are stellar large scale motions in the longitudinal and the meridional directions, respectively. Thanks to helioseismology, the differential rotation in the Sun has been measured with great accuracy (Schou *et al.* 1998). The profile shown in Fig. 1 (bottom-left panel) is a long term average of the angular velocity. Its main characteristics are: fast equator, slow poles with conical isocontours. Latitudinal differential rotation goes from the surface down to the tachocline at $\sim 0.71 R_\odot$, from where the rotation is rigid. At the surface there is a thin layer of negative shear called the near-surface shear layer (NSSL). The solar DR suffers periodic speeds-up and slows-down called torsional oscillations (e.g., Howard & Labonte 1980; Kosovichev & Schou 1997; Antia & Basu 2001; Vorontsov *et al.* 2002).

The meridional circulation, observed in the solar photosphere with different techniques, shows poleward flows on each hemisphere. Its amplitude of, about 20 m s^{-1}, also fluctuates with the solar cycle. The helioseismic signal of the meridional flow from the solar interior is weak. For that reason, despite that few attempts have been conducted, its profile has not yet been accurately measured. Zhao *et al.* (2013); Schad *et al.* (2013); Böning *et al.* (2017) inferred meridional circulation patterns with multiple radial cells. Liang *et al.* (2018) reported evidences for single cell circulation.

Global simulations aim to reproduce the observed properties of these mean-flows. However, explaining their peculiarities is still challenging. Simulations from diverse numerical techniques and codes have consistently found a transition from anti-solar DR (with slow equator and fast poles) for slow rotating models, to solar-like DR for rotationally constrained turbulent convection (Gilman 1977; Käpylä *et al.* 2011; Guerrero *et al.* 2013; Gastine *et al.* 2014). This transition seems to occur when the Rossby number, measuring the ratio between the rotational period and the convective turnover time (Ro $= P_{\rm rot}/\tau_c$), is about one†. This point marks the transition between strong and fast convection and slow convective motions influenced by the Coriolis force. Self consistently developed tachoclines have been found in recent simulations (e.g., Guerrero *et al.* 2013; Brun *et al.* 2017). Similarly, the formation of the NSSL has been studied in recent papers (Matilsky *et al.* 2019).

From the simulation results it is possible to explore how this flows are sustained. Differential rotation may be explained from the distribution of angular momentum $\mathcal{L} = \rho\varpi\overline{u}_\varphi$, where $\overline{u}_\varphi$ is the mean, longitudinally averaged, zonal flow and $\varpi = r\sin\theta$ is the lever arm. In the HD case its evolution equation is

$$\frac{\partial\mathcal{L}}{\partial t} = -\nabla\cdot\left(\varpi\left[\rho(\overline{u}_\varphi + \varpi\Omega_0)\overline{u}_{\rm m} + \rho\overline{u}'_\varphi u'_{\rm m}\right]\right), \tag{3.1}$$

where, $\overline{u}_{\rm m}$, $u'_{\rm m}$ are the mean and turbulent meridional (r and θ) components of the velocity field, respectively. All these terms can be computed from the global simulations. For solar-like differential rotation, the second term on the RHS, namely the Reynolds stresses, dominate over the meridional circulation terms, first term on the RHS. A mostly positive latitudinal Reynolds stress component, $\rho\varpi\overline{u}'_\phi u'_\theta$, transporting angular momentum towards the equator, is a robust feature of global simulations of the Sun (Brun & Toomre 2002; Guerrero *et al.* 2013; Featherstone & Miesch 2015; Passos *et al.* 2017). This result is in agreement with the Λ-effect theory (Ruediger 1989) where the Reynold stresses are parametrized as $R_{ij} = R^\Lambda_{ij} + R^\nu_{ij}$. Here, the first and second terms correspond to non-diffusive and diffusive parts of the Λ-effect, respectively. Positive values of the non diffusive part are required to sustain solar-like DR (Kitchatinov 2013).

For anti-solar DR the meridional circulation terms dominate, advecting angular momentum towards higher latitudes. Featherstone & Miesch (2015) argue that this meridional flow is driven by the transport of angular momentum through the so-called gyroscopic pumping mechanism which results from Eq. (3.1) and considering steady

† Note, however, that there are several definitions of the Rossby number and this value might change.

state ($\partial \mathcal{L}/\partial t = 0$). Under this scenario, negative values of $-\nabla \cdot \varpi \rho \overline{u'_\varphi u'_m}$ will induce a strong meridional flow away from the rotation axis at lower latitudes. Because of the closed boundary conditions, this flow transports angular momentum to the high latitudes increasing the angular velocity. At the equator it will result in a differential rotation that decreases away from the rotation axis (Miesch & Hindman 2011).

This description is completed with the equation of the thermal wind balance (TWB) which in its HD form reads,

$$\varpi \frac{\partial \overline{\Omega}^2}{\partial z} = \frac{g}{r} \frac{\partial}{\partial \theta} \left(\frac{\Theta'}{\Theta_s} \right) , \tag{3.2}$$

where z is the height in cylindrical coordinates. If the DR has isocontours which are not aligned with the rotation axis, the term on the LHS implies gyroscopic pumping. The term on the RHS is the baroclinicity of the system. Models with anti-solar differential rotation result in warm equator and cold poles, favoring then a counterclockwise meridional motion in the northern hemisphere. cell (Featherstone & Miesch 2015). The same, gyroscopic pumping mechanism might explain the, more complex, multi cell pattern of MC typical of simulations with solar-like differential rotation (Guerrero et al. 2016b).

An alternative explanation to the gyroscopic pumping dominance in defining the meridional circulations comes from the Λ-effect theory. Mean-field models under this approximation explain MC from departures of the TWB (Eq. 3.2). Since the TWB is likely lost at the boundaries of the convection zone, viscous forces might drive the meridional flows (Kitchatinov 2013). In contrast with global simulations where all the driving forces arise naturally, in Λ-effect mean-field simulations these forces are parametrized by theoretical turbulence models. Thus, including this viscous force in the equation for the vorticity results in single circulation cells for both fast and slow rotating stars (Karak et al. 2014). This disagrees with the results of HD global simulations where single cell circulation is obtained for slow rotation and multiple cells show up in fast rotation simulations. (See, however, Pipin & Kosovichev (2018) for Λ-effect mean-field simulations with double cell MC.)

In the MHD case, simulations have shown that the magnetic field quenches the convective flux affecting the distribution of the Reynolds stresses and, therefore, of angular momentum. It allows for slow rotating models to develop solar-like DR (Fan & Fang 2014; Karak et al. 2015; Guerrero et al. 2019b). The HD and MHD balance of angular momentum and thermal wind for solar-like DR are discussed in detail by Passos et al. (2017). The transition from solar-like to anti-solar DR for HD and MHD simulations was explored by Karak et al. (2015). It was found that the transition is smooth for dynamo simulations and occurs at larger values of the Rossby number. In the EULAG-MHD simulations presented in Guerrero et al. (2019b) this transition is not observed for Rossby numbers up to ~ 3 (63 days rotational period). This can be seen in Fig. 3 depicting the DR (left of each panel) and MC (right) of some simulations in that publication. Note also that in all the cases the MC has multiple cells in radius at lower latitudes.

Differential rotation has been observed in other stars by different means including tracking magnetic features, photometric variability and asteroseismology (e.g., Reinhold & Gizon 2015; Kővári et al. 2017; Benomar et al. 2018). The results indicate DR consistent with faster equator for a large number of observed stars. Anti-solar DR has not been yet confirmed. Reinhold & Gizon (2015); Kővári et al. (2017) found a shear parameter, $\alpha^* = \Delta\Omega/\Omega_{eq}$, increasing with the rotational period of the stars with values between 10^{-3} and 0.5. The recent asteroseismology analysis of Benomar et al. (2018) does not show a clear defined trend and found shear values larger than unity. In the Sun $\Delta\Omega/\Omega_{eq}$, with $\Delta\Omega = \Omega_{eq} - \Omega_{45}$, is about 0.1.

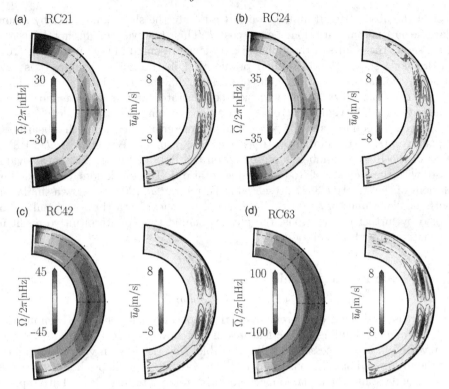

Figure 3. Differential rotation in the inertial frame, $\overline{\Omega} - \Omega_0$, and meridional circulation of characteristic simulations of Guerrero *et al.* (2019b). In the differential rotation panels the color contours show regions of iso-rotation. In the meridional circulation panels the continuous (dashed) lines represent clockwise (counterclockwise) circulation. The color contours show the mean latitudinal velocity, $\overline{u}_\theta$. Adapted from Guerrero *et al.* (2019b).

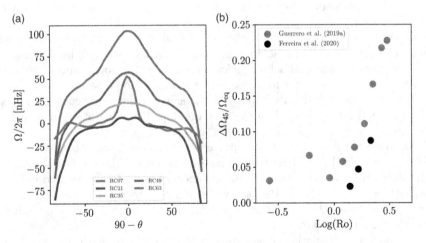

Figure 4. (a) Latitudinal differential rotation in the inertial frame, $\overline{\Omega} - \Omega_0$, of the simulations presented in Guerrero *et al.* (2019b); (b) relative shear, $\Delta\Omega/\Omega_{\mathrm{eq}}$, where $\Delta\Omega = \Omega_{\mathrm{eq}} - \Omega_{45}$, as a function of the Rossby number (in logarithmic scale). The red and black dots correspond to the simulations of Guerrero *et al.* (2019b) and Ferreira *et al.* (2020), respectively.

Most of the observational findings on DR estimate the shear parameter by assuming a solar-like latitudinal variation, i.e., $\sim \cos^2 \theta$. This, however, might not be the case. Figure 4 shows the results of the global simulations presented in Guerrero et al. (2019b). The LHS panel shows the latitudinal profile of $\overline{\Omega}$ in the inertial frame. It is evident that the latitudinal shear has different profiles for different rotational periods. The fast rotating simulations show strong variation concentrated at the equator and the poles. Intermediate cases have smooth profiles while slow rotating simulations develop a strong shear between the equator and the poles. The parameter α^*, measuring the shear between the equator and 45° (to compare with astroseismic results of Benomar et al. 2018), as a function of the Rossby number is presented on the RHS of Fig. 4. The shear evidently increases with the increase of Ro (decreasing rotation). The black dots, corresponding to simulations of the star HD43587 presented in Ferreira et al. (2020), follow a similar trend yet with smaller values of α^*. If the increase of the shear with the rotational period is confirmed by further observations, the physics behind this counterintuitive result must be elucidated through global models.

4. Large-scale magnetic field

4.1. *The solar dynamo*

For the reasons discussed in previous sections, performing global MHD simulation of the solar dynamo, including all the features of the convection zone, and all the relevant scales involved in the processes, is still unfeasible. Nevertheless, in view of the ambiguities of the mean-field simulations, global MHD modeling is perhaps the most promising method to understand the dynamo mechanism in stellar interiors. Several attempts have been done over the last decades with the aim of capturing some relevant physics. In the solar context, Gilman & Miller (1981); Gilman (1983) presented pseudo-spectral simulations of rotating convection in the Boussinesq approximation. They were able to obtain dynamo action and even magnetic field reversals. The anelastic approximation was used in the early works of Glatzmaier (1984, 1985a,b). With the advent of parallel supercomputer architectures, simulations with improved resolutions were performed by Brun et al. (2004) with the ASH code. They obtained amplification of the magnetic field without a large-scale dynamo.

Large-scale organized magnetic fields were found in the simulations performed by Browning et al. (2006). They achieve mean-field dynamo action by forcing a shear layer, i.e., mimicking the tachocline, below the convection zone. Nevertheless, Brown et al. (2010) showed that the formation of organized global structures do not need strong tachocline shear but only rapidly rotating turbulent convection. Their simulations, rotating ~ 3 times faster than the Sun, developed steady antisymmetric magnetic wreaths.

Periodic magnetic fields were first reported by Ghizaru et al. (2010) with the ILES formulation of the EULAG-MHD code. The lower values of the numerical viscosity due to the implicit SGS scheme, together with the energy description presented in §2.6 led to a self-consistent development of the tachocline and a solar-like differential rotation. The resulting dynamo mode does not show, however, clear latitudinal migration and reverses with a cycle period of 30 yrs.

Fully compressible simulations of spherical wedges performed with the pencil-code also found cyclic dynamo solutions (Käpylä et al. 2010). Also, Käpylä et al. (2012) reported solutions where the toroidal magnetic field migrates towards the equator resembling the solar observations. Considering slope-limited diffusion for diffusing momentum, and eddy values for the heat transport coefficients and the magnetic diffusivity, Augustson et al.

(2015) obtained periodic field reversals with the ASH code. These cases simulate only in the convection zone and the cycle periods are between 3 to 6 yrs.

Masada *et al.* (2013), with a code based on second order finite differences and using the Yin-Yang grid explored the role of penetrative convection in models with and without a fraction of the radiative interior. They found that larger magnetic fields develop when the stable layer is present. Nevertheless, there is not a well defined tachocline in their results. Guerrero *et al.* (2016a) performed the same comparison with EULAG-MHD and explored simulations with rotation rates of $2\Omega_\odot$, $\Omega_\odot$ and $\Omega_\odot/2$. The results showed striking differences between the two cases. While the resulting latitudinal shear was roughly the same, the presence of the radial shear at the tachoclines led to a different behavior of the strong magnetic field developed at and below these shear regions. As a consequence, the cycle period, of about 2 yrs in simulations of the convection zone only, resulted of the order of decades.

Figure 5 shows the time-latitude butterfly diagram of a recent EULAG-MHD simulation of the solar analog HD43587 presented in Ferreira *et al.* (2020). The color contours represent the strength of the mean toroidal field, $\overline{B}_\phi$. The continuous (dashed) contour lines show the positive (negative) levels of the mean radial magnetic field, $\overline{B}_r$. The panels (a), (b) and (c) correspond to different depths, as indicated. At $0.85R_*$, where $R_* = 1.19R_\odot$ is the radius of the star, the butterfly diagram shows qualitative similarities with the solar one, i.e., there is strong magnetic field migrating equatorward at lower latitudes and poleward migration at higher latitudes. This configuration is likely the result of magnetic field emerging from the tacholine at lower latitudes, (see also the upper-left panel of Fig. 5 which shows a snapshot of the magnetic field configuration in the meridional, $\theta - r$, plane), followed by poleward advection at the upper part of the domain. At $0.95R_*$, upper boundary of the simulation, only the polar branch is evident (this stage corresponds to the lower-left panel of Fig. 5).

4.2. *Mean-field interpretation*

Similar to the theory of the Λ-effect, a mean-field model can be used to describe the evolution of the large-scale magnetic field, $\overline{B}$ in stellar interiors (Parker 1955; Steenbeck *et al.* 1966). After separation of the turbulent and the large scales, the induction equation (Eq. 2.4) becomes,

$$\frac{\partial \overline{B}}{\partial t} = \nabla \times (\overline{u} \times \overline{B}) + \nabla \times (\alpha \overline{B}) - \nabla \times (\eta \nabla \times \overline{B}), \qquad (4.1)$$

where $\overline{B} = (\overline{B}_r, \overline{B}_\theta, \overline{B}_\phi)$ and $\overline{u} = (\overline{u}_r, \overline{u}_\theta, \overline{u}_\phi)$ are the averaged magnetic and velocity fields. In the second term on the RHS, the α term stems for the α-effect which has kinetic and magnetic contributions, namely $\alpha = \alpha_k + \alpha_m$. These terms are the first order correlations of the turbulent electromotive force,

$$\overline{\varepsilon} = \overline{u' \times B'} \simeq \alpha \overline{B} - \eta_t \mu_0 \nabla \times \overline{B}, \qquad (4.2)$$

where u' and B' are the turbulent velocity and magnetic fields and η_t is the turbulent magnetic diffusivity. The α terms induce the large-scale magnetic field from small scale helical motions and currents. Under suitable closure models (e.g., Pouquet *et al.* 1976; Moffatt 1978), the it may be expressed as:

$$\alpha = \alpha_k + \alpha_m = -\frac{\tau_c}{3}\langle \omega' \cdot u' \rangle + \frac{\tau_c}{3}\langle j' \cdot B' \rangle/\rho, \qquad (4.3)$$

where $\omega' = \nabla \times u'$ and $j' = \nabla \times B'$ are the small scale vorticity and current, respectively, and τ_c is the convective turnover time of the turbulent motions.

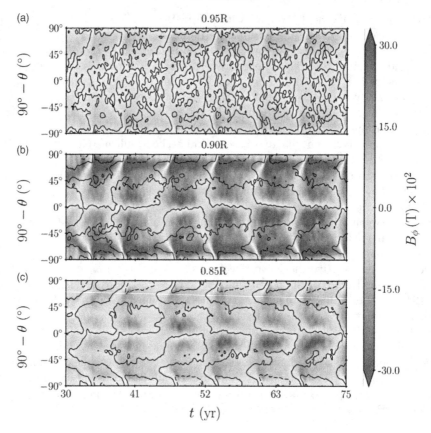

Figure 5. Time-latitude butterfly diagram of one simulation of the star HD43587. The color contours show the toroidal mean magnetic field, $\overline{B}_\phi$. The continuous (dashed) contour lines show the positive (negative) mean radial magnetic field, $\overline{B}_r$.

In the third term in Eq. (4.1), $\eta = \eta_m + \eta_t$ is the sum of the molecular and the turbulent magnetic diffusivities. The turbulent contribution is dominant by several orders of magnitude and under the first order smoothed approximation (FOSA) may be computed as

$$\eta_t = \frac{1}{3}\tau_c u'^2 . \qquad (4.4)$$

Either the terms in the electromotive force (Eq. 4.2) or those in Eqs. (4.3 and 4.4) may be estimated from the global simulations output allowing to search for a theoretical interpretation of complex results. This has been done for the EULAG-MHD simulations of Ghizaru *et al.* (2010) by using a method for multidimensional regression (Racine *et al.* 2011). Their results indicate that the process is consistent with an $\alpha^2\Omega$ dynamo (see also Simard *et al.* 2013). For pencil-code dynamo results Warnecke *et al.* (2014) computed the dynamo coefficients, α and Ω, and proved that the evolution of the magnetic field follows the Parker (1955)-Yoshimura (1975) sign rule. More recently Warnecke *et al.* (2018) used the test-field method (Schrinner *et al.* 2005; Brandenburg *et al.* 2008) to perform a fiducial prediction of the electromotive force coefficients. For the profiles and amplitude of the coefficients, they conclude that the dominant dynamo in pencil-code simulations is of $\alpha\Omega$ type with a localized α^2 dynamo.

Figure 6. Snapshots presenting half magnetic cycle in the EULAG-MHD simulation P25 presented in Ferreira *et al.* (2020). On the left of each snapshot, the LIC representation depicts the distribution of the poloidal field lines with the color indicating the magnitude of the mean radial field, $\bar{B}_r(t, \theta, r)$. The colored contours on the right quadrants correspond to the azimuthal mean magnetic field, $\bar{B}_\varphi(t, \theta, r)$.

The FOSA coefficients of Eqs. (4.3 and 4.4) were computed by Guerrero *et al.* (2019b) for EULAG-MHD simulations including both, radiative and a convective layer; having solar-like ambient state and rotating at different rates. The results showed that for fast rotation the dominant dynamo is of $\alpha\Omega$ type and operates at the convection zone (Fig. 7(a)). For intermediate and slow rotation rates, the dominant dynamo operates locally at the tachocline and is of $\alpha^2\Omega$ type (Fig. 7(b and c)). The α-effect contributes to the amplification of the toroidal field at intermediate to high latitudes while the rotational shear generates toroidal field at the equator. The results indicated that the appearance of cycles in tachocline dynamos is related to the strength of the equatorial shear. The most striking result observed in the simulations is that the α-effect, at and below the tachocline, has magnetic origin, i.e., the inductive contribution of small-scale current helicity (second term in Eq. 4.3) is dominant over the induction of magnetic field by small-scale helical eddies.

The magnetic loop observed in the deep seated $\alpha^2\Omega$ dynamos described above is illustrated in Fig. 8. Turbulent convection is necessary to generate and sustain the differential rotation (left panel). It supplies the strong shear necessary for the Ω-effect to develop a

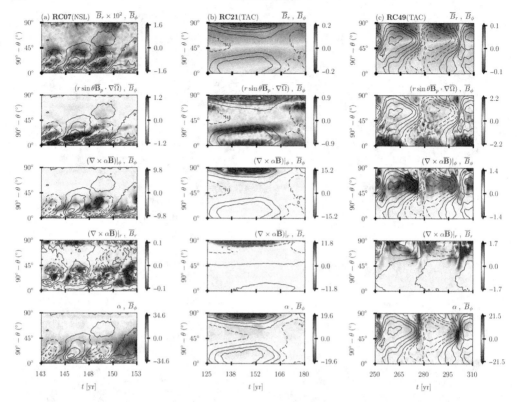

Figure 7. Mean-field sources of large-scale magnetic fields in some characteristic simulations presented in Guerrero *et al.* (2019b). The columns correspond to simulations with a cycle period of 7 (a), 21 (b) and 49 (c) days, respectively. All the quantities are averaged in longitude and over a radial extent in the near-surface layer (a) or in the tachocline (b and c). From top to bottom the panels display the radial, $\bar{B}_r$ (colored contours) and toroidal, $\bar{B}_\phi$ (contour lines) mean magnetic fields; the sources of the toroidal field, $r \sin\theta (\bar{\mathbf{B}}_p \cdot \nabla)\bar{\Omega}$ and $(\nabla \times \alpha\bar{\mathbf{B}})|_\phi$ (colored contours), both compared with $\bar{B}_\phi$; the source of the radial field, $(\nabla \times \alpha\bar{\mathbf{B}})|_r$ (color), compared with $\bar{B}_r$ (contour lines); and $\alpha = \alpha_k + \alpha_m$ (colored contours), compared with $\bar{B}_\phi$. In the color map the dimensions are [T] for the magnetic field, 10^{-8} [T/s] for the source terms, and [m/s] for the α-effect, respectively. Adapted from Guerrero *et al.* (2019b).

large-scale toroidal magnetic field, $\bar{B}_\phi$ (top). The generated layer of toroidal field is unstable to the shear present in the region and decays in non-axisymmetric modes ($m \neq 0$), whose collective contribution, in turn, develop an axisymmetric magnetic α-effect (right). This α-effect contributes to the generation of both, toroidal and poloidal mean magnetic fields. Between maxima and minima the developed magnetic tension speeds-up and slows-down the azimuthal motions creating a pattern of torsional oscillations. The mechanism generating this pattern is described in detail in Guerrero *et al.* (2016b).

Several instabilities are simultaneously operating in the cycle depicted in Fig. 8. Disentangle them is a rather difficult task, thus, simple simulations are necessary to understand the processes separately. The most intriguing mechanism in this dynamo is perhaps the generation of small scale helicities in the stable layer because of instabilities of the magnetic field. We have claimed that it is consistent with a magnetoshear instability where a decaying toroidal field produces helical currents that allow the growing of the poloidal field. This kind of instabilities was first proposed by (Tayler 1973) and studied further by several authors (e.g., Pitts & Tayler 1985; Bonanno & Urpin 2012, 2013). In the presence of shear the instability was explored in detail by Cally (2003);

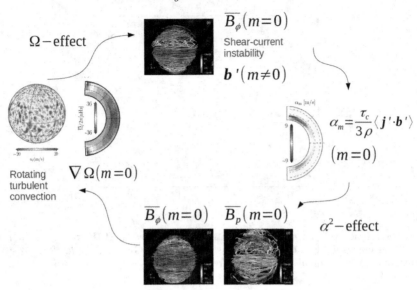

Figure 8. Dynamo loop for oscillatory $\alpha^2\Omega$ dynamos operating in the simulations of Guerrero *et al.* (2019b).

Miesch *et al.* (2007). They reported that the shear-current instability for a broad band of toroidal magnetic field results in modes where the field lines open in the form of a clamshell. When the initial field has two bands of opposite polarity, the axis of the bands tilts in latitude. Similarly to the Tayler instability, in these cases the fastest growing mode is $m = 1$. The clamshell instability is observed in the simulations presented in (Guerrero *et al.* 2019b, see their Fig. 10). A cyclic dynamo based on Tayler-like instabilities has been previously proposed by Spruit (2002); Zahn *et al.* (2007); Bonanno (2013), however, the global EULAG-MHD simulations of Guerrero *et al.* (2019b) were the first ones to develop and sustain the dynamo.

To fully understand how these instabilities are the source of an α-effect, Guerrero *et al.* (2019b) explored the stabilizing properties of gravity to the Tayler instability using the EULAG-MHD code. The results showed that for weak gravitational stratification the instability has radial dependence, while for larger stratification the unstable modes develop in horizontal layers and might have oscillatory behavior. The growth rate depends on the local value of the Brunt-Väisäla (BV) frequency (in turn depending on the gravity force). An extension to this work is currently being carried out including rotation and shear (Monteiro *et al.* 2020, *in preparation*).

4.3. *Stellar cycles*

The study of magnetism in stars different from the Sun is necessary to get a broad understanding of the physics in stellar interiors as well as the interaction between stars and planetary systems. It may also provide hints and constraints to unveil the details of the solar dynamo. From the Mount Wilson data several stars were found to have activity cycles (Baliunas *et al.* 1995; Saar & Brandenburg 1999) evident in the Ca II H and K spectral lines. Further studies correlated the magnetic cycle period to the rotational period (Noyes *et al.* 1984; Brandenburg *et al.* 1998; Böhm-Vitense 2007; Brandenburg *et al.* 2017). Two main branches, classified as active (A) or inactive (I) depending also in the strength of the magnetic field, have been canonically discussed in the literature (Brandenburg *et al.* 1998; Böhm-Vitense 2007; Brandenburg *et al.* 2017). Both of them

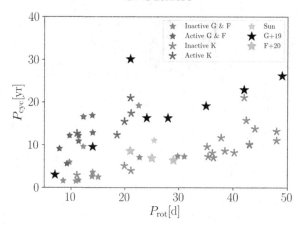

Figure 9. Magnetic cycle period of rotation, $P_{\rm cyc}$ as a function of the rotational period $P_{\rm rot}$. The black stars correspond to the simulations presented in Guerrero *et al.* (2019b), the green stars correspond to the simulations of the star HD43586 presented in Ferreira *et al.* (2020).

show a positive linear correlation between $P_{\rm cyc}$ and $P_{\rm rot}$. The same two branches appear parallel when $\log(P_{\rm rot}/P_{\rm cyc})$ is compared with $\log\langle R'_{HK}\rangle$, where $\langle R'_{HK}\rangle = F'_{HK}/F_{\rm bol}$ is the mean fractional Ca II H and K flux normalized with the stellar bolometric flux, $F_{\rm bol}$ (Brandenburg *et al.* 1998, 2017). Olspert *et al.* (2018) re-visited the Mount Wilson data computing cycle periods with sophisticated techniques for analysis of periodic and quasi-periodic time series. Their results confirmed the existence of the two branches, A and I. They found, however, that the trend of the active branch is not positive but broadly distributed and connects with a third transitional branch of activity (Lehtinen *et al.* 2016).

Over the last years few attempts have been done to explore the relation between magnetic cycles and rotation by using global convective dynamo simulations. Strugarek *et al.* (2017) performed ILES simulations with the EULAG-MHD code. They found a negative correlation between $P_{\rm cyc}$ and $P_{\rm rot}$ and argue that the cycle period is established by the back-reaction of the magnetic field over the differential rotation (Strugarek *et al.* 2018). Warnecke (2018) reported a similar relation in pencil-code wedge simulations. They attributed, however, the magnetic cycles to $\alpha\Omega$ dynamos operating at the base of the convection zone. The resulting cycle periods follow the linear mean-field dynamo theory (see e.g., Stix 1976):

$$P_{\rm cyc} \propto (C_\alpha C_\Omega)^{-1/2}\,, \tag{4.5}$$

where C_α and C_Ω are non-dimensional quantities comparing the inductive effects of the α and the Ω effects, respectively, with the magnetic turbulent diffusion. Warnecke (2018) argues that their simulations are reproducing the transitional branch. Another set of simulations performed with the pencil-code cover the entire domain in the φ direction but without reaching the poles (Viviani *et al.* 2018). They explore fast convective dynamos with rotations from 1 to 30 $\Omega_\odot$. Since rotation breaks the convective cells, these simulations require high resolution. They found that the slow rotating cases fall in the I branch but having a negative tilt. On the other hand, the fast rotating cases result in non-axisymmetric dynamos with cycles falling in a different branch of activity representative of superactive stars.

All these cyclic dynamo simulations have in common the absence of the strong shear at the tachocline. Guerrero *et al.* (2016a) found that long-lived strong magnetic fields develop in models that include the tachocline and a stable layer underneath. A longer set of simulations was recently presented in Guerrero *et al.* (2019b) with dynamo solutions for rotational periods between 7 and 63 days. The results revealed two different types of dynamos: $\alpha\Omega$ dynamos operating at the bulk of the convection zone for fast rotating simulations, and depth seated $\alpha^2\Omega$ dynamos for intermediate and slow rotating simulations. In between, there is a bifurcation point where deep seated oscillatory dynamo at lower latitudes coexists with a steady dynamo at the poles. This point marks the level of equatorial shear necessary to generate tachocline cyclic dynamos. In both types of dynamos P_{cyc} increases with P_{rot}. The ones operating in the convection zone fall close to the active branch whereas the tachocline dynamos fall in between the two branches (see black stars in Fig. 9). Although the deep seated magnetic cycles are consistent with $\alpha^2\Omega$ dynamos (as seen in §4.2), the cycle periods do not agree with the mean-field prediction (Eq. 4.5). The cycle is established once a balance between a non-linear dynamo and tachocline instabilities is reached. A better understanding of this mechanism is necessary to be able of predicting cycle lengths as a function of stellar parameters. Furthermore, the numerical experiments presented in Ferreira *et al.* (2020) show that the simulations are rather sensitive to the profile o the ambient state, Θ_{amb}, described in section §2.3.

In one hand, the level of superadiabaticity of the convection zone determines the strength of the convection, and therefore the profile of differential rotation. On the other hand, the level of subadiabaticity in the stable layer determines the Brunt-Väisäla frequency and therefore the development of gravity waves in the stable layer. These waves might be fundamental for the development of the instability of toroidal fields in radiative zones (Guerrero *et al.* 2019a). We have implemented these ideas in the simulations of the solar analog HD43568 (Ferreira *et al.* 2020). The stratification profile in the convection zone is set to fit the convective velocity obtained by the mixing length theory (MLT); in the stable layer it fits the BV frequency of the same structural model. In addition, we changed the magnetic boundary condition at the bottom of the domain from radial field to perfect conductor. The results agree with the spectroscopic analysis indicating an activity cycle of ~ 10 years and low magnetic activity levels. Simulations for 21, 25 and 29 days results in cycles which fall in the inactive branch (see green stars in Fig. 9). Further simulations with this improved ambient state configuration are currently being performed for the Sun and other stars of the Mount Wilson sample.

5. Conclusions

In this work we described the modeling of solar and stellar dynamos through self-consistent global simulations. We argue that these are the most relevant tools to understand the development and sustain of stellar magnetic fields. Special focus is given to the recent ILES simulations performed with the anelastic solver EULAG-MHD. A short discussion on the need of including SGS modeling in the current global modeling was also presented.

In spite of the recent advances, the current models still are unable to reproduce the main characteristics of the solar dynamo nor the empirical relations between magnetic cycle periods and magnetic field strength with stellar parameters like the Rossby number. It is worth mentioning, however, that these relations still need to be revised with a large number of samples and longer observational missions.

Regarding the development of mean flows, the result of global simulations with different numerical schemes show good agreement. Consequently, great advances have been done in the understanding of the buildup of these flows. While the role of the Reynolds stresses is unanimously recognized as the responsible for the sustain of the differential rotation, the formation of meridional circulation has not yet a closed theoretical interpretation. It is attributed to the gyroscopic pumping mechanism resulting from the thermal wind balance (Miesch & Hindman 2011), or to deviations of this equilibrium given mostly by viscous forcing at boundary layers (Kitchatinov 2013). Most of global models tend to support the first alternative (e.g., Featherstone & Miesch 2015; Guerrero *et al.* 2016b).

With respect to the development of mean magnetic fields, several questions still remain to be answered. Perhaps the most important are, first, what global simulations have missed until now to reproduce solar-like patterns? Second, what physical processes are occurring in the global simulations? And third, how relevant are the tachoclines in the dynamo process? About the first question, there is still room for exploration. For instance, the role of the stratification, compressibility and a realistic upper magnetic boundary are yet to be studied. In addition, simulations with higher Reynolds numbers would likely be possible over the starting decade. With respect to the second question, the work until now has proven that the FOSA approximation might allow to distinguish what type of dynamo is occurring. However, computing the full set of elements in the electromotive force tensors seems to be the appropriate alternative to fully understand the dynamo process (e.g., Warnecke *et al.* 2018). The third question has been an issue of debate over the last years. Doubts about the relevance of the layers of strong radial shear started from the observational results of Wright & Drake (2016). They reported that stars without tachoclines follow the same scaling law between the magnetic field strength and Ro that stars with this layer. On the other hand, recent observations of the magnetic field topology in solar-type stars with different ages, show a clear change at the age of the formation of the radiative zone (Gregory *et al.* 2012). The structural change corresponds to the shift from a mainly poloidal configuration to a magnetic field dominated by high order modes. This is a clear indication of a change in the dynamo regime at the age when the tachoclines appear. Whatever is the case, for the time being global models with and without tachoclines are able to develop large-scale magnetic fields with polarity reversals. As these models are better understood, further observational evidence will provide a clearest scenario of stellar magnetism.

Acknowledgments

The author is thankful to the IAU and the Physics Department of UFMG for funding his participation in the symposium. I'm grateful to all the collaborators who contributed to the publications discussed in this work. Special thanks to Rafaella Barbosa and Alexander Kosovichev for providing some of the figures. The simulations were performed in the NASA cluster Pleiades and Brazilian super computer SDumont of the National Laboratory of Scientific Computation (LNCC).

References

Antia, H. M. & Basu, S. 2001, *ApJL*, 559, L67
Augustson, K., Brun, A. S., Miesch, M., & Toomre, J. 2015, *ApJ*, 809, 149
Baliunas, S. L., Donahue, R. A., Soon, W. H., *et al.* 1995, *ApJ*, 438, 269
Benomar, O., Bazot, M., Nielsen, M. B., *et al.* 2018, *Science*, 361, 1231
Böhm-Vitense, E. 2007, *ApJ*, 657, 486
Bonanno, A. 2013, *Sol. Phys.*, 287, 185
Bonanno, A. & Urpin, V. 2012, *ApJ*, 747, 137
Bonanno, A. & Urpin, V. 2013, *ApJ*, 766, 52
Böning, V. G. A., Roth, M., Jackiewicz, J., & Kholikov, S. 2017, *ApJ*, 845, 2
Brandenburg, A., Mathur, S., & Metcalfe, T. S. 2017, *ApJ*, 845, 79
Brandenburg, A., Rädler, K. H., & Schrinner, M. 2008, *A&A*, 482, 739
Brandenburg, A., Saar, S. H., & Turpin, C. R. 1998, *ApJL*, 498, L51
Brown, B. P., Browning, M. K., Brun, A. S., Miesch, M. S., & Toomre, J. 2010, *ApJ*, 711, 424
Browning, M. K., Miesch, M. S., Brun, A. S., & Toomre, J. 2006, *ApJL*, 648, L157
Brun, A. S., Miesch, M. S., & Toomre, J. 2004, *ApJ*, 614, 1073

Brun, A. S., Miesch, M. S., & Toomre, J. 2011, *ApJ*, 742, 79

Brun, A. S., Strugarek, A., Varela, J., *et al.* 2017, *Astrophys. J.*, 836, 192

Brun, A. S. & Toomre, J. 2002, *ApJ*, 570, 865

Cally, P. S. 2003, *MNRAS*, 339, 957

Cossette, J.-F., Charbonneau, P., Smolarkiewicz, P. K., & Rast, M. P. 2017, *ApJ*, 841, 65

Elliott, J. R. & Smolarkiewicz, P. K. 2002, International Journal for Numerical Methods in Fluids, 39, 855

Fan, Y. & Fang, F. 2014, *ApJ*, 789, 35

Featherstone, N. A. & Miesch, M. S. 2015, *ApJ*, 804, 67

Ferreira, R. R., Barbosa, R., Castro, M., *et al.* 2020, Submitted to *A&A*, doi:10.1051/0004-6361/201937219

Gastine, T., Yadav, R. K., Morin, J., Reiners, A., & Wicht, J. 2014, *MNRAS*, 438, L76

Germano, M., Piomelli, U., Moin, P., & Cabot, W. H. 1991, Physics of Fluids, 3, 1760

Ghizaru, M., Charbonneau, P., & Smolarkiewicz, P. K. 2010, ApJL, 715, L133

Gilman, P. A. 1977, Geophysical and Astrophysical Fluid Dynamics, 8, 93

Gilman, P. A. 1983, *Astrophys. J.s*, 53, 243

Gilman, P. A. & Miller, J. 1981, *ApJ*, 46, 211

Glatzmaier, G. A. 1984, Journal of Computational Physics, 55, 461

Glatzmaier, G. A. 1985a, *ApJ*, 291, 300

Glatzmaier, G. A. 1985b, Geophysical and Astrophysical Fluid Dynamics, 31, 137

Gregory, S. G., Donati, J.-F., Morin, J., *et al.* 2012, *ApJ*, 755, 97

Grinstein, F., Margolin, L., & Rider, W. 2007, Implicit Large Eddy Simulation: Computing Turbulent Fluid Dynamics (Cambridge University Press)

Guerrero, G., Del Sordo, F., Bonanno, A., & Smolarkiewicz, P. K. 2019a, *MNRAS*, 490, 4281

Guerrero, G., Smolarkiewicz, P. K., de Gouveia Dal Pino, E. M., Kosovichev, A. G., & Mansour, N. N. 2016a, *ApJ*, 819, 104

Guerrero, G., Smolarkiewicz, P. K., de Gouveia Dal Pino, E. M., Kosovichev, A. G., & Mansour, N. N. 2016b, *ApJL*, 828, L3

Guerrero, G., Smolarkiewicz, P. K., Kosovichev, A. G., & Mansour, N. N. 2013, *ApJ*, 779, 176

Guerrero, G., Zaire, B., Smolarkiewicz, P. K., *et al.* 2019b, *ApJ*, 880, 6

Haugen, N. E., Brandenburg, A., & Dobler, W. 2004, *Phys. Review E*, 70, 016308

Haugen, N. E. L. & Brandenburg, A. 2006, Physics of Fluids, 18, 075106

Held, I. M. & Suarez, M. J. 1994, Bulletin of the American Meteorological Society, 75, 1825

Hotta, H., Rempel, M., & Yokoyama, T. 2016, *Science*, 351, 1427

Howard, R. & Labonte, B. J. 1980, *ApJL*, 239, L33

Kaneda, Y., Ishihara, T., Yokokawa, M., Itakura, K., & Uno, A. 2003, Physics of Fluids, 15, L21

Käpylä, P. J., Korpi, M. J., Brandenburg, A., Mitra, D., & Tavakol, R. 2010, Astronomische Nachrichten, 331, 73

Käpylä, P. J., Mantere, M. J., & Brandenburg, A. 2012, ApJL, 755, L22

Käpylä, P. J., Mantere, M. J., Guerrero, G., Brandenburg, A., & Chatterjee, P. 2011, *A&A*, 531, A162

Karak, B. B., Käpylä, P. J., Käpylä, M. J., *et al.* 2015, *A&A*, 576, A26

Karak, B. B., Kitchatinov, L. L., & Choudhuri, A. R. 2014, *ApJ*, 791, 59

Kővári, Z., Oláh, K., Kriskovics, L., *et al.* 2017, Astronomische Nachrichten, 338, 903

Kitchatinov, L. L. 2013, in IAU Symposium, Vol. 294, Solar and Astrophysical Dynamos and Magnetic Activity, ed. A. G. Kosovichev, E. de Gouveia Dal Pino, & Y. Yan, 399–410

Kosovichev, A. G. & Schou, J. 1997, *ApJL*, 482, L207

Lehtinen, J., Jetsu, L., Hackman, T., Kajatkari, P., & Henry, G. W. 2016, *A&A*, 588, A38

Liang, Z.-C., Gizon, L., Birch, A. C., Duvall, T. L., & Rajaguru, S. P. 2018, *A&A*, 619, A99

Margolin, L. G. 2019, Shock Waves, 29, 27

Margolin, L. G. & Rider, W. J. 2002, International Journal for Numerical Methods in Fluids, 39, 821

Masada, Y., Yamada, K., & Kageyama, A. 2013, *ApJ*, 778, 11

Matilsky, L. I., Hindman, B. W., & Toomre, J. 2019, *ApJ*, 871, 217

Miesch, M. S., Brun, A. S., & Toomre, J. 2006, *ApJ*, 641, 618

Miesch, M. S., Gilman, P. A., & Dikpati, M. 2007, *ApJ*, 168, 337

Miesch, M. S. & Hindman, B. W. 2011, *ApJ*, 743, 79

Moffatt, H. K. 1978, Magnetic field generation in electrically conducting fluids

Noyes, R. W., Weiss, N. O., & Vaughan, A. H. 1984, *ApJ*, 287, 769

Olspert, N., Lehtinen, J. J., Käpylä, M. J., Pelt, J., & Grigorievskiy, A. 2018, *A&A*, 619, A6

Parker, E. N. 1955, *ApJ*, 122, 293

Passos, D., Miesch, M., Guerrero, G., & Charbonneau, P. 2017, *A&A*, 607, A120

Pipin, V. V. & Kosovichev, A. G. 2018, *ApJ*, 854, 67

Pitts, E. & Tayler, R. J. 1985, *MNRAS*, 216, 139

Pouquet, A., Frisch, U., & Leorat, J. 1976, Journal of Fluid Mechanics, 77, 321

Prusa, J. M., Smolarkiewicz, P. K., & Wyszogrodzki, A. A. 2008, Comput. Fluids, 37, 1193

Racine, É., Charbonneau, P., Ghizaru, M., Bouchat, A., & Smolarkiewicz, P. K. 2011, *ApJ*, 735, 46

Reinhold, T. & Gizon, L. 2015, *A&A*, 583, A65

Ruediger, G. 1989, Differential rotation and stellar convection. Sun and the solar stars

Saar, S. H. & Brandenburg, A. 1999, *ApJ*, 524, 295

Schad, A., Timmer, J., & Roth, M. 2013, *ApJL*, 778, L38

Schou, J., Antia, H. M., Basu, S., *et al.* 1998, *ApJ*, 505, 390

Schrinner, M., Rädler, K.-H., Schmitt, D., Rheinhardt, M., & Christensen, U. 2005, Astronomische Nachrichten, 326, 245

Simard, C., Charbonneau, P., & Bouchat, A. 2013, *ApJ*, 768, 16

Smagorinsky, J. 1963, Monthly Weather Review, 91, 99

Smolarkiewicz, P. K. 2006, International Journal for Numerical Methods in Fluids, 50, 1123

Smolarkiewicz, P. K. & Charbonneau, P. 2013, J. Comput. Phys., 236, 608

Spitzer, L. 1962, Physics of Fully Ionized Gases

Spruit, H. C. 2002, *A&A*, 381, 923

Steenbeck, M., Krause, F., & Rädler, K.-H. 1966, Zeitschrift Naturforschung Teil A, 21, 369

Stix, M. 1976, in IAU Symposium, Vol. 71, Basic Mechanisms of Solar Activity, ed. V. Bumba & J. Kleczek, 367

Strugarek, A., Beaudoin, P., Charbonneau, P., & Brun, A. S. 2018, *ApJ*, 863, 35

Strugarek, A., Beaudoin, P., Charbonneau, P., Brun, A. S., & do Nascimento, J.-D. 2017, *Science*, 357, 185

Tayler, R. J. 1973, *MNRAS*, 161, 365

Vidotto, A. A., Gregory, S. G., Jardine, M., *et al.* 2014, *MNRAS*, 441, 2361

Viviani, M., Warnecke, J., Käpylä, M. J., *et al.* 2018, *A&A*, 616, A160

Vorontsov, S. V., Christensen-Dalsgaard, J., Schou, J., Strakhov, V. N., & Thompson, M. J. 2002, *Science*, 296, 101

Warnecke, J. 2018, *A&A*, 616, A72

Warnecke, J., Käpylä, P. J., Käpylä, M. J., & Brandenburg, A. 2014, *ApJL*, 796, L12

Warnecke, J., Rheinhardt, M., Tuomisto, S., *et al.* 2018, *A&A*, 609, A51

Wright, N. J. & Drake, J. J. 2016, *Nature*, 535, 526

Wright, N. J., Drake, J. J., Mamajek, E. E., & Henry, G. W. 2011, *ApJ*, 743, 48

Yoshimura, H. 1975, *ApJ*, 201, 740

Zahn, J.-P., Brun, A. S., & Mathis, S. 2007, *A&A*, 474, 145

Zhao, J., Bogart, R. S., Kosovichev, A. G., Duvall, Jr., T. L., & Hartlep, T. 2013, *ApJL*, 774, L29

Discussion

BRANDENBURG: The cycle to rotation frequency is found to increase with stellar activity. Do you agree that there is currently no model that reproduces this? How can we hope to reproduce the Sun if this essential feature is not reproduced. Could you speculate on what might be missing in the models?

GUERRERO: As a matter of fact no current model reproduces this relation. Nevertheless, cycle dynamos in global simulations are relatively new. There is still a broad parameter space to explore in the simulations, specially concerning the resolution and the energy transport issue. I'm optimistic that as soon as these issues are clarified the models will be able to reproduce better the solar and stellar magnetism. Note also that observations are still providing new results.

STRUGAREK: The trend of differential rotation seems to increase as rotation rate diminishes. This is at odds with other 3D simulations with and without underlying stable layer. How is differential rotation sustained in your models?

GUERRERO: The latitudinal shear indeed increases when the rotational rate decreases. This relation, not included in the presentation, is shown in Fig. 4 of this proceeding. The results, however, are not completely at odds with other simulations. See, for instance, the results presented in Gastine *et al.* (2014) including global simulations from several groups. As rotation decreases the shear increases before changing sing for anti-solar differential rotation. In Guerrero *et al.* (2016b) we presented the angular momentum balance of one particular simulation, a detailed analysis of all the simulations is still on its way. This will clarify the reason for this relation.

LUHMANN: There is a well-known observed relationship between surface polar fields and cycle size. Do the simulations reveal a physical reason for that relationship?

GUERRERO: This property has not yet been studied in the results of global models. The reason is that unfortunately there is not yet a model reproducing well all the properties of the solar cycle.

Solar and Stellar Magnetic Fields: Origins and Manifestations
Proceedings IAU Symposium No. 354, 2019
A. Kosovichev, K. Strassmeier & M. Jardine, ed.
doi:10.1017/S1743921320000897

3D Modeling of the Structure and Dynamics of a Main-Sequence F-type Star

Irina N. Kitiashvili[1]⬤ and Alan A. Wray[2]

[1]NASA Ames Research Center
Moffett Field, MS 258-6, Mountain View, USA
email: irina.n.kitiashvili@nasa.gov

[2]NASA Ames Research Center
Moffett Field, MS 258-6, Mountain View, USA
email: alan.a.wray@nasa.gov

Abstract. Current state-of-the-art computational modeling makes it possible to build realistic models of stellar convection zones and atmospheres that take into account chemical composition, radiative effects, ionization, and turbulence. The standard 1D mixing-length-based evolutionary models are not able to capture many physical processes of the stellar interior dynamics. Mixing-length models provide an initial approximation of stellar structure that can be used to initialize 3D radiative hydrodynamics simulations which include realistic modeling of turbulence, radiation, and other phenomena.

In this paper, we present 3D radiative hydrodynamic simulations of an F-type main-sequence star with 1.47 solar mass. The computational domain includes the upper layers of the radiation zone, the entire convection zone, and the photosphere. The effects of stellar rotation is modeled in the f-plane approximation. These simulations provide new insight into the properties of the convective overshoot region, the dynamics of the near-surface, highly turbulent layer, and the structure and dynamics of granulation. They reveal solar-type differential rotation and latitudinal dependence of the tachocline location.

Keywords. convection; hydrodynamics; methods: numerical; stars: general, horizontal-branch, interiors, rotation, fundamental parameters

1. Introduction

A dramatic increase in observational data from NASA's Kepler, K2, and TESS missions and supporting ground-based observatories has opened new opportunities to investigate the internal structure, dynamics, and evolution of stars and their atmospheres. However, analysis and interpretation of the observations are challenging, especially for characterizing the structure and dynamics of the outer stellar convection layers and atmospheres. Until recently, only models based on mixing-length theory (MLT) were used to investigate the interior structure and oscillations.

Numerous observed phenomena such as superflares, activity cycles, starspots, and oscillation properties all reflect the properties and dynamics of turbulent convection in the outer layers of stars (e.g. Carroll & Strassmeier 2014; Balona 2015; Antoci *et al.* 2019; Mathur *et al.* 2019; Strassmeier *et al.* 2019). Rapidly growing computational capabilities have enabled 3D stellar hydrodynamic simulations on local scales (e.g. Beeck *et al.* 2015; Kitiashvili *et al.* 2012; Kitiashvili 2016; Salhab *et al.* 2018).

In this paper, we present results of 3D radiative hydrodynamic simulations of an F-type main-sequence star of $1.47 M_\odot$, with and without rotation. We discuss the physical

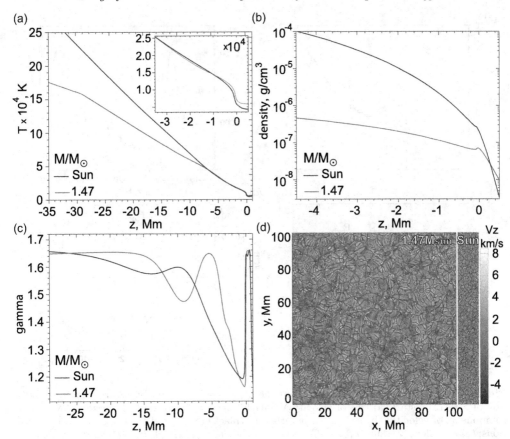

Figure 1. The mixing-length 1D model of the 1.47 $M_\odot$ F-type star used as the initial conditions to perform the 3D radiative realistic modeling (red curves): a) temperature, b) density, and c) gamma. The black curves show the corresponding mixing-length model of the Sun for comparison. Panel d) shows a snapshot of the vertical velocity at the stellar photosphere. A vertical strip on the right shows the simulations of the solar surface for comparison of the surface granulation scales.

properties of the stellar convection and its interactions with the radiative zone in the overshoot layer.

2. Numerical setup

The modeling is performed using the StellarBox code (Wray *et al.* 2018). This code performs modeling from first principles, taking into account the realistic chemical composition, equation of state, radiative transfer, and the effects of turbulence. Starting from a 1D mixing-length model of the internal structure (Fig. 1a-c) obtained with CESAM code (Morel 1997; Morel & Lebreton 2008), we produced a series of 3D dynamical simulations of an F-spectral type main-sequence star of 1.47$M_\odot$ both with rotation and without rotation. The computational domain covers about 20% of the stellar radius (or 51 Mm in depth), which includes the entire convection zone and the upper layers of the radiative zone. The simulations include a 1 Mm high atmospheric layer. The horizontal size of the computational domain is 102.4 Mm with a resolution of 100 km; the vertical resolution increases with depth from 25 km at the atmosphere to 183 km near the bottom boundary. In the StellarBox code, stellar rotation is implemented using the f-plane approximation.

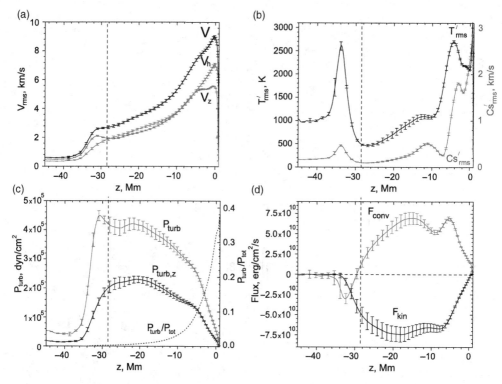

Figure 2. Vertical profiles, obtained from the 3D numerical radiative hydrodynamic simulation of a 1.47M$_\odot$ star: (a) rms of velocity V (black), vertical Vz (red) and horizontal Vh (blue) components of velocity; (b) rms of temperature T' (black) and sound-speed $c_{S'}$ (blue) perturbations; (c) turbulent pressure P_{turb} (blue curve), turbulent pressure of the vertical motions $P_{turb,z}$ (black), and the ratio of turbulent pressure P_{turb} to total pressure $P_{turb} + P$; and (d) convective energy flux F_{conv} (blue curve) and kinetic energy flux F_{kin} (black), calculated according to the formulation of Nordlund & Stein (2001). Vertical dashed lines indicate the bottom boundary of the convection zone of the corresponding 1D stellar model: $z_{cz} = -28.5$ Mm.

3. Thermodynamic structure of non-rotating stars with shallow convection zones

The turbulent dynamics of stars with relatively shallow convection zones attracts our interest because it allows us to investigate the properties of turbulent convection across the whole convection zone and the transition into the convectively stable radiative zone with a high degree of realism.

Our simulation results have revealed that a characteristic feature of the stellar photosphere of F-type stars is the co-existence of several granulation scales (Kitiashvili 2016). The convection pattern represents well-defined 'small' and 'large' granules self-organized into clusters (Fig. 1d). In deeper layers of the convection zone, the scale of the convection pattern gradually increases. In the intergranular lanes, some downdrafts penetrate through the entire convection zone, reaching velocities of more than 20 km/s. These downdrafts penetrate into the radiative zone, form an overshoot layer and cause local heating (Fig. 2b). This increases the local sound speed and initiates density fluctuations that are a source of internal gravity oscillations (g-modes).

It is informative to consider mean vertical profiles of the fluctuations for various turbulence quantities and the convective and kinetic energy fluxes from the stellar photosphere down to the radiative zone (Fig. 2). The strength of the horizontal flows gradually decreases from the photosphere to the bottom of the convection zone. In the overshoot region, the amplitude of the horizontal velocities increases slightly due to the splashing of

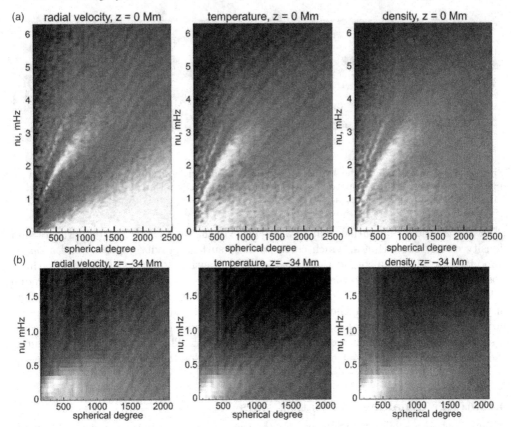

Figure 3. Power spectra of the radial velocity, temperature, and density, represented as functions of angular spherical harmonic degree and frequency for: a) the stellar surface and b) the radiative zone just below the tachocline layer at depth 34 Mm below the photosphere.

downdrafts (Fig. 2a). The rms vertical velocity profile shows a strong deviation from the horizontal flows near the surface layer. In particular, a broad bump in horizontal velocity located from 0 – 5 Mm below the surface corresponds to one of the characteristic scales of the granulation layer. Below 10 Mm, the fluctuations of vertical velocity are stronger than the horizontal ones, probably due to the penetrating high-speed downdrafts. In the overshoot layer, the velocity strength sharply decreases.

The distribution of temperature fluctuations with depth (Fig. 2b, black curve) shows significant variations near the surface layers and in the overshoot region. The near-surface fluctuations (peaked at near –5 Mm) are mostly related to strong radiative cooling in the intergranular lanes and are associated with downflows that determine the granulation scales. A sharp increase in temperature fluctuations from –10 to –5Mm corresponds to the He II ionization zone. In this region, the rms of vertical velocity and the convection energy flux (Fig. 2a,c) also increase, indicating an enhancement of turbulent convection in the ionization zone. Temperature fluctuations at $z \approx -34$Mm, which is associated with the overshoot region, reflect the intense local heating in these layers due to plasma compression by downdrafts that penetrate through the convection zone. The rms sound-speed perturbations (Fig. 2b, blue curve) have a similar shape, but their amplitude in the overshoot region is much smaller than in the subsurface layers.

The stellar oscillations excited in the convection zone are shown in the form of the power spectra ($l - \nu$ diagrams) near the photosphere (Fig. 3a) and the overshoot region

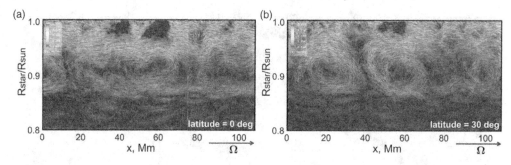

Figure 4. Comparison of the rotating convection zone with period of rotation of 1 day at different latitudes: a) 0° (equator), and b) 30°. The dynamical structure of the convective patterns is visualized by the particle tracing method.

(Fig. 3b). At the stellar surface, the power spectra of the radial velocity, temperature, and density clearly show ridges, which correspond to surface gravity f-modes and acoustic p-modes. An additional broad ridge visible in the density power spectrum below 1mHz possibly corresponds to internal gravity g-modes. However, the power spectra of radial velocity and temperature do not show this ridge. The power spectra of the upper layers of the radiative zone, near the overshoot region, do not show any acoustic-type oscillations (Fig. 3b), probably because of the flow amplitude. However, all quantities show a signature of g-modes, excited in the overshoot region. This may explain why g-mode oscillations are not observed even for stars with shallow convection zones, where the surface convective turbulent flows are strong.

4. Interior dynamics of rotating stars

Links between stellar rotation and stellar thermodynamic properties are a hot topic in various fields (e.g. Law 1981; MacGregor *et al.* 2007; Eggenberger *et al.* 2010; Brito & Lopes 2019). Therefore, understanding how stellar internal dynamics and structure depends on the rotation period is critical for interpreting observational data. To study rotation effects, we performed simulations for rotational periods of 1 and 14 days using the f-plane approximation, in which the Coriolis force is assumed constant in horizontal planes. This approximation is applicable for local simulations because the simulation domain covers only $\sim 5.6°$ in latitude. The full radiative hydrodynamic simulations are performed for three latitudes: 0° (equator), 30°, and 60° for both rotation rates.

The simulation results show that stellar rotation does not affect the structure and dynamics of the surface layers. However, in deeper layers of the convection zone, two distinct layers can be identified (Fig. 4). In the outer layer of the convection zone, large-scale flow patterns are transported in the direction opposite to stellar rotation. Such a distinct structure of the flow in the upper layers of the convection zone represents a subsurface shear flow, similar to the one observed in the Sun. Profiles of the azimuthal velocity, meridional circulation, and the mean vertical velocity (Fig. 5a-c) show that the subsurface shear layer extends up to 10.5 Mm into the convection zone. The inner layers of the convection zone reveal cellular or roll-like circular motions. The structure and scale of these large-scale eddies depend on latitude. At the equator (Fig. 4a), they are very elongated and often can split into smaller and weaker patterns. At higher latitudes, the eddies are more compact and stable with sizes of about 25 Mm at 30° latitude for $P_{rot} = 1$day (Fig. 4b).

The ability to model the effects of stellar rotation at different latitudes allows us to study the properties of differential rotation and meridional circulation for various periods of rotation and latitudes (Fig. 5a,b). The differential rotation shows faster rotation at

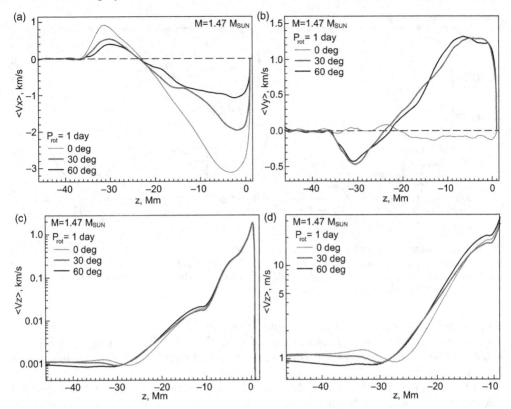

Figure 5. Effect of stellar rotation on the mean velocity profiles for a rotation period of 1 day at three latitudes: 0° (red curves), 30° (blue curves), and 60° (black curves). Panel a) shows radial profiles of the azimuthal velocity, representing differential rotation. Panel b): Radial profiles of the mean velocity field corresponding to the meridional circulation at different latitudes. Panels c) and d) show the mean radial profiles for the vertical velocity component across the whole domain (panel c), and a zoom into the tachocline layer, d). All profiles are averages in horizontal planes over one stellar hour.

the equator and slower at 60° latitude, indicating solar type rotation (Fig. 5a). In the tachocline layer, reverse azimuthal flows (corresponding to deceleration of the stellar rotation) are present. The relative strength of the return flows correlates with subsurface shear flows at different latitudes. Thus, the reverse flows are weaker at higher latitude, and the subsurface shear flows are also weaker in comparison with the flows at the equator. The strongest subsurface and reverse flow location depends on the latitude. In particular, the maximum of the subsurface shear flow is located closer to the stellar photosphere in comparison with the flows at the equator. The maximum of the reverse flows is shifted into the deeper layers of the stellar interiors in comparison with the flows at higher latitudes. In the case of slower rotation, $P_{rot} = 14$ days, these trends also take place but are weaker and develop slowly, requiring a significantly longer simulation run.

Another component of the large-scale flows, perpendicular to the stellar rotation, can be interpreted as meridional circulation. Because of the periodic lateral boundary conditions, the resulting models are able to reproduce only a narrow range of latitudes of the meridional circulations. Figure 5b shows the radial profiles of the meridional circulation at the equator, 30°, and 60° latitudes. The meridional component of the velocity at the equator fluctuates around zero, meaning that there is no significant mean cross-equator meridional flow, which is not surprising. A weak negative flow velocity in the convection

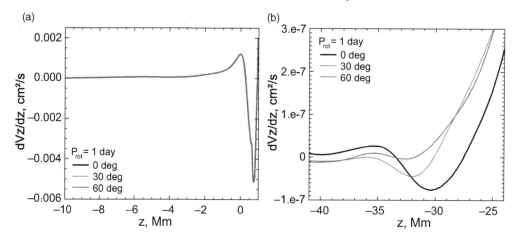

Figure 6. Radial profiles of the mean vertical velocity time-derivative, dVz/dt, for the whole computational domain (panel a) and near the tachocline layer. The profiles were obtained for a rotating main-sequence 1.47 $M_\odot$ star with a period of 1 day at 3 latitudes: $0°$ (thick black lines), $30°$ (thin blue lines), and $60°$ (red curves).

zone is likely due to only 1-hour averaging, but potentially may reflect the flow variations on longer time-scales. This requires additional study. The meridional circulation at $30°$ and $60°$ latitudes reveals a consistent meridional circulation structure in the flow variations.

The overall mean distribution of the vertical flows is shown in Figure 5c. The vertical flow distribution is the same in the layers from the photosphere down to 9.5Mm below the photosphere. In the deeper layers, the vertical velocities start to deviate at different latitudes (Fig. 5c,d). The fastest decrease of the velocity is at the equator, where it reaches a local minimum around 27 Mm, which corresponds to the top of the tachocline layer. After that, the vertical velocity increases until the bottom of the tachocline layer. In the radiative zone, the velocities slowly decrease with depth. Similar behavior of the vertical flows takes place at $30°$ latitude, with a minimum at a depth of 30 Mm below the surface. At $60°$, there is no prominent minimum. The vertical velocity derivative reveals a clearer picture (Fig. 6). In particular, it shows the velocity variations near the stellar surface and in deeper layers. Near the tachocline layer, the derivative minimum at the equator is located at 30 Mm below the stellar surface, at 32 Mm for $30°$, and 32.5 Mm for $60°$ latitude. This provides evidence that the depth of the tachocline varies with latitude.

5. Conclusion

Despite the availability of advanced observational data from modern space and ground instruments, investigation of the dynamics and structure of the surface and subsurface layers of stars is quite challenging. We performed a series 3D radiative hydrodynamic simulations of an F-type star with mass 1.47 $M_\odot$, in which the whole convection zone and upper layers of the radiative zone were included in the computational domain. The simulation results reveal the formation of an overshoot layer and also multi-scale populations and clustering of the surface granulation. High-speed convective downdrafts of $20 - 25$km/s penetrate through the convection zone, form an overshoot layer, and contribute to excitation of internal gravity waves (g-modes). These waves are identified near the overshoot layer. At the stellar photosphere, these modes are hidden among strong turbulent convective flows, and only f- and p-modes are clearly displayed in the simulated power spectra.

Simulating of effects of stellar rotation, for rotational periods of 1 and 14 days at different latitudes, allowed us to identify the formation of a subsurface shear flow and roll-like convective patterns in the deep layers of the convection zone. The radial profiles of the differential rotation indicate that it is of solar type. The subsurface shear flow velocity peaks closer to the photosphere at higher latitudes. The meridional circulation profiles do not show a significant difference at 30° and 60° latitudes. The simulation results show that the tachocline layer is located deeper and is less prominent at higher latitudes.

Our future plans include expansion of the computational domain higher into the stellar atmosphere and taking into account magnetic fields, as well as developing these types of numerical models for more massive stars.

Acknowledgment

The work is supported by NASA Astrophysics Theory Program.

References

Antoci, V., Cunha, M. S., Bowman, D. M., *et al.* 2019, *MNRAS*, 490, 4040

Balona, L. A. 2015, *MNRAS*, 447, 2714

Beeck, B., Schüssler, M., Cameron, R. H., & Reiners, A. 2015, *A&A*, 581, A42

Brito, A. & Lopes, I. 2019, *MNRAS*, 488, 1558

Carroll, T. A. & Strassmeier, K. G. 2014, *A&A*, 563, A56

Eggenberger, P., Meynet, G., Maeder, A., *et al.* 2010, *A&A*, 519, A116

Kitiashvili, I. N., Guzik, J. A., Kosovichev, A. G., *et al.* 2012, Astronomical Society of the Pacific Conference Series, Vol. 462, Radiation Hydrodynamics Simulations of Turbulent Convection for Kepler Target Stars, ed. H. Shibahashi, M. Takata, & A. E. Lynas-Gray, 378

Kitiashvili, I. N., Kosovichev, A. G., Mansour, N. N., & Wray, A. A. 2016, *ApJL*, 821, L17

Law, W. Y. 1981, *A&A*, 102, 178

MacGregor, K. B., Jackson, S., Skumanich, A., & Metcalfe, T. S. 2007, *ApJ*, 663, 560

Mathur, S., García, R. A., Bugnet, Lisa andSantos, Â. R. G., Santiago, N., & Beck, P. G. 2019, Frontiers in Astronomy and Space Sciences, 6, 46

Morel, P. 1997, *A&AS*, 124, 597

Morel, P. & Lebreton, Y. 2008, *AP&SS*, 316, 61

Nordlund, Å. & Stein, R. F. 2001, *ApJ*, 546, 576

Salhab, R. G., Steiner, O., Berdyugina, S. V., *et al.* 2018, *A&A*, 614, A78

Strassmeier, K. G., Carroll, T. A., & Ilyin, I. V. 2019, *A&A*, 625, A27

Wray, A. A., Bensassiy, K., Kitiashvili, I. N., Mansour, N. N., & Kosovichev, A. G. 2018, Realistic simulations of Stellar Radiative MHD. In Book: Variability of the Sun and Sun-like Stars: from Asteroseismology to Space Weather, ed. E. B. J.-P. Rozelot (EDP Sciences), 39–62

Solar and Stellar Magnetic Fields: Origins and Manifestations
Proceedings IAU Symposium No. 354, 2019
A. Kosovichev, K. Strassmeier & M. Jardine, ed.
doi:10.1017/S1743921320000630

Helioseismic insights into the generation and evolution of the Sun's internal magnetic field

Anne-Marie Broomhall[ID] and René Kiefer

Centre for Fusion, Space and Astrophysics, Department of Physics,
University of Warwick, CV4 7AL
email: a-m.broomhall@warwick.ac.uk

Abstract. Properties of helioseismic acoustic oscillations (p modes) are modified by flows and magnetic fields in the solar interior, with frequencies, amplitudes and damping rates all varying systematically through the solar cycle. Crucially, now, we have a long enough baseline of helioseismic data to compare of the different activity cycles. We review recent efforts along these lines, from the impact of near-surface magnetic fields on p-mode frequencies to the evolution of the torsional oscillation and meridional circulation. We show that each activity cycle for which we have helioseismic data is slightly different in terms of the relationship between p mode frequencies and atmospheric proxies of activity, and in terms of the rotation and meridional circulation flows. However, many challenges remain, crucially including our ability to constrain flows and magnetic fields in the deep solar interior.

Keywords. Sun: activity, Sun: helioseismology, Sun: interior, Sun: magnetic fields, Sun: oscillations, Sun: rotation

1. Introduction

Helioseismology uses the Sun's resonant oscillations to infer conditions beneath the solar surface. At any point in time there are thousands of acoustic oscillations travelling throughout the solar interior. These oscillations sample different but overlapping regions of the solar interior and so by studying the properties of these oscillations, and using models, we can build up a profile of the inside of the Sun. For example, we can learn about how temperature and rotation vary with depth and we can even use heliosiesmology to learn about the composition of the Sun. The acoustic oscillations, upon which we will focus here, are referred to as p modes as the main restoring force is a pressure differential.

We describe helioseismic oscillations in terms of spherical harmonics, which are charactised by three main "quantum" numbers. The first of these is the spherical harmonic degree, or ℓ, which specifies the total number of nodes present at the surface. As an oscillation travels inwards, the temperature, and therefore the sound speed, increases so, unless the oscillation is travelling perfectly radially, it will be refracted until it returns to the surface. The depth a mode travels to before being completely refracted depends on the angle it's travelling inwards at near the surface and this varies as a function of ℓ: the lowest-ℓ modes travel radially, or nearly radially, and so penetrate deeper in the solar interior than high-ℓ modes, which only sample the near surface regions. It is this difference in the depth of the "lower-turning point" that allows us to build up profiles of the solar interior.

The second number we use to describe the spherical harmonic structure of the modes is the azimuthal degree or m, and this describes the number of nodes round the equator. Again showing similarities to quantum mechanics, if the Sun were completely symmetric, each one of the m components of a mode would have the same frequency. However asymmetries, like rotation and other flows, split the frequencies of the different m components, with the magnitude of that splitting providing information about the asymmetries responsible.

The final number used to described the modes is the radial degree, n. As the name suggests, n describes the number of nodes in the radial direction, with the frequency of the oscillations increasing as a function of n.

As already mentioned, the modes, are refracted in the solar interior until they return towards the surface. When the modes reach the surface again they are reflected by the sharp drop in density and the radius at which reflection takes place depends on the length scale of the mode in comparison to the density scale height and, therefore, on mode frequency. More specifically, the high-frequency modes are reflected further out than low-frequency modes. Another way of thinking of this is that, for a particular radius, there is a specific acoustic cut-off frequency, where all modes below that frequency have already been reflected and all modes above that frequency are transmitted.

Helioseismic observations can be made using both Doppler velocity and intensity. Then there are two main branches of helioseismology: global and local. Global helioseismology studies the natural resonant oscillations that cause the Sun to oscillate as a whole and itself can be split into two sub-branches. There is unresolved, Sun-as-a-star helioseismology, whose observations are only sensitive to low-ℓ modes or those with the largest horizontal length scales. This is because, for unresolved observations, the forwards and backwards motions associated with higher-ℓ modes cancel each other out, meaning the modes are not visible. Sun-as-a-star helioseismic data can easily observe oscillations with $\ell \leq 3$, and modes with $\ell = 4$ and 5 are just about visible in some data sets (e.g. Chaplin *et al.* (1996); Lund *et al.* (2014)). However, it is worth remembering that these low-ℓ modes are the ones that travel deepest in the solar interior. A number of long-baseline Sun-as-a-star datasets now exist, which readily allow solar-cycle helioseismic studies to be performed. These include the Birmingham Solar Oscillations Network (BiSON), which has been making Doppler velocity observations since the 1970s, the Global Oscillations at Low Frequency (GOLF) instrument, which also makes Doppler velocity observations and has been operational since its host spacecraft, the Solar and Heliospheric Observatory (SoHO) was launched in 1995, and the Variability of Solar Irradiance and Gravity Oscillations (VIRGO) instrument, which is also onboard SoHO and makes intensity observations. Resolved observations use spatial filters to allow much higher-ℓ modes to be observed. There are also long-baseline resolved observations of global modes available, which are suitable for solar cycle studies, including those from the Global Oscillations Network Group (GONG), which produces Doppler velocity data for global modes with $0 \leq \ell \leq 200$ going back to 1996. Other commonly used instruments include the Michelson Doppler Imager (MDI), which observed from the launch of SoHO until 2011, and the Helioseismic and Magnetic Imager (HMI), which is onboard the Solar Dynamics Observatory (SDO) and was launched in 2010. Using the overlap to scale the data, it is common practice to combine MDI and HMI results when studying solar cycle variations in the p modes. Further details on global helioseismology can be found in Basu (2016).

Local helioseismology studies waves in localised patches of the surface and is able to infer conditions and parameters, such as those used to describe localised flows, say beneath a sunspot. This too has different branches and methodologies based on the different observational techniques. Ring-diagram analysis is most akin to global helioseismology, only observations are made on small "tiles" and Fourier transforms are performed in 3 dimensions, giving power spectra as a function of two horizontal wavenumbers, k_x and k_y, as well as frequency. The power appears in rings, hence the name, that are shifted and distorted, in the x and y directions by flows and in frequency by localised

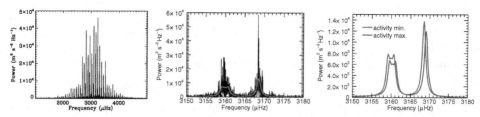

Figure 1. Left: spectrum obtained from 365 d of BiSON Sun-as-a-star Doppler velocity data. Middle: close up of an $\ell = 2$ and an $\ell = 0$ mode in this spectrum. The red line is the fitted Lorentzian profiles. Right: comparison of the profiles fitted to these two modes at activity maximum and activity minimum.

changes in the wave speed. The second technique frequently used in helioseismology is known as time-distance helioseismology. Here, the travel time of a wave between two surface locations is determined using cross-correlation techniques. Inhomogeneities, such as flows, magnetic fields and anisotropies in the sound speed affect the travel time and inversions are performed to determine the properties of the inhomogeneity. GONG, MDI and HMI are all frequently used in local helioseismic studies. Further details on local helioseismology can be found in Gizon & Birch (2005).

We now move on to describe how the parameters of the global p modes vary through the solar cycle (in Section 2) before a discussion on flows in the solar interior that play important roles in many dynamo models (in Section 3) and finally we will describe how these flows vary through the solar cycle.

2. Solar cycle variations in p mode parameters

When we observe global p modes we tend to make relatively long observations of the Sun, of the order of months to years, and then perform some form of Fourier-like transform to make a power spectrum. The left panel of Figure 1 shows an example of one such power spectrum obtained from 365 d of Sun-as-a-star Doppler velocity observations, made by BiSON (Davies *et al.* (2014); Hale *et al.* (2016)). Each one of the peaks you can see is a different mode of oscillation. If we zoom in on just two modes, as shown in the middle panel of Figure 1, we can see that the mode peaks have structure because the modes behave like damped harmonic oscillators. We can, therefore, fit asymmetric Lorentzian profiles to the peaks in order to determine mode parameters, such as frequencies, powers and damping rates (which are indicated by the width of the peaks). These parameters all vary through the Sun's magnetic activity cycle. In the right panel of Figure 1, we can see a comparison between the mode profiles that were fitted to the two p modes shown in the middle panel at solar maximum and minimum. The variations in frequency and maximum power of the profiles are easily visible. Less easy to discern is the variation in the width of these profiles, but it is there nonetheless.

Figure 2 shows the average shift in frequency of p modes as a function of time. A scaled and shifted version of the 10.7 cm flux is also plotted for comparison. While the zero point is somewhat arbitrary, the 11 yr solar cycle is clearly visible in frequency shifts. The p-mode frequencies increase as the solar magnetic field increases, with $\ell = 0$ p modes at about 3000 μHz experiencing a shift of about 0.4 μHz between cycle minimum and maximum. The causes of these variations can broadly be split into two categories: direct and indirect effects. The direct effects involve the Lorentz force, which provides an additional restoring force, thereby increasing the frequency of the modes. The indirect effects involve changes in the properties of the cavity in which the modes are trapped, for example, the size of the acoustic cavity.

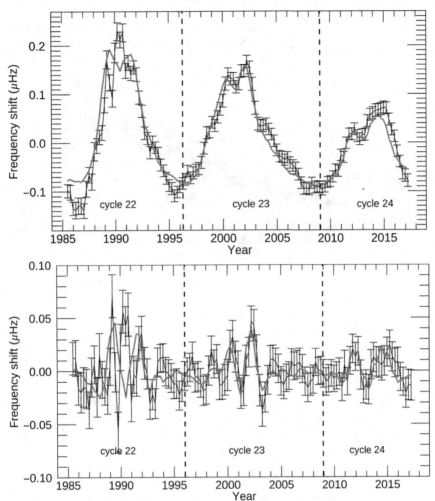

Figure 2. Top: average shift in p-mode frequencies between cycle minimum and cycle maximum (black data points). Red line shows a scaled version of the 10.7 cm flux for comparison. Bottom: frequency-shift residuals that remain once a smoothed version of the frequency shifts in the top panel has been subtracted (black data points). Red line shows the 10.7 cm flux residuals, obtained by the same process.

The frequency-shift residuals plotted in the bottom panel of Figure 2 were obtained by removing a smooth 11 yr variation from the frequency-shift data (obtained using a box-car smoothing with a 3 yr window). There is evidence for the "quasi-biennial oscillation (QBO)" in the residuals, which are highly correlated with those seen in atmospheric proxies of solar activity, including the 10.7 cm flux. In fact, the QBO is seen in a large number of activity proxies Bazilevskaya *et al.* (2014). We note that the amplitude of the observed QBO varies with time, being at a maximum close to solar maximum. This too is a feature of the QBO seen in other activity proxies. The presence of the QBO in helioseismic data provides a link between the signal seen in atmospheric activity proxies with the magnetic field in the solar interior. The challenge now is to use this signal to gain insights into the origin and structure of the magnetic field responsible for the QBO.

It is also interesting to compare the frequencies observed at different times in the solar cycle. Figure 3 shows the difference in p-mode frequencies observed at cycle minimum

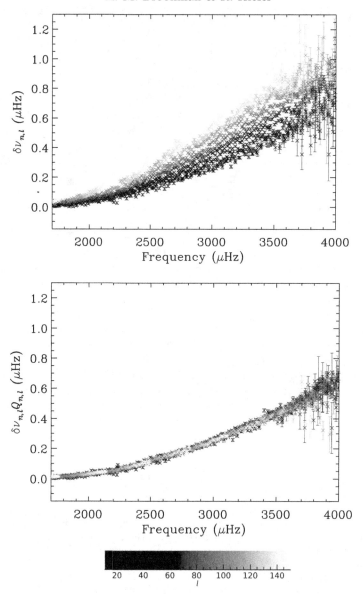

Figure 3. Top: frequency shifts observed when comparing observations from cycle maximum and cycle minimum. Bottom: inertia corrected frequency shifts observed when comparing observations from cycle maximum and cycle minimum. Modified from Broomhall *et al.* (2017).

and cycle maximum. The size of the frequency shift is dependent on both the degree and frequency of the mode. The degree dependence can be understood in terms of mode inertia Christensen-Dalsgaard & Berthomieu (1991), which is lower for the high-ℓ modes than the low-ℓ modes, meaning that the high-ℓ modes are more easily perturbed. Normalizing by the mode inertia removes the degree dependence, leaving something that is dependent on frequency alone (as can be seen in Figure 3). The strong frequency dependence, with little or no dependence on degree, is taken to imply that most of frequency changes are caused by effects confined to near-surface layers (Gough (1990); Libbrecht & Woodard (1990)). The higher the mode frequency, the higher the upper turning point of the mode (defined as the radius at which the modes are reflected back into the solar interior).

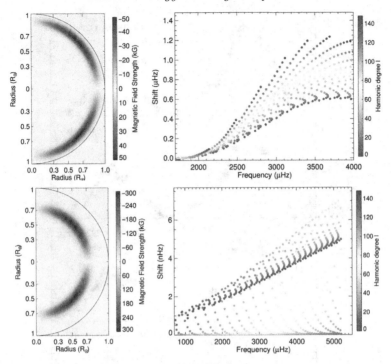

Figure 4. Top, left: dipolar, near-surface magnetic field, used to simulate the impact of a magnetic field on the mode frequencies at cycle maximum. To simulate the impact of a magnetic field on the modes at cycle minimum the maximum field strength was reduced by 10 kG. Top, right: shift in frequency between the simulated cycle maximum and cycle minimum. First published in Kiefer & Roth (2018). Bottom, left: Strong dipolar magnetic field introduced at the base of the convection zone. Bottom, right: Frequency shift caused by introduction of field shown in bottom, left panel.

Therefore, if the magnetic perturbation occurs in the near-surface region, a region in which low-frequency modes may not penetrate but high-frequency modes do, it stands to reason that the perturbation has a larger impact on high-frequency modes than low-frequency modes. Latitudinal inversions, such as those performed in Howe *et al.* (2002), show that the latitudinal structure of the perturbation very closely follows that of the well-known butterfly diagram.

Kiefer *et al.* (2017) and Kiefer & Roth (2018) demonstrated that the observed frequency shift can be replicated by modelling direct and indirect effects: A toroidal magnetic field is constructed and, by varying its strength but maintaining the same structure, used to simulate the solar cycle. These magnetic fields are then used to perturb the mode frequencies and the difference in mode frequency between the two states is determined. Figure 4 shows the impact of a change in near-surface field strength of 10 kG, when the field is dipolar. It is clear that this figure closely follows the form seen in the real data (i.e. as seen in Figure 3). Kiefer & Roth (2018) also tried putting a very strong dipolar field of 300 kG at the base of the convection zone. As can be seen in Figure 4, this produced a tiny shift of the order of nHz, which can be compared to the µHz shift produced by the much smaller near-surface field. This indicates that it is going to be extremely difficult to detect the impact of any field at the base of the convection zone. It is, however, a worthy goal, given that many believe this region to be the seat of the solar dynamo.

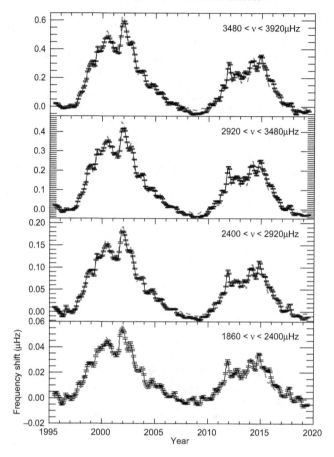

Figure 5. Frequency shifts as a function of time for $0 \leq \ell \leq 150$ modes observed by GONG. The frequency shift data (black data points) have been averaged over four different frequency ranges. For comparison, scaled and shifted versions of the 10.7 cm flux have been included.

2.1. *Solar cycle comparisons*

Arguably one of the most interesting, and least understood, features of the solar cycle is its variability: each solar cycle is different to the next. Even though it is often referred to as the the 11 yr solar cycle, the length of each cycle varies, usually lying somewhere in the region of $9-14$ yr Hathaway (2015). The amplitude of an activity cycle also varies from one cycle to the next, making accurate predictions of the strength and timing of solar maximum notoriously difficult. Although relatively new on the scene, compared to say sunspot records, helioseismology now has continuous observations spanning decades, meaning we even have data covering a full Hale cycle (which consists of two solar cycle and accounts for the fact that there is a polarity reversal after each 11 yr cycle). This means that we can begin to compare activity cycles to determine if we can discern what causes the cycle-to-cycle variability.

It is common, when looking at solar cycle variation in p-mode parameters to average the observed variation over a range of frequencies. Figure 5 shows two full cycles of frequency shifts as seen with GONG, where the frequency shifts are averaged over four frequency ranges. This was also the approach taken by Basu *et al.* (2012), Salabert *et al.* (2015), and Howe *et al.* (2017), when looking at frequency shifts in low-ℓ modes, seen in Sun-as-a-star data. Each of these authors showed that there is a change in the relationship

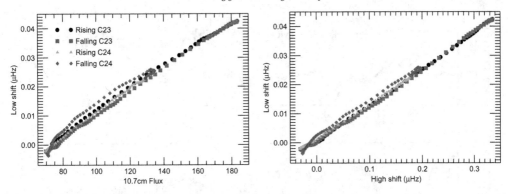

Figure 6. Left: frequency shift in the low-frequency band from Figure 5, which extends between $1860 \leq \nu \leq 2400\,\mu\text{Hz}$, as a function of 10.7 cm flux (measured in RFU). Right: Frequency shift in the low-frequency band as a function of the frequency shift in the high-frequency band, which extends between $2920 \leq \nu \leq 3480\,\mu\text{Hz}$. The difference colours indicate the different phases of different cycles, as indicated in the legend.

between low-frequency, low-ℓ modes and atmospheric activity proxies between cycles 23 and 24, which resulted in far smaller changes in frequency through the solar cycle then expected and a far lower correlation with atmospheric proxies than seen in previous cycles. However, as can be seen in the bottom panel of Figure 5, a definite solar cycle variation is observed for the low-frequency medium-ℓ modes. Nevertheless, as shown in Howe *et al.* (2018a), there is still evidence for a small but systematic and significant cycle-to-cycle variation in intermediate-ℓ frequency shifts, with solar cycle 24 showing a shift that was around 10% larger for the same change in atmospheric activity proxy as was observed in cycle 23. This cycle-to-cycle variation is also evident in Figure 6, which directly compares the intermediate-ℓ low-frequency-range frequency shifts, as shown in the bottom panel of Figure 5, with the 10.7 cm flux. The activity cycles are offset from one another, as are the rising and falling phases from each individual cycle, indicating the change in relationship between atmospheric and interior activity proxies. Interestingly, this same behaviour can be seen when comparing the low-frequency-range shifts with the frequency shifts observed in a higher frequency band (e.g. here the $2920 < \nu < 3480\,\mu\text{Hz}$ range was used). This is likely to be related to the upper turning point of the modes, which is further out for higher mode frequencies. Thus we can speculate that the cycle-to-cycle change is occurring in the layer containing the upper turning points of the low-frequency modes. For example, one explanation, proposed by Basu *et al.* (2012), is that the layer responsible for the magnetic perturbation has become thinner between cycles 23 and 24, meaning that in cycle 24 the low-frequency modes penetrate less far into the perturbation, if at all, and are thus less affected by the magnetic field.

2.2. *Variations in other p-mode parameters*

As already mentioned, it is not just the mode frequencies that vary with time: mode powers and damping rates also show solar cycle variations (see Kiefer *et al.* (2018), and references therein). The top row of Figure 7 shows the variation in time of mode widths, which are defined as the widths of the Lorentzian profiles like those plotted in Figure 1. These results were obtained using GONG data but the results are consistent with those obtained using Sun-as-a-star data. The mode widths are proportional to mode damping rates and are observed to vary in phase with the solar activity, meaning the damping rates are highest at solar maximum. The powers, defined as the integrated area under the Lorentzian curves plotted in Figure 1, vary in anti-phase with the solar cycle, meaning the

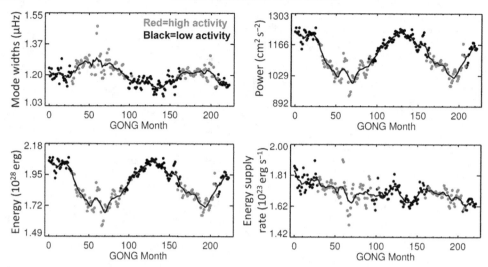

Figure 7. Top, left: variation in mode width, which is proportional to damping rate, as a function of time. Top, right: variation in mode power as a function of time. Bottom, left: variation in mode energy as a function of time. Bottom, right: variation in the rate at which energy is supplied to modes as a function of time. Adapted from Kiefer *et al.* (2018).

powers are at a minimum at solar maximum. The mode energies are directly proportional to the mode powers and so they, too, vary in anti-phase with the solar cycle (e.g. Goldreich *et al.* (1994)). The mode energies and damping rates can then be combined to determine the rate at which energy is supplied to the mode (e.g. Goldreich *et al.* (1994)), which is found to be approximately constant with time. We note that the results plotted in Figure 7 require numerous corrections to the data to account for e.g. the fill of the data, but the results are consistent with those found by Keith-Hardy *et al.* (2019), who perform an alternative correction.

3. Flows in the solar interior and their relationship with magnetic activity

As already mentioned, the separation in frequency of the different m components can be used to determine the internal rotation profile of the solar interior. Helioseismology revealed that the differential rotation pattern observed on the surface of the Sun, whereby the equator rotates faster than the poles, extends throughout the outer $\sim 30\%$ of the solar interior, down to the base of the convection zone (see e.g. (Basu 2016, Basu, 2016)). At the base of the convection zone there is a switch to solid body rotation, resulting in a narrow shear layer known as the tachocline. The tachocline is important in many dynamo models of the Sun, as it is often regarded as the location where the magnetic field is generated and maintained (e.g. Charbonneau (2010)).

The rotation profile of the solar interior does not remain constant through the solar cycle. One such temporal variation is known as the torsional oscillation, which is highlighted by subtracting the mean rotation profile from the rotation profile observed at a specific time. This can be done for different depths in the solar interior allowing a profile of the variation to be inferred. Figure 8 shows the torsional oscillation at $0.99\,R_\odot$. Coherent bands of faster than average and slower than average rotation are clearly visible. The bands of faster than average rotation coincide with the inner edge of the wings on a traditional butterfly diagram but can be traced back to well before a cycle starts and possibly covering the outer edge of the previous cycle. We note that this flow was substantially faster in cycle 24 than cycle 23. The equatorward branch of the new activity

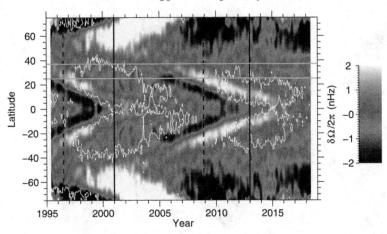

Figure 8. Torsional oscillations observed at $0.99\,R_\odot$, obtained using regularized least squares inversions of GONG, MDI and HMI data. The vertical dashed line indicates solar minimum, while the solid vertical lines indicate solar maximum. The white contours, which indicate the locations of the surface magnetic flux at 10% of the maximum level, allow comparison with the butterfly diagram. Originally published in Howe *et al.* (2018b).

cycle can just about be seen, but was still relatively weak at the time this figure was published. A poleward branch of faster than average rotation can also be seen from the beginning of cycle 23. Although hard to see in Figure 8, a similar band is present in cycle 24 but it is far weaker. Kosovichev & Pipin (2019) reveal zones of deceleration of the torsional oscillation that originate at high latitudes near the base of the convection zone and migrate towards the surface. They suggest that the deceleration is caused by magnetic field and thus the results are consistent with magnetic dynamo waves, as predicted by Parker's dynamo theory Parker (1955).

Another flow that is important in dynamo models is the meridonal flow. As the name suggests, a meridional flow is the circulation of plasma in the meridional plane or perpendicular to rotation. The meridional flow is an important component of flux-transport dynamo models as it is presumed to carry flux from old active regions towards the poles, where it is subducted to the base of the convection zone and transported back towards the equator (see e.g. Charbonneau (2010), for a review). Furthermore, the strength and timescale of the meridional flow is believed to determine the strength and timescale of the Sun's activity cycle, although many models disagree on the exact impact of these parameters. Detecting the near-surface flow is relatively straightforward (e.g. Giles *et al.* (1997)). However, as soon as we try and go deeper the uncertainties increase substantially (e.g. Mitra-Kraev & Thompson (2007)): The modes used to make these inversions simply are not sensitive enough to the deeper layers of the Sun because of the high sound speed there. Of course mass conservation indicates that there must be a return flow but results describing the return flows remain controversial.

Figure 9 shows three examples of recently published results on meridional flows (we note that these are not the only three published recently but have been selected because they have plotted their results in a similar manner, making comparison easier). Each uses a slightly different methodology and each produces a different result. Mandal *et al.* (2018) find a single cell meridional flow pattern, with a return flow at a depth of $0.78\,R_\odot$. Chen & Zhao (2017), on the other hand, find a double cell structure, with an equatorward flow found between about 0.82 and $0.91\,R_\odot$ for low latitude areas and between about 0.85 and $0.91\,R_\odot$ for higher latitude areas. This equatorward flow is sandwiched between two poleward flows. While Böning *et al.* (2017) demonstrate that observations are consistent

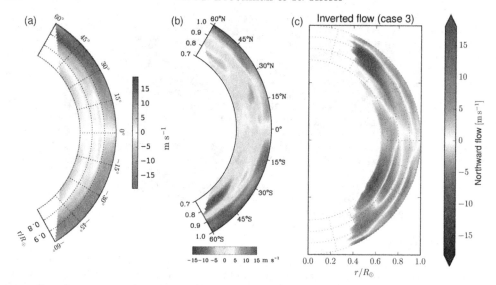

Figure 9. (a) Meridional flow inferred by Mandal *et al.* (2018). (b) Meridional flow inferred by Chen *et al.* (2017). (c) Meridional flow inferred by Böning *et al.* (2017).

with both single- and multi-cell meridional flow profiles, but find a much shallower return flow than that observed by Mandal *et al.* (2018). We note that both Chen *et al.* (2017) and Mandal *et al.* (2018) use HMI data, albeit covering slightly different epochs, while Böning *et al.* (2017) use GONG data. All use relatively long data sets, covering six years or more, in order to improve precision at the greater depths needed to study the equatorward meridional flow. However, shorter time series can be used to study solar cycle variations in the near-surface meridional flow.

For example, a recent study by Komm *et al.* (2018) demonstrated that there are evolving bands of faster than average and slower than average meridional flow (just as the torsional oscillation comprises of faster and slower than average rotation). One such band of faster than average meridional flow was observed to migrate towards the equator between the maxima of cycles 23 and 24, meaning that these faster than average meridional flows can be regarded as precursors to the appearance of magnetic field on the surface. In cycle 23, the fast flows were observed at 30 degrees approximately 2 yr before the surface activity from cycle 24 was observed. A new band of fast meridional flow reached 30 degrees in late 2016 – early 2017, implying that we may expect to observe the new cycle in late 2019 – early 2020. The fast flow is sandwiched between bands of slower than average rotation, each of which can be associated with a particular activity cycle, and the first hint of the slow meridional flow associated with cycle 25 appeared in 2016. Komm et al. also compare each fast flow with the slow flow preceding it and find that the difference is larger by around a factor of two for cycle 24 compared to the upcoming cycle 25. Liang *et al.* (2018) observed a significant reduction in travel time shifts in cycle 24 but only in the northern hemisphere, this implies a rapid decrease in poleward flows with increasing depth. This north-south asymmetry was not present in cycle 23 and so could provide a hint as to why the cycle 24 was so weak.

4. Summary

Numerous challenges remain in helioseismology and these challenges impact far beyond the Sun. For example, with the current growth of asteroseismology, it will become increasingly important to understand the inner workings of our closest star. Helioseismic results can provide important constraints not only for models of solar and stellar structure and evolution but also solar and stellar dynamo models. Helioseismology has already provided information vital for the construction of solar dynamo models including, but not limited to, the depth and extent of the tachocline, the rotation profile of the solar interior and the near-surface meridional circulation profile. However, perhaps one area where we can only now make substantial progress is in the comparison of different activity cycles. Can we determine why cycles are so different and what role do changes in the internal flow speeds play?

Dynamo models would greatly benefit from further constraints on the deep solar interior. The deep component of the meridional flow also remains uncertain, with seemingly small changes in methodology resulting in important changes in the inferred rotation profile. Resolving this issue is certainly one of the biggest current challenges facing helioseismologists and yet we are limited by the inherent insensitivity of p modes to the deep solar interior. For the same reason the impact on p modes of magnetic fields at the base of the convection zone is expected to be minuscule (although we should remember that novel analysis techniques have provided a hint in this regard, e.g. Baldner & Basu (2008)). Perhaps gravity modes will play an important role in constraining our understanding of the deep solar interior, and yet the quest to unequivocally detect gravity modes remains frustratingly ongoing.

References

Baldner, Charles S. & Basu, Sarbani 2008, *ApJ*, 686, 1349

Basu, Sarbani, Broomhall, Anne-Marie, Chaplin, William J. & Elsworth, Yvonne 2012, *ApJ*, 758, 43

Basu, Sarbani 2016, *LRSP*, 13, 2

Bazilevskaya, G., Broomhall, A.-M., Elsworth, Y. & Nakariakov, V. M. 2014, *Space Sci. Rev.*, 186, 359

Böning, Vincent G. A., Roth, Markus, Jackiewicz, Jason & Kholikov, Shukur 2017, *ApJ*, 845, 2

Broomhall, A. -M. 2017, *Sol. Phys.*, 292, 67

Chaplin, W. J., Elsworth, Y., Howe, R., Isaak, G. R., McLeod, C. P., Miller, B. A. & New, R. 1996, *MNRAS*, 280, 1162

Charbonneau, Paul 2010, *LRSP*, 7, 3

Chen, Ruizhu & Zhao, Junwei 2017, *ApJ*, 849, 144

Christensen-Dalsgaard, Jørgen & Berthomieu, Gabrielle 1991, In: *Solar interior and atmosphere*, Tucson, AZ, University of Arizona Press, 401

Davies, G. R., Chaplin, W. J., Elsworth, Y. P. & Hale, S. J. 2014, *MNRAS*, 441, 3009

Giles, P. M., Duvall, T. L., Scherrer, P. H., & Bogart, R. S. 1997, *Nature*, 390, 52

Goldreich, P., Murray, N. & Kumar, P. 1994, *ApJ*, 424, 466

Gough, D.O. 1990, In: Osaki, Y., Shibahashi, H. (eds.) *Progress of Seismology of the Sun and Stars, Lecture Notes in Physics* 367, Springer, Berlin, 283

Gizon, Laurent & Birch, Aaron C. 2005, *LRSP*, 2, 6

Hale, S. J., Howe, R., Chaplin, W. J., Davies, G. R. & Elsworth, Y. P. 2016, *Sol. Phys.*, 291, 1

Hathaway, David H. 2015, *LRSP*, 12, 4

Howe, R., Komm, R.W., & Hill, F. 2002, *ApJ*, 580, 1172

Howe, R., Davies, G. R., Chaplin, W. J., Elsworth, Y., Basu, S., Hale, S. J., Ball, W. H. & Komm, R. W. 2017, *MNRAS*, 470, 1935

Howe, R., Chaplin, W. J., Davies, G. R., Elsworth, Y., Basu S., & Broomhall, A. -M. 2018, *MNRAS*, 480, L79

Howe, R., Hill, F., Komm, R., Chaplin, W. J., Elsworth, Y., Davies, G. R., Schou, J. & Thompson, M. J. 2018, *ApJL*, 865, L5

Keith-Hardy, J. Z., Tripathy, S. C., Jain, K. 2019, *ApJ*, 877, 148

Kiefer, René, Schad, Ariane & Roth, Markus 2017, *ApJ*, 846, 162

Kiefer, René & Roth, Markus 2018, *ApJ*, 854, 74

Kiefer, René, Komm, Rudi, Hill, Frank, Broomhall, Anne-Marie & Roth, Markus 2018, *Sol. Phys*, 293, 151

Komm, R., Howe, R. & Hill, F. 2018, *Sol. Phys*, 293, 145

Kosovichev, Alexander G. & Pipin, Valery V. 2019, *ApJL*, 871, L20

Libbrecht, K.G. & Woodard, M.F. 1990, *Nature*, 345, 779

Liang, Zhi-Chao, Gizon, Laurent, Birch, Aaron C., Duvall, Thomas L. & Rajaguru, S. P. 2018, *A&A*, 619, 99

Lund, Mikkel Nørup, Kjeldsen, Hans, Christensen-Dalsgaard, Jørgen, Handberg, Rasmus & Silva Aguirre, Victor 2014, *ApJ*, 782, 2

Mandal, K., Hanasoge, S. M., Rajaguru, S. P. & Antia, H. M. 2018, *ApJ*, 863, 39

Mitra-Kraev, U. & Thompson, M. J. 2007, *AN*, 328, 1009

Parker E. N. 1955, *ApJ*, 122, 293

Salabert, D., García, R. A. & Turck-Chièze, S. 2015, *A&A*, 578, 137

Discussion

KOSOVICHEV: The p-mode frequency shift varies similarly to the radio flux but there is a significant difference in cycle 24. What can be learned from this difference?

BROOMHALL: There does appear to be a change in relationship between the radio flux and the frequency shifts from one cycle to the next. If we plot one against the other, as is done in Figure 6, we can readily see this change. However, we only really have two cycles to compare (three if we use BiSON data) and so, at the moment, it is difficult to determine exactly what causes this change. This behaviour is seen if we compare the frequency shifts to other proxies as well, such as sunspot area, and, similarly, if you compare two non-helioseismic activity proxies. One potential explanation comes in terms of the extent of the near-surface layer responsible for perturbing the mode frequencies. This layer becoming thinner between cycles 23 and 24 would be consistent with the observations. However, I believe more will be revealed as the next cycle progresses.

Solar and Stellar Magnetic Fields: Origins and Manifestations
Proceedings IAU Symposium No. 354, 2019
A. Kosovichev, K. Strassmeier & M. Jardine, ed.
doi:10.1017/S1743921320001416

Resolving Power of Asteroseismic Inversion of the Kepler Legacy Sample

Alexander G. Kosovichev[1]ⓘ and Irina N. Kitiashvili[2]ⓘ

[1]New Jersey Institute of Technology, Newark, NJ 07102, U.S.A.
email: alexander.g.kosovichev@njit.edu

[2]NASA Ames Research Center
Moffett Field, CA 94035, U.S.A.
email: irina.n.kitiashvili@nasa.gov

Abstract. The Kepler Asteroseismic Legacy Project provided frequencies, separation ratios, error estimates, and covariance matrices for 66 Kepler main sequence targets. Most of the previous analysis of these data was focused on fitting standard stellar models. We present results of direct asteroseismic inversions using the method of optimally localized averages (OLA), which effectively eliminates the surface effects and attempts to resolve the stellar core structure. The inversions are presented for various structure properties, including the density stratification and sound speed. The results show that the mixed modes observed in post-main sequence F-type stars allow us to resolve the stellar core structure and reveal significant deviations from the evolutionary models obtained by the grid-fitting procedure to match the observed oscillation frequencies.

Keywords. stars: interiors, stars: late-type, stars: oscillations, methods: data analysis, stars: individual (KIC 10162436, KIC 5773345)

---

1. Introduction

The Kepler mission (Borucki *et al.* 2010) provided a wealth of stellar oscillation data enabling asteroseismic investigation of the internal structure and rotation of many stars across the HR diagram. A primary tool of asteroseismology employed for interpretation of observed oscillation frequencies used a method of grids of stellar models. In this method, the asteroseismic calibration of stellar models is performed by matching the observed oscillation frequencies to theoretical mode frequencies calculated for a grid of standard evolutionary models. In combination with spectroscopic constraints, this approach provided estimates of stellar radius, composition, and age with unprecedented precision (see Chaplin *et al.* 2010; Metcalfe *et al.* 2010, and references therein). In addition, the high accuracy measurements of oscillation frequencies opened a new opportunity for asteroseismic inversions which allows us to reconstruct the internal structure and test evolutionary stellar models. A similar approach has been used in helioseismology for more than two decades.

A specific feature of asteroseismology data is that only low-degree oscillations can be observed, typically for modes of spherical harmonic degrees, $\ell = 0, 1, 2$, and sometimes, $\ell = 3$. In this situation, only the structure of stellar cores can be resolved by inversion techniques. An additional difficulty is caused by uncertainties in the mass and radius of distant stars. Gough and Kosovichev (1993a) showed that this difficulty can be overcome by an additional condition in the inversion procedure, which constraints the frequency scaling factor, $q = M/R^3$, where M and R are the stellar mass and radius. They showed

Table 1. Characteristics of the stellar models that are used for inversion of the observed oscillation frequencies.

KIC	M/M$_\odot$	lg(R/R$_\odot$)	Age(Gyr)	lg(L/L$_\odot$)	Ysurf	Zsurf	Teff	alpha
7206837	1.298	0.1980	2.90	0.5523	0.2800	0.0220	6320	1.791
1435467	1.382	0.2436	2.56	0.6694	0.2637	0.0197	6414	1.692
10162436	1.461	0.3082	2.51	0.8110	0.2460	0.0173	6494	1.684
9353712	1.516	0.3261	2.03	0.9008	0.2603	0.0180	6665	1.713
5773345	1.579	0.3074	2.07	0.7931	0.2593	0.0306	6400	1.715
12069127	1.588	0.3403	1.76	0.9385	0.2669	0.0216	6700	1.672

that if the low-ℓ mode frequencies are measured with precision of 0.1 μHz then the structure of the stellar core can reconstructed even when the stellar mass and radius are not known. When the measurement precision is relatively low (e.g. 1 μHz), the localization of the inversion accuracy is degraded. For such cases, Gough and Kosovichev (1993b) suggested a procedure of calibration of the averaging kernels to estimate averaged properties of the stellar core. The effectiveness of this procedure was demonstrated by applying it to the low-degree solar oscillation frequencies observed by the IPHIR instrument that measured the total solar irradiance onboard the PHOBOS spacecraft(Toutain and Froehlich 1992). For analysis of the Kepler asteroseismology data, the inversion technique has been recently used by Bellinger *et al.* (2019), and the kernel calibration method was applied by Buldgen *et al.* (2017) to estimate integrated properties of stellar structure.

In this paper, we show that the detection of mixed modes in the oscillation spectra of F-type stars allows us to resolve the structure of the inner stellar cores. The mixed modes have properties of internal gravity waves (g-modes) in the convectively stable helium core and properties of acoustic modes outside the core. The oscillation frequencies of these modes may be quite sensitive to the properties of the core. Including them in the inversion procedure allows us to localize the averaging kernels in the core region. This opens a unique opportunity for testing the stellar evolution theory for subgiant stars where the energy release is in a hydrogen shell surrounding a helium core.

2. Target selection. Evolutionary models.

From the Kepler Asteroseismic Legacy Sample (Silva Aguirre *et al.* 2017) we selected 6 stars in the mass range from about 1.3 to 1.6 solar masses, and, using the MESA stellar evolutionary code (Paxton *et al.* 2011), we calculated models of the internal structure, matching as close as possible the stellar parameters determined by grid fitting methods. Specifically, we chose the stellar mass, chemical composition, and the mixing parameter from the grid pipeline YMCM (Silva Aguirre *et al.* 2015) and evolved starting from a premain sequence phase to the age estimated from the grid fitted models. Basic parameters of the calculated models are shown in Table 1.

In this paper, we present results for two models: KIC 10162436 and KIC 5773345, the radial profiles of the density, sound speed, and Brunt-Väisälä frequency are shown in Figure 1. The Brunt-Väisälä frequency displays a sharp peak at the helium core outer boundary, located at about 0.05 $R_\odot$ in KIC 10162436, and at 0.07 $R_\odot$ in KIC 577334.

Figure 2 shows the difference between the observed and modeled frequencies plotted as a function of the radius of the acoustic inner turning points, r_t, calculated from the asymptotic relation: $r_t/c_S(r_t) = (l + 1/2)/\omega_{nl}$, where $c_S(r)$ is the sound-speed profile, ℓ is the angular degree, and ω_{nl} is the mode frequency. The acoustic turning points form three branches corresponding to the modes of angular degree $l = 0, 1$, and 2. Behavior of the frequency deviations for both models is similar, indicating a systematic difference of the evolutionary models from the real stellar structure.

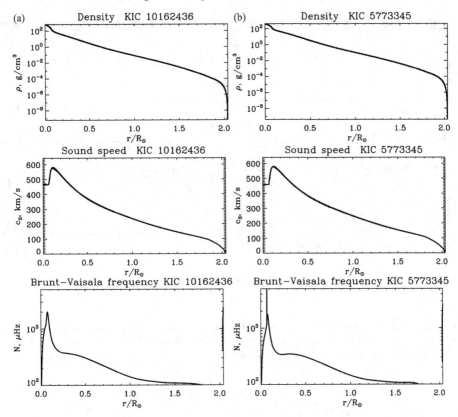

Figure 1. Distributions of the density, the sound speed and the Brunt-Väisälä frequency in the stellar models of: a) KIC 10162436, and b) KIC 5773345.

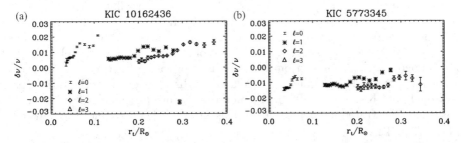

Figure 2. The differences between the observed and modeled frequencies as a function of the acoustic inner turning points for: a) KIC 10162436 and b)KIC 5773345.

3. Sensitivity kernels

The acoustic turning points of some $\ell = 0$ modes are located in the stellar core. Nevertheless, the sensitivity of these modes to the core structure is low. Some of the observed low-frequency non-radial modes of $\ell = 1$ and 2 represent mixed acoustic-gravity modes which have properties of internal gravity (g) modes in the stellar cores and acoustic (p) modes outside the core. The oscillation frequencies of these modes are very sensitive to the core properties.

Using explicit formulations for the variational principle, frequency perturbations can be reduced to a system of integral equations for a chosen pair of independent variables; e.g. for (ρ, γ):

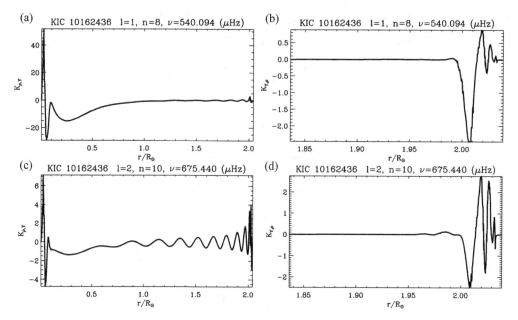

Figure 3. Sensitivity kernels of oscillation frequencies for two mixed modes of $\ell = 1$ (panels $a - b$) and $\ell = 2$ $(c - d)$ to relative variations in stellar density $(a - c)$, and variations in the abundance of helium $(b - d)$, for KIC 10162436).

$$\frac{\delta\omega^{(n,l)}}{\omega^{(n,l)}} = \int_0^R K_{\rho,\gamma}^{(n,l)} \frac{\delta\rho}{\rho} dr + \int_0^R K_{\gamma,\rho}^{(n,l)} \frac{\delta\gamma}{\gamma} dr,$$

where $K_{\rho,\gamma}^{(n,l)}(r)$ and $K_{\gamma,\rho}^{(n,l)}(r)$ are sensitivity (or 'seismic') kernels. These are calculated using the initial solar model parameters, ρ_0, P_0, γ, and the oscillation eigenfunctions for these model, $\vec{\xi}$ (for an explicit formulation, see e.g. Kosovichev 1999). The sensitivity for various pairs of solar parameters, such the sound speed, Brunt-Väisälä frequency, temperature, and chemical abundances, can be obtained by using the relations among these parameters, which follow from the equations of solar structure ('stellar evolution theory'). These 'secondary' kernels are then used for direct inversion of the various parameters (Gough and Kosovichev 1988). A general procedure for calculating the sensitivity kernels can be illustrated in operator form (Kosovichev 2011). Consider two pairs of solar variables, $\vec{X}$ and $\vec{Y}$, e.g.

$$\vec{X} = \left(\frac{\delta\rho}{\rho}, \frac{\delta\gamma}{\gamma} \right) ; \quad \vec{X} = \left(\frac{\delta\rho}{\rho}, \frac{\delta Y}{Y} \right),$$

where Y is the helium abundance. The linearized structure equations (the hydrostatic equilibrium equation and the equation of state) that relate these variables can be written symbolically as

$$A\vec{X} = \vec{Y}.$$

Let $\vec{K}_X$ and $\vec{K}_Y$ be the sensitivity kernels for X and Y; then the frequency perturbation is:

$$\frac{\delta\omega}{\omega} = \int_0^R \vec{K}_X \cdot \vec{X} dr \equiv \left\langle \vec{K}_X \cdot \vec{X} \right\rangle,$$

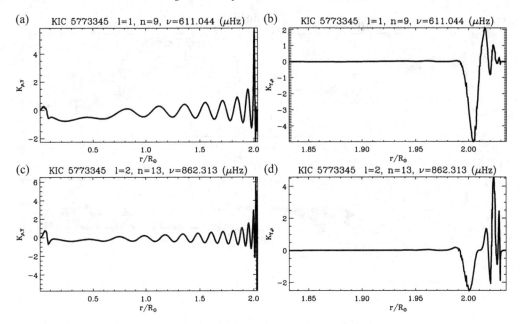

Figure 4. The same as Fig. 3 for KIC 5773345.

where $< \cdot >$ denotes the inner product. Similarly,

$$\frac{\delta\omega}{\omega} = \left\langle \vec{K}_Y \cdot \vec{Y} \right\rangle.$$

Then from the stellar structure equation $A\vec{X} = \vec{Y}$:

$$\left\langle \vec{K}_Y \cdot \vec{Y} \right\rangle = \left\langle \vec{K}_Y \cdot A\vec{X} \right\rangle = \left\langle A^*\vec{K}_Y \cdot \vec{X} \right\rangle,$$

where A^* is an adjoint operator. Thus: $\left\langle A^*\vec{K}_Y \cdot \vec{X} \right\rangle = \left\langle \vec{K}_X \cdot \vec{X} \right\rangle$. This is valid for any $\vec{X}$ if $A^*\vec{K}_Y = \vec{K}_X$. This means that the equation for the sensitivity kernels is adjoint to the stellar structure equations. An explicit formulation in terms of the stellar structure parameters and mode eigenfunctions was given by Kosovichev (1999).

Examples of the sensitivity kernels to density and helium-abundance deviations for the mixed acoustic-gravity modes of $\ell = 1$ and $\ell = 2$ for KIC 10162436, and KIC 5773345 are shown in Figures 3 and 4, respectively. In the case of KIC 10162436, the sensitivity of the mode frequencies to the density stratification is high in the central core. In the case of KIC 5773345, the sensitivity of the observed mixed modes in the core is not that high but still quite significant. In such cases the mixed oscillation modes open a unique opportunity for probing the stellar cores by direct structure inversion. The sensitivity kernels for helium abundance are concentrated in the helium and hydrogen ionization zones.

4. Inversion procedure

In the inversion procedure it is important to take into account potential systematic uncertainties in the stellar mass and radius. Because the oscillation frequencies are scaled linearly with the factor $q = M/R^3$, then, following (Gough and Kosovichev 1993a), the mode frequencies ω_i can be expressed in terms of their relative small difference $\delta\omega_i^2/\omega_i^2$ from those of a standard reference model of similar mass and radius according to the linearized expression

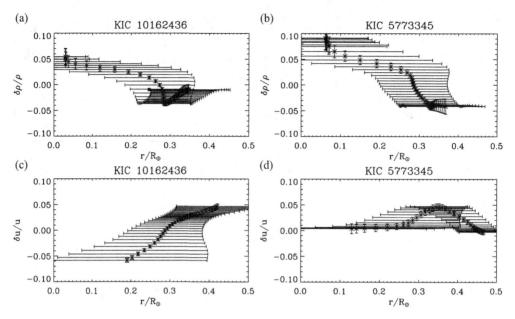

Figure 5. The relative deviations of the stellar structure from the evolutionary models of density $(a-b)$ and parameter $u = P/\rho$ $(c-d)$ as a function of radius for KIC 10162436 and KIC 5773345, obtained by inversion of the observed mode frequencies. Crosses show the center locations the localized averaging kernels, the horizontal bars show the spread of the averaging kernels, and the vertical bars show the uncertainties calculated using error estimates of the observed frequencies.

$$\delta\omega_i^2/\omega_i^2 = \int_0^1 \left(K_{f,Y}^i \frac{\delta f}{f} + K_{Y,f}^i \delta Y \right) dx - I_q^i \delta q,$$

where $x = r/R$, $q = M/R^3$, Y is the helium abundance, and f can be any function of p and ρ (Däppen $et\ al.$ 1991); M and R are stellar mass and radius, in solar units, $K_{f,Y}$ and $K_{Y,f}$ are appropriate kernels, and I_q is an integral over the reference model. In this paper we consider two cases: $f = \rho$, and $f = u \equiv P/\rho$ (Dziembowski $et\ al.$ 1991).

These constraints can provide localized averages of $\delta \ln f$ and estimates of δY and δq of the kind

$$\overline{\delta \ln f} \equiv \int_0^1 \sum_i a_i(x_0) K_{f,Y}^i \delta \ln f dx \equiv \int_0^1 A_{f,Y}(x, x_0) \delta \ln f dx = \sum_i a_i(x_0) \frac{\delta\omega_i^2}{\omega_i^2}$$

by minimizing over the coefficients $a_i(x_0)$ the functional (Backus and Gilbert 1968):

$$\int_0^1 A_{f,Y}^2(x, x_0) J_f dx + \lambda_1 \int_0^1 \left(\sum_i a_i K_{Y,f}^i \right)^2 J_Y dx + \lambda_2 \left(\sum_i a_i I_q^i \right)^2 + \alpha \sum_i a_i^2 \epsilon_i^2$$

for tradeoff parameters λ_1, λ_2 and α, where $J_f = (x - x_0)^2$, and ϵ_i are standard relative errors in the data. The tradeoff parameters are chosen empirically, by selecting a sufficiently smooth solution and using the L-curve criterion (Hansen 1992).

The non-adiabatic frequency shift (aka 'surface effect') can be approximated by a smooth function of frequency, $F(\omega)$ scaled with the factor, $Q \equiv I(\omega)/I_0(\omega)$, where $I(\omega)$ is the mode inertia, and $I_0(\omega)$ is the mode inertia of radial modes ($l = 0$), calculated at frequency ω; that is:

$$\frac{\delta\omega^2_{nonad,i}}{\omega_i^2} = F(\omega_i)/Q(\omega_i).$$

Function $F(\omega)$ can be approximated by a polynomial function of degree K (Dziembowski *et al.* 1990):

$$F(\omega_i) = \sum_{k=0}^{K} c_k P_k(\omega_i),$$

where P_k are the Legendre polynomials of degree k. Then the influence of nonadiabatic effects is reduced by applying $K+1$ additional constraints for a_i:

$$\sum_{i=1}^{N} a_i P_k(\omega_i)Q(\omega_i) = 0, \quad k = 0, ..., K.$$

5. Inversion results

The inversion results obtained by the Optimally Localized Averaging (OLA) procedure described in the previous section are shown in Figure 5. They show the relative deviations in stellar structure from the evolutionary models of density and parameter $u = P/\rho$ as a function of radius for KIC 10162436 and KIC 5773345. Crosses show the center locations the localized averaging kernels, the horizontal bars show the spread of the averaging kernels, and the vertical bars show the uncertainties calculated using the observational error estimates (for the definitions of these properties see, Kosovichev 1999). It is remarkable that the mixed modes observed in the oscillation spectra of these stars allow us to obtain constraints on the density stratification in the central helium core and the hydrogen-burning shell. The averaging kernels for the parameter u (or, equivalently, the sound speed) are localized only outside the core. Therefore, it is very important to choose the pairs of inversion variables that provide the best resolution. These variables can be different for different regions of a star. In our cases only the density inversions provide the optimally localized averaging kernels centered in the stellar cores. The sound-speed inversions are incapable of resolving the core.

For both stars, the inversion results show that the density of the core and the surrounding shell are about 5% higher than in the stellar models, but lower outside the energy-release shell. The boundary of the helium core is located at 0.05 $R_\odot$ in KIC 10162436, and at 0.07 $R_\odot$ in KIC 5773345. Outside the helium cores, the nuclear energy production shells extend 0.3 $R_\odot$, with the peak rate at ~ 0.08 $R_\odot$ in both models. Perhaps these stellar regions involve physical processes that are not described by the evolutionary models. For understanding these deviations it will be beneficial to perform more detailed structure inversion studies for a large sample of post-main sequence stars with hydrogen-burning shells.

6. Conclusion

High accuracy measurements of oscillation frequencies for a large number stars as well as identification of the observed oscillations in terms of normal modes for stellar models by the grid-fitting techniques open possibilities for performing asteroseismic inversions and for testing evolutionary models. The discovery of the oscillation modes with mixed g- and p-mode characteristics in post-main sequence stars allows us to reconstruct the density stratification in the stellar helium cores and hydrogen burning shell.

We applied the previously developed asteroseismic inversion method (Gough and Kosovichev 1993a) to two F-type stars from the Kepler Asteroseismic Legacy Project

(Silva Aguirre *et al.* 2017) and performed inversions for the density and squared isothermal sound-speed parameter. The background models were calculated using the MESA code, and the mass, composition, and age were previously determined by the model-grid fitting method. The inversion technique takes potential discrepancies in the estimated mass and radius from the actual properties, as well as the potential frequency shifts due non-adiabatic near-surface effects, and constructs optimally localized averaging kernels following the (Backus and Gilbert 1968) method.

The inversion results for both stars showed that the density in the helium core and the inner part of the hydrogen shell may be about 5% higher than in the evolutionary models, and, in the outer part of the shell, lower. This suggests that the differences may be due to physical processes not described by the evolutionary models. However, more detailed inversion studies for a large sample of stars are needed for quantifying the deviations more precisely.

Acknowledgments

The work was partially supported by the NASA Astrophysics Theory Program and grants: NNX14AB7CG and NNX17AE76A.

References

Backus, G. and Gilbert, F.: 1968, *Geophysical Journal* 16(2), 169

Bellinger, E. P., Basu, S., Hekker, S., and Christensen-Dalsgaard, J.: 2019, *Astrophys. J.* 885(2), 143

Borucki, W. J., Koch, D., Basri, G., Batalha, N., Brown, T., Caldwell, D., Caldwell, J., Christensen-Dalsgaard, J., Cochran, W. D., DeVore, E., Dunham, E. W., Dupree, A. K., Gautier, T. N., Geary, J. C., Gilliland, R., Gould, A., Howell, S. B., Jenkins, J. M., Kondo, Y., Latham, D. W., Marcy, G. W., Meibom, S., Kjeldsen, H., Lissauer, J. J., Monet, D. G., Morrison, D., Sasselov, D., Tarter, J., Boss, A., Brownlee, D., Owen, T., Buzasi, D., Charbonneau, D., Doyle, L., Fortney, J., Ford, E. B., Holman, M. J., Seager, S., Steffen, J. H., Welsh, W. F., Rowe, J., Anderson, H., Buchhave, L., Ciardi, D., Walkowicz, L., Sherry, W., Horch, E., Isaacson, H., Everett, M. E., Fischer, D., Torres, G., Johnson, J. A., Endl, M., MacQueen, P., Bryson, S. T., Dotson, J., Haas, M., Kolodziejczak, J., Van Cleve, J., Chandrasekaran, H., Twicken, J. D., Quintana, E. V., Clarke, B. D., Allen, C., Li, J., Wu, H., Tenenbaum, P., Verner, E., Bruhweiler, F., Barnes, J., and Prsa, A.: 2010, *Science* 327(5968), 977

Buldgen, G., Reese, D., and Dupret, M.-A.: 2017, in *European Physical Journal Web of Conferences*, Vol. 160 of *European Physical Journal Web of Conferences*, p. 03005

Chaplin, W. J., Appourchaux, T., Elsworth, Y., García, R. A., Houdek, G., Karoff, C., Metcalfe, T. S., Molenda-Żakowicz, J., Monteiro, M. J. P. F. G., Thompson, M. J., Brown, T. M., Christensen-Dalsgaard, J., Gilliland, R. L., Kjeldsen, H., Borucki, W. J., Koch, D., Jenkins, J. M., Ballot, J., Basu, S., Bazot, M., Bedding, T. R., Benomar, O., Bonanno, A., Brandão, I. M., Bruntt, H., Campante, T. L., Creevey, O. L., Di Mauro, M. P., Doğan, G., Dreizler, S., Eggenberger, P., Esch, L., Fletcher, S. T., Frandsen, S., Gai, N., Gaulme, P., Handberg, R., Hekker, S., Howe, R., Huber, D., Korzennik, S. G., Lebrun, J. C., Leccia, S., Martic, M., Mathur, S., Mosser, B., New, R., Quirion, P. O., Régulo, C., Roxburgh, I. W., Salabert, D., Schou, J., Sousa, S. G., Stello, D., Verner, G. A., Arentoft, T., Barban, C., Belkacem, K., Benatti, S., Biazzo, K., Boumier, P., Bradley, P. A., Broomhall, A. M., Buzasi, D. L., Claudi, R. U., Cunha, M. S., D'Antona, F., Deheuvels, S., Derekas, A., García Hernández, A., Giampapa, M. S., Goupil, M. J., Gruberbauer, M., Guzik, J. A., Hale, S. J., Ireland, M. J., Kiss, L. L., Kitiashvili, I. N., Kolenberg, K., Korhonen, H., Kosovichev, A. G., Kupka, F., Lebreton, Y., Leroy, B., Ludwig, H. G., Mathis, S., Michel, E., Miglio, A., Montalbán, J., Moya, A., Noels, A., Noyes, R. W., Pallé, P. L., Piau, L., Preston, H. L., Roca Cortés, T., Roth, M., Sato, K. H., Schmitt, J., Serenelli, A. M., Silva Aguirre, V., Stevens, I. R., Suárez, J. C., Suran, M. D., Trampedach, R., Turck-Chièze, S., Uytterhoeven, K., Ventura, R., and Wilson, P. A.: 2010, *Astrophys. J. Lett.* 713(2), L169

Däppen, W., Gough, D. O., Kosovichev, A. G., and Thompson, M. J.: 1991, *A New Inversion for the Hydrostatic Stratification of the Sun*, Vol. 388, p. 111

Dziembowski, W. A., Pamiatnykh, A. A., and Sienkiewicz, R. .: 1991, *Mon. Not. Roy. Astron. Soc.* 249, 602

Dziembowski, W. A., Pamyatnykh, A. A., and Sienkiewicz, R.: 1990, *Mon. Not. Roy. Astron. Soc.* 244, 542

Gough, D. O. and Kosovichev, A. G.: 1988, in E. J. Rolfe (ed.), *Seismology of the Sun and Sun-Like Stars*, Vol. 286 of *ESA Special Publication*, pp 195–201

Gough, D. O. and Kosovichev, A. G.: 1993a, in W. W. Weiss and A. Baglin (eds.), *IAU Colloq. 137: Inside the Stars*, Vol. 40 of *Astronomical Society of the Pacific Conference Series*, p. 541

Gough, D. O. and Kosovichev, A. G.: 1993b, *Seismic Analysis of Stellar P-Mode Spectra*, Vol. 42 of *Astronomical Society of the Pacific Conference Series*, p. 351

Hansen, P. C.: 1992, *SIAM Review* 34(4), 561

Kosovichev, A. G.: 1999, *Journal of Computational and Applied Mathematics* 109(1), 1

Kosovichev, A. G.: 2011, *Advances in Global and Local Helioseismology: An Introductory Review*, In: *The Pulsations of the Sun and the Stars*, pp 3–84, Springer Berlin Heidelberg, Berlin, Heidelberg

Metcalfe, T. S., Monteiro, M. J. P. F. G., Thompson, M. J., Molenda-Żakowicz, J., Appourchaux, T., Chaplin, W. J., Doğan, G., Eggenberger, P., Bedding, T. R., Bruntt, H., Creevey, O. L., Quirion, P. O., Stello, D., Bonanno, A., Silva Aguirre, V., Basu, S., Esch, L., Gai, N., Di Mauro, M. P., Kosovichev, A. G., Kitiashvili, I. N., Suárez, J. C., Moya, A., Piau, L., García, R. A., Marques, J. P., Frasca, A., Biazzo, K., Sousa, S. G., Dreizler, S., Bazot, M., Karoff, C., Frandsen, S., Wilson, P. A., Brown, T. M., Christensen-Dalsgaard, J., Gilliland, R. L., Kjeldsen, H., Campante, T. L., Fletcher, S. T., Hand berg, R., Régulo, C., Salabert, D., Schou, J., Verner, G. A., Ballot, J., Broomhall, A. M., Elsworth, Y., Hekker, S., Huber, D., Mathur, S., New, R., Roxburgh, I. W., Sato, K. H., White, T. R., Borucki, W. J., Koch, D. G., and Jenkins, J. M.: 2010, *Astrophys. J.* 723(2), 1583

Paxton, B., Bildsten, L., Dotter, A., Herwig, F., Lesaffre, P., and Timmes, F.: 2011, *ApJS* 192(1), 3

Silva Aguirre, V., Davies, G. R., Basu, S., Christensen-Dalsgaard, J., Creevey, O., Metcalfe, T. S., Bedding, T. R., Casagrande, L., Handberg, R., Lund, M. N., Nissen, P. E., Chaplin, W. J., Huber, D., Serenelli, A. M., Stello, D., Van Eylen, V., Campante, T. L., Elsworth, Y., Gilliland, R. L., Hekker, S., Karoff, C., Kawaler, S. D., Kjeldsen, H., and Lundkvist, M. S.: 2015, *Mon. Not. Roy. Astron. Soc.* 452(2), 2127

Silva Aguirre, V., Lund, M. N., Antia, H. M., Ball, W. H., Basu, S., Christensen-Dalsgaard, J., Lebreton, Y., Reese, D. R., Verma, K., Casagrande, L., Justesen, A. B., Mosumgaard, J. R., Chaplin, W. J., Bedding, T. R., Davies, G. R., Handberg, R., Houdek, G., Huber, D., Kjeldsen, H., Latham, D. W., White, T. R., Coelho, H. R., Miglio, A., and Rendle, B.: 2017, *Astrophys. J.* 835(2), 173

Toutain, T. and Froehlich, C.: 1992, *A&A* 257(1), 287

Solar and Stellar Magnetic Fields: Origins and Manifestations
Proceedings IAU Symposium No. 354, 2019
A. Kosovichev, K. Strassmeier & M. Jardine, ed.
doi:10.1017/S1743921319010007

Cycle times of early M dwarf stars: mean field models versus observations

Manfred Küker[1], Günther Rüdiger[1], Katalin Oláh[2] and Klaus G. Strassmeier[1]

[1]Leibniz-Institut für Astrophysik Potsdam,
An der Sternwarte 16, 14482 Potsdam, Germany
emails: mkueker@aip.de, gruediger@aip.de, kstrassmeier@aip.de

[2]Konkoly Observatory,
Budapest Hungary
email: olahkatalin5@gmail.com

Abstract. Observations of early-type M stars suggest that there are two characteristic cycle times, one of order one year for fast rotators ($P_{\rm rot} < 1$ day) and another of order four years for slower rotators. For a sample of fast-rotating stars, the equator-to-pole differences of the rotation rates up to 0.03 rad d^{-1} are also known from Kepler data. These findings are well-reproduced by mean field models. These models predict amplitudes of the meridional flow, from which the travel time from pole to equator at the base of the convection zone of early-type M stars can be calculated. As these travel times always exceed the observed cycle times, our findings do not support the flux transport dynamo.

Keywords. Stars: late-type – stars: magnetic field – stars: activity – magnetohydrodynamics (MHD) – turbulence

---

1. Introduction

The solar cycle is widely believed to be the result of a flux transport dynamo in which the toroidal magnetic field is advected towards the equator at the bottom of the convection zone by the large-scale meridional flow. The cycle time is then determined by the flow speed (Choudhuri *et al.* 1995; Dikpati & Charbonneau 1999; Küker *et al.* 2001; Bonanno *et al.* 2002). In a traditional $\alpha\Omega$ shell dynamo, on the other hand, the cycle time in the linear regime is

$$\tau_{\rm cyc} \simeq c_{\rm cyc} \frac{R_* D}{\eta_T}, \tag{1.1}$$

where R_* is the stellar radius, D the thickness of the convective layer, and η_T the turbulent magnetic diffusivity, and $c_{\rm cyc}$ a scaling factor (Roberts 1972). For the Sun, $\eta_T \simeq 10^{12}$ cm^2 s^{-1} is required to reproduce the cycle time of eleven years. This value agrees with what is found from the decay of large active regions and from cross helicity $\langle \vec{u} \cdot \vec{b} \rangle$. Interestingly, Equation 1.1 does not contain the rotation rate.

As the radii of M dwarfs are smaller than the solar radis, we expect shorter cycles unless the magnetic diffusivity coefficient is substantially smaller than 10^{12}cm^{2}s^{-1}. Indeed, cycle times derived from light curves lie in the range from 300 d to 2700 d, cf. Fig. 1 . However, there seems to be a dependence on the rotation period, as very rapidly rotating stars show shorter cycle times.

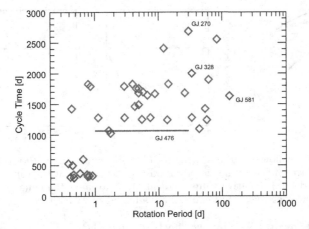

Figure 1. Cycle times vs. rotation period (both in days) for early M stars (diamonds). The red diamonds indicate stars observed by the KEPLER spacecraft, the blue ones observations with the STELLA telescope. The blue bar indicates GJ 476, for which the rotation period has not been determined. Data from Vida *et al.* (2013, 2014); Distefano *et al.* (2016); Suárez Mascareño *et al.* (2016); Wargelin *et al.* (2017); Küker *et al.* (2019)

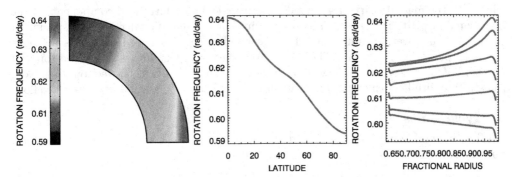

Figure 2. Differential rotation of an M star with $M = 0.6 M_\odot$ rotating with a period of 10 d. Left: Color contour plot of the angular velocity. Center: Surface rotation as a function of latitude. Right: Angular velocity vs. fractional stellar radius for different latitudes. From top to bottom: 0°, 15°, 30°, 45°, 60°, 75°, and 90°.

2. Differential rotation and meridional flow

Meridional flows are hard to observe even on the Sun. For stars we have to rely on theoretical models. We have therefore applied the mean field model of Küker *et al.* (2011), which reproduces the solar differential rotation very well, to a sample of three stellar models. The masses were chosen to be 0.4, 0.6, and 0.66 solar masses in order to cover the range of spectral types in the sample of M dwarfs with observed cycle times. The stellar models were computed with the Mesa stellar evolution code of Paxton *et al.* (2011) assuming solar metallicity. Figure 2 shows the resulting differential rotation patterns for 0.6 solar masses and a rotation period of ten days. The rotation is solar-type, i.e. more rapid rotation at the equator than at the poles. However, the isocontours are more cylinder-shaped than what both our model and helioseismology find for the Sun. The latitudinal shear is larger at the surface than at the bottom of the convection zone. The radial shear is positive at the equator and negative at the poles.

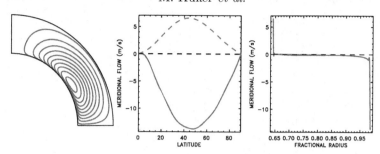

Figure 3. Meridional flow of an M star with $M = 0.6 M_\odot$ rotating with a period of 10 d. Left: isocontours of the stream function. Center: meridional flow speed at the top (blue solid line) and bottom (red dashed line). Positive values indicate flow towards the equator, negative towards the pole. Right: flow speed vs. fractional stellar radius at mid-latitudes.

The meridional flow is driven by two effects, namely the gradient of the angular velocity parallel to the rotation axis, $\partial \Omega^2 / \partial z$, and the latitudinal gradient of the entropy, $\partial s / \partial \theta$. Fig. 3 shows the flow pattern for the same case as Fig. 2. The stream function ψ is defined through

$$\rho u_r = \frac{1}{r^2 \sin \theta} \frac{\partial \psi}{\partial \theta}, \qquad \rho u_\theta = \frac{1}{r \sin \theta} \frac{\partial \psi}{\partial r},$$

where ρ is the mass density and u_r and u_θ the vertical and horizontal components of the meridional flow. Its isocontours are stream lines of the meridional flow. The spacing of the isolines, however, does not directly correspond to the flow speed, as is illustrated by the right panels in the figure. There are a thin layers of fast flow at both the top and bottom boundaries while the flow is much slower in the bulk of the convection zone. Note that the amplitude of the return flow at the bottom is about half that of the surface flow.

3. Cycle times

With the meridional flows computed above, we can now estimate the cycle times for a flux transport dynamo. We define the travel time,

$$\tau = \frac{\pi R_{\text{bot}}}{2 u_m}$$

where R_{bot} and u_m radius and the average flow speed at the bottom of the convection zone. As one would have to integrate $1/u_\theta$ over latitude and that integral diverges when taken from zero to $\pi/2$, we define $u_m = 0.4 \max(u_\theta)$ as the average flow speed. This choice reproduces the observed eleven year cycle time when applied to our model for the Sun and corresponds to integrating from $8°$ to $82°$.

The left panel in Fig. 4 shows u_m for all three models for a range of rotation periods from 0.2 to 30 days. For each model, the variation with the rotation period is rather moderate. The lines of 0.66 $M_\odot$ and 0.6 $M_\odot$ models are almost identical while the 0.4 $M_\odot$ for the model with 0.4 $M_\odot$ the flow amplitude is about half that of the models with larger masses. All cases show an increase of the flow speed with increasing rotation rate.

The right panel of Fig. 4 shows the travel times computed with the flow speeds shown in the left panel and the observed cycle times from Fig. 1. Despite the differences in flow speed, particularly between the $0.4 M_\odot$ case and the $0.6 M_\odot$ and $0.66 M_\odot$ models, the three curves are practically identical, i.e. the differences in flow speed and radius cancel. The sun is also very close to the three curves, despite being much more massive and luminous.

A comparison between the observed cycle times and our model travel times shows that while there are three cases where both are the same, the vast majority of the stars in

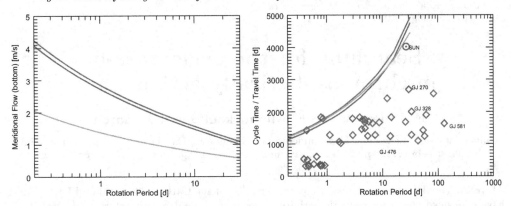

Figure 4. Left: Meridional flow speed (left) and the resulting travel time at the bottom of the convection zone for stars of $M = 0.66\,M_\odot$ (green line), $M = 0.60\,M_\odot$ (red line), and $M = 0.40\,M_\odot$ (yellow line). The right panel also shows the observed rotation periods and cycle times from Fig. 1 (diamods) and the Sun ($\odot$).

the sample shows cycle times that are substantially shorter than the travel times. The discrepancy is particularly large for the rapidly rotating stars observed by KEPLER, but also for some of the slow rotators.

4. Conclusions

In the flux transport dynamo, the flow speed that sets the cycle time. As the gas in the solar convection zone is not a perfect conductor, the magnetic field must actually be expected to move a bit more slowly than the gas. That makes our estimated travel time a lower estimate for the cycle time, i.e. all stars in Fig. 4 should lie on or above the green, yellow, and red lines. As almost all the stars in the sample lie below the lines, our findings do not support the flux transport dynamo as the mechanism behind the activity of these stars.

References

Bonanno, A., Elstner, D., Rüdiger, G., & Belvedere, G. 2002, *A&A*, 390, 673

Choudhuri, A. R., Schüssler, M., & Dikpati, M. 1995, *A&A*, 303, L29

Dikpati, M. & Charbonneau, P. 1999, *ApJ*, 518, 508

Distefano, E., Lanzafame, A. C., Lanza, A. F., Messina, S., & Spada, F. 2016, *A&A*, 591, A43

Küker, M., Rüdiger, G., & Kitchatinov, L. L. 2011, *A&A*, 530, A48

Küker, M., Rüdiger, G., Olah, K., & Strassmeier, K. G. 2019, *A&A*, 622, A40

Küker, M., Rüdiger, G., & Schultz, M. 2001, *A&A*, 374, 301

Paxton, B., Bildsten, L., Dotter, A., *et al.* 2011, *ApJ*, 192, 3

Roberts, P. H. 1972, Philosophical Transactions of the Royal Society of London Series A, 272, 663

Suárez Mascareño, A., Rebolo, R., & González Hernández, J. I. 2016, *A&A*, 595, A12

Vida, K., Kriskovics, L., & Oláh, K. 2013, Astronomische Nachrichten, 334, 972

Vida, K., Oláh, K., & Szabó, R. 2014, *MNRAS*, 441, 2744

Wargelin, B. J., Saar, S. H., Pojmański, G., Drake, J. J., & Kashyap, V. L. 2017, *MNRAS*, 464, 3281

Solar and Stellar Magnetic Fields: Origins and Manifestations
Proceedings IAU Symposium No. 354, 2019
A. Kosovichev, K. Strassmeier & M. Jardine, ed.
doi:10.1017/S1743921320000885

Searching for the cycle period in chromospherically active stars

F. Villegas[ID], R. E. Mennickent[ID] and J. Garcés

Universidad de Concepción, Departamento de Astronomía, Casilla 160-C, Concepción, Chile
emails: fabrivillegas@udec.cl, rmennick@udec.cl, jgarcesletelier@gmail.com

Abstract. The detection and analysis of line emission of the CaII, H(396.8nm) and K(393.3nm) have confirmed the chromospheric activity of some single and binaries stars. This activity is associated to the presence of magnetic fields which in turn are produced by internal convective flows along with stellar rotation producing a long-term photometric cycle length related to the apparition and vanishing of superficial stellar spots. We present a photometric study of stars of the type RS CVn, Rotationally variable Star and BY Dra, that have shown evidence of chromospheric activity. The analysis of these measurements has allowed us to delimit periods of rotation. In addition, we have detected and measured the cycle length in some cases. It allows us to complement previous investigations and in some cases to determine for the first time the presence of a long photometric cycle, contributing to complement the link between rotation and magnetic cycles.

Keywords. Activity, binaries: eclipsing, spots.

1. Introduction

Chromospheric activity (CA) is an interesting phenomenon present in different types of stars such as RS Canum Venaticorum (RS CVn) and BY Draconis variables (BY Dra) which are one of the best and most complete astrophysical laboratories to study stellar activity according to Montes (1995). Since the latter are binaries (RS CVn), the phenomenon of synchronization between P_{rot} and P_{orb} is present due to its proximity, which at the same time generates a fast rotation.

Wilson (1978) was the pioneer in finding evidence of cyclic variation (P_{cyc}) by tracking spectroscopic indicators such as the emission lines of Ca II H (3968 Å) & K (3933 Å), and subsequent studies have allowed to establish that stars exhibit different levels of cromopheric activity. From this discovery and to establish the link between P_{rot} and P_{cyc}, the classification in the known $P_{rot}-P_{cyc}$ plane arises, in which two Active (A) and Inactive (I) sequences are present according to Saar & Brandenburg (1999).

The search for P_{cyc} has not only been addressed in the spectroscopic field, long term photometric missions such as those carried out by Phillips & Hartmann (1978) or Baliunas & Vaughan (1985) found evidence of cyclic activity in low mass stars. In addition, using a photometric analysis, authors such as Böhm-Vitense (2007) and Messina & Guinan (2002) find changes in the rotational period, associated with differential rotation.

Other variation in the light curve associate to the CA is the O'Connell Effect investigated by O'Connell (1951) this phenomenon is produced by asymmetries in the maxima of the light curve.

We present a new photometric study of 5 RS CVn and BY Dra stars, finding new cycle periods (not previously observed) and other indicators associated with CA, for example secondary peaks in the Fourier periodograms (associated with differential rotation) and morphological changes in the light curves.

Table 1. Results of the analysis of light-curves obtained, described in Villegas (2019).

Target	V_{ASAS} [mag]	P_{rot}(Pub.) [day]	P_{rot}(Med.) [day]	Error [day]	P_{cyc} [day]	Error [day]	P_{cyc}/P_{rot} -
YZ Men	7.770	19.58	19.401	0.009	~827	-	43.54
CK Ant	8.506	29.83	29.5605	0.0049	3203.67	95.07	108.26
V1380	11.593	3.091	3.0921	0.0001	334.97	19.65	108.31
V1382	9.165	38.03	38.03	0.010	2623.1	38.6	69.84
YY CMa	11.262	11.22	5.625	0.010	-	-	-

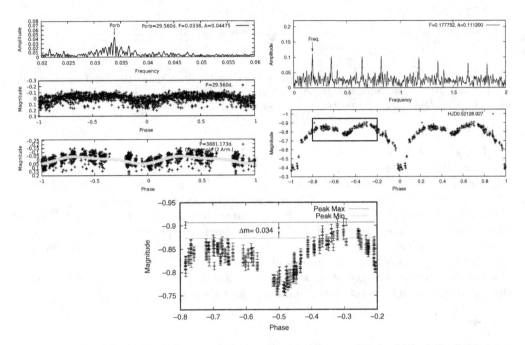

Figure 1. Up left the Periodogram (relation between the amplitude "A" of the light curve and the frequency "F"), rotational and long-cycle phase diagram obtained to the CK Ant star. (The yellow line represent the fit used to apply the disentangling) Top right: is similar but includes only the rotational phase and the periodogram to the YY CMa star, where is possible see secondary peaks associated with alias of the a frequency (for example 2F, 4F) and Down: is a zoom to the region where the O'Connell effect is identified.

2. Overview

Using photometric data, we obtained the final light curve from TAROT telescope, a 25-cm telescope located in La Silla Observatory, Chile with g,r,i and c filters, (data were reduced by a standard way using IRAF tasks ccdred and phot) and photometric observations provided by the All Sky Automated Survey described in Pojmanski (2003), these observations consist of simultaneous photometry in filter V through five apertures.

We disentangled the light curve into an orbital and long-cycle part with the aid of a Fourier decomposition algorithm described by Mennickent *et al.* (2012).

An overview of the five new long-cycle periods is shown in Table 1. The orbital and long-cyclic light curves for one system (CK Ant) are shown in Fig 1. In addition, the detection of O'Connell effect in YY CMa, indicator of Chromospheric Activity is also shown.

In the Figure 2, we include different published databases and the used fit, shown in Vida *et al.* (2013), Boro Saikia *et al.* (2018), Vida *et al.* (2014) and Oláh *et al.* (2016), incorporating our data we make a new adjustment shown with the blue line.

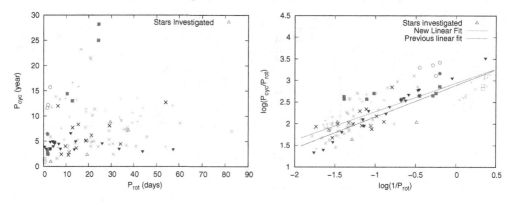

Figure 2. Different relation between orbital and magnetic cycle, using the database shown in Vida *et al.* (2013), Boro Saikia *et al.* (2018), Vida *et al.* (2014) and Oláh *et al.* (2016), the new fit is given by: $log(P_{cyc}/P_{rot}) = (0.66 \pm 0.04)log(1/P_{rot}) + (2.94 \pm 0.5)$.

3. Implications

- Using about 4500 CCD images obtained from our own TAROT observations and together with ASAS database, we find in 4 of the 6 stars analyzed, for the first time the presence of a long photometric cycle associated with chromospheric activity, for YY CMa confirm the activity through the presence of the O'Connell effect.
- When inspecting the $P_{cyc}-P_{rot}$ plane, unlike initially shown by Böhm-Vitense (2007), we observe the disappearance of possible I and A sequences.
- A lineal fit of the type $f(x) = ax + b$ was made including our new data, to determine the relationship between the magnetic and rotation period.
- We identify the presence of secondary peaks in the periodograms, showing possible signals related to stellar spots and the differential rotation.

References

Baliunas, S. L. & Vaughan, A. H. 1985, *ARA&A*, 23, 379
Böhm-Vitense, E. 2007, *ApJ*, 657, 486
Boro Saikia, S., Marvin, C. J., Jeffers, S. V., Reiners, A., Cameron, R., Marsden, S. C., Petit, P., Warnecke, J. & Yadav, A. P. 2018, *A&A*, 616A, 108B
Mennickent, R. E., Djurašević, G., Kołaczkowski, Z., & Michalska, G., 2012, *MNRAS*, 421, 862
Montes D. 1995, Doct en Ciencias Físicas, Universidad Complutense de Madrid, Madrid
Messina, S. & Guinan, E. F. 2002, *A&A*, 393, 225
O'Connell, D. 1951, *Publ. Riverview College Obs.*, 2, 85
Oláh, K., Kövári, Zs., Petrovay, K., Soon, W., Baliunas, S., Kolláth, Z., & Vida, K. 2016, *A&A*, 590, A133
Phillips, M. J. & Hartmann, L. 1978, *ApJ*, 224, 182–184
Pojmanski, G. 2003, *Acta Astronomica*, 53, 341
Saar, S. H. & Brandenburg, A. 1999, *ApJ*, 524, 295
Vida, K., Kriskovics, L., & Oláh, K. 2013, *AN*, 334, 972
Vida, K., Oláh, K., & Szabó, R. 2014, *MNRAS*, 441, 2744
Villegas, F. 2019, M.Sc. thesis , U. de Concepción
Wilson, O. 1978, *ApJ*, 226, 379

Solar and Stellar Magnetic Fields: Origins and Manifestations
Proceedings IAU Symposium No. 354, 2019
A. Kosovichev, K. Strassmeier & M. Jardine, ed.
doi:10.1017/S1743921320001428

Are there local dynamo in solar polar region?

Chunlan Jin[iD]

Key Laboratory of Solar Activity, National Astronomical Observatories,
Chinese Academy of Sciences, Beijing 100101, China
email: cljin@nao.cas.cn

Abstract. Polar magnetic field, as a component produced by the global dynamo, is thought to be the remant of toroidal magnetic field transported poleward from Sun's active belt. With the improvement of instruments, more and more observations are challenging the viewpoint. Recently, we identify the bipolar magnetic emergences (BMEs) in the polar region, and find that the distribution of the magnetic axes for these BMEs shows random state, which does not follow the Joy's law of active region. The result implies the possible existence of local dynamo in the solar polar region.

Keywords. Sun: atmosphere, Sun: magnetic fields, Sun: photosphere, sunspots

---

1. Introduction

Observations of the magnetic field in solar polar region are very important in understanding the long-term variation and the origin of solar magnetism. The polar field was firstly observed by the ground-based magnetograms (Babcock & Babcock 1955): a general, predominantly dipolar field. The average flux density in the polar region is about several G (Babcock & Babcock 1955; Tang & Wang 1991; Deng *et al.* 1999; Tsuneta *et al.* 2008).

The vector magnetic field in the polar region was firstly quantitatively measured by Deng *et al.* (1999). It was found that the polar field is an inclined field, and deviate from the normal of the surface by about 40 degree. The result is confirmed by Sun *et al.* (2015) based on the SDO/HMI vector magnetic field observations in the polar region. The polar field contributes about 5.0e22 Mx flux (Deng *et al.* 1999). More observations with high spatial resolution and polarization sensitivity come from the SOT aboard Hinode. Many magnetic flux tubes with kilo-Gauss field strength is distributed in polar region, and they have same polarity, consistent with the global polarity in polar region (Tsuneta *et al.* 2008). The polar region is different from the quiet region of the Sun: a larger area of kilo-Gauss magnetic concentration in the polar region than those of the quiet Sun (Ito *et al.* 2010).

The magnetic field of polar region is thought to be the direct manifestation of the global poloidal field. It is the seed field of global dynamo, and produces the toroidal field which results in the formation of sunspot and active regions. The source of polar magnetic field is explained by the solar cycle model: with away from solar minimum, magnetic field of decaying active region dispersed into the polar region owing to differential rotation, meridional flow and turbulent diffusion (Wang *et al.* 1991; Sheeley 1992; Sheeley 2005) and finally makes the polarity of the polar region reverse near solar maximum. However, more and more observations have challenged the solar global dynamo models in explaining the magnetic origin for the polar region (Severny 1971; Lin 1994; Benevolenskaya 2004; Shiotaet *et al.* 2012; Jin *et al.* 2020).

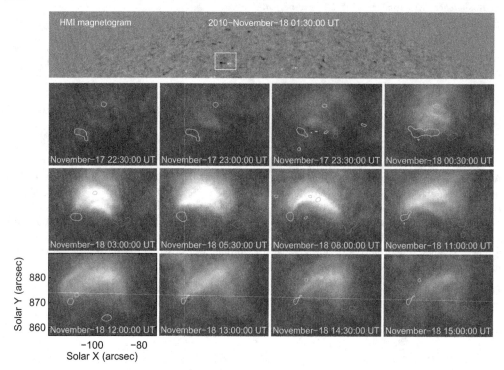

Figure 1. An identified BME example. Top panel: the HMI magnetic observation. Bottom panels: the atmospheric response in AIA 211 Å waveband during the evolution of BME. The green and red lines mean the −30 G and 30 G magnetic contours.

2. Challenge of global dynamo explaining polar magnetic origin

In low-resolution magnetogram, the polar region may appear unipolar, but the polar magnetic field is dominated by small-scale magnetic elements with mixed polarities (Severny 1971). Lin (1994) found that during sunspot maximum period, the polar regions were occupied by about equal numbers of positive and negative magnetic elements. Jin & Wang (2011) study the vector magnetic field in a polar region during solar minimum years, and they pointed that the ratio of magnetic flux in the minority polarity to the dominant polarity reaches 0.5, i.e., the mixed-polarities magnetic elements provides about 2/3 unsigned magnetic flux in the polar region.

Based on the magnetic observations during the maximum and decreasing phase of cycle 22, i.e., from early 1991 to mid-1993, Lin (1994) found that as the solar cycle develops toward sunspot minimum, the average magnetic field of the dominant polarity in the polar region increases, and the average magnetic field of the minority polarity decreases. However, Benevolenskaya (2004) uses the observations from the SOHO/MDI in 1996–2003, and reveals an interesting results that the total polar magnetic flux does not change during the polarity reversal in both hemispheres, but the positive and negative parts of the total flux change. By detecting the magnetic flux per patch in the solar polar region, Shiotaet *et al.* (2012) found that almost all large patches ($\geqslant 1.0e18$ Mx) have the same polarity, while smaller patches have a balance of both polarities. During the increasing phase of cycle 24, the net magnetic flux of the polar region clearly decreases, while the total magnetic flux of these smaller patches ($< 1.0e18$ Mx) does not change.

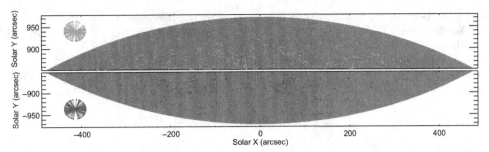

Figure 2. The distribution of BMEs magnetic axes in the polar region. The length and direction of each arrow in the polar region means the maximum separation and the magnetic axis direction for each BME. The arrows of the concentric circle in the top left and bottom left corners display the magnetic axis distribution of BMEs with the same distance of the opposite polarities in the corresponding polar region.

3. Random distribution of BME magnetic axis in the polar region

In our recent study, we try to explore the bipolar magnetic emergences (BMEs) in the polar region by considering the simultaneous magnetic evolutions and imaging observations in extreme ultraviolet waveband in the interval from 2010 June to 2011 May. More than 300 BMEs are definitely identified by two criterions. Firstly, a pair of emerging flux with opposite polarities in the LOS magnetogram appears quasi-instantaneously in the solar surface. Secondly, the brightening loop in extreme ultraviolet images between the opposite polarities appears with the increasing magnetic flux. An example of identified BME is shown in Figure 1. According to the Joy's law of active region, if these BMEs come from the global dynamo, their magnetic tilt should be consistent with the Joy's law of active region. We determine the magnetic barycenter position of both polarities for each BME, and obtain the direction of BME tilt. The distribution of magnetic tilt is shown in Figure 2. It can be found that the direction of magnetic tilt for BMEs is random either in the southern or northern polar regions, and does not obey the Joy's law of active regions (Jin *et al.* 2020).

Our observation result confirmed the possible existence of local dynamo in the solar polar region besides global dynamo. More careful and serious efforts need to be made to further explore the magnetic origin of polar magnetic field.

Acknowledgements

This work is supported by the National Natural Science Foundations of China (11533008, 11573038 and 11873059) and Basic Frontier Scientific Research Programs of CAS(ZDBS-LY-SLH013)

References

Babcock, H. W. & Babcock, H. D. 1955, *Astrophys. J.*, 121, 349
Benevolenskaya, E. E. 2004, *A&A*, 428, L5
Deng, Y. Y., Wang, J. X., & Ai, G. X. 1999, *Science in China*, 42, 10
Ito, H., Tsuneta, S., Shiota, D., Tokumaru, M., & Fujiki, K. 2010, *Astrophys. J.*, 719, 131
Jin, C. L. & Wang, J. X. 2011, *Astrophys. J.*, 732, 4
Jin, C. L., Zhou, G. P., Zhang, Y. Z., & Wang, J. X. 2020, *Astrophys. J. Lett.*, 889, L26
Lin, H., Varsik, J., & Zirin, H. 1994, *Sol. Phys.*, 155, 243
Severny, A. B. 1971, *IAUS*, 43, 675
Sheeley, N. R., Jr. 1992, *ASPC*, 27, 1
Sheeley, N. R., Jr. 2005, *LRSP*, 2, 5

Shiota, D., Tsuneta, S., Shimojo, M., Sako, N., *et al.* 2012, *Astrophys. J.*, 753, 157
Sun, X., Hoeksema, J. T., Liu, Y., Norton, A. A., *et al.* 2015, *American Geophysical Union, Fall Meeting, SH23A-2429*
Tang, F. & Wang, H. M. 1991, *Sol. Phys.*, 132, 247
Tsuneta, S., Ichimoto, K., Katsukawa, Y., Lites, B. W., *et al.* 2008, *Astrophys. J.*, 688, 1374
Wang, Y. M., Sheeley, N. R., Jr., & Nash, A. G. 1991, *Astrophys. J.*, 383, 431

Solar and Stellar Magnetic Fields: Origins and Manifestations
Proceedings IAU Symposium No. 354, 2019
A. Kosovichev, K. Strassmeier & M. Jardine, ed.
doi:10.1017/S1743921320000071

A Clock in the Sun?

C. T. Russell[1], J. G. Luhmann[2]🄾 and L. K. Jian[3]

[1]Earth, Planetary and Space Sciences, University of California, Los Angeles, CA 90095, USA

[2]Space Sciences Laboratory, University of California, Berkeley, CA 94720, USA

[3]Goddard Space Flight Center, Greenbelt, MD, 20771

Abstract. The sunspot cycle is quite variable in duration and amplitude, yet in the long term, it seems to return to solar minimum on schedule, as if guided by a clock with an average period of close to 11.05 years for the sunspot number cycle and 22.1 years for the magnetic cycle. This paper provides a brief review of the sunspot number cycle since 1750, discusses some of the processes controlling the solar dynamo, and provides clues that may add to our understanding of what controls the cadence of the solar clock.

Keywords. solar cycle, solar dynamo, sunspot number

1. Introduction

In the center of the Sun is a nuclear furnace that produces the heat that eventually radiates into space, producing a very habitable zone near 1 AU. The nuclear-fusion produced heat is radiated upward inside the Sun to about 0.7 solar radii, where convection contributes to the outward transport of the energy that is ultimately radiated into space. The rotation of the convection zone affects the circulation of its magnetized plasma. It enforces rotational columns of fluid similar to those that occur within a rotating sphere (Taylor 1922), but the presence of the magnetic field and the convection zone boundary introduce differences that are essential to the solar activity cycle.In particular, magnetic flux ropes produced near the bottom of the convection zone rise upward to pervade the convection zone, manifesting themselves on the surface of the Sun, as sunspots, active regions, coronal mass ejections, and a myriad of phenomena collectively known as solar activity. Figure 1a shows a cut-away drawing of the Sun's interior with an idealistic convection pattern, together with some of the related features that appear on the observed photosphere. Figure 1b shows a cross section of the Sun and contours of the period of the motion of plasma in the convection zone, here depicted as symmetric with respect to the solar rotation axis. This figure assumes that the rotation periods in the north and south are precisely identical, whereas, in fact, the sun often has two quite independent north and south hemispheres, especially in the regions above the poles. Helioseismology informs us of these motions, but only recently (Komm *et al.* 2018) has it been possible to separate the flow structure in the north and south, even though we know the north and south hemispheres can magnetically be quite different. Finally, Figure 1c shows the rotation period of the Sun versus radial distance. The core and the radiative zone are thought to share the same rotation period, but in the convection zone, this period becomes a function of latitude.

Two important phenomena associated with this circulation are manifested on the photosphere as distinctive surface features: the rush to the poles and the torsional oscillation. The global appearance of these, which arise from a combination of zonal and meridional flow features, are sketched in Figure 2a on the surface of the photosphere, and 2b in the cross section, while Figures 2c and 2d show the surface manifestation of these

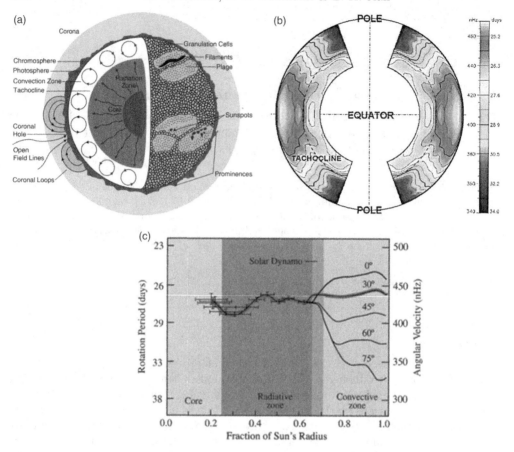

Figure 1. The structure of the Sun. The interior structure of the Sun is on the left, while the phenomena on the surface are shown on the right. Sunspots are seen equally in both hemispheres, but with opposite magnetic symmetries. Leading and trailing polarities are opposite in the two hemispheres and reverse every sunspot cycle (Russell *et al.* 2015). Internal structure of the Sun's rotation, showing the location of the tachocline where rigid rotation gives way to differential rotation throughout the convection zone. The color scales are for solar differential rotation frequency (nHz) and the corresponding rotation period (day). Figure is adapted from NASA MSFC (https://solarscience.msfc.nasa.gov/images/internal_rotation_mjt.jpg), based on the original results from helioseismology published by Thompson *et al.* (2003). Internal structure of the Sun's rotation, showing the rotation period as a function of radius at different latitudes. The latitude where the zonal velocity transitions from effective superrotation at low latitudes, to subrotation at mid and high latitudes, occurs at around 30 degrees. The torsional oscillation (see text) tends to initially appears around 55 degrees, while the associated surface magnetic activity appears with it after it has migrated to ∼30 degrees. The reasons for this behavior have yet to be determined. Courtesy of K.R. Lang, Tufts University (https://ase.tufts.edu/cosmos/print_images.asp?id=25). This figure is an adaptation of the original results from helioseismology published by the National Solar Observatory (NSF).

motions over the double sunspot cycle, in a magnetogram and a zonal velocity-gram (from Kosovichev & Pipin 2019).

Figure 2e, showing the zonal velocity at two instants of time during the solar cycle, emphasizes that panel 2b does not show material motion to the pole or to the equator, but rather results from the location of bands of fast flows moving poleward and equatorward during the cycle. Figures 2c and 2d show that the solar magnetic cycle is truly 22.1 years long (on average) and that two successive magnetic cycles are always in operation with

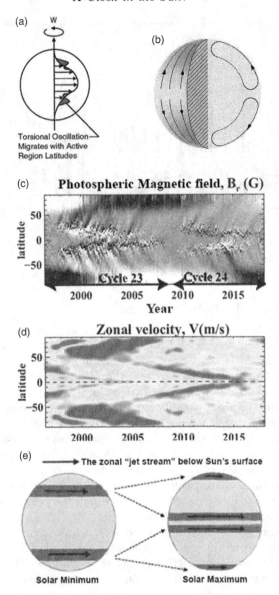

Figure 2. Illustrations of the torsional oscillations (top) and meridional circulation (bottom) described in the text. Figure adapted from Kosovichev & Pipin (2019), illustrating the cyclic behavior of the torsional oscillation velocity in the lower panel (d), compared with the solar magnetic cycle as seen in the longitudinally averaged radial surface magnetic field (c). The similarity of these patterns, with both showing poleward and equatorward branching features, suggests they are related. The zonal velocity at two instants of time during the solar cycle, emphasizes that material motion is not to the pole or to the equator, but the locations of fast flow move poleward and equatorward during the cycle.

each in a different phase. Solar cycles are nested. Just counting sunspots may give an erroneous impression of the independence of successive cycles. In fact, successive cycles 'always' coexist over half their durations with the preceding and then the following cycle overlapping, as demonstrated by the zonal oscillation pattern that requires 22 years to complete.

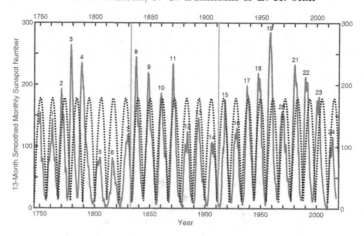

Figure 3. Sunspot record since 1749, solid line (based on Clette & Lefèvre 2016). The dotted line is an 11.05 yr period cycle with amplitude of 178.7, the average maximum SSN of cycles 1-24, for the current sunspot maxima. The 11.05-yr period is the average period for cycles 1-23.

2. The Irregularity of the Solar Cycle

Figure 3 shows the time series of sunspot numbers since 1750 (see Clette & Lefèvre 2016), as recently checked and revised for consistency across the centuries (Hathaway 2015; Svalgaard 2013). A dotted line of constant amplitude (177.8 spots) and period (11.05 yr) provides a chronometer to illustrate the variation of the amplitude and phase of the solar sunspot cycle. Two variabilities are worthy of note: the amplitude can strongly exceed the average, and the phase can differ greatly during the solar cycles shown, forming 'grand' cycles. The amplitudes are not random. They are quite variable, but they do define these 'grand cycles' quite clearly. The phase also has a large-scale variation. The sunspot number phase can change more rapidly than the chronometer phase, but when the sunspot numbers drop, the phase of the sunspot can quickly return to being in phase with our chronometer (e.g. B.J.I. Bromage, 2009, personal communication). This phase reset seems to be associated with a multiple-cycle dip in the peak cycle sunspot number. Our solar clock is irregular but not random.

Robert Robert Dicke (1978) long ago noted that despite first appearances, the Sun was an accurate clock. His metric was that the length of a solar cycle remained constant over the long term and varied in length for only short periods. The Sun keeps track of time internally, and reveals the correct time on the photosphere only occasionally.

3. What Is the Sun Trying to Tell Us?

Well before Dicke's seminal paper, Waldmeier noted that the faster the sunspot number rose, the higher the sunspot maximum became (Waldmeier 1935). This is illustrated in Figure 4a. This relation is obeyed in both hemispheres separately and quite independently, as shown in Figure 4b for north and south hemispheres separately. This relationship has a very simple interpretation if the Sun's convection layer is an electrically conducting medium in which the magnetic field is supported by local currents that are dissipative. The existence of the strong magnetic fields of sunspots depends on 'local' currents, and the longer they are present , the more their magnetic fields weaken. Solar cycles end with a decrease in the sunspot number, due to continued decay of the sunspot fields together with a cutoff of the supply of strong fields from below. We can use these observations to give us some insight into the workings of the solar dynamo. In Figure 5, we show the sunspot number maximum for each of cycles 1 to 24 as a function of

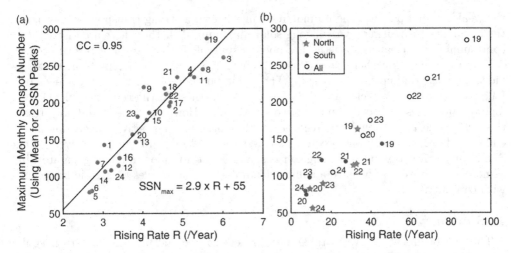

Figure 4. (a) (Left) Waldmeier effect for the solar cycles 1-25 using latest calculations. (b) (Right) Waldmeier effect for north and south hemispheres separately for cycles 19-24 when data for north and south hemispheres were available separately. Star: Northern hemisphere. Closed circle: Southern. Open circle: Combined sunspot numbers.

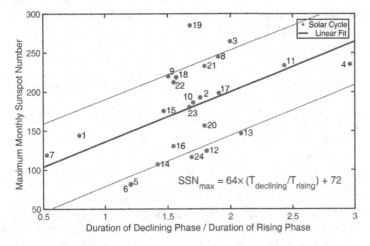

Figure 5. Maximum monthly sunspot numbers for cycles 1 to 24, versus the ratios of the duration of the declining phase divided by the duration of the rising phase.

the ratio of the duration of the declining phase when the cycle is ending, to the duration of the rising phase when the sunspot maximum is being created. This plot shows some interesting behavior. This ratio varies over a factor of 6 from 0.5 to 3, and it defines 3 lines with distinctly different activity. There are the weak sunspot cycles 5, 6, 12, 13, 14, 16, and 24. There are the 'normal' or medium-activity cycles 1, 2, 4, 7, 10, 11, 15, 23, and the high activity cycles 3, 8, 9, 21, and 22, with cycle 19 being a clear outlier above all the rest. The low activity cycle balancing the high cycle 19 could be the Maunder minimum (not shown) which for the purposes of this discussion is a statement that the Sun does not have to produce sunspots.

A linear fit to these data is given by the center line. The two parallel lines are equally spaced parallel lines that approximate the weak maxima and the strong maxima. This gives the impression that the Sun may have different levels of internal behavior: the Maunder minimum behavior when the transport/production of magnetic flux is minimal;

the weak minima pattern, the normal/median pattern, a strong transport scenario, and the capability of creating a very strong solar maximum. Whether these are discrete or a continuum is not clear as statistics are poor in spite of the long sunspot record used.

There is also a hint here of predictive behavior somewhat different from that provided by the Waldmeier relationship. The point farthest to the right is cycle 4. It is followed by two weak cycles 5 and 6. The second farthest to the right is cycle 11. It is followed by cycles 12 and 13, another two weak cycles. However, cycle 23 did not presage the weak cycle 24, although the recovery of the phase of solar activity occurring at the end of cycle 23, certainly was predictive, as was the phase shift at the end of cycle 4. The Sun clearly is functioning according to rules, and is not a random number generator, but as noted above we do not have a sufficient number of solar cycles to decode these rules with certainty.

4. Summary

The Sun begins with a very stable interior heat source and produces a very irregular magnetic envelope. The sunspots vary in number and strength. Rotation and convection play together to produce a very complex magnetic field. However, there are patterns in the circulation and the sunspot number that provide clues to what is happening within the Sun. While the combination of all of these factors produces complexity, it is clear that the Sun itself is keeping an accurate measure of time. For a more detailed discussion of this problem, see "The Solar Clock", a recent paper by these authors in Reviews of Geophysics (Russell et al. 2019).

References

Clette, F. & Lefèvre, L. 2016. The New Sunspot Number: assembling all corrections, *Solar Physics, 291.* doi:10.1007/s11207-016-1014-y
Dicke, R. H. 1978. Is there a chronometer hidden deep in the Sun? *Nature, 276,* 676–680. doi:10.1038/276676b0
Hathaway, D. H. 2015. The Solar Cycle. *Living Reviews in Solar Physics, 12.* https://doi.org/10.1007/lrsp-2015-4
Komm, R., Howe, R., Hill, F. *et al.* 2018. Subsurface zonal and meridional flow during cycles 23 and 24. *Solar Physics, 293.* doi:10.1007/s11207-018-1365-7
Kosovichev, A. G. & Pipin, V. V. 2019. Dynamo wave patterns inside the Sun revealed by torsional oscillations. *Astrophys. J. Lett., 871.* doi:10.3847/2041-8213/aafe82.d
Russell, C. T., Jian, L. K., & Luhmann, J. G. 2019. The Solar Clock. *Rev. Geophys.,* 57
Russell, C. T., Luhmann, J. G., & Strangeway, R. J. 2015. Space Physics: An Introduction, Cambridge University Press, 479
Svalgaard, L. 2013. Solar activity-Past, present, future. *J. Space Weather and Space Climate, 3.* https://doi.org/10.1051/swsc/2013046
Taylor, G. I. 1922. The motion of a sphere in a rotating liquid. *Proceedings of the Royal Society A, 102, 715.* doi:10.1098/rspa.1922.0079
Thompson, M. J., Christensen-Dalsgaard, J., Miesch, M. S., Toomre, J. 2003. The Internal Rotation of the Sun. *Annu. Rev. Astron. Astrophys., 41,* 599–643
Waldmeier, M. (1935). *Astron. Mitt. Zurich,* 14, 105

IAU 354: Question and Answer

Paper: A Clock in the Sun? by C.T. Russell, L.K. Jian and J.G. Luhmann

Question: Please explain in more detail why the polar, surface magnetic fields of the Sun are decoupled from the lower latitude photospheric fields.

–Chia-Hsien Lin

Answer: This effect in the Sun's magnetic dynamo is a result of the large non-convective radiative zone in its interior and the rotation of the Sun. The radiative zone contributes, at most, mimimally to the solar cycle, and acts mainly to restrict the generation of the magnetic field to the convective, electrically conducting outer 30% of the Sun. Rotating fluids, whether they are cylinders in the laboratory or spherical planetary and stellar bodies in space, form cylinders of rotating fluids, more or less parallel to the rotation axis of the body as they conduct heat to the surface of the body. This effect, combined with the spherical interior non-convecting region, divides the convection zone into three regions that weakly communicate: the northern polar zone, the mid- and low-latitude zone, and the southern polar zone. As a result, solar cycles can begin and end and behave quite independently in these three regions. The fact that they do not ever become totally uncorrelated indicates there is always some small coupling between them.

Solar and Stellar Magnetic Fields: Origins and Manifestations
Proceedings IAU Symposium No. 354, 2019
A. Kosovichev, K. Strassmeier & M. Jardine, ed.
doi:10.1017/S1743921319009888

Various scenarios for the equatorward migration of sunspots

Detlef Elstner[ID], Yori Fournier and Rainer Arlt

Leibniz-Institute for Astrophysics Potsdam, An der Sternwarte 16,
D-14482 Potsdam, Germany
email: delstner@aip.de

Abstract. The profile of the differential rotation together with the sign of the alpha-effect determine the dynamo wave direction. In early models of the solar dynamo the dynamo wave often leads to a poleward migration of the activity belts. Flux transport by the meridional flow or the effect of the surface shear layer are possible solutions. In a model including the corona, we show that various migrations can be obtained by varying the properties of the corona. A new dynamo of Babcock-Leighton type also leads to the correct equatorward migration by the non-linear relation between flux density and rise time of the flux.

Keywords. solar dynamo, solar corona

---

1. Introduction

The butterfly diagram of the solar magnetic field is still challenging dynamo theory. The simple explanation by a dynamo wave fails because of the increasing angular velocity at lower latitudes. Including the effect of a meridional circulation leads to the flux transport dynamos, which can explain many features of the solar cycle under the assumption of low turbulent diffusivities (eg., Choudhuri *et al.* 1995, Dikpati & Charbonneau 1999, Bonanno *et al.* 2002). Most of these models use the mathematically convenient but physically questionable vacuum boundary condition at the solar surface. A better choice is probably a force free magnetic field extension into the atmosphere (Bonanno 2016). The role of the corona for the dynamo is still poorly understood. Dynamical models including parts of a corona with high viscosity and diffusivity were investigated by Warnecke *et al.* (2013). We consider kinematic models with various assumptions for the mean rotation and the turbulent diffusivity in the next section. A second possibility for the occurrence of a solar butterfly diagram by nonlocal and finite time correlation effects of the turbulence is given in the last section.

2. The role of the Corona

We present simple 3-dimensional $\alpha^2\Omega$-dynamos with a solar rotation law neglecting the surface shear layer and meridional flow. We solve the induction equation

$$\frac{\partial \mathbf{B}}{\partial t} = \text{curl}(\mathbf{u} \times \mathbf{B} + \alpha \circ \mathbf{B} - \eta_{\text{T}}\text{curl}\mathbf{B}), \tag{2.1}$$

in spherical coordinates (r, θ, φ). The mean flow $\mathbf{u} = (0, 0, r\sin(\theta)\,\Omega)$ is a solar type rotation within the convection zone similar to Dikpati & Charbonneau (1999).

For the rotation of the atmosphere we consider 3 different cases:

(1) no radial shear at the stellar surface, same latitudinal dependence as in the convection zone

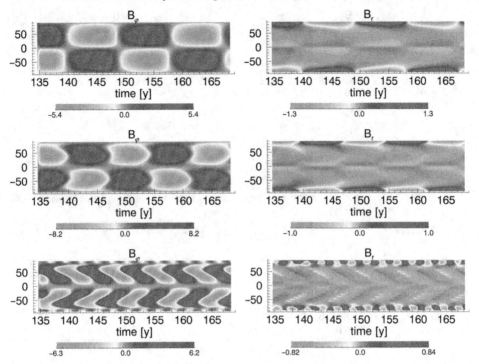

Figure 1. Time-latitude diagrams for halo diffusivities $\eta_h = 10\eta_c$ (upper), $\eta_h = \eta_c$ (middle) and $\eta_h = 0.1\eta_c$ (lower). The halo rotates with the same latitudinal dependence as the convection zone.

(2) rigid rotation same as the core

(3) rigid rotation same as the pole at stellar surface

The α-tensor has only diagonal components $\alpha_{ii} = \alpha_0 \cos(\theta)$ in the convection zone independent of solar radius with $\alpha_0 = 5$ cm/s and is locally quenched with magnetic energy density. The diffusivity η_c in the convection zone is 5×10^{11}cm^2/s. The inner radial boundary at 0.7 solar radius is a perfect conductor. We add a solar atmosphere up to 1.5 $R_\odot$. There we set a pseudo vacuum boundary condition (radial magnetic field only) and consider diffusivities in the atmosphere $\eta_h = 10\eta_c$, $\eta_h = \eta_c$ and $\eta_h = 0.1\eta_c$.

We show butterfly diagrams for the toroidal and radial field beneath the solar surface for a corona rotating with the angular velocity of the solar surface in Fig. 1. These models have no radial shear at the solar surface but latitudinal shear in the corona. For a rigidly rotating corona we present two cases, corotation with the core and corotation with the pole in Fig. 2 and Fig. 3, respectively. All the nine models show amazingly different behavior what underlines the role of the boundary condition for the dynamo. The models with no radial shear at the stellar surface show no latitudinal migration for the higher diffusivity cases in the atmosphere, but also an equatorward migration for the low diffusivity. But this model has a mixed mode solution of an axisymmetric oscillation and a strong non-axisymmetric polar field. Therefore the spots at the pole in the radial butterfly diagram show the double period caused by the rotation of the mode with azimuthal wave number $m = 1$. Best models appear for the high diffusivity case of the rigidly rotating atmospherical layer. Rigid halo rotation with the pole (3) shows a nice butterfly diagram for the toroidal field but the equatorward migration of the radial field is better in agreement with the observations of Hathaway (2010) in case (2) of rigid rotation with the core.

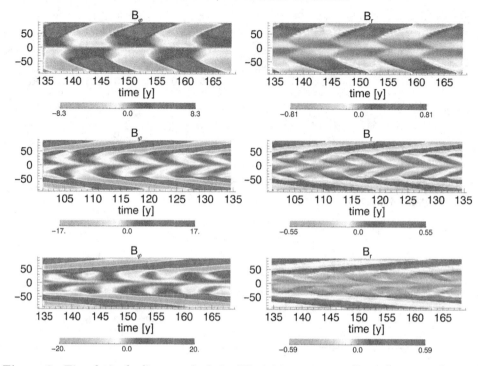

Figure 2. Time-latitude diagrams for halo diffusivities $\eta_h = 10\eta_c$ (upper), $\eta_h = \eta_c$ (middle) and $\eta_h = 0.1\eta_c$ (lower). The halo rotates with the core.

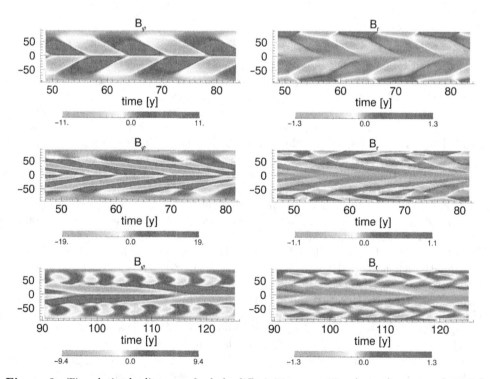

Figure 3. Time-latitude diagrams for halo diffusivities $\eta_h = 10\eta_c$ (upper), $\eta_h = \eta_c$ (middle) and $\eta_h = 0.1\eta_c$ (lower). The halo rotates as the pole.

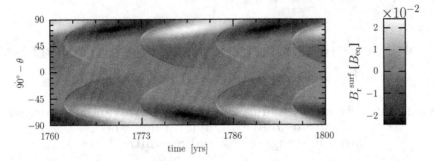

Figure 4. Butterfly diagram for the radial field of a subcritical Babcock-Leighton dynamo.

Dynamos with usually taken standard vacuum boundary conditions have no magnetic Poynting flux through the solar surface, whereas it should be different from zero for the magnetic heating of the corona. It is still open how far the large scale field contributes to the Poynting flux from the photosphere into the chromosphere, which can not be seen in models with a potential field boundary. Because the large scale dynamo is a global phenomenon it depends crucially on its boundary condition as it is demonstrated with our simple models. How far the rotation in the corona depends on the dynamo can only be found by a dynamical approach (Warnecke *et al.* 2016). Common models of stellar dynamos with their corona are also necessary to explain stellar activity in dependence on rotation.

3. A dynamo with field dependent memory effect

In Fournier *et al.* (2018) we found for a Babcock-Leighton type dynamo equatorward migration in the diffusive regime independent of the meridional flow (see Fig. 4). Magnetic flux tubes have a finite rise time scaling with rotation period of the star and the ratio between buoyant force and Coriolis force modified by the magnetic tension. For the sun we have $\tau_{delay} = \tau_0/\sin(\theta)|B_\varphi/B_{eq}|^q$. It is a non-local α-effect in space and time, where the non locality depends on the field strength in a non-linear way for q between -0.91 and -2.0. The toroidal flux can accumulate at the surface during the cycle. It is the subcritical regime, which leads to the equatorial migration at low latitudes. The system automatically saturates for strong fields, where the accumulation of flux is impossible because of the fast rise time. No additional quenching of the source term is needed.

References

Bonanno, A., Elstner, D., Rüdiger, G. 2002, *A&A*, 390, 673
Bonanno, A. 2016, *ApJ*, 833, L22
Choudhuri, A. R., Schüssler, M., Dikpati, M. 1995, *A&A*, 303, L29
Dikpati, M. & Charbonneau, P. 1999, *ApJ*, 518, 508
Fournier, Y., Arlt, R., Elstner, D. 2018, *A&A*, 620, 135
Hathaway, D. H. 2010, *LRSP*, 7, 1
Warnecke, J., Käpylä, P. J., Mantere, M. J., Brandenburg, A. 2013, *ApJ*, .778, 41
Warnecke, J., Käpylä, P. J., Käpylä, M. J., Brandenburg, A. 2016, *A&A*, 596, A115

Solar and Stellar Magnetic Fields: Origins and Manifestations
Proceedings IAU Symposium No. 354, 2019
A. Kosovichev, K. Strassmeier & M. Jardine, ed.
doi:10.1017/S1743921320003993

A solar cycle 25 prediction based on 4D-var data assimilation approach

Allan Sacha Brun[1], **Ching Pui Hung**[1,2], **Alexandre Fournier**[2],
Laurène Jouve[3], **Olivier Talagrand**[4],
Antoine Strugarek[1] **and Soumitra Hazra**[1]

[1]DAp/AIM, CEA Paris-Saclay, 91191 Gif-sur-Yvette, France

[2]IPGP, Université de Paris, UMR 7154 CNRS, F-75005 Paris, France

[3]Université de Toulouse, UPS-OMP, IRAP, 31028 Toulouse Cedex 4, France

[4]LMD, UMR 8539, Ecole Normale Supérieure, Paris Cedex 05, France

Abstract. Based on our modern 4D-var data assimilation pipeline *Solar Predict* we present in this short proceeding paper our prediction for the next solar cycle 25. As requested by the Solar Cycle 25 panel call issued on January 2019 by NOAA/SWPC and NASA, we predict the timing of next minimum and maximum as well as their amplitude. Our results are the following: the minimum should have occured within the first semester of year 2019. The maximum should occur in year 2024.4 ± 6 months, with a value of the sunspot number equal to 92 ± 10. This is in agreement with the NOAA/NASA consensus published in April 2019. Note that our prediction errors are based on 1-σ measure and do not consider all the systematics, so they are likely underestimated. We will update our prediction and error analysis regularly as more data becomes available and we improve our prediction pipeline.

Keywords. Sun, dynamo, 11-yr cycle, solar cycle prediction, data analysis, data assimilation

1. Introduction

The Sun possesses an intense surface activity modulated by its 11-yr magnetic dynamo cycle (Brun & Browning 2017). Over the last four centuries it has become clear that the period of the so-called Hale solar cycle is not perfectly stable, varying between 9 and 13 years typically (Clette & Lefèvre 2012). Likewise its amplitude has varied significantly, from being weak or even null (grand minima phase) to being very strong as in cycle 19. Hence it has become crucial to be able to predict the solar activity in order to anticipate and ideally mitigate the impact of our fierce Sun and its highly variable activity. It is of course very difficult to predict the solar activity cycle given the high degree of nonlinearity of the solar dynamo. Still there seems to be some order in this otherwise chaotic behavior and we can attempt to capture it to the best we can. To this end several groups have developed various ways of predicting the strength and timing of the next solar cycle, see for instance (Hathaway 2015) and (Petrovay 2019) for recent updates and summaries. Here we briefly present our own solar cycle predicting pipeline based on a novel 4-D var data assimilation method (Talagrand 2010), coupled to a 2.5D mean field dynamo model (see details in Jouve *et al.* 2011; Hung *et al.* 2015, 2017). In this short paper we present a summary of our answer to the Solar Cycle 25 - Call for Predictions issued in January 2019 by the NOAA Space Weather Prediction Center and NASA.

In §2 we briefly present the methodology behind our solar cycle prediction 4D-var tool *Solar Predict*. In §3 we present the solar data used to perform our prediction for solar cycle 25 and the *hindcasting* of cycles 22, 23 and 24. In §4 we validate our procedure using

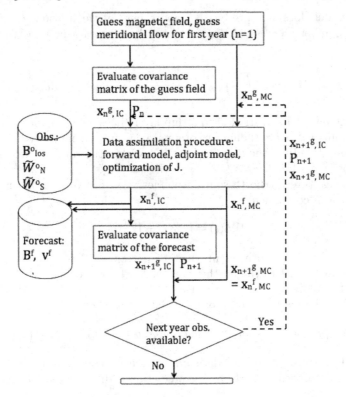

Figure 1. A schematic diagram illustrating data assimilation procedure. Integer subscripts refer to discrete time indices.

solar data from past solar cycles in order to test our prediction skills before performing our data assimilation (DA) analysis for cycle 25 and discussing our results in §5.

2. Brief presentation of the *Solar Predict* Data Assimilation pipeline

Our *Solar Predict* tool developed and used to perform inversion and prediction of the solar dynamics and cycle is based on a 4-D Var data assimilation (DA) method (Talagrand 2010), coupling sunspot number (SSN) and butterfly diagram (e.g. temporally stacked $B_{los}(\theta, t)$) time series data within a 2.5-D mean field axisymmetric dynamo model (depending on radius r and colatitude θ, but solving for all 3 components (B_r, B_θ, B_ϕ) of the magnetic field). This versatile tool can assimilate various lenght of solar data time series and invert different physical ingredients using our DA algorithm. Here, we focus on predicting solar cycle 25, using solar cycle 24 data (so-called "recent climatology"). An extensive description of *Solar Predict* can be found in (Jouve *et al.* 2011; Hung *et al.* 2015, 2017) and an image summarizing the pipeline is shown in Figure 1.

2.1. *Observational Proxy*

In order to make our prediction we need a model to generate time series of physical variables that can be directly compared to the observations. To do so, we use a mean field Babcock-Leighton dynamo model as described in (Jouve & Brun 2007; Sanchez *et al.* 2014; Hung *et al.* 2017), with small modifications to the parameters and with a slightly more complex resistivity profile, with a 2-step profile in radial direction.

Since the model does not produce sunspots per se, we introduce a proxy for the total sunspot number (SSN^f) of the model, in the form of a pseudo-Wolf number $\tilde{W}^f$ defined by

$$\tilde{W}^f(t) = \int_0^\pi \int_{0.7}^{0.71} \left[B_\phi^f(r, \theta, t) \right]^2 r^2 \sin \theta \quad dr \, d\theta. \tag{2.1}$$

In radius, the integral is restricted to a thin layer (between 0.70 and 0.71 solar radius) where toroidal flux tubes are thought to originate. We further multiply $\tilde{W}^f$ by a constant c_{SSN} that allows to adjust the amplitude of $\tilde{W}^f$ within the range of values of real solar sunspot number record, e.g. $SSN^f = c_{SSN} * \tilde{W}^f$ (cf. Fig. 2).

If we wish to capture the north-south asymmetry, we can further decompose $\tilde{W}^f$ into its north and south components

$$\tilde{W}^f(t) = \tilde{W}_N^f(t) + \tilde{W}_S^f(t), \tag{2.2}$$

in which the north (resp. south) component $\tilde{W}_N^f$ (resp. $\tilde{W}_S^f$) is computed by restricting the integration in Eq. 2.1 to the northern (resp. southern) hemisphere.

The other class of data will consist of time series of the line-of-sight component of the magnetic field at the model surface for all co-latitudes θ, $B(\theta, t)_{los}^f$, defined as

$$B_{los}^f(\theta, t) = B_r^f(r = 1, \theta, t) \sin \theta = (\cos \theta + \sin \theta \partial_\theta) A_\phi^f(r = 1, \theta, t). \tag{2.3}$$

assuming $r = 1$ at the surface. Such data will be directly compared to solar butterfly-like data. The solar data used to perform our prediction is discussed in the next section.

2.2. *Objective function $\mathcal{J}$*

To succesfully perform our prediction, we aim to minimize an objective function defined in terms of the differences between the observations and our dynamo model trajectory,

$$\mathcal{J} = \sum_{t_i}^{N_t} \left\{ \sum_{\theta_j}^{N_\theta} \frac{(B_{i,j}^f - B_{i,j}^o)^2}{x / \sin^2 \theta} + \frac{(SSN_i^f - SSN_i^o)^2}{\sigma_{SSN,i}^2} \right\}, \tag{2.4}$$

where $\mathcal{J}$ is the objective function, B is the surface radial field, SSN is the sunspot number (see equations (2.2) and (2.3) for a description of how we compute these quantities from our dynamo model). Superscripts o and f denote observed and forecast (model-based) values respectively. The misfit of the surface radial field is normalized with $1/\sin^2 \theta$, as the uncertainty of observation increases with latitude. The relative weighting of misfit in the surface field and SSN is controlled by the factor x. The misfit in SSN is normalized with the variance of SSN^o. The misfits are summed over the number of observations in time (N_t) and latitudes (N_θ, for the surface field).

We then assimilate the magnetic observations for the nth sunspot cycle, and get an estimate of the average flow and the magnetic configuration on the meridional plane, at the end of the cycle. Finally, we can obtain a preliminary guess of the maximum of the $n + 1^{th}$ cycle by extrapolation of the model beyond the assimilation window, based on the forecast flow and magnetic configuration estimated from the observations of the nth cycle. The first maximum of the extrapolated dynamic trajectory of the modeled SSN is our guess of the $n + 1^{th}$ maximum.

As mentioned above, the observations used for the experiment are compared to our magnetic sunspot proxy (Equation 2.1) and with the surface line of sight magnetic field B_{los}^o (Equation 2.3).

Table 1. Guessed maxima of the $(n+1)^{th}$ cycle by extrapolation of the forecast model, from the results of assimilation of the observations of the nth cycle.

x	$n=21$		$n=22$		$n=23$	
	$t_{max,n+1}$	$SSN^f_{max,n+1}$	$t_{max,n+1}$	$SSN^f_{max,n+1}$	$t_{max,n+1}$	$SSN^f_{max,n+1}$
0.1	1990.8	190	2000.7	208	2012.5	135
1.0	1991.1	222	2000.7	239	2013.0	115
5.0	1991.4	202	2001.1	211	2013.5	90.4
10	1991.5	192	2001.2	196	2013.8	76.2
50	1991.7	174	2001.7	152	2014.1	68.9
observed	1990.5	212 ± 11	2001.1	177 ± 9	2013.7	106 ± 8

Table 2. Predicted minima between cycle 24 and 25 by extrapolation of the forecast model, from the results of assimilation of the observations of cycle 24.

	minimum $n=24/25$ forecast
x	$t_{min,24-25}$
0.1	2019.09 ± 0.10
0.3	2019.16 ± 0.10
0.5	2019.1 ± 0.10
1	2019.08 ± 0.10
10	2019.12 ± 0.10

Table 3. Predicted maxima of the cycle 25 by extrapolation of the forecast model, from the results of assimilation of the observations of cycle 24.

	maximum cycle $n=25$ forecast	
x	$t_{max,25}$	$SSN^f_{max,25}$
0.1	2024.3 ± 0.40	97 ± 6
0.3	2024.3 ± 0.40	97 ± 6
0.5	2024.3 ± 0.40	96 ± 6
1.0	2024.3 ± 0.35	94 ± 6
10	2024.5 ± 0.40	88 ± 6

2.3. *Overall Data assimilation procedure*

In Figure 1 we represent the overall data assimilation procedure. We use both direct and adjoint (tangent linear) 2.5D mean field dynamo models to assess the variation (gradients) of the objective function $\mathcal{J}$ (again see details in: Jouve *et al.* (2011); Hung *et al.* (2015, 2017)). Thanks to the coupling of the direct and adjoint dynamo models in our DA pipeline we are able to efficiently minimize $\mathcal{J}$. This results on the misfit between the observations and the model observation proxies to tend to zero. We perform this over about one solar cycle prior to letting the forecast (e.g. the model with observationally constrained magnetic field configurations for $A_\phi(\vec{r}, t)$ and $B_\phi(\vec{r}, t)$ and the meridional circulation) evolve freely. We repeat this procedure for as many x control parameters as needed (typically 5 or 6 values). We then systematically perform stochastic perturbations of each average trajectories to further assess the error bars of the predictions (see Tables 2 & 3). So overall our procedure is divided in two steps: a) data assimilation of the existing solar data over about one solar cycle to obtain good initial magnetic and meridional states to be used in step b) as initial conditions of the dynamo model that we let go unconstrained. The trajectory then obtained in step b) constitutes a prediction of the magnetic state of the Sun.

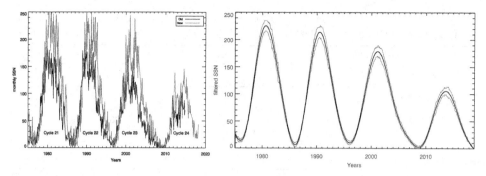

Figure 2. (Left) Monthly averaged SSN time series (old and new in red Clette & Lefèvre (2018)) from SIDC/SILSO data base and (Right) filtered version SSN^o time series used to perform our data assimilation procedure. On the filtered curve the +/- 1-σ curve are being shown as well.

3. Solar data set used

We base our prediction on two types of solar data a) sunspot number time series and b) butterfly diagram of solar activity (see for instance Hathaway (2015)), from which we derive the line of sight magnetic field B^o_{los} magnetic field. Historically, sunspot series have started in 1749 and for instance are available at the Solar Influences Data Analysis Center (SIDC) http://www.sidc.be/silso/datafiles. As indicated in the NOAA/NASA call of January 2019, there are 2 time sunspot number series (SSN), the old and the revised ones (Clette & Lefèvre 2018). For the predictions presented in this note we will make use of the new SSN time series up to December 2018. We start from the monthly smoothed one as shown on Figure 2 and apply a filter on that data (see below).

Daily magnetograms of the surface magnetic field of the Sun have been available since the 2nd half of the 20th century thanks to facilities such as Kitt Peak and Wilcox observatories or more recently from space probes such as SoHO or SDO or the ground network GONG. There are now easily accessible via the NSO web site: https://www.nso.edu.

To create the time-latitude butterfly diagram used in our assimilation and prediction pipeline, we start from synoptic maps also provided as a "by-product" on NSO website. After applying an azimuthal average on the synoptic maps for every single Carrington Rotation (CR) available in the data bases, we generate one time snapshot "slice" as a function of latitude (or sine of latitude) and stack them in time. This procedure is similar and inspired by the one used by Dr. David Hathaway to generate his blue-yellow butterfly diagram (http://solarcyclescience.com/solarcycle.html). We can generate the butterfly diagram with any source of synoptic maps data, including synchronous maps if necessary.

The magnetic observations used to make the butterfly diagram shown on Figure 3 are obtained from Kitt Peak (KPVT), SOHO (MDI), GONG and SOLIS 1-degree synoptic maps data from 1976 up to December 2018. For the most recent data we use the NISP (NSO Integrated Synoptic Program) web server and use SOLIS maps when available and replace them by MDI or GONG maps when necessary. We use Carrington Rotation synpotic maps from CR 1625 until CR 2211, hence covering the last 4 solar cycles or so.

We further processed the observations with a low-pass Butterworth filter of order 4 with the cut-off frequency $1/T$ set at $T = 10$ years for the surface radial field and project the latitudinal spatial structures into Legendre polynomial $P_\ell(\cos\theta)$, using a cut-off of $\ell_{max} = 10$. For the SSN time series we also apply a filter (using again a 4th order Butterworth filter) whose cut-off is at $T = 5$ years. The latest data point includes observations up to December 2018. We display the original and filtered solar data used in this study in Figures 2 and 3.

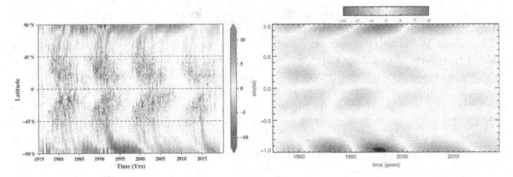

Figure 3. (Left) Full resolution butterfly diagram using Carrington Rotation synoptic maps (see text for details). The color contour plots are scaled between +/- 10 Gauss. (Right) Filtered version $B^o_{los}(\theta, t)$ used in the DA pipeline.

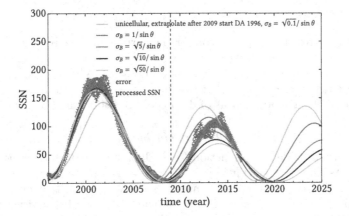

Figure 4. An example of hindcasting for cycle 23: Data assimilation of the magnetic observations for the whole cycle 23, and extrapolation of the modeled SSN based on the estimate of the average flow (recall that one of our control parameter is the meridional circulation amplitude and profile). We conduct the tests with different weighting factor x, namely, x = 0.1 (brown), 1 (red), 5 (green), 10, (black) and 50 (light blue) (see objective function $\mathcal{J}$ in eq. 2.4 for the definition of x). The 13-month mean and (low pass) filtered observations are shown with black and blue error bars, respectively.

4. Hindcasting of the previous solar cycles 22, 23 and 24

In order to validate our prediction of the next solar cycle 25, we have used our pipeline on cycles 22, 23 and 24, using real solar data from the previous cycles 21, 22, and 23 (see section 3 for details and Figures 2 and 3). This has allowed us to assess the best range of parameters in our 4D-Var assimilation procedure and the accuracy of our prediction and its associated 1-σ error bars. These parameters and error bar information are then used in the next section 5 to perform our solar cycle 25 prediction. We display in Figure 4 one realization of our *hindcasting* validation procedure on cycle 23 and provide in table 1 a summary for all three past cycles. On Figure 4, we note that the model tracks the solar data very well in the data assimilation part, here from 1996 until 2009. Various trajectories of the *Solar Predict* pipeline are being plotted against real solar data. The parameter x has been varied such that we give more or less weight to the SSN time series over the butterfly diagram. All models except the one represented with a mustard curve track are within the observation error bars. In the extrapolation part beyond 2009, for which the *Solar Predict* pipeline was run in its prediction mode as if there were no solar data available, we can assess how the various trajectories evolve in time. Since solar data for cycle 24 is mostly available, we can directly compare these extrapolated trajectories with reality. We note that our set of trajectories contain the trajectory that the Sun really evolved to after 2009. This gives us confidence that our *Solar Predict* pipeline is able to guess educate the near future of the Sun's magnetic state. We further see on table 1 that the range of the x parameter between 0.1 and 10 generally makes a good job at predicting the next solar cycle timing and amplitude. For each of the three test cycles (22, 23, 24) we are able to predict the timing and the amplitude of the cycle within the observations error bars. Hence, in the next section we will use this range of value of x to bracket the future magnetic state of the Sun.

5. Equatorially Symmetric Prediction of Solar Cycle 25

In this section we present in detail our prediction for the next solar cycle 25. We first consider the equatorially symmetric case.

5.1. *Next Minimum*

On Table 2, we provide our prediction for the current minimum between cycle 24 and 25. As we can see our model predicts that we were close to reaching the minimum in the late part of the first semester of 2019. This minima seems to have extended into fall 2019 but did not deepen further. Recall that this prediction used data up to December 2018, not November 2019, 11 more months of data could of course modify our prediction.

5.2. *Next Maximum*

On Table 3, we provide our prediction (as of Janauary 2019) for the next solar cycle 25 maximum. As we can see our model predicts that we will reach the next maximum in the middle of 2024 between the first trimester and the 3rd and that cycle 25 should be comparable in amplitude with cycle 24. We provide 1-σ error bar and hence our timing for the next cycle seems quite precise with only a 6 months window. This may be a bit too optimistic with respect to the real precision of our pipeline and future work is needed to assess further the precision and accuracy of our solar cycle prediction.

Our SSN proxy predicts mean values for the sunspot number ranging between 88 and 97. If we further take into account ensemble forecasting (based on stochastic perturbations of the trajectories) and systematic error bars, we predict a sunspot number for cycle 25 as low as 82 or as high as 103.

We also display on Figure 5 our various realizations and associated time series, for which we have changed the parameter x that controls the relative weighting of the SSN vs the butterfly diagram data set while assimilating the data and minimizing the objective function $\mathcal{J}$. We see that during the time period when we have data of solar cycle 24 available, the model curves follow closely the SSN time series and that even if we vary x to have less weight on the SSN data series and instead favor the butterfly diagram one, this does not make a large difference. Indeed all curves track the data well, as expected from advanced DA procedure as soon as the objective function $\mathcal{J}$ is kept small.

We also note that our DA pipeline indicates a maximum horizon of predictability of one solar cycle. Making a prediction from minimum state is empirically more favorable than near the maximum. This feature is well known in the solar forecasting community (Cameron & Schüssler 2008). By starting from a state close to the minimum of cycle 24, our horizon of predictibility for the next cycle reaches about one cycle period. However, near the maximum this horizon is shortened to less than a cycle period. The initial state and the control of the growth of error is key in performing predictions (Sanchez *et al.* 2014).

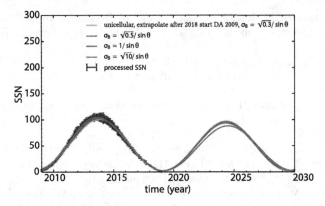

Figure 5. An example of forecasting for cycle 25: Extrapolation beyond December 2018 from the estimated average unicellular flow and magnetic fields, from the forecast based on assimilation of magnetic observations in cycle 24. The magnitude of scaling factor of the misfit in surface radial field is the parameters to be studied, with brown, green, red and violet curves for x = 0.3; 0.5, 1.0 and 10.0, respectively (see objective function $\mathcal{J}$ in eq. 2.4). The observed monthly mean SSN values (to which we applied a 5-yr filtered) are shown with blue error bars.

5.3. *Discussion of our predictions and caveats*

Since our prediction was submitted to the NOAA/NASA cycle 25 panel in early February 2019, the panel has published a consensus prediction (see for instance Weber *et al.* 2019, contribution in this conference proceedings). The NOAA/NASA consensus (version 2 as of April 2019) is: a SSN between 95 and 130 and a maximum reached between 2023 and 2026. So our prediction falls well into the consensus and we can note that our own error bars are certainly too optimistic (small). There are some caveats in analyzing the raw results of our solar cycle 25 predictions. For instance, our current dynamo model does not yet include large asymmetries between rising and declining phases of the cycle nor a time derivative of the meridional circulation state (we assume constant flow for the time being but can perform time dependent inversions as demonstrated in (Hung *et al.* 2017)). Further the Waldmeier effect is only captured by adapting the diffusivity profile at the base of the convection zone as done by (Karak & Choudhuri 2011). Such a modelling shortcoming induces biaises that are hard to quantify, and they do not appear in our uncertainty analysis, which should therefore be considered with caution, as it likely underestimates the systematic errors impacting our prediction. Our pipeline can also perform north and south hemispheres specific predictions and will report on them in a future paper.

Acknowledgments

We acknowledge financial support through IRS SpaceObs project of Paris-Saclay, SolarGeoMag project of Labex UnivEarths, the ERC Synergy grant 810218 WHOLE SUN and ERC Proof of Concept grant 640997 SolarPredict, CNES Solar Orbiter and Space Weather grants and INSU/PNST funding.

References

Brun, A. S. 2007. Towards using modern data assimilation and weather forecasting methods in solar physics. *Astronomische Nachrichten*, 328, 329

Brun, A. S. & Browning, M. K. 2017, Magnetism, Dynamo Action and the Solar-Stellar Connection, Living Reviews in Solar Physics, 14, 4

Cameron, R. & Schüssler, M. 2008 A Robust Correlation between Growth Rate and Amplitude of Solar Cycles: Consequences for Prediction Methods. *The Astrophysical Journal*, 685, 1291

Clette, F. & Lefèvre, L. 2012, Are the sunspots really vanishing?. Anomalies in solar cycle 23 and implications for long-term models and proxies. *Journal of Space Weather and Space Climate*, 2, A06

Clette, F. & Lefèvre, L. 2018, The new Sunspot Number: continuing upgrades and possible impacts, IAU Symposium, 17

Hathaway, D. H. 2015. The Solar Cycle. *Living Reviews in Solar Physics*, 12, 4

Hung, C. P., Jouve, L., Brun, A. S., Fournier, A., Talagrand, O., *et al.* 2015. Estimating the Deep Solar Meridional Circulation Using Magnetic Observations and a Dynamo Model: A Variational Approach. *The Astrophysical Journal*, 814, 151

Hung, C. P., Brun, A. S., Fournier, A., Jouve, L., Talagrand, O., Zakari, M., *et al.* 2017. Variational Estimation of the Large-scale Time-dependent Meridional Circulation in the Sun: Proofs of Concept with a Solar Mean Field Dynamo Model. *The Astrophysical Journal*, 849, 160

Jouve, L. & Brun, A. S. 2007. On the role of meridional flows in flux transport dynamo models. *Astronomy and Astrophysics*, 474, 239–250

Jouve, L., Brun, A. S., Talagrand, O., *et al.* 2011. Assimilating Data into an $\alpha\Omega$ Dynamo Model of the Sun: A Variational Approach. *The Astrophysical Journal*, 735, 31

Karak, B. B. & Choudhuri, A. R. 2011. The Waldmeier effect and the flux transport solar dynamo. *Monthly Notices of the Royal Astronomical Society*, 410, 1503–1512

Petrovay, K. 2019, Solar Cycle prediction, arXiv e-prints, arXiv:1907.02107

Sanchez, S., Fournier, A., & Aubert, J. 2014, The Predictability of Advection-dominated Flux-transport Solar Dynamo Models. *The Astrophysical Journal*, 781, 8

Talagrand, O. 2010, in Data Assimilation: Making Sense Of Observations, ed. W. Lahoz, B. Khattatov & R. Menard (Berlin: Springer)

Weber, M., Upton, L. Biesecker, D., *et al.* 2019, Solar Cycle 25 Prediction, https://www.swpc.noaa.gov/news/solar-cycle-25-forecast-update

Solar and Stellar Magnetic Fields: Origins and Manifestations
Proceedings IAU Symposium No. 354, 2019
A. Kosovichev, K. Strassmeier & M. Jardine, ed.
doi:10.1017/S174392132000071X

Global evolution of solar magnetic fields and prediction of activity cycles

Irina N. Kitiashvili🄳

NASA Ames Research Center, Moffett Field, MS 258-6, Mountain View, USA
email: irina.n.kitiashvili@nasa.gov

Abstract. Prediction of solar activity cycles is challenging because physical processes inside the Sun involve a broad range of multiscale dynamics that no model can reproduce and because the available observations are highly limited and cover mostly surface layers. Helioseismology makes it possible to probe solar dynamics in the convective zone, but variations in differential rotation and meridional circulation are currently available for only two solar activity cycles. It has been demonstrated that sunspot observations, which cover over 400 years, can be used to calibrate the Parker-Kleeorin-Ruzmaikin dynamo model, and that the Ensemble Kalman Filter (EnKF) method can be used to link the modeled magnetic fields to sunspot observations and make reliable predictions of a following activity cycle. However, for more accurate predictions, it is necessary to use actual observations of the solar magnetic fields, which are available only for the last four solar cycles. In this paper I briefly discuss the influence of the limited number of available observations on the accuracy of EnKF estimates of solar cycle parameters, the criteria to evaluate the predictions, and application of synoptic magnetograms to the prediction of solar activity.

Keywords. Sun: interior, magnetic fields, sunspots; stars: activity; methods: data analysis, statistical

---

1. Introduction

Physics-based solar activity forecasts require knowledge of subsurface and surface evolution of both large-scale flows and magnetic fields. However, at present we have access to measurements of subsurface flow and surface magnetic fields only for the last few solar cycles, which makes calibration of existing dynamo models, as well as the development of new models, very challenging. There have been many attempts to develop models and data analysis techniques to understand the nature of the global dynamics of the Sun and improve the accuracy of activity forecasts (e.g. Dikpati & Gilman 2007; Cameron & Schüssler 2007; Choudhuri *et al.* 2007; Kitiashvili & Kosovichev 2008; Jouve *et al.* 2011; Karak & Choudhuri 2013; Upton & Hathaway 2018; Jiang & Cao 2018; Macario-Rojas *et al.* 2018; Covas *et al.* 2019; Labonville *et al.* 2019). Sunspot number data, available for over 400-years, qualitatively capture variations in the toroidal magnetic field component. This property allows us to use the sunspot number as a proxy of the toroidal field and to calibrate a low-dimensional dynamo model (Kitiashvili & Kosovichev 2009).

Because of inaccuracy in both models and observations, a data assimilation approach has been used to correct model solutions according to corresponding observational data, to estimate uncertainties, to obtain an improved description of the current state of solar activity, and to build a prediction of future states. This analysis is performed using the Ensemble Kalman Filter method (Evensen 1997), which is one of the best methods for applying data assimilation to non-linear problems (Kalnay 2002). It has been applied to assimilate the available synoptic magnetic field observations into the dynamo model and to made estimates for the upcoming Cycle 25 (Kitiashvili 2020b).

2. Ensemble Kalman Filter method for solar activity prediction

The Ensemble Kalman Filter method performs a statistical analysis of possible activity states (so-called ensemble members) and allows us to estimate magnetic field evolution more accurately by taking into account uncertainties in observations as well as potential model errors. To perform the data assimilation analysis, we use a low-mode approximation (Weiss *et al.* 1984) to the mean-field Parker-Kleeorin-Ruzmaikin (PKR) dynamo model in the form of a non-linear dynamical system (Kitiashvili & Kosovichev 2009; Kitiashvili 2020b):

$$\frac{dA}{dt} = DB - A,$$
$$\frac{dB}{dt} = iA - B, \tag{2.1}$$
$$\frac{d\alpha_m}{dt} = -\nu\alpha_m - D\left[B^2 - \lambda A^2\right],$$

where A, B, α_m, and t are non-dimensional variables for poloidal field vector-potential, toroidal field strength, magnetic helicity, and time respectively; D is a non-dimensional dynamo number, $\lambda = \mathrm{Rm}^{-2}$, where Rm is an effective magnetic Reynolds number, and ν is the ratio of characteristic turbulence time-scales (Kleeorin & Ruzmaikin 1982). The reduced dynamo model describes the evolution of the mean global toroidal and poloidal field components and the magnetic helicity, for a set of parameters D, ν, and λ, and the initial conditions for the initial model state. We assume a relationship between the toroidal magnetic field and the sunspot number in the form suggested by Bracewell (1988), $W = CB^{3/2}$, where W is the sunspot number, and C is a normalization constant.

The EnKF method has been used for correcting the periodic solution of the model given by (2.1) to predict Solar Cycle 24 (SC24) using the observational data up to the preceding minimum of the solar activity (Kitiashvili & Kosovichev 2008). In addition, Kitiashvili (2016) investigated the potential for early forecast of the next solar cycle using data up to the time of the polar field reversals during the preceding cycle. However, there are several uncertainties in this method. For instance, because of the statistical approach used to estimate the global activity states, the size of the statistical ensemble can affect the resulting solutions. Figure 1 shows reconstruction of Solar Cycle 23 (SC23) starting from the preceding activity minimum and assuming different sizes of ensemble. In the case of a small ensemble size, the reconstructed sunspot number variations for Cycles 21 and 22 (blue curves) and predictions for Cycle 23 (red curves) are significantly noisier than for larger ensembles. As the size of the ensemble increases, the noise of the solutions decreases. Testing ensemble size effects for several solar cycles allowed us to conclude that, for this problem, using 300 ensemble members is optimal for describing stochastic variations relative to the mean annual sunspot numbers.

3. Influence of short series of observational data on prediction accuracy

Modern observational data (such as magnetograms) are available only for the last few cycles. Therefore, it is important to investigate the influence of short time-series on prediction accuracy. Investigation of limited observational time series was performed for solar cycles 19 – 24 using the annual sunspot number data. Initially, data assimilation was performed using observations for nine cycles, and then the observational data was sequentially removed cycle-by-cycle and the ability of the procedure to reconstruct a

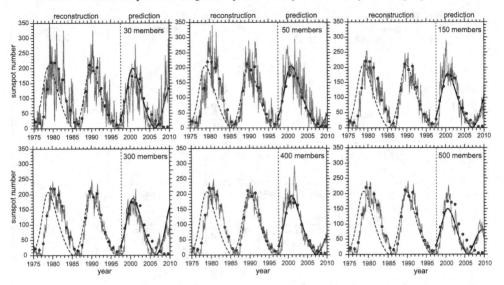

Figure 1. Test prediction of Solar Cycle 23 for 30, 50 150, 300, 400, and 500 ensemble members. The black curves show the periodic dynamo solution (dashed curves), corrected according to the available observations, to obtain new initial conditions. The exact model solution calculated using the updated initial conditions for the prediction phase is shown by black curves. Blue curves show the ensemble mean model solution corrected according to observations using the EnKF method assuming different sizes of the ensemble. Red curves show the estimated 'prediction' of SC23. Black dots show the actual observational data.

target cycle was evaluated. Figure 2 shows test predictions for cycles 19, 21, and 23, when observations of only two preceding cycles were used in the EnKF analysis, performed for 300 ensemble members. The resulting model solutions have been evaluated by using the following criteria: 1) the signs of the last available observation (for toroidal field) and the corresponding model solution should be same; 2) the exact model solution for the prediction phase must be consistent with the model solution for the reconstruction phase (no solution flattening, jumps or 'bumps', but the solution may shift according to new initial condition); 3) the corrected solution (first guess estimate) at the initial moment of time during the prediction phase should not be greater than the best-estimate variations of the toroidal field; and 4) the phase discrepancy between the exact model solution and observations should not be greater than 2 years (Kitiashvili 2020a). After the evaluation, we compared the predicted evolution with the actual observations. It was found that the first two criteria are the most important for accuracy estimation. The third criterion has a weaker correlation with the accuracy and is primarily used to make a choice among different solutions when the other criteria are satisfied. We consider this criterion only for the toroidal magnetic field because only sunspot observations are used in the analysis.

4. Application of synoptic magnetograms to predict global solar activity

4.1. *Magnetic field observations*

Our tests with short sunspot number series showed the ability of the EnKF method based on the PKR dynamo model to make, in some cases, a reliable prediction even when observations are available for only two preceding cycles. This makes it reasonable to apply the data assimilation methodology to synoptic magnetograms available from Kitt Peak Observatory (Harvey *et al.* 1980; Worden & Harvey 2000), the SOLIS instrument (Keller *et al.* 2003), and SOHO/MDI and SDO/HMI instruments (Scherrer *et al.* 1995, 2012) for

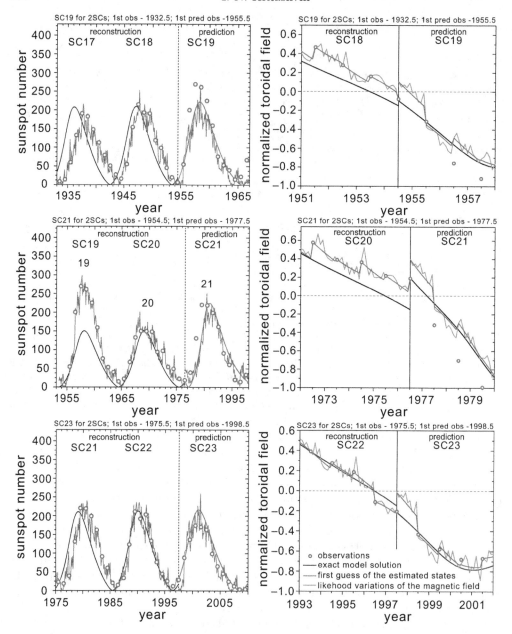

Figure 2. Predictions of Solar Cycles 19, 21, and 23 using only the annual sunspot number observations for the two preceding solar cycles. Transition between the reconstruction and prediction phases is shown for the toroidal magnetic field component in the right panels.

the last four solar cycles, i.e., from 1976 (Carrington rotation 1645) to 2019 (Carrington rotation 2216).

Decomposition of the synoptic magnetograms into toroidal and poloidal field components is challenging due to the difficulty of finding a unique solution, especially from only a line-of-sight magnetic field component. To simplify the magnetic field decomposition problem we assume that the high-latitude magnetic field (above the active latitudes)

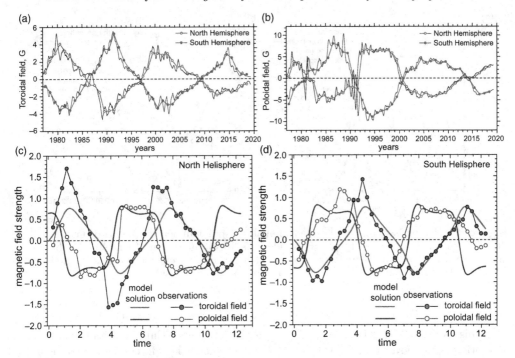

Figure 3. Temporal variations of the toroidal (panel a) and poloidal fields (panel b) in the northern (blue curves) and southern hemispheres (red curves). Panels c and d: Time-series of the annual toroidal (red dots) and poloidal (blue) field observations calibrated to the corresponding periodic dynamo solutions (thick curves) for the northern (panel a) and southern hemispheres (b). The magnetic fields and time units are non-dimensional.

characterizes the poloidal field component and that the unsigned flux in the active latitudes corresponds to the toroidal field. This assumption is acceptable for the 1D model with some level of uncertainty, because it requires estimates of the relative behavior of the field components. To account for toroidal field reversals, the sign of the estimated toroidal field is prescribed according to the Hale polarity law. Figure 3a shows the variation with time in the magnitude of the estimated toroidal field for each hemisphere. The time-series of the estimated toroidal and poloidal fields are averaged over 1-year intervals and are shown by circles for each hemisphere in Figure 3b. Thin curves show the unsmoothed variations of the field components for reference.

The resulting annual observations have been normalized to match the model periodic solutions for the toroidal and poloidal fields (Fig. 3 c, d) for each hemisphere. Normalization for the poloidal field was chosen for best agreement with the field variation amplitude. For the toroidal field, the normalization is performed relative to the last observed solar cycle, following the approach of Kitiashvili & Kosovichev (2008). Figure 3c shows an example of the toroidal component of the magnetic field calibration in the model solutions for the prediction of SC25. Traditionally, solar activity cycles are characterized by the sunspot number; the toroidal field can be converted to the sunspot number with a corresponding normalization. A comparison of the observed hemispheric sunspot number and that estimated from the synoptic magnetograms is shown in Figure 4. It is important to note that discrepancies between the observed sunspot numbers and those estimated from magnetograms can increase the uncertainty in estimates of the sunspot cycle strength.

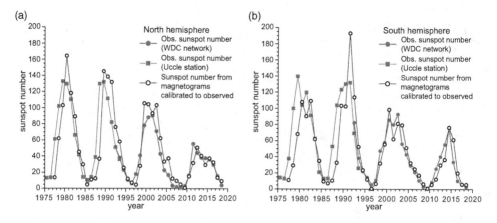

Figure 4. Comparison of the observed annual sunspot number time-series from the WDC-SILSO network (blue circles, http://www.sidc.be/silso/datafiles), and Uccle station (blue rectangles) with the calibrated annual sunspot number estimated from the synoptic magnetograms (green circles) for: a) northern hemisphere and b) southern hemisphere.

4.2. *Solar activity prediction based on the synoptic magnetograms*

In this section we examine the predictive capabilities of the EnKF method as applied to synoptic magnetograms for Solar Cycles 23 and 24 and make a forecast for the upcoming Cycle 25. The forecasting procedure has been tested for SC23 using synoptic magnetograms obtained during two previous cycles, SC21 and SC22 from 1977.5 to 1996.5. Because the synoptic observations started in the rising phase of SC21, we added two synthetic observations for 1976.5 and 1975.5 corresponding to the solar minimum between SC20 and SC21. Using the EnKF procedure, the periodic model solution is corrected according to the annual observations of the toroidal and poloidal fields for the corresponding hemispheres. The additional model variables, e.g. magnetic helicity, for which observations are not available, are generated from the model solution with an imposed 10% noise. Comparison of the model prediction with the actual toroidal field variations shows good agreement for both hemispheres up to the SC23 maximum. After the maximum, the predicted toroidal field in the northern hemisphere quickly deviates from the observed evolution. In the southern hemisphere, deviations of the predicted toroidal field become significant two years after the SC23 maximum. The predicted evolution of the poloidal field in both hemispheres quickly deviates from the actual data after the first prediction year.

The test prediction for Solar Cycle 24 was performed by using assimilation of observational data for the previous two and three solar cycles (Kitiashvili 2020b). In the three cycle case, we used the available magnetic field measurements from 1977.5 to 2008.5 to predict SC24 (Fig. 5). The predicted evolution of the toroidal field is in good agreement for both hemispheres, although there are some discrepancies during the solar maximum in the northern hemisphere. Increasing discrepancies between the predicted and observed toroidal fields during the decay phase of solar activity in the southern hemisphere are expected because of the step-like variations of the predicted toroidal field evolution (red thin curve, Fig. 5a) and the sunspot number (red thin curve, Fig. 5d). This behavior indicates accumulation of errors during the analysis and, in general, gives us a warning that the forecast quality is potentially low. This effect previously was discussed by Kitiashvili & Kosovichev (2008) for assimilation of the sunspot-number time-series. Accuracy of the poloidal field prediction (Fig. 5b) is good for up to 3 years and provides a correct prediction for the time of the polar field reversals. After this, the predicted and observed

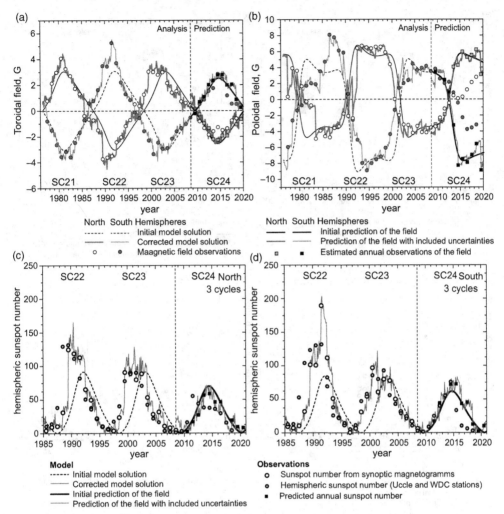

Figure 5. Predictions for the mean toroidal (panel a) and poloidal (b) fields and sunspot number variations (panels c and d) in the northern and southern hemispheres during Cycle 24. Vertical dashed lines indicate the prediction start time.

field components quickly diverge. The sunspot number estimates (Fig. 5) show good agreement with the actual data for both hemispheres. Some deviations in the shape of the predicted activity cycles are expected, and this reflects restrictions of the dynamo model formulation. The total sunspot number maximum is slightly overestimated, but in general the prediction results show a good agreement for the whole solar cycle.

To perform a prediction of upcoming Solar Cycle 25, all four solar cycles of synoptic magnetic field data from 1977 to 2019 have been used. Figure 6 shows the predictions for the toroidal (panel a) and poloidal (panel b) fields and the sunspot number (panel c and d) in both hemispheres. As expected, the forecast for the toroidal fields (and the sunspot number) is more accurate for the northern hemisphere than for the southern hemisphere because of smaller discrepancies between the model solution and observations at the end of Cycle 24. The model solutions show strong variation in the toroidal fields near and after 2026.5 (red curves, Figs 6a, c, d). These strong variations indicate that prediction uncertainties significantly increase after 2026.5. Thus, the sunspot number prediction for

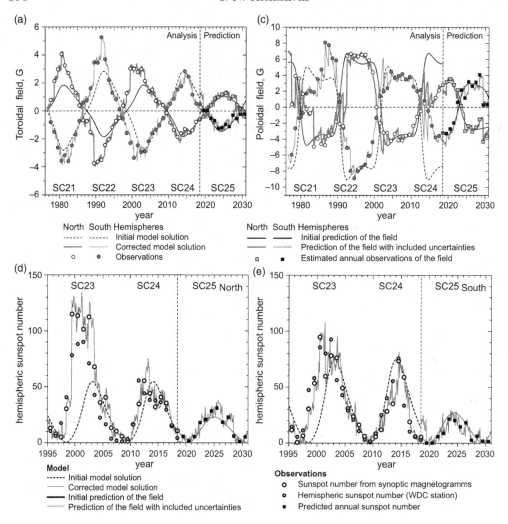

Figure 6. Predictions for Solar Cycle 25 of the mean toroidal (panel a) and poloidal (b) fields and sunspot number variations (panels c and d) in the northern and southern hemispheres based on field observations for four solar cycles. Vertical dashed lines indicate the prediction start time.

SC25 in the northern hemisphere is about 30 (that is $\sim 50\%$ weaker than SC24) with an error of $15-20\%$ and about 25 for southern hemisphere ($\sim 65\%$ weaker than SC24) with error $25-30\%$. The solar maximum is expected during $2024-2026$ in the northern hemisphere, and during $2024-2025$ in the southern hemisphere.

5. Discussion and conclusions

With very limited information about the evolution and structure of the magnetic fields and flows in the interior, reconstruction of the current state of global solar activity and prediction of future activity is a challenging task. The limited observational data restrict our ability to build accurate global models. The long series of sunspot number observations only provide a rough estimate of the global toroidal magnetic field, and, though the synoptic magnetograms carry more information about the toroidal and poloidal field evolution, they are available for only four cycles.

Nevertheless, correlations between the surface magnetic fields and the sunspot number variations allow us to test ideas of how to improve long-term solar activity predictions by developing new models and data analysis techniques and invoking data assimilation and machine learning approaches. Using the sunspot time series we were able to calibrate the low-order mean-field PKR dynamo model (Kitiashvili & Kosovichev 2009) using an approximate relationship between sunspot number and the global toroidal magnetic field strength, identify criteria to evaluate the prediction quality (Kitiashvili 2020a), and demonstrate the potential of the EnKF data assimilation method to make a reliable prediction of future solar activity using the data for only three preceding sunspot cycles. In some cases, the derived criteria gave us a warning that the prediction results may be not accurate. More work needs to be done for developing quantitative criteria for the forecasting accuracy.

Application of the Ensamble Kalman Filter method to predict solar cycle variations using synoptic magnetograms shows certain limitations because the magnetograms are available for only four activity cycles. In addition, there is no accurate procedure to uniquely decompose the line-of-sight magnetic field measurements onto poloidal and toroidal field components. Also, there is no one-to-one correspondence between the observed sunspot numbers (Fig. 4) and those estimated from magnetograms; this adds some bias in the interpretation of the prediction results. Nevertheless, the data assimilation approach combined with synoptic magnetogram data allowed us to make predictions for the next solar cycle. According to this analysis, Solar Cycle 25 will be weaker than the current cycle and will start after an extended solar minimum during 2019 – 2021. The maximum of activity will occur in 2024 – 2025 with a sunspot number at the maximum of about 50 ± 15 (for the v2.0 sunspot number series) with an error estimate of 30%. The Solar Cycle will start in the southern hemisphere in 2020 and reach maximum in 2024 with a sunspot number of ~ 28 ($\pm 10\%$). Solar activity in the northern hemisphere will be delayed for about 1 year (with an error of ± 0.5 year) and reach maximum in 2025 with a sunspot number of $\sim 23 \pm 5$ ($\pm 21\%$).

The presented results encourage future development of the data assimilation methodology for more detailed dynamo models and more complete data sets, and, in particular, development of cross-analysis of different data sources to characterize the global dynamics of the Sun.

Acknowledgment

The work is supported by NSF grant AGS-1622341.

References

Bracewell, R. N. 1988, *MNRAS*, 230, 535

Cameron, R. & Schüssler, M. 2007, *ApJ*, 659, 801

Choudhuri, A. R., Chatterjee, P., & Jiang, J. 2007, *Physical Review Letters*, 98, 131103

Covas, E., Peixinho, N., & Fernandes, J. 2019, *Sol. Phys.*, 294, 24

Dikpati, M. & Gilman, P. A. 2007, *New Journal of Physics*, 9, 297

Evensen, G. 1997, Data Assimilation: The Ensemble Kalman Filter (Springer)

Harvey, J., Gillespie, B., Miedaner, P., & Slaughter, C. 1980, NASA STI/Recon Technical Report N, 81

Jiang, J. & Cao, J. 2018, *Journal of Atmospheric and Solar-Terrestrial Physics*, 176, 34

Jouve, L., Brun, A. S., & Talagrand, O. 2011, *ApJ*, 735, 31

Kalnay, E. 2002, Atmospheric Modeling, Data Assimilation and Predictability (Cambridge University Press), 364

Karak, B. B. & Choudhuri, A. R. 2013, *Research in Astronomy and Astrophysics*, 13, 1339

Keller, C. U., Harvey, J. W., & Giampapa, M. S. 2003, in *Proc. SPIE*, Vol. 4853, Innovative
    Telescopes and Instrumentation for Solar Astrophysics, ed. S. L. Keil & S. V. Avakyan,
    194–204

Kitiashvili, I. & Kosovichev, A. G. 2008, *ApJ*, 688, L49

Kitiashvili, I. N. 2016, *ApJ*, 831, 15

Kitiashvili, I. N. 2020a, arXiV:2001.09376

Kitiashvili, I. N. 2020b, *ApJ*, 890, 36

Kitiashvili, I. N. & Kosovichev, A. G. 2009, Geophysical and Astrophysical Fluid Dynamics,
    103, 53

Kleeorin, N. I. & Ruzmaikin, A. A. 1982, *Magnetohydrodynamics*, 18, 116

Labonville, F., Charbonneau, P., & Lemerle, A. 2019, *Sol. Phys.*, 294, 82

Macario-Rojas, A., Smith, K. L., & Roberts, P. C. E. 2018, *MNRAS*, 479, 3791

Scherrer, P. H., Bogart, R. S., Bush, R. I., *et al.* 1995, *Sol. Phys.*, 162, 129

Scherrer, P. H., Schou, J., Bush, R. I., *et al.* 2012, *Sol. Phys.*, 275, 207

Upton, L. A. & Hathaway, D. H. 2018, *Geophys. Res. Lett.*, 45, 8091

Weiss, N. O., Cattaneo, F., & Jones, C. A. 1984, Geophysical and Astrophysical Fluid Dynamics,
    30, 305

Worden, J. & Harvey, J. 2000, *Sol. Phys.*, 195, 247

Solar and Stellar Magnetic Fields: Origins and Manifestations
Proceedings IAU Symposium No. 354, 2019
A. Kosovichev, K. Strassmeier & M. Jardine, ed.
doi:10.1017/S1743921320000563

Solar Open Magnetic Flux Migration Pattern over Solar Cycles

Chia-Hsien Lin[1]⬛, Guan-Han Huang[1] and Lou-Chuang Lee[2]

[1]Graduate Institute of Space Science, National Central University, Taiwan
email: chlin@jupiter.ss.ncu.edu.tw

[2]Institute of Earth Sciences, Academia Sinica, Taiwan

Abstract. The objective of this study is to investigate the solar-cycle variation of the areas of solar open magnetic flux regions at different latitudes. The data used in this study are the radial-field synoptic maps from Wilcox Solar Observatory from May 1970 to December 2014, which covers 3.5 solar cycles. Our results reveal a pole-to-pole trans-equatorial migration pattern for both inward and outward open magnetic fluxes. The pattern consists of the open flux regions migrating across the equator, the regions generated at low latitude and migrating poleward, and the regions locally generated at polar regions. The results also indicate the destruction of open flux regions during the migration from pole to equator, and at low latitude regions. The results have been published in Scientific Reports (Huang *et al.* 2017)

Keywords. Sun: activity, Sun: corona, Sun: magnetic fields

---

1. Introduction

How and why the solar magnetic fields change polarity every 11 years is still not fully understood. Coronal holes are the regions with "open" magnetic fields, which are the fields with field lines extending far away from the Sun. Such fields are the largest-scale global magnetic fields of the Sun. Therefore, coronal holes are good tracers for the change of global solar magnetic field, and many studies have examined the variations of different properties of the coronal holes (e.g., Obridko & Shelting 1999, Bilenko 2002, Hess Webber *et al.* 2014, Karna *et al.* 2014, Karachik *et al.* 2010, Bilenko & Tavastshema 2016). In this study, we identify the coronal holes as the open magnetic flux (OMF) regions, and examine the temporal and spatial variations of the area and magnetic polarities of the open magnetic flux regions over three and half solar cycles from 1976 to 2014.

2. Identification of coronal holes

We use the radial-field synoptic maps from the Wilcox Solar Observatory (WSO) from May 1976 to December 2014, corresponding to Carrington rotation number 1642 to 2158, for this work. To identify the regions with open magnetic fields, we first applied the Potential Field Source Surface model (Schatten *et al.* 1969) to construct the three-dimensional magnetic field between the solar surface and an upper boundary (source surface), where all field lines are assumed to have become radial. The source surface is placed at $2.5R_\odot$ from the solar center, following earlier studies (Obridko & Shelting 1999, Wang & Sheeley 1990). Next, the magnetic field lines are traced from the source surface to the solar surface, and the footpoints of open field lines are identified as the OMF regions.

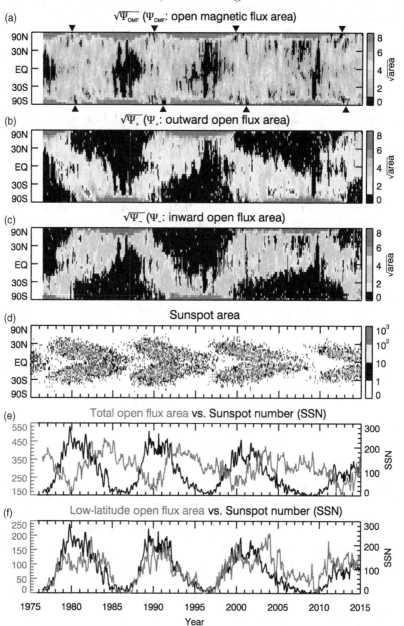

Figure 1. The upper three panels are the time maps of (a) unsigned OMF; (b) outward OMF; (c) inward OMF. Panel (d) shows the time map of sunspot areas. Panel (e) and (f) compare the total sunspot number (black line) with the total open flux area and the low-latitude open flux area, respectively. (adapted from Huang *et al.* 2017)

3. Solar cycle variation of coronal holes

The time maps of outward (Ψ_+), inward (Ψ_-), and unsigned OMF ($\Psi_{OMF} = \Psi_+ + \Psi_-$) areas are plotted in the upper three panels in Fig 1 (adapted from Huang *et al.* 2017). As a comparison, we also constructed the time map of sunspot area by using the data from the Royal Greenwich Observatory. The sunspot area map is placed in Fig 1(d).

Figure 1(a) shows that the variation of OMF area is approximately symmetric at the two poles. Fig. 1(b) and (c) reveals that the outward and inward fluxes are mostly concentrated in the opposite polar regions during the quiet period, spread to lower latitude during the rising phase of solar activity, cross the equator around the solar maximum, and reach the opposite pole during the decreasing phase of sunspot number, leading to the polarity reversal in the two hemispheres. The average migration rate estimated from the plots is $\approx 10.3 \pm 2.5$m s^{-1}, slightly slower than the surface poleward meridional flow speed ($\approx 15 - 20$ m s^{-1}). The temporal variations of the OMF areas are compared with that of the sunspot number in panels (e) and (f). The comparison shows that the total OMF area is negatively correlated with the sunspot number while the low-latitude OMF area is positively correlated with the sunspot number.

The cause for the pole-to-pole trans-equatorial (PPTE) pattern of the OMF regions is investigated by comparing the total areas within different latitude ranges during the evolution of a solar cycle. The analysis indicates that the PPTE pattern consists of four components: (1) majority of the polar open fluxes are locally generated in the polar regions; (2) some open flux regions are locally generated and dissipated without migrating to higher latitudes; (3) some open flux regions migrate across the equator; (4) some open flux regions migrate to higher latitudes.

References

Bilenko, I. A. 2002, *Astronomy and Astrophysics*, 396, 657

Bilenko, I. A. & Tavastsherna, K. S. 2016 *Solar Physics*, 291, 2329

Hess Webber, S. A., Karna, N., Pesnell, W. D. & Kirk, M. S. 2014, *Solar Physics*, 289, 4047

Huang, G.-H., Lin, C.-H., & Lee, L. C. 2017, *NatSR*, 7, 9488

Karna, N., HessWebber, S. A. & Pesnell,W. D 2014, *Solar Physics*, 289, 3381

Karachik, N. V., Pevtsov, A. A. & Abramenko, V. I. 2010, *ApJ*, 714, 1672

Obridko, V. N. & Shelting, B. D. 1999 *Solar Physics*, 187, 185

Schatten, K. H., Wilcox, J. M. & Ness, N. F. 1969, *Solar Physics*, 6, 442

Wang, Y. -M. & Sheeley, N. R., Jr. 1990 *ApJ*, 365, 372

Solar and Stellar Magnetic Fields: Origins and Manifestations
Proceedings IAU Symposium No. 354, 2019
A. Kosovichev, K. Strassmeier & M. Jardine, ed.
doi:10.1017/S1743921320000757

Probing solar-cycle variations of magnetic fields in the convection zone using meridional flows

Chia-Hsien Lin[1]📷 and Dean-Yi Chou[2]

[1]Graduate Institute of Space Science, National Central University, Taiwan
email: chlin@jupiter.ss.ncu.edu.tw

[2]Department of Physics, National Tsing-Hua University, Taiwan
email: chou@phys.nthu.edu.tw

Abstract. Solar magnetic fields are believed to originate from the base of convection zone. However, it has been difficult to obtain convincing observational evidence of the magnetic fields in the deep convection zone. The goal of this study is to investigate whether solar meridional flows can be used to detect the magnetic-field effects. Meridional flows are axisymmetric flows on the meridional plane. Our result shows that the flow pattern in the entire convection zone changes significantly from solar minimum to maximum. The changes all centered around active latitudes, suggesting that the magnetic fields are responsible for the changes. The results indicate that the meridional flow can be used to detect the effects of magnetic field in the deep convection zone.

The results have been published in the Astrophysical Journal (Lin & Chou 2018).

Keywords. Sun: helioseismology, Sun: magnetic fields, Sun: oscillations

1. Introduction

The solar magnetic fields are the main driver for most of observed solar activities and phenomena. However, how and where they are generated are still not fully understood. While it is generally accepted that they are generated by a dynamo mechanism at the base of the convection zone and brought up by magnetic buoyancy, there has been no unambiguous observational evidence for the existence of magnetic field in the deep convection zone. It is because the effects of the magnetic fields on the properties of waves are too weak to be distinguished from noise by current observation and analysis capability.

Solar meridional flows are axisymmetric flows on the meridional planes, and penetrate the entire convection zone. They play an important role in transporting magnetic flux and energy, and can, in turn, be affected by the magnetic fields. Liang & Chou (2015) applied the time-distance analysis method (see, e.g., Kosovichev 1996, Giles 1999, Zhao & Kosovichev 2004) to examine the solar-cycle variation of the travel-time difference of waves due to the meridional flow in the convection zone. Their results show that the pattern of the travel-time difference changes with solar cycle. This indicates that the meridional flows are sensitive to the variation in the solar magnetic activity. In this study, we apply a helioseismic inversion procedure to the travel-time difference data from Liang & Chou (2015) to infer the meridional flow patterns during the solar minimum and maximum, and examine whether the difference in the patterns are related to the magnetic fields.

2. Travel-time difference Data

Liang & Chou (2015) used the full-disk Doppler images taken by Michelson Doppler Imager (MDI) on board Solar and Heliospheric Observatory (SOHO) spacecraft (Scherrer *et al.* 1995) to measure the travel-time difference $\delta\tau$ of different travel distances Δ_n, at different latitudes L_m, and at different times t from May 1996 to November 2010, which includes two solar minima and one maximum. To reduce the error caused by telescope pointing, they kept only the anti-symmetric component of $\delta\tau$ relative to the equator. Their results indicate that the patterns of the travel-time difference of the two minima are similar, but are significantly different from that of the maximum.

In this study, we averaged their measured $\delta\tau(L_m, \Delta_n, t)$ over the two minimum periods (May 1996 to December 1997, and January 2008 to December 2009) to represent the travel-time difference of the solar minimum, $\delta\tau^{(\mathrm{min})}(L_m, \Delta_n)$, and averaged the travel-time difference over the one maximum period (January 2000 to December 2001) to represent the travel-time difference for the solar maximum, $\delta\tau^{(\mathrm{max})}(L_m, \Delta_n)$. The travel-time difference data used for this study are shown in Fig. 1 (adapted from Lin & Chou 2018). The panels (A), (C), and (E) are the patterns of $\delta\tau^{(\mathrm{min})}$, $\delta\tau^{(\mathrm{max})}$, and the difference between the two $\delta\tau^{(\mathrm{diff})} = \delta\tau^{(\mathrm{max})} - \delta\tau^{(\mathrm{min})}$. Positive $\delta\tau$ corresponds to northward flow, and vice versa. Their respective errors are shown in panels (B), (D), and (F). The exact ranges of the travel distances and the latitude locations used in our study to estimate the meridional flow are 89 travel distances in the range $7.2° \leqslant \Delta_n \leqslant 60°$, and 79 latitudes in the range $-39° \leqslant L_m \leqslant 39°$ for the shallowest layer.

3. Inversion methods

In this study, the travel-time difference is related to the meridional flow by the ray path approximation, which assumes that the propagation of acoustic waves can be represented by their ray paths. Although the measured $\delta\tau$ consists of the contributions from the horizontal and radial components of the meridional flow, Giles (1999) showed that the contribution from the radial flow component is much smaller than the contribution from the horizontal flow component. Therefore, we neglected the radial flow contribution to $\delta\tau$ (Kosovichev 1996, Giles 1999, Kosovichev *et al.* 2000):

$$\delta\tau(L_m, \Delta_n) = -2 \int_{\Gamma(L_m, \Delta_n)} \frac{\mathbf{U} \cdot \mathbf{n}}{c^2} \, ds \tag{3.1}$$

$$\approx \int_{\Gamma(L_m, \Delta_n)} \frac{V_{\mathrm{gh}}}{c^2 V_{\mathrm{gr}}} U_{\mathrm{h}} \, ds \,, \tag{3.2}$$

$$\approx \sum_{i,j} K(L_m, \Delta_n; r_i, \theta_j) U_{\mathrm{h}}(r_i, \theta_j) \,, \tag{3.3}$$

where c is the sound speed, $\mathbf{U}$ is the flow velocity, $\mathbf{n}$ is the unit vector along the ray path Γ, which is specified by L_m and Δ_n, U_h is the horizontal (latitudinal) component of the flow velocity, and V_{gh} and V_{gr} are the horizontal and radial components of the group velocity of the acoustic wave, respectively. The last equation is the discretized form of the integration, where r_i and θ_j are the radial and latitudinal coordinates of the grid points, and K is called the sensitivity kernel and can be computed from the standard solar model of Christensen-Dalsgaard *et al.* (1996).

To determine U_h, we implemented the Subtractive Optimally Localized Averages (SOLA) inversion method (Pijpers & Thompson 1994). The basic idea of SOLA is to superpose the sensitivity kernels corresponding to different ray paths to form a localized averaging kernel around a target point $(r_{i'}, \theta_{j'})$. The weighting coefficients C_{mn} of the

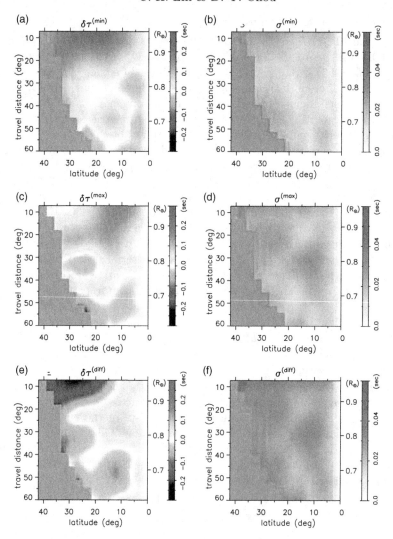

Figure 1. Measured travel-time difference at the solar minimum (panel A), solar maximum (panel C), and their difference (panel E). The three right panels are the corresponding measurement errors. The left- and right-side axis labels of each panel are the travel distance and the corresponding radius of the lower turning point, respectively.

superposition is determined by minimizing the difference between the averaging kernel and a target function which peaks at the target point and small else where:

$$\sum_{mn} C_{mn}^{i'j'} \delta\tau(L_m, \Delta_n) = \sum_{mn}\sum_{ij} C_{mn}^{i'j'} K(L_m, \Delta_n; r_i, \theta_j) U_{\mathrm{h}}(r_i, \theta_j)$$

$$+ \sum_{mn} C_{mn}^{i'j'} \sigma_{mn} \tag{3.4}$$

$$\approx \sum_{ij} \bar{K}^{i'j'}(r_i, \theta_j) U_h(r_i, \theta_j) \tag{3.5}$$

$$\approx \sum_{ij} T^{i'j'}(r_i, \theta_j) U_h(r_i, \theta_j) \tag{3.6}$$

$$\approx \langle U_h \rangle_{i'j'} \tag{3.7}$$

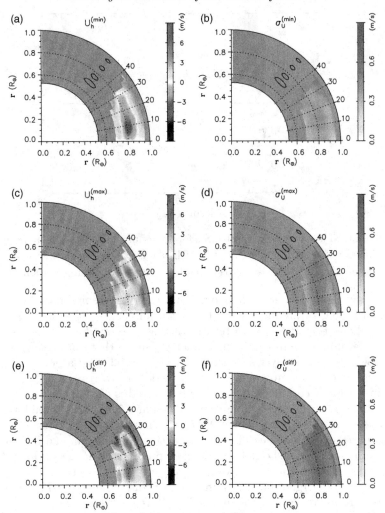

Figure 2. Estimated horizontal component of the meridional flow at the minimum (panel A), maximum (panel C), and their difference (panel E). The three right panels show the corresponding errors. The contours represent the half-maximum widths of the averaging kernels at four different depths.

where the superscripts i' and j' correspond to the subscripts of target point $(r_{i'}, \theta_{j'})$, σ_{mn} is the measurement error of travel-time difference, $\bar{K}^{i'j'}$ and $T^{i'j'}$ are the averaging kernel and the target function of the target point (i', j'), respectively, and $\langle U_h \rangle_{i'j'}$ is an estimate of U_h from the inversion with an estimated error of $\sqrt{\sum_{mn}(C_{mn}^{i'j'}\sigma_{mn})^2}$. To compute $\{C_{mn}^{i'j'}\}$, we applied the Singular Value Decomposition (SVD) method (Press *et al.* 1992).

4. Results and discussion

The inversion results using measured $\delta\tau(L_m, \Delta_n)$ are shown in Fig. 2 (adapted from Lin & Chou 2018). Panels (A), (C) and (E) show the horizontal flow speed at the minimum, $U_h^{(\mathrm{min})}$, maximum, $U_h^{(\mathrm{max})}$, and the difference between the two, $U_h^{(\mathrm{diff})}$, respectively. The three right panels (B), (D) and (F) are their corresponding errors. The half-maximum

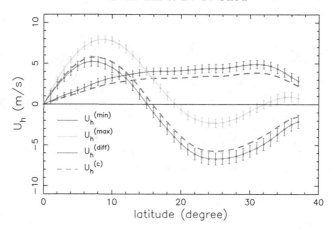

Figure 3. Estimated horizontal flow at $0.96R_\odot$ for the minimum (blue solid line), the maximum (green solid line), and their difference (red solid line). The blue dashed line is $\alpha U_h^{(min)}$ using $\alpha = 0.79$, and the red dashed line is the resulting $U_h^{(c)}$, computed from Equation (4.1).

widths of the averaging kernels at four different depths are represented by the contours in the high-latitude gray area where no data are available.

The figure shows that the flow pattern in the entire convection zone changes significantly from solar minimum to maximum. Panel (A) shows that during the minimum, the horizontal flow has a three-layer structure in $0.67 \leqslant r \leqslant 0.96R_\odot$: a poleward flow in the upper convection zone, an equator-ward in the middle convection zone, and a poleward flow again in the lower convection zone. The flow speed is close to zero within the error bar around the base of the convection zone, located at $0.7R_\odot$. This three-layer structure is similar to the result of Chen & Zhao (2017), which uses the HMI data in the period of 2010.05 – 2017.04, cycle 24. The difference between the maximum and minimum can be seen clearly in panel (E). It shows that the most prominent changes from the minimum to maximum flow are the appearance of a convergent flow above $0.9R_\odot$, another around $0.8R_\odot$, and a weak divergent flow near the base of convection zone. The signal-to-noise ratio of this weak divergent flow is about 3–4. It is interesting to note that all these changes are centered around the active latitudes $\approx 15 - 17°$.

To compare our results with the direct measurements of the surface meridional flow by previous studies, we plotted the the result of our shallowest layer ($0.96R_\odot$) in Fig. 3 (adapted from Lin & Chou 2018). The latitudinal dependence of $U_h^{(min)}$ (blue solid line) is similar to the typical sine-shape distribution, but $U_h^{(max)}$ (green solid line) is significantly different from $U_h^{(min)}$. The difference $U_h^{(diff)}$ (red solid line) changes sign at the active latitude $\approx 15°$ (Liang & Chou 2015). The pattern of $U_h^{(diff)}$ indicates that at the solar maximum, an additional convergent flow toward the active latitudes is generated relative to the flow at the solar minimum. To approximate the pattern of $U_h^{(max)}$, we can combine a reduced $U_h^{(min)}$ and a convergent flow:

$$U_h^{(max)}(L) = \alpha \, U_h^{(min)}(L) + U_h^{(c)}(L), \qquad (4.1)$$

where α is the suppression coefficient describing the reduction in flow magnitude, and $U_h^{(c)}(L)$ is the the convergent flow. With the simplification assumptions that α is a constant and the peaks of $|U_h^{(c)}(L)|$ above and below the active latitude are equal, we determined the value of α to be 0.79 using measured $U_h^{(min)}(L)$ and $U_h^{(max)}(L)$. The blue dashed line in Fig. 3 represents $\alpha U_h^{(min)}(L)$ using $\alpha = 0.79$, and the resulting $U_h^{(c)}(L)$ is plotted as the red dashed line. $U_h^{(c)}(L)$ is approximately anti-symmetric with respect to latitude 16°, with a peak speed of 5.8 m s^{-1} at about 7° and 25°.

Earlier study by Hathaway & Rightmire (2010) reported a reduction of about 30% in flow speed at the maximum on the surface. In comparison, our analysis suggests a reduction of about 21% at $0.96R_\odot$. The convergent flow around the active latitudes at the maximum has also been reported in previous studies. Hathaway & Rightmire (2011) reported a convergent flow of about several m s^{-1} on the surface at the maximum. The time-distance analysis by Zhao & Kosovichev (2004) reported a magnitude of the convergent flow about $2 - 8$ m s^{-1} at $0.987R_\odot - 0.996R_\odot$. The ring-diagram analysis by Haber *et al.* (2002) obtained a magnitude of about several m s^{-1} at $0.99R_\odot$. In our study, the magnitude of the convergent flow is 5.8 m s^{-1} at $0.96R_\odot$.

5. Conclusion

The objective of this study is to investigate the solar-cycle variation of the meridional flow and whether the variation can be used to probe the magnetic fields.

The results show that the horizontal flow during minimum has three layers: a poleward flow in the upper and lower convection zon, and an equator-ward flow in the middle convection zone. The flow changes significantly from the minimum to maximum. The main differences are the reduction in flow magnitude and the appearance of two convergent flows and a weak divergent flow at the maximum. The convergent flows and the divergent flow are all centered around the active latitudes, suggesting that the changes are related to the magnetic fields. The results indicate that the solar-cycle variation of the meridional flow can be used to probe the magnetic fields in the deep convection zone.

References

Chen, R. & Zhao, J. 2017, *ApJ*, 849, 144

Christensen-Dalsgaard, J., Dappen, W., Ajukov, S. V., *et al.* 1996, *Science*, 272, 1286

Giles, P. M. 1999, *PhD thesis, Stanford Univ.*

Haber, D. A., Hindman, B. W., Toomre, J., Bogart, R., & Larsen, R. 2002 *ApJ*, 570, 855

Hathaway, D. H. & Rightmire, L. 2010, *Science*, 327, 1350

Hathaway, D. H. & Rightmire, L. 2011, *ApJ*, 729, 80

Kosovichev, A. G. 1996, *ApJL*, 461, L55

Kosovichev, A. G., Duvall, T. J., Jr., & Scherrer, P. H. 2000 *SoPh*, 192, 159

Liang, Z.-C. & Chou, D.-Y. 2015, *ApJ*, 809, 150

Lin, C.-H. & Chou, D.-Y. 2018, *ApJ*, 860, 48

Pijpers, F. P. & Thompson, M. J. 1994, *A&A*, 281, 231

Press, W. H., Teukolsky, S. A., Vetterling, W. T., & Flannery, B. P. 1992, *Numerical Recipes in C (2nd eg.): The Art of Scientific Computing (New York: Cambridge Univ. Press)*

Scherrer, P. H., Bogart, R. S., Bush, R. I., *et al.* 1995, *SoPh*, 162, 129

Zhao, J. & Kosovichev, A. G. 2004, *ApJ*, 603, 776

Discussion

ALEXANDER KOSOVICHEV ASKED: Did you find a N-S asymmetry of the meridional flow?

CHIA-HSIEN LIN REPLIED: There is no asymmetric information in the data because only the anti-symmetric component is kept.

IRINA KITIASHVILI ASKED: How does meridional flow speed vary with solar cycle?

CHIA-HSIEN LIN REPLIED: The meridional flow has a 3-layer pattern during minimum, and becomes more complicated during the maximum.

Chapter 4. Stellar rotation and magnetism

Lauren Doyle

Solar and Stellar Magnetic Fields: Origins and Manifestations
Proceedings IAU Symposium No. 354, 2019
A. Kosovichev, K. Strassmeier & M. Jardine, ed.
doi:10.1017/S1743921320001404

Magnetic field evolution in solar-type stars

Axel Brandenburg[1,2,3,4,5]

[1]Nordita, KTH Royal Institute of Technology and Stockholm University,
Roslagstullsbacken 23, SE-10691 Stockholm, Sweden

[2]Department of Astronomy, Stockholm University, SE-10691 Stockholm, Sweden

[3]JILA and Laboratory for Atmospheric and Space Physics, Univ. Colorado, Boulder, USA

[4]McWilliams Center for Cosmology, Carnegie Mellon University, Pittsburgh, PA 15213, USA

[5]Faculty of Natural Sciences and Medicine, Ilia State University, 0194 Tbilisi, Georgia
email: brandenb@nordita.org

Abstract. We discuss selected aspects regarding the magnetic field evolution of solar-type stars. Most of the stars with activity cycles are in the range where the normalized chromospheric Calcium emission increases linearly with the inverse Rossby number. For Rossby numbers below about a quarter of the solar value, the activity saturates and no cycles have been found. For Rossby numbers above the solar value, again no activity cycles have been found, but now the activity goes up again for a major fraction of the stars. Rapidly rotating stars show nonaxisymmetric large-scale magnetic fields, but there is disagreement between models and observations regarding the actual value of the Rossby number where this happens. We also discuss the prospects of detecting the sign of magnetic helicity using various linear polarization techniques both at the stellar surface using the parity-odd contribution to linear polarization and above the surface using Faraday rotation.

Keywords. (magnetohydrodynamics:) MHD, turbulence, techniques: polarimetric, Sun: magnetic fields, stars: magnetic fields

1. Introduction

The purpose of this paper is to discuss recent results relevant for understanding the connection between observations and simulations of magnetic fields in solar-like stars. We focus on observational measures of stellar activity, the occurrence of activity cycles in other stars, and new ways of interpreting stellar surface magnetic fields in terms of linear polarization measurements. We also address the possibility of a radial sign reversal of the star's magnetic helicity some distance above the stellar surface.

The Sun's magnetic field exhibits remarkable regularity in space and time, as is demonstrated by Maunder's butterfly diagram (Maunder 1904). Although comparable diagrams are only now beginning to become possible for other stars (see, e.g., Alvarado-Gómez *et al.* 2018), there is clear evidence that cyclic chromospheric variability is ubiquitous. This became clear after Wilson (1963, 1978) selected a set of stars that were then monitored for the next three decades at Mount Wilson (Baliunas *et al.* 1995).

Much of our knowledge on stellar magnetic activity comes from understanding the Sun's magnetic field. The occurrence of a fairly regular 11-year activity cycle is of course one of its main characteristics. A cycle as clear as that of the Sun has not been observed for any of the other stars monitored so far. The perhaps best observed cycle is that of HD 81809, but that star is a binary and the cyclic component is not a main sequence star, but a subgiant; see Egeland (2018) for a detailed discussion of this interpretation. The lack of equally clear cycles makes one wonder whether the Sun is perhaps a special case. Other evidence in favor of such thinking is the fact that in a diagram of cycle period versus rotation period, the Sun lies right between two different branches, the high and low activity branches (Böhm-Vitense 2007). There are a few other aspects suggesting that the Sun is special. Some of them are discussed below.

In the earlier work of Brandenburg *et al.* (1998), which also showed two distinct activity branches, the Sun appeared closer to the inactive branch and not really between two branches. One reason why the Sun was closer to the inactive branch was the fact that they used the 10 year cycle period obtained by Baliunas *et al.* (1995) for the time interval for which their solar $\langle R'_{\rm HK} \rangle$ data were determined from nightly moonlight observations. Using instead the 11 year cycle period determined for the full record since the time of Schwabe (1844) and before (since the time of the Maunder minimum, as was established by Eddy 1976) yields a position of the Sun that is now further away from the inactive branch (Brandenburg *et al.* 2017), although it is still not really between the two branches. Regarding the clear cyclicity, it should also be kept in mind that the Sun is relatively old compared with many of the stars for which cycles have been obtained. It is therefore possible that there is some sort of selection effect for why only the Sun has such a well defined cycle.

We begin by discussing first the relation between rotation period and stellar activity. Also this relation suggests that the Sun's location in that diagram is between two different modes of behavior, which is when the activity attains a minimum as a function of rotation periods. Both for faster and for slower rotation, the activity increases relative to that of the Sun, at least approximately; see the work Brandenburg & Giampapa (2018), who offered an interpretation in terms of the stellar differential rotation changing from solar-like to antisolar-like differential rotation right at the Sun's rotation rate. So, again, the Sun appears to take a special position among the many other stars. We finish by discussing new ideas for determining solar and stellar magnetic helicity from linear polarization measurements. This technique, however, has so far only been applied to the Sun.

2. Activity versus rotation

Stellar variability is usually characterized by the chromospheric Ca II H+K line emission, normalized by the bolometric flux and corrected for photospheric contributions to give $R'_{\rm HK}$. This quantity is believed to be a good measure of the mean magnetic field normalized by the equipartition field strength, which is defined based on the kinetic energy as $B_{\rm eq} = \sqrt{4\pi\rho}u_{\rm rms}$, where $u_{\rm rms}$ is the rms value of the turbulent velocity and ρ is the local gas density. Schrijver *et al.* (1989) estimated that $R'_{\rm HK}$ is related to the rms magnetic field strength, $B_{\rm rms}$, through

$$R'_{\rm HK} \propto (B_{\rm rms}/B_{\rm eq})^{0.5}, \tag{2.1}$$

where $B_{\rm rms}$ is expressed in terms of a filling factor f and the typical field strength in spots $B_{\rm spot}$, which can be estimated by the photospheric pressure $p_{\rm phot}$ as $B_{\rm spot} = \sqrt{8\pi p_{\rm phot}}$. Saar & Linsky (1985) thus proposed

$$B_{\rm rms} = f\, B_{\rm spot}. \tag{2.2}$$

As stars become more active, f increases up to the point when the entire stellar surface is covered with spots, leading to saturation as f cannot increase beyond unity ($f \leq 1$ by definition).

Stellar rotation with angular velocity Ω affects the dynamo process through the Coriolis force, $2\Omega \times u$, and its strength is characterized by the Coriolis number, $\mathrm{Co} = 2\Omega\tau$, where τ is some measure of the turnover time. One often quotes the Rossby number, $\mathrm{Ro} = P_{\rm rot}/\tau$, where $P_{\rm rot} = 2\pi/\Omega$ is the rotation period. The two parameters are then related to each other through $\mathrm{Ro} = 4\pi/\mathrm{Co}$, although Brandenburg *et al.* (1998) defined the Rossby

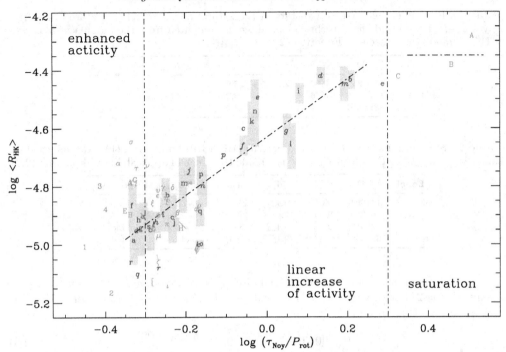

Figure 1. Activity versus Rossby number, showing saturation for $\log(\tau_{\rm Noy}/P_{\rm rot}) > 0.3$ (see Table 1) and enhanced activity for many stars in the range $\log(\tau_{\rm Noy}/P_{\rm rot}) < -0.3$ (see also Table 2).

number simply as Co^{-1}. There are also other reasons why a statement about the Rossby number should be taken with care.

To know the actual values of Ro or Co, one has to agree on a good definition for τ. From an observational point of view, all that matters is that τ is a monotonically decreasing function of stellar mass, or, in practice, a monotonically increasing function of the color $B - V$. But even then there can be significant uncertainties. In the relevant color range of $B - V$ between 0.6 and 0.75, the turnover times of Barnes & Kim (2010) are about 2.4 times longer than those of Noyes *et al.* (1984a). In the work of Brandenburg & Giampapa (2018), for example, the turnover times of Noyes *et al.* (1984a) were adopted. For definitiveness, we refer to those times as $\tau_{\rm Noy}$. The original calculation of $\tau_{\rm Noy}$ was based upon stellar mixing length models of Gilman (1980), where $\tau_{\rm Noy} = \ell/u_{\rm rms}$ was based on the local mixing length ℓ and the rms velocity $u_{\rm rms}$ about one scale height above the bottom of the convection zone.

It has been known for some time that $R'_{\rm HK}$ scales with Co (Noyes *et al.* 1984a; Vilhu 1984) and is therefore inversely proportional to Ro, until saturation is reached for rapid rotation. As alluded to above, saturation is often interpreted in terms of the filling factor of the magnetic field on the stellar surface approaching unity. However, none of the stars with cycles are anywhere near saturation, so they obey $\langle R'_{\rm HK}\rangle \propto \mathrm{Co}$, but there is some leveling off for non-cyclic stars with $\log(\tau_{\rm Noy}/P_{\rm rot}) > 0.3$. This is shown in Figure 1, where we have combined data from the three stars of Table 1 (see the orange labels A–C for HD 17925, 131156A, and 131156B, respectively) with data from Brandenburg *et al.* (2017), red characters for K dwarfs and blue ones for G and F dwarfs, and Brandenburg & Giampapa (2018), green symbols, for the stars of the open cluster M 67 with rotation periods estimated from gyrochronology based on an estimated age of 4 Gyr; see Brandenburg & Giampapa (2018) for details. The orange numbers 1–4 for small values of

Table 1. Parameters of three active stars in the saturation range. The data are taken from Noyes *et al.* (1984a) and the gyrochronological age is computed from the formula of Mamajek & Hillenbrand (2008); see also Equation (2.3).

Label	HD	B−V	τ [d]	P_{rot} [d]	$\log\langle R'_{HK}\rangle$	Age [Gyr]
A	17925	0.87	21.3	6.6	−4.28	0.2
B	131156A	0.76	17.8	6.2	−4.38	0.3
C	131156B	1.17	24.3	11.5	−4.42	0.5

Table 2. Stars in the range of enhanced activity. HD 187013 and HD 224930 do indeed show enhanced activity. For HD 187013, $B - V$ is outside the range of applicability of Equation (2.3).

Label	HD	B−V	τ [d]	P_{rot} [d]	$\log\langle R'_{HK}\rangle$	Age [Gyr]
1	141004	0.60	9.1	25.8	−5.00	5.6
2	161239	0.65	12.0	29.2	−5.16	5.5
3	187013	0.47	3.1	8.0	−4.79	old
4	224930	0.67	13.1	33.0	−4.88	6.4

$\log(\tau/P_{rot})$ are for other field stars discussed already in Brandenburg & Giampapa (2018); see also Table 2. For those stars, we have computed their ages from Equations (12)–(14) of Mamajek & Hillenbrand (2008) as

$$t = \left\{ P_{rot}/[0.407\,(B - V - 0.495)^{0.325}] \right\}^{1.767}, \tag{2.3}$$

provided $B - V > 0.495$; see also Brandenburg *et al.* (2017).

Some of the stars of M67 show increasing activity with decreasing τ_{Noy}/P_{rot} or decreasing rotation rate, which suggests that the activity rises again as they slow down further. This has been interpreted by Brandenburg & Giampapa (2018) as possible evidence for anti-solar differential rotation, which is theoretically expected for very slow rotation (Gilman 1977). The absolute differential rotation in the regime of antisolar differential rotation is known to be stronger than that in the regime of solar-like differential rotation (Gastine *et al.* 2014; Käpylä *et al.* 2014), and also leads to larger magnetic activity for stars with smaller τ_{Noy}/P_{rot} (Karak *et al.* 2015), thus explaining their enhanced activity.

However, when comparing with simulations, the situation is still somewhat puzzling. In fact, numerical simulations of convectively driven dynamo action show that there is another transition, where the large-scale magnetic field becomes predominately nonaxisymmetric with an $m = 1$ azimuthal modulation (Viviani *et al.* 2018). Present models show that these two transitions happen more or less at the same value of Co, but observations shows that the two transitions happen at different values: anti-solar differential rotation for Ro ≳ 2.0 (Brandenburg & Giampapa 2018) and nonaxisymmetric large-scale field for Ro ≲ 0.5...1.0; see Table 5 of Viviani *et al.* (2018), who refer to data of Lehtinen *et al.* (2016). In fact, recent work of Lehtinen *et al.* (2020) suggests that the transition from nonaxisymmetric to axisymmetric magnetic fields might be accompanied a change in the slope of the chromospheric activity versus Rossby number diagram, which becomes particularly clear when they combine data of main sequence stars with those of subgiants and giants. The lack of corresponding features in data from numerical simulations is an important shortcoming of current simulations. It could be that some sort of "renormalization" is required when comparing numerical simulations with observational data. Indeed, simulations are long known to yield cyclic behavior only for rotation rates that exceed the solar value by about a factor of three (Brown *et al.* 2011).

Not all stars necessarily slow down with age. Their Sun-like activity cycles may just disappear, but they would still spin rapidly (Metcalfe & van Saders 2017). If their magnetic field topology develops predominantly small scales, as has now been demonstrated

Table 3. Selected stars with well defined cycles using data from Brandenburg *et al.* (2017).

Label	HD	B–V	τ [d]	$P_{\rm rot}$ [d]	$\log\langle R'_{\rm HK}\rangle$	Age [Gyr]	$P_{\rm cyc}$ [yr]	comp
a (blue)	Sun	0.66	12.6	25.4	−4.90	4.6	11.0	6.6
c (blue)	10476	0.84	20.6	35.2	−4.91	4.9	9.6	9.3
f (blue)	26965	0.82	20.1	43.0	−4.87	7.2	10.1	10.9
m (blue)	160346	0.96	22.7	36.4	−4.79	4.4	7.0	8.5

by Metcalfe *et al.* (2019), magnetic breaking would become progressively inefficient. This idea emerged when discrepancies between helioseismic ages and gyrochronological ages became apparent (van Saders *et al.* 2016). Brandenburg & Giampapa (2018) speculated that this could still be reconciled with the possibility of antisolar differential rotation if there is a bifurcation into two possible scenarios: stars that make the transition to antisolar differential rotation as a result of a sufficiently chaotic evolution, and others that just change their field topology and remain rapidly spinning. Demonstrating this with actual models would clearly be a next important step.

3. Cycle frequency versus activity

A systematic dependence of cycle frequency $\omega_{\rm cyc}$ $(=2\pi/P_{\rm cyc})$ on rotation rate Ω was first found by Noyes *et al.* (1984b) based on the early analyses of the sample of Wilson (1963, 1978) measured at Mount Wilson. They found $\omega_{\rm cyc}\propto\Omega^{1.25}$. It is important to emphasize that *the exponent is larger than unity*, which has long been a theoretical difficulty to explain. An early theoretical analysis by Kleeorin *et al.* (1983) based on the fastest growing linear eigenmode yielded promising results with $\omega_{\rm cyc}\propto\Omega^{4/3}$, but very different solutions were obtained for nonlinear saturated dynamos (Tobias 1998). This was also emphasized by Brandenburg *et al.* (1998), who proposed that spatial nonlocality could be strong enough so that only solutions with the lowest wavenumber would exist. This seems to be the best explanation even today.

As already mentioned in the introduction, there are only very few stars that show well defined cycles that are nearly as clean as that of the Sun. We have listed the properties of these stars in Table 3, where we also list the cycle periods, $P_{\rm cyc}$, as well as those *computed* from the formula of Brandenburg *et al.* (2017) that assumes that the stars lie exactly on the long-period branch. The stars cover the full range in $\log(\tau_{\rm Noy}/P_{\rm rot})$ from −0.3 to 0.2; see Figure 1. The ages of those stars are in the range from 4.4 to 4.9 Gyr, except for HD 26965, which is 7.2 Gyr. This shows that all stars with well defined cycles are old stars.

Furthermore, the analysis of many cycle data by Baliunas *et al.* (1995) suggested the existence of multiple cycles. Their reality remains debated even today (Boro Saikia *et al.* 2016; Olspert *et al.* 2018). The work of Brandenburg *et al.* (1998) and Brandenburg *et al.* (2017) suggested two nearly parallel branches of values of $\omega_{\rm cyc}/\Omega$ versus $\langle R'_{\rm HK}\rangle$. The lower branch has cycle periods that are about six times longer than those on the regular (upper) branch, where also the Sun was thought to be located, if we adopted the 10 yr period; see the discussion above. The two branches were originally called active and inactive branches, because they were also well separated with respect to the vertical line $\lg\langle R'_{\rm HK}\rangle=-4.65$. Brandenburg *et al.* (2017) suggested, however, that (i) the branches are now well overlapping and that (ii) stars younger than 3 Gyr might exhibit both shorter and longer cycles simultaneously. It should be remembered that these stars tend to be rapid rotators, whose large-scale magnetic field is expected to be nonaxisymmetric (Viviani *et al.* 2018). Such a magnetic field is similar to that of a dipole lying in the equatorial plane and with opposite polarities at longitudes that are 180° away from each other.

Nonaxisymmetric magnetic fields have long been predicted based on mean-field models with an anisotropic α effect, and a tensor α_{ij} whose diagonal components do not all have the same value. Since the early work of Rüdiger (1978), it was known that at rapid rotation, α_{ij} attains an additional piece proportional to $\Omega_i\Omega_j$, so the tensor approaches the form

$$\alpha_{ij} \to \alpha_0 \left(\delta_{ij} - \Omega_i\Omega_j/\Omega^2 \right), \tag{3.1}$$

showing that the component of α in the $\mathbf{\Omega}$ direction vanishes. In other words, if the $\mathbf{\Omega}$ direction corresponds to the z direction in Cartesian coordinates, α_{ij} is proportional to diag $(\alpha_{xx}, \alpha_{yy}, 0)$. In general, there can also be off-diagonal components, but those are not important for the present discussion. The main point is that with $\alpha_{zz} = 0$, the dynamo-generated magnetic field is, in Cartesian geometry, always horizontal. In a sphere, it then corresponds to a dipole lying in the equatorial plane; see Fig. 3(a) of Moss & Brandenburg (1995).

A simple example of such a field is that generated by the Roberts flow I (the first of four flows I–IV studied by Roberts 1972). This flow takes the form

$$\mathbf{u} = \mathbf{\nabla} \times \psi\hat{\mathbf{z}} + k_\mathrm{f}\psi\hat{\mathbf{z}}, \tag{3.2}$$

which is a prototype example for modeling magnetic fields generated by an α effect. Here, the α tensor is indeed of the form of Equation (3.1) with $\alpha_{ij} = \mathrm{diag}\,(\alpha_0, \alpha_0, 0)$ and some coefficient α_0. It is also a model of the magnetic field generated in the Karlsruhe dynamo experiment, which, in turn, is an idealized model of the geodynamo (Rädler et al. 2002; Rädler & Brandenburg 2003). The Coriolis number of the geodynamo is extremely large—much larger than that of any of the observed stars. It therefore tends to exaggerate the effects of rotation.

Observationally, nonaxisymmetric magnetic fields have been inferred from light curve modeling and Doppler imaging (see, e.g., Kochukhov et al. 2017). However, we also know that stars with nonaxisymmetric magnetic field can exhibit what is known as flip-flop phenomenon (Jetsu et al. 1994). This means that the two opposite polarities alternate in strengths in a cyclic fashion.

It is generally expected that such variations correspond to a mixed parity solution of the type originally investigated by Rädler et al. (1990). Subsequent work of Moss et al. (1995) found it difficult to obtain such solutions with their more realistic simulations. Qualitative discussions have also been offered by Elstner & Korhonen (2005).

Another related question concerns the cycle period observed for stars on the active or long-period branch discussed above. Guerrero et al. (2019a) suggest that some sort of magnetic shear instability might be responsible for cycle periods comparable to those of the Sun. In this connection, there is also the question whether the Tayler instability might play a role; see Guerrero et al. (2019b). An important question concerns the surface appearance of magnetic fields from the two branches of short and long cycle periods, especially for young and rapidly rotating stars. Are they really nonaxisymmetric and how can we understand the observed occurrence of multiple cycles, i.e., the occurrence of multiple branches with the same stars on both of them? Modeling this convincingly would be a major step forward in understanding the truth behind these two branches.

4. New twists to polarimetric measurements

Zeeman Doppler Imaging (ZDI) provides a powerful tool for characterizing the actual magnetic field structure and its temporary changes. Both in solar and stellar physics, one tends to display the results directly in terms of the full magnetic field vector. Those results are in general subject to the 180° ambiguity, which is also sometimes referred to as the π ambiguity. In the solar context, this π ambiguity might be an important source of error in calculating the sign of the Sun's magnetic helicity at large length scales. It may therefore be advantageous to work directly with the Stokes parameters (Brandenburg *et al.* 2019; Brandenburg 2019; Prabhu *et al.* 2020).

Magnetic helicity is a quantity that characterizes the handedness of the magnetic field. Its sign would change if one looked at the star through a mirror. In this connection, it is important to realize that the Stokes Q and U parameters can directly be expressed in terms of a quantity that characterizes the sense of handedness. This technique is routinely employed in cosmology, where one expresses Q and U in terms of what is known as the parity even E and the parity odd B polarizations. To obtain E and B, one expands Q and U not in terms of the ordinary spherical harmonics, but in terms of spin-2 spherical harmonics, $_2Y_{\ell m}(\theta, \phi)$; see Goldberg *et al.* (1967). One then obtains E and B as the real and imaginary parts of the transformed quantity in the form (Kamionkowski *et al.* 1997; Seljak & Zaldarriaga 1997; Zaldarriaga & Seljak 1997; Durrer 2008; Kamionkowski & Kovetz 2016)

$$E + iB \equiv R = \sum_{\ell=2}^{N_\ell} \sum_{m=-\ell}^{\ell} \tilde{R}_{\ell m} Y_{\ell m}(\theta, \phi), \tag{4.1}$$

where the $\tilde{R}_{\ell m}$ are given by

$$\tilde{R}_{\ell m} = \int_{4\pi} (Q + iU) \, _2Y_{\ell m}^*(\theta, \phi) \, \sin\theta \, d\theta \, d\phi. \tag{4.2}$$

In spectral space, we then define $\tilde{E}_{\ell m} = (\tilde{R}_{\ell m} + \tilde{R}_{\ell, -m}^*)/2$ as the parity-even part and $\tilde{B}_{\ell m} = (\tilde{R}_{\ell m} - \tilde{R}_{\ell, -m}^*)/2i$ as the parity-odd part, where the asterisk means complex conjugation. This has recently been done for the Sun's magnetic field using synoptic vector magnetograms, which yield a global map. It turns out that E is indeed even about the equator and B is odd about the equator. Therefore, $\tilde{E}_{\ell m}$ has contributions mainly from even values of ℓ and $\tilde{B}_{\ell m}$ has mainly contributions from odd values of ℓ. As a useful proxy, one can therefore employ the correlators

$$K_\ell^+ = \tilde{E}_\ell \tilde{B}_{\ell+1}^* \quad \text{and} \quad K_\ell^- = \tilde{E}_\ell \tilde{B}_{\ell-1}^*, \tag{4.3}$$

respectively, for different values of ℓ. There have also been attempts to determine the handedness of magnetic fields in the solar neighborhood of the interstellar medium (Bracco *et al.* 2019).

To illustrate the decomposition further, we now discuss the corresponding Cartesian decomposition. It reads (Durrer 2008)

$$\tilde{R}(k_x, k_y) = -(\hat{k}_x - i\hat{k}_y)^2 \tilde{P}(k_x, k_y), \tag{4.4}$$

where

$$\tilde{P}(k_x, k_y) = \int P(x, y) e^{-i\mathbf{k}\cdot\mathbf{x}} d^2x \tag{4.5}$$

is the Fourier transform of $P = Q + iU$, and $\mathbf{k} = (k_x, k_y)$ $\mathbf{x} = (x, y)$ are two-dimensional vectors. The return transform, here written for R, is given by

$$R(x, y) = \int \tilde{R}(k_x, k_y) e^{i\mathbf{k}\cdot\mathbf{x}} d^2x/(2\pi)^2. \tag{4.6}$$

To generate a two-dimensional vector field, it suffices to combine the vector $\mathbf{b} = (b_x, b_y)$ into a single complex field, $\mathcal{B}(x, y) = b_x + ib_y$. Next, to generate a periodic pattern, we use the complex wavenumber $\mathcal{K} \equiv k_x + ik_y$ and generate a pattern in Fourier space as the simplest possible nontrivial analytic function, $\tilde{\mathcal{B}}(\mathcal{K}) = \mathcal{K}$, along with some phase factor

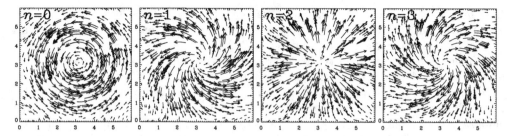

Figure 2. Plots of $e^{i\pi n/4}(k_x + ik_y)/k^3$ for $n = 0$, 1, 2, and 3. Different values of the phase sample E patterns ($n = 0$ and 2) and B patterns ($n = 1$ and 3) in a continuous fashion.

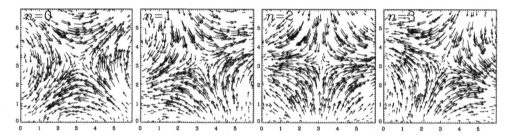

Figure 3. Plots of $e^{i\pi n/4}(k_x - ik_y)/k^3$ for $n = 0$, 1, 2, and 3. Unlike the case with $(k_x + ik_y)$, shown in Figure 2, these patterns do not correspond to the usual E or B polarizations.

$e^{i\phi}$ with phase ϕ; see Figure 2. We also adopt its complex conjugate, $\mathcal{K}$; see Figure 3. In both cases we normalize by $|\mathcal{K}|^3$ to reduce the values further away from the origin.

We see that the different patterns of $\tilde{\mathcal{B}}(\mathcal{K}) = e^{i\phi}\mathcal{K}$ correspond to qualitatively different patterns. For $n = 0$ and 2, we obtain parity-even patterns (E polarizations with different signs of E), whereas for $n = 1$ and 3, we obtain parity-odd patterns (B polarizations with different signs of B). For $e^{i\phi}\mathcal{K}^*$, on the other hand, the patterns are quite different, and they correspond to just the inverse of the former ones in the sense that if one multiplies the $\tilde{\mathcal{B}}(\mathcal{K})$ of Figure 2 with those of Figure 3, one obtain just the phase of $R = E + iB$, so its decomposition into real and imaginary parts returns as the E and B polarizations, as expected from Equation (4.4).

Before closing, let us mention here one more aspect regarding polarized emission. Normally, Faraday rotation leads to Faraday depolarization, that is, the cancellation of polarized emission from the line-of-sight integration of the intrinsic polarized emission of different orientations. Thus, in galactic magnetic field measurements with radio telescopes, Faraday rotation was long regarded as something "bad". For a helical field, however, Faraday rotation can also bring some benefits in that it allows us to estimate the sign of magnetic helicity and the approximate length scales of helical magnetic fields. This property was exploited originally in the galactic context (Brandenburg & Stepanov 2014), but it can probably also be applied to coronal magnetic fields (Brandenburg *et al.* 2017).

Studying the helicity of coronal magnetic fields further will potentially be extremely useful in order to assess the possibility of a sign reversal of magnetic helicity some distance above the surface of the Sun and of other stars. Such a sign reversal was first noted in the solar wind (Brandenburg *et al.* 2011), where, far away from the Sun, the typical sign of magnetic helicity was found to be opposite to what it is at a solar surface. This was then later also found in simulations of simple models of global dynamos with a conducting exterior (Warnecke *et al.* 2011, 2012). Studying and understanding this phenomenon further will be an important aspect of future studies.

5. Discussion

In this work, we have offered some speculation on how stars like the Sun may have evolved from the times they were born to the time when they reached an age beyond that of the present Sun. We expect that the Sun is shortly before changing its rotation from solar-like to antisolar-like differential rotation where the equator rotates slower than the poles; see also Karak *et al.* (2019) for recent mean-field modeling of such stars. Stars with antisolar differential rotation are also potential candidates for displaying superflares (Katsova *et al.* 2018). Unfortunately, there is currently no explicit observation of this phenomenon, except for giants (Kővári *et al.* 2015, 2017). For main sequence stars, this idea is solely based on numerical simulations. It must therefore be hoped that future observations can give us more explicit evidence for this suggestion. Helioseismic techniques might provide one such approach and has already been partially successful (Benomar *et al.* 2018). Other techniques could involve the measurement of light curves, as has been proposed by Reinhold & Arlt (2015).

We also discussed the need for a better understanding of the cyclic nature of slowly and rapidly rotating stars. Observations are consistent with a magnetic dipole lying in the equatorial plane, but there is also a long-term cyclic variation that could be compatible with the two poles alternating in their relative strengths. But this is not yet borne out by reasonably realistic simulation data.

A promising long-term record of simulation data has been assembled by (Käpylä *et al.* 2016). Those simulations display a huge variety of different behaviors, including poleward and equatorward migratory patterns, as well as short and long cycle periods, all in one and the same run. It would therefore be useful to produce similar data for stars with different Rossby numbers in an attempt to better understand the various observation signatures, including the various transitions and the proper position of the Sun in this vast parameter space.

Finally, we turned attention to more direct inspection techniques using linear polarization data. This is motivated by the possibility that standard inversion techniques to obtain the magnetic field might be severely flawed by the fact that no safe π disambiguation technique exists that tells us whether the magnetic field vector points for forward or backward. This problem results from the fact that polarization "vectors" are not proper vectors as they do not have neither head nor tail. In fact, just to obtain the sign of magnetic helicity, it is, under some conditions, not even necessary to disambiguate the polarization vector. The sign of handedness can in fact be obtained directly from the linear polarization—without resorting to the magnetic field.

An intermediate approach here is to first determine the magnetic field, but then to make it ambiguous again by estimating $Q + iU$ from $(b_x + ib_y)^2$. This sounds somewhat odd, but it has the advantage that one does then not need to worry about wavelength dependencies of the line spectra of Stokes Q and U. This has been discussed in detail in the work of A. Prabhu (private communication).

A different approach to using Stokes Q and U directly is in connection with the determination of magnetic helicity in solar and stellar coronae. This idea was originally proposed for edge-on galaxies (Brandenburg & Stepanov 2014), but it can equally well if he applied to the Sun and other stores. The work of Brandenburg *et al.* (2017) suggests, however, that one may have a better chance of exploiting this technique by using millimeter wavelengths rather than infrared. In any case, it is necessary to measure polarized intensity over a broad range of different wave lengths. This has now become possible with the emergence of more refined detector technology. In the context of galactic polarization measurements, this technological advance is what led to the development of Faraday tomography (Brentjens & de Bruyn 2005). This is based on only work of Burn (1966), who recognized that the line-of-sight integral of linear polarization is the same as a

Fourier integral and can therefore be inverted, provided one covers a sufficiently broad range of wavelengths. This was exactly the problem that was difficult to overcome in the early days of radio astronomy, where observations could only be carried out in a small number of frequency bands

A reliable measurement of magnetic helicity in the solar corona is greatly helped by the possibility to use the moon as a perfect coronograph. The total eclipse during the IAU symposium has not yet been utilized for that purpose, but this would hopefully change in the near future.

Acknowledgements

This work was supported by the National Science Foundation under the grant AAG-1615100 and the Swedish Research Council under the grant 2019-04234. We acknowledge the allocation of computing resources provided by the Swedish National Allocations Committee at the Center for Parallel Computers at the Royal Institute of Technology in Stockholm.

References

Alvarado-Gómez, J. D., Hussain, G. A. J., Drake, J. J., Donati, J.-F., Sanz-Forcada, J., Stelzer, B., Cohen, O., Amazo-Gómez, E. M., Grunhut, J. H., Garraffo, C., Moschou, S. P., Silvester, J., Oksala, M. E. 2018, *MNRAS, 473,* 4

Baliunas, S. L., Donahue, R. A., Soon, W. H., Horne, J. H., *et al.* 1995, *ApJ, 438,* 269

Barnes, S. A., & Kim, Y.-C. 2010, *ApJ, 721,* 675

Benomar, O., Bazot, M., Nielsen, M. B., Gizon, L., Sekii, T., Takata, M., Hotta, H., Hanasoge, S., Sreenivasan, K. R., & Christensen-Dalsgaard, J. 2018, *Science, 361,* 1231

Böhm-Vitense, E. 2007, *ApJ, 657,* 486

Boro Saikia, S., Jeffers, S. V., Morin, J., Petit, P., Folsom, C. P., Marsden, S. C., Donati, J.-F., Cameron, R., Hall, J. C., Perdelwitz, V., Reiners, A., & Vidotto, A. A. 2016, *A&A, 594,* A29

Bracco, A., Candelaresi, S., Del Sordo, F., & Brandenburg, A. 2019, *A&A, 621,* A97

Brandenburg, A. 2019, *ApJ, 883,* 119

Brandenburg, A., & Giampapa, M. S. 2018, *ApJ, 855,* L22

Brandenburg, A., & Stepanov, R. 2014, *ApJ, 786,* 91

Brandenburg, A., Ashurova, M. B., & Jabbari, S. 2017, *ApJ, 845,* L15

Brandenburg, A., Bracco, A., Kahniashvili, T., Mandal, S., Roper Pol, A., Petrie, G. J. D., & Singh, N. K. 2019, *ApJ, 870,* 87

Brandenburg, A., Mathur, S., & Metcalfe, T. S. 2017, *ApJ, 845,* 79

Brandenburg, A., Saar, S. H., & Turpin, C. R. 1998, *ApJ, 498,* L51

Brandenburg, A., Subramanian, K., Balogh, A., & Goldstein, M. L. 2011, *ApJ, 734,* 9

Brentjens, M. A., & de Bruyn, A. G. 2005, *A&A, 441,* 1217

Brown, B. P., Miesch, M. S., Browning, M. K., Brun, A. S., & Toomre, J. 2011, *ApJ, 731,* 69

Burn, B. J. 1966, *MNRAS, 133,* 67

Durrer, R. 2008, *The Cosmic Microwave Background, Chapter 5* (Cambridge University Press, Cambridge, United Kingdom, 2008)

Eddy, J. A. 1976, *Science, 286,* 1198

Egeland, R. 2018, *ApJ, 866,* 80

Elstner, D., & Korhonen, H. 2005, *Astron. Nachr., 326,* 278

Gastine, T., Yadav, R. K., Morin, J., Reiners, A., & Wicht, J. 2014, *MNRAS, 438,* L76

Giampapa, M. S., Brandenburg, A., Cody, A. M., Skiff, B. A., & Hall, J. C. 2017, *ApJ,* submitted http://www.nordita.org/preprints, no. 2017-121

Gilman, P. A. 1977, *Geophys. Astrophys. Fluid Dyn., 8,* 93

Gilman, P. A. 1980, in *Stellar turbulence; Proceedings of the Fifty-first Colloquium, London, Ontario, Canada, August 27-30, 1979,* ed. Gray, D. F. & Linsky, J. L. (Berlin and New York, Springer-Verlag), 19

Goldberg, J. N., Macfarlane, A. J., Newman, E. T., Rohrlich, F., & Sudarshan, E. C. G. 1967, *JMP, 8,* 2155

Guerrero, G., Zaire, B., Smolarkiewicz, P. K., de Gouveia Dal Pino, E. M., Kosovichev, A. G., Mansour, N. N. 2019a, *ApJ, 880,* 6

Guerrero, G., Del Sordo, F., Bonanno, A., & Smolarkiewicz, P. K. 2019b, *MNRAS, 490,* 4281

Jetsu, L., Tuominen, I., Grankin, K. I., Mel'nikov, S. Yu., & Shevenko, V. S. 1994, *A&A, 282,* L9

Kamionkowski, M., Kosowsky, A., & Stebbins, A. 1997, *Phys. Rev. Lett., 78,* 2058

Kamionkowski, M., & Kovetz, E. D. 2016, *ARA&A, 54,* 227

Käpylä, M. J., Käpylä, P. J., Olspert, N., Brandenburg, A., Warnecke, J., Karak, B. B., & Pelt, J. 2016, *A&A, 589,* A56

Käpylä, P. J., Käpylä, M. J., & Brandenburg, A. 2014, *A&A, 570,* A43

Karak, B. B., Käpylä, M. J., Käpylä, P. J., Brandenburg, A., Olspert, N., & Pelt, J. 2015, *A&A, 576,* A26

Karak, B. B., Tomar, A., & Vashishth, V. 2019, *MNRAS, 491,* 3155

Katsova, M. M., Kitchatinov, L. L., Livshits, M. A., Moss, D. L., Sokoloff, D. D., & Usoskin, I. G. 2018, *Astron. Rep., 95,* 78

Kleeorin, N. I., Ruzmaikin, A. A., & Sokoloff, D. D. 1983, *Ap&SS, 95,* 131i

Kochukhov, O., Petit, P., Strassmeier, K. G., Carroll, T. A., Fares, R., Folsom, C. P., Jeffers, S. V., Korhonen, H., Monnier, J. D., Morin, J., Rosén, L., Roettenbacher, R. M., & Shulyak, D. 2017, *Astron. Nachr., 338,* 428

Kővári, Z., Kriskovics, L., Künstler, A., Carroll, T. A., Strassmeier, K. G., Vida, K., Oláh, K., Bartus, J., Weber, M. 2015, *A&A, 573,* A98

Kővári, Z., Strassmeier, K. G., Carroll, T. A., Oláh, K., Kriskovics, L., Kővári, E., Kovács, O., Vida, K., Granzer, T., & Weber, M. 2017, *A&A, 606,* A42

Lehtinen, J., Jetsu, L., Hackman, T., Kajatkari, P., & Henry, G. W. 2016, *A&A, 588,* A38

Lehtinen, J. J., Spada, F., Käpylä, M. J., Olspert, N., & Käpylä, P. J. 2020, *Nat. Astron., 4,* 658

Mamajek, E. E., & Hillenbrand, L. A. 2008, *ApJ, 687,* 1264

Maunder, E. W. 1904, *MNRAS, 64,* 747

Metcalfe, T. S., & van Saders, J. 2017, *Solar Phys., 292,* 126

Metcalfe, T. S., Kochukhov, O., Ilyin, I. V., Strassmeier, K. G., Godoy-Rivera, D., & Pinsonneault, M. H. 2019, *ApJ, 887,* L38

Moss, D., & Brandenburg, A. 1995, *Geophys. Astrophys. Fluid Dyn., 80,* 229

Moss, D., Barker, D. M., Brandenburg, A., & Tuominen, I. 1995, *A&A, 294,* 155

Noyes, R. W., Hartmann, L., Baliunas, S. L., Duncan, D. K., & Vaughan, A. H. 1984a, *ApJ, 279,* 763

Noyes, R. W., Weiss, N. O., & Vaughan, A. H. 1984b, *ApJ, 287,* 769

Olspert, N., Lehtinen, J. J., Käpylä, M. J., Pelt, J., & Grigorievskiy, A. 2018, *A&A, 619,* A6

Prabhu, A., Brandenburg, A., Käpylä, M. J., & Lagg, A. 2020, *A&A,* doi:10.1051/0004-6361/202037614, arXiv:2001.10884

Rädler, K.-H., Wiedemann, E., Brandenburg, A., Meinel, R., & Tuominen, I. 1990, *A&A, 239,* 413

Rädler, K.-H., Rheinhardt, M., Apstein, E., & Fuchs, H. 2002, *Nonl. Processes Geophys., 38,* 171

Rädler, K.-H., & Brandenburg, A. 2003, *Phys. Rev. E, 67,* 026401

Reinhold, T., & Arlt, R. 2015, *A&A, 576,* A15

Roberts, G. O. 1972, *Phil. Trans. Roy. Soc. London A, 271,* 411

Rüdiger, G. 1978, *Astron. Nachr., 299,* 217

Saar, S. H., & Linsky, J. L. 1985, *ApJ, 299,* L47

Schrijver, C. J., Cote, J, Zwaan, C., Saar, S. H. 1989, *ApJ, 337,* 964

Schwabe, H. 1844, *Astron. Nachr., 21,* 233

Seljak, U., & Zaldarriaga, M. 1997, *Phys. Rev. Lett., 78,* 2054

Tobias, S. 1998, *MNRAS, 296,* 653

van Saders, J. L., Ceillier, T., Metcalfe, T. S., Silva Aguirre, V., Pinsonneault, M. H., García, R. A., Mathur, S., & Davies, G. R. 2016, *Nature, 529,* 181

Vilhu, O. 1984, *A&A, 133,* 117

Viviani, M., Warnecke, J., Käpylä, M. J., Käpylä, P. J., Olspert, N., Cole-Kodikara, E. M., Lehtinen, J. J., & Brandenburg, A. 2018, *A&A, 616,* A160

Warnecke, J., Brandenburg, A., & Mitra, D. 2011, *A&A, 534,* A11

Warnecke, J., Brandenburg, A., & Mitra, D. 2012, *J. Spa. Weather Spa. Clim., 2,* A11

Wilson, O. C. 1963, *ApJ, 138,* 832

Wilson, O. C. 1978, *ApJ, 266,* 379

Zaldarriaga, M. & Seljak, U. 1997, *Phys. Rev. D, 55,* 1830

Discussion

KLAUS STRASSMEIER: It is indeed true that we have found no good evidence for anti-solar differential rotation in main sequence stars, but only in giants and subgiants. The stars with anti-solar differential rotation were not solar-like stars when on the main sequence, but may have been some sort of Ap stars without a sign of an outer convection zone. Do we see differential rotation at all in Ap stars? Or, if seen in solar-like main sequence stars, what process switches differential rotation from solar- to anti-solar?

AXEL BRANDENBURG: The process causing this switching from solar-like to anti-solar differential rotation is an emerging dominance of meridional circulation over the Reynolds stress. Subgiants could also have meridional circulation in a thin outer convective layer causing anti-solar differential rotation.

CHRISTOPHER KAROFF: What would happen to the topography of the magnetic field as the stars change to anti-solar differential rotation?

AXEL BRANDENBURG: According to the simulations, it should be the same, so the magnetic field would still be poleward migrating. This is because in the solar-like differential rotation regime, it has been very difficult to reproduce solar-like equatorward migration. Therefore, poleward migration has been obtained in both cases. In reality, however, this might not be true, and so poleward migration might emerge only with the moment that the star begins to display antisolar differential rotation.

MOIRA JARDINE: One of the ways to confuse a periodogram is to have spots whose lifetime is less than the rotation period of the star. Can you say something about any trends in spot lifetime with accuracy?

AXEL BRANDENBURG: Sunspots are known to have a large range of lifetimes from half a day to three months. Their decay times scale with their surface area, so for the Sun, the larger and more dominant spots do have lifetimes longer than the rotation period.

Solar and Stellar Magnetic Fields: Origins and Manifestations
Proceedings IAU Symposium No. 354, 2019
A. Kosovichev, K. Strassmeier & M. Jardine, ed.
doi:10.1017/S1743921319009864

Magnetic field and prominences of the young, solar-like, ultra-rapid rotator AP 149

Tianqi Cang⬤, Pascal Petit, Colin Folsom and Jean-Francois Donati

Institut de Recherche en Astrophysique et Planétologie, Université de Toulouse,
CNRS, CNES, 31400 Toulouse, France

Abstract. Young solar analogs reaching the main sequence experience very strong magnetic activity, directly linked to their angular momentum loss through wind and mass ejections. We investigate here the surface and chromospheric activity of the ultra-rapid rotator AP 149 in the young open cluster alpha Persei. With a time-series of spectropolarimetric observations gathered over two nights with ESPaDOnS, we are able to reconstruct the surface distribution of brightness and magnetic field using the Zeeman-Doppler-Imaging (ZDI) method. Using the same data set, we also map the spatial distribution of prominences through tomography of H-alpha emission. We find that AP 149 shows a strong cool spot and magnetic field closed to the polar cap. This star is the first example of a solar-type star to have its magnetic field and prominences mapped together, which will help to explore the respective role of wind and prominences in the angular momentum evolution of the most active stars.

Keywords. stars: individual: AP 149, stars: magnetic field, stars: solar-type, stars: rotation, stars: spots, stars: winds

---

1. Introduction

Magnetic activity of solar-type stars is an important ingredient in their early evolutionary phases. Young Suns tend to have high rotation rates since they have conserved most of the angular momentum acquired during the stellar formation process, and this rapid rotation plays a key role in their evolution (see the review of Bouvier 2013). High rotation rates are responsible for the efficient amplification of internal magnetic fields, through the action of global stellar dynamos.

Tomographic mapping is a powerful tool to characterize the surface magnetic field of rapid rotators. Since its first application to HR 1099 (Donati *et al.* 1992), this approach has been applied to several dozens of cool active stars on the main sequence (Folsom *et al.* 2018). Early investigations dedicated to young Suns (e.g. Donati *et al.* 2003, Petit *et al.* 2008) provide unique information about magnetic geometries and differential rotation probing a range of stellar parameters (mass, age, rotation age).

Prominences are among the most spectacular manifestations of stellar magnetic activity. For rapidly rotating stars, prominence systems become much more massive and extended than on the Sun (see review by Collier Cameron 1999). Prominences and stellar winds remove the angular momentum of stars and become a critical contributor to the early evolution of active stars.

In a first step towards the exploration of the magnetic field and prominences of ultra-rapidly rotating young suns, we investigated the G type star AP 149 (also named V530 Per), which is a young member of the α Persei open cluster.

Table 1. Parameters of AP 149

Parameter	Value	Ref.	Parameter	Value	Ref.
Distance	167.7 ± 15 pc	1,2	Mass	$1.03 \pm 0.13\ M_\odot$	1
Age	33^{+10}_{-7} Myr	1	$log\ L_x$	31.2	5
$T_{\rm eff}$	5281 ± 96 K	1	Rossby number	0.0128 ± 0.0019	1
$log\ g$	4.1 ± 0.19	1	Co-rotation radius	$1.92\ R_*$	1
$[Fe/H]$	0	1	$v \sin i$	105.6 km s^{-1}	1
m_V^{max}	11.657 ± 0.13	4	Eq. rot. period	0.3205 ± 0.0002 d	1
m_K	9.422 ± 0.019	4	$d\Omega$	0.043 ± 0.007 rad d^{-1}	1
m_J	10.08 ± 0.019	4	Inclination angle	$35 \pm 0.2°$	1
Luminosity	$0.76 \pm 0.18\ L_\odot$	1	Radial velocity	-0.96 ± 0.04 km s^{-1}	1
Radius	$1.04 \pm 0.11\ R_\odot$	1			

References: 1. This work 2. Yen *et al.* (2018) 3. Stauffer *et al.* (1999) 4. Zacharias *et al.* (2012) 5. Pillitteri *et al.* (2013).

2. Observations

We obtained a time-series of spectropolarimetric observations of AP 149 in late 2006 (29 Nov and 05 Dec). The data were collected by the ESPaDOnS spectropolarimeter (Donati *et al.* 2006a), mounted at the Canada-France-Hawaii Telescope (CFHT). The observation was optimized to have full coverage of the rotational phase and normalized, reduced 56 Stokes I, and 14 Stokes V spectra are extracted. The LSD method (Donati *et al.* 1997) is used to combine a list of selected spectral lines to produce a pseudo line profile with improved S/N.

3. Fundamental Parameters

By combining literature measurements and ZDI results, we derive fundamental parameters of AP 149 and locate it on the H-R diagram. Details of the parameters are shown in Table 1.

4. Brightness imaging

By applying the Doppler Imaging method (DI, Vogt *et al.* 1987) using the ZDI code of Folsom *et al.* (2018), we recovered the surface brightness distribution of AP 149 in Fig 1. The reconstruction includes a solar-like differential rotation law, with a roughly solar shear level optimizing our model.

5. Magnetic field

Zeeman Doppler Imaging technique (ZDI, Semel 1989; Donati *et al.* 2006b; Folsom *et al.* 2018) is applied to model the two nights of Stokes V data, in order to reconstruct the 2D magnetic field distribution of AP 149. The map of the three components is shown in Fig.1. According to the map, the average surface magnetic field is 150G, with local peaks slightly above 1kG. About 2/3 of the magnetic energy belongs to the toroidal field component. Moreover, the complexity of the magnetic field is very high, with only 1.1% of the magnetic energy stored in the dipolar component.

6. Differential Rotation

The very dense phase coverage of our time-series constitutes a very good basis for studying the short term evolution of photospheric brightness, especially under the action of differential rotation. In our model, the rotation rate Ω is assumed to vary with the latitude (θ), following a simple surface dependence: $\Omega(\theta) = \Omega_{eq} - d\Omega \sin^2 \theta$. A χ^2 minimizing method is used to derive the differential rotation $d\Omega$ and the rotational rate of the equator Ω_{eq} (Petit et al. 2002), shown in Figure 3. The quantitative results can be found in Table 1, and highlight a shear level roughly solar in magnitude.

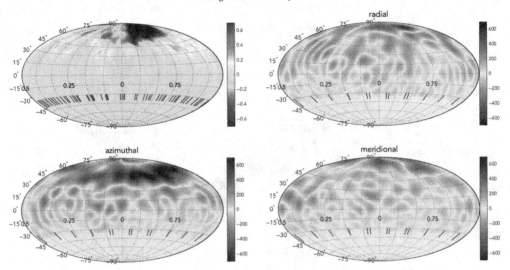

Figure 1. Brightness and Magnetic Map of AP 149. Logarithmic normalized brightness map (upper left) of AP 149 reconstructed assuming a combination of dark (blue) and bright (yellow) spots. A large, dark spot is identified close to the pole, as well as a complex spot distribution at lower latitudes. Since the inclination is small, most of the southern hemisphere could not be observed and modeled. The other three panels show different field components in a spherical projection with the unit in Gauss. For the radial (upper right) and meridional (lower right) components, the strongest magnetic spots are found close to the pole. For the azimuthal (lower left) component, a ring of negative field is reconstructed above a latitude of $45°$.

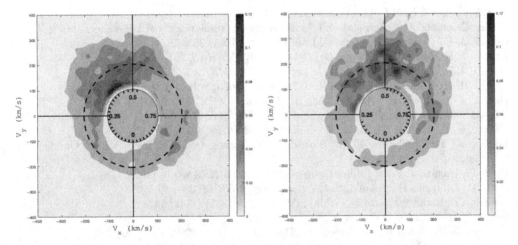

Figure 2. Dynamic Spectra showing the $H\alpha$ line of AP 149. The upper and lower panels are for 29 Nov and 05 Dec, respectively. From left to right, we display the observations, the outcome of the tomographic model, and the residuals. $H\alpha$ mapping from the first night leads to a reduced $\chi^2 = 6.8$, while the second night provides us with a reduced $\chi^2 = 7.5$.

7. Prominence System

For active stars, prominences are constituted of cool, dense gas trapped in closed magnetic loops, which always be connected with stellar wind. The prominence structure produces strong, rotationally-modulated, double peak $H\alpha$ emission. The line profiles of $H\alpha$ are always seen in emission throughout the observing sequence of AP 149, and we

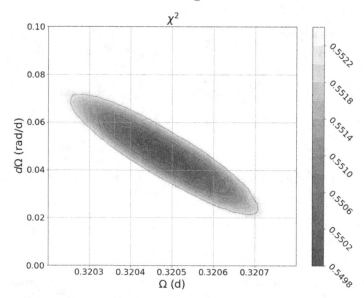

Figure 3. Reduced χ^2 map for the shear parameter $d\Omega$ and the equatorial rotation rate Ω_{eq}. The three red solid lines illustrate the 1σ, 2σ, and 3σ confidence intervals.

reconstructed prominence maps of AP 149 for the two observational nights (see Fig. 2), unveiling clues of short-term variations.

References

Bouvier, J. 2013, EAS Publications Series, 143

Collier Cameron, A. 1999, Solar and Stellar Activity: Similarities and Differences, 146

Donati, J.-F., Brown, S. F., Semel, M., *et al.* 1992, *A&A*, 265, 682

Donati, J.-F., Semel, M., Carter, B. D., *et al.* 1997, *MNRAS*, 291, 658

Donati, J.-F., Collier Cameron, A., & Petit, P. 2003, *MNRAS*, 345, 1187

Donati, J.-F., Catala, C., Landstreet, J. D., *et al.* 2006a, Solar Polarization 4, 362a

Donati, J.-F., Howarth, I. D., Jardine, M. M., *et al.* 2006b, *MNRAS*, 370, 629

Folsom, C. P., Bouvier, J., Petit, P., *et al.* 2018, *MNRAS*, 474, 4956

Folsom, C. P., Petit, P., Bouvier, J., *et al.* 2016, *MNRAS*, 457, 580

Lodieu, N., McCaughrean, M. J., Barrado Y Navascues, D., *et al.* 2005, VizieR Online Data Catalog, J/A+A/436/853

Petit, P., Donati, J.-F., & Collier Cameron, A. 2002, *MNRAS*, 334, 374

Petit, P., Dintrans, B., Solanki, S. K., *et al.* 2008, *MNRAS*, 388, 80

Pillitteri, I., Remage Evans, N., Wolk, S. J., *et al.* 2013, *AJ*, 145, 143

Semel, M. 1989, *A&A*, 225, 456

Stauffer, J. R., Barrado y Navascués, D., Bouvier, J., *et al.* 1999, *ApJ*, 527, 219

Vidotto, A. A., Gregory, S. G., Jardine, M., *et al.* 2014, *MNRAS*, 441, 2361

Vogt, S. S., Penrod, G. D., & Hatzes, A. P. 1987, *ApJ*, 321, 496

Yen, S. X., Reffert, S., Schilbach, E., *et al.* 2018, *A&A*, 615, A12

Zacharias, N., Finch, C. T., Girard, T. M., *et al.* 2012, VizieR Online Data Catalog, I/322A

Solar and Stellar Magnetic Fields: Origins and Manifestations
Proceedings IAU Symposium No. 354, 2019
A. Kosovichev, K. Strassmeier & M. Jardine, ed.
doi:10.1017/S1743921319009803

Dipolar stability in spherical simulations: The impact of an inner stable zone

Bonnie Zaire[ID] and Laurène Jouve

IRAP, Université de Toulouse, CNRS / UMR 5277, CNES, UPS, 14 avenue E. Belin,
Toulouse, F-31400 France
email: bzaire@irap.omp.eu

Abstract. Magnetic fields vary in complexity for different stars. The stability of dipolar magnetic fields is known to depend on different quantities, e.g., the stellar rotation, the stratification, and the intensity of convective motions. Here, we study the dipolar stability in a system with an inner stable zone. We present preliminary results of dynamo simulations using the Rayleigh number as a control parameter. The stiffness of the stable zone is accordingly varied to keep a constant ratio of the Brunt-Väisälä frequency to the angular velocity. Similarly to the completely convective spherical shell, we find that a transition exists between a regime where the magnetic field is dipolar to a multipolar regime when the Rossby number is increased. The value of the Rossby number at the transition is very close to the one of the fully convective case.

Keywords. Star: magnetic field; Stellar interiors: stably-stratified; Dynamo: simulation.

1. Introduction

Fully convective simulations show that the stability of dipolar dynamos depends on the Rossby number (Gastine *et al.* 2012). Two regimes exist: one for low Rossby number, which results in strong dipolar fields, and another with higher Rossby numbers that lead to complex magnetic field configurations. It is not clear if the same trend exists in the presence of an inner stable zone, a configuration that describes stars with a radiative core.

Helioseismology analysis reveals the presence of a shear layer at the interface of the stable zone with the convective zone in the Sun. This layer is known to be a source of strong toroidal fields through the so-called $\Omega-$effect. Numerical simulations possessing a stable zone indicate that complex dynamics operate (Guerrero *et al.* 2016). Although the shear is expected to impact the magnetic field generation, no parametric study using the Rayleigh number to force the intensity of the convection has been carried out for this setup using 3D simulations.

In this work, we probe the influence of the inner stable zone in the magnetic field properties. We simulate fully convective spherical shells and partially convective shells (radiative zone + convective envelope). In Section 2 we describe the setup of the simulations and in Section 3 we present our magnetic results.

2. Models

We perform 3D MHD simulations in spherical geometry. Our computational domain covers a full shell with inner radius r_i and outer radius r_o. Two different setups are considered in this work: **(i)** a fiducial fully convective setup, which consists of a spherical shell with aspect ratio $r_i/r_o = 0.6$; and **(ii)** a partially convective setup with aspect ratio

$r_i/r_o = 0.4$. The setup of the latter is essentially the convective zone of model **(i)** with an additional inner stable zone.

Next, we show the governing equations that describe our system and we define the reference states that capture the physical conditions of each setup.

2.1. *Governing equations and non-dimensional parameters*

We solve the MHD equations for a stratified fluid in a spherical shell that rotates with angular velocity Ω about the axis $\hat{\mathbf{e}}_z$. We adopt a dimensionless formulation where r_o is the reference lengthscale, the viscous diffusion time r_o^2/ν is the timescale, and $r_o|d\bar{s}/dr|_{r_o}$ is the entropy scale. Gravity, density, and temperature are normalised by their outer radius value.

We use the anelastic version of the code MagIC (Gastine & Wicht 2012) to solve the non-dimensional equations that govern convective motions, magnetic field generation, and entropy fluctuations:

$$\mathrm{E}\left[\frac{\partial \vec{\mathbf{u}}}{\partial t} + (\vec{\mathbf{u}} \cdot \boldsymbol{\nabla})\vec{\mathbf{u}}\right] = -\boldsymbol{\nabla}\left(\frac{p}{\bar{\rho}}\right) + \frac{\mathrm{Ra}\,\mathrm{E}}{\mathrm{Pr}}gs'\hat{\mathbf{e}}_r - 2\hat{\mathbf{e}}_z \times \vec{\mathbf{u}} + \frac{1}{\mathrm{Pm}\,\bar{\rho}}(\boldsymbol{\nabla} \times \vec{\mathbf{B}}) \times \vec{\mathbf{B}} + \frac{\mathrm{E}}{\bar{\rho}}\boldsymbol{\nabla} \cdot S,$$

$$(2.1)$$

$$\frac{\partial \vec{\mathbf{B}}}{\partial t} = \boldsymbol{\nabla} \times \left(\vec{\mathbf{u}} \times \vec{\mathbf{B}}\right) - \frac{1}{\mathrm{Pm}}\boldsymbol{\nabla} \times \left(\boldsymbol{\nabla} \times \vec{\mathbf{B}}\right),$$

$$(2.2)$$

$$\bar{\rho}\bar{T}\left[\frac{\partial s'}{\partial t} + (\vec{\mathbf{u}} \cdot \boldsymbol{\nabla})s' + u_r\frac{d\bar{s}}{dr}\right] = \frac{1}{\mathrm{Pr}}\boldsymbol{\nabla} \cdot (\bar{\rho}\bar{T}\boldsymbol{\nabla}s') + \frac{\mathrm{Pr}\,\mathrm{Di}}{\mathrm{Ra}}Q_\nu + \frac{\mathrm{Pr}\,\mathrm{Di}}{\mathrm{Pm}^2\,\mathrm{E}\,\mathrm{Ra}}(\boldsymbol{\nabla} \times \vec{\mathbf{B}})^2,$$

$$(2.3)$$

$$\boldsymbol{\nabla} \cdot (\bar{\rho}\vec{\mathbf{u}}) = 0, \quad \boldsymbol{\nabla} \cdot \vec{\mathbf{B}} = 0,$$

$$(2.4)$$

where S and Q_ν are, respectively, the strain-rate tensor and viscous heating. The equations above are expressed in terms of five dimensionless control parameters, which are: the Ekman number (E), Rayleigh number (Ra), Prandtl number (Pr), magnetic Prandtl number (Pm), and dissipation number (Di). These non-dimensional numbers are defined as follows:

$$\mathrm{E} = \frac{\nu}{\Omega r_o^2}, \quad \mathrm{Ra} = \frac{g_o r_o^4}{c_p \kappa \nu}\left|\frac{d\bar{s}}{dr}\right|_{r_o}, \quad \mathrm{Pr} = \frac{\nu}{\kappa}, \quad \mathrm{Pm} = \frac{\nu}{\lambda}, \quad \text{and} \quad \mathrm{Di} = \frac{g_o r_o}{c_p T_o}, \quad (2.5)$$

where, ν is the viscosity, κ is the thermal diffusivity, and λ is the magnetic diffusivity. The dimensionless gravity profile adopted is $g(r) = -\frac{7.36r}{r_o} + \frac{4.99r^2}{r_o^2} + \frac{3.71r_o}{r} - \frac{0.34r_o^2}{r^2}$.

2.2. *Reference state*

Thermodynamical quantities in equations 2.1 to 2.3 were expressed in terms of a static (reference) state and fluctuations around it. This reference state, indicated by an overbar, is assumed to be an ideal gas nearly adiabatic given by

$$\frac{1}{\bar{T}}\frac{\partial \bar{T}}{\partial r} = \epsilon_s\frac{d\bar{s}}{dr} - \mathrm{Di}\,\alpha_o g(r) \quad \text{and} \quad \frac{1}{\bar{\rho}}\frac{\partial \bar{\rho}}{\partial r} = \epsilon_s\frac{d\bar{s}}{dr} - \frac{\mathrm{Di}\,\alpha_o}{\Gamma}g(r), \quad (2.6)$$

where the condition $\epsilon_s \ll 1$ is necessary to ensure that the governing equations still hold near adiabaticity.

Here, $d\bar{s}/dr$ is a prescribed non-adiabaticity that controls the radial stratification in the simulation domain. Stably-stratified regions occur whenever $d\bar{s}/dr > 0$, while negative gradients set convectively-unstable regions. In model **(i)** convection is set by imposing

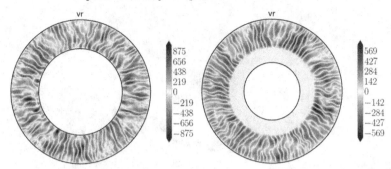

Figure 1. Equatorial view of the radial velocity, v_r, for runs with $Ra = 1.56 \times 10^8$. Left panel shows the result using the setup **(i)** and right panel shows the result using the setup **(ii)**. Units are the inverse of the Ekman number, E^{-1}.

$d\bar{s}/dr = -1$ in the entire radial domain, i.e, $r \in (0.6, 1.0)r_o$. On the other hand, in model **(ii)** the non-adiabaticity is given by

$$\frac{d\bar{s}}{dr} = \begin{cases} \left(\frac{N}{\Omega}\right)^2 \frac{Pr}{Ra\,E^2}, & r < 0.6r_o, \\ -1, & r \geq 0.6r_o, \end{cases}$$

a profile that creates a stably-stratified layer for $r < 0.6r_o$. The amplitude of this stable layer is a function of the non-dimensional numbers and the ratio of the Brunt-Väisälä frequency (N) to the angular velocity.

In the next section we compare the magnetic configurations achieved on both setups. In the full set of simulations $E = 1.6 \times 10^{-5}$, $Pr = 1$, $Pm = 5$, and $Di = 2.7$, where the choice of Di sets a density contrast in the convective region equivalent to 4.8 in both models. Furthermore, we chose to keep $N/\Omega = 2$ in this first analysis of the stably-stratification impact on the magnetic field topology.

3. Magnetic results

The set of simulations present here uses the Rayleigh number as a control parameter. We cover four different forcing levels: $Ra = 4.77 \times 10^7$, 6.25×10^7, 7.81×10^7, and 1.56×10^8. Pair of simulations were run for each one of the forcing levels in order to access the impact of the stable zone in the magnetic configuration. These pairs of simulations correspond to the two different setups introduced in Section 2: **(i)** fully convective and **(ii)** partially convective.

Equatorial views of the radial velocity field are depicted in Fig. 1 for the case with $Ra = 1.56 \times 10^8$. Left and right panels correspond respectively to setups **(i)** and **(ii)**. The snapshot evidences the existence of the stable zone in setup **(ii)**. Following Christensen & Aubert (2006), we define the local Rossby number as

$$Ro_l = \frac{u_{rms}}{\Omega r_o} \frac{\bar{l}}{\pi},$$

where $\bar{l}$ is the mean spherical harmonics degree in the kinetic energy spectrum. A comparison between both runs shows that the intensity of the convective flows is diminished in the presence of the stable zone, thus lowering the Rossby number. This damping behavior occurs for all partially convective simulations, however it is more prominent in the run with higher forcing.

Figure 2 shows the complexity of the magnetic field in our simulations as a function of the local Rossby number. We characterise the complexity of the surface magnetic field

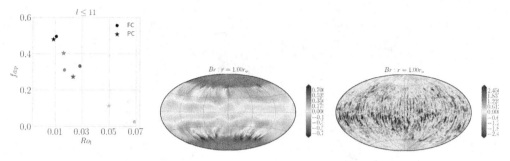

Figure 2. Left panel shows dipolar fraction in our set of simulations. Symbols denote simulations with different setups, where circles represent the setup **(i)** and stars represent the setup **(ii)**. Colours distinguish simulations by its forcing with black, red, blue, and orange standing for $Ra = 4.77 \times 10^7$, 6.25×10^7, 7.81×10^7, and 1.56×10^8 respectively. Right most panels depict the magnetic topology for the partially convective setup at the lowest and the highest Rossby numbers.

through the fraction of dipole. We express it as the relative energy in the axial dipole to the total energy in spherical harmonics degrees up to 11 at the surface:

$$f_{\mathrm{dip}}(r = r_o) = \frac{\vec{\mathbf{B}}^2_{(l=1, m=0)}(r = r_o)}{\sum_{l=0}^{11} \sum_{m=0}^{l} \vec{\mathbf{B}}^2_{(l,m)}(r = r_o)}. \tag{3.1}$$

Magnetic fields were consistently generated in our set of simulations. The transition from dipolar to multipolar fields was observed for both setups considered. Runs with $Ra = 4.77 \times 10^7$ (black color in Fig. 2), $Ra = 6.25 \times 10^7$ (red) and $Ra = 7.81 \times 10^7$ (blue) produced strong axial dipolar fields, whereas the case with $Ra = 1.56 \times 10^8$ (orange) resulted in complex surface magnetic fields. For both setups the case with $Ra = 7.81 \times 10^7$ (blue) showed a reversal dipole, a characteristic behavior for the simulations with Rossby number at the transition for multipolar solutions.

4. Conclusions and perspectives

We performed numerical simulations to access the stability of dipolar solutions in the presence of a stable zone. Two different setups were considered, one fully convective and another partially convective. Our set of simulations corresponds to a parametric search using the Rayleigh number, which translates into a variation of Rossby number, i.e. the influence of rotation on convection.

For both setups dipolar and multipolar solutions were achieved. This preliminary results indicate that the dipolar transition occurs in the same region for simulations with and without stable zone. Additional runs are necessary to identify at which Rossby value the transition occurs and if other regimes exist where the dipolar solution would be favoured.

References

Christensen, U. R. & Aubert, J. 2006, *Geophysical Journal International*, 166(1), 97–114
Gastine, T. & Wicht, J. 2012, *Icarus*, 219(1), 428–442
Gastine, T.,Duarte, L., Wicht, J., et al. 2012, *Astronomy & Astrophysics*, 546, A19
Guerrero, G., Smolarkiewicz, P. K., de Gouveia Dal Pino, E. M., Kosovichev, A. G., Mansour, N. N., et al. 2016, *The Astrophysical Journal*, 819:104

Solar and Stellar Magnetic Fields: Origins and Manifestations
Proceedings IAU Symposium No. 354, 2019
A. Kosovichev, K. Strassmeier & M. Jardine, ed.
doi:10.1017/S1743921319009785

A large rotating structure around AB Doradus A at VLBI scale

J. B. Climent[1]📖, J. C. Guirado[1,2], R. Azulay[1] and J. M. Marcaide[1]

[1]Departament d'Astronomia i Astrofísica, Universitat de València, C. Dr. Moliner 50,
46100 Burjassot, València, Spain
email: j.bautista.climent@uv.es

[2]Observatori Astronòmic, Universitat de València, Parc Científic, C. Catedrático
José Beltrán 2, 46980 Paterna, València, Spain

Abstract. We report the results of three VLBI observations of the pre-main-sequence star AB Doradus A at 8.4 GHz. With almost three years between consecutive observations, we found a complex structure at the expected position of this star for all epochs. Maps at epochs 2007 and 2010 show a double core-halo morphology while the 2013 map reveals three emission peaks with separations between 5 and 18 stellar radii. Furthermore, all maps show a clear variation of the source structure within the observing time. We consider a number of hypothesis in order to explain such observations, mainly: magnetic reconnection in loops on the polar cap, a more general loop scenario and a close companion to AB Dor A.

Keywords. stars: pre–main-sequence, stars: rotation, stars: imaging, stars: magnetic fields, stars: flare, radio continuum: stars

1. Introduction

About 15 pc away, AB Doradus is a pre-main-sequence (PMS) system formed by two pairs of stars separated by 9″, AB Dor A/C and AB Dor Ba/Bb (Close *et al.* 2005; Guirado *et al.* 2006), giving name to the AB Doradus moving group (AB Dor-MG). The main star of this system, the K0 dwarf AB Dor A is a PMS star (Herbst & Shevchenko 1999) with a rotation period of 0.5 days and presents strong emission at all wavelengths, from radio to X-rays. It has been well studied by the Hipparcos satellite (Lindegren & Kovalevsky 1995; Lestrade *et al.* 1995) and very-long-baseline-interferometry (VLBI) arrays (Lestrade *et al.* 1995; Guirado *et al.* 1997; Azulay *et al.* 2017). This joined effort revealed the presence of AB Dor C, a low-mass companion with 0.090 $M_\odot$, orbiting AB Dor A at an average angular distance of 0.2″. The pair AB Dor A/C has also been observed by different near-infrared instruments at the VLT (Close *et al.* 2005, 2007; Boccaletti *et al.* 2008) allowing independent photometry of AB Dor C which, along with the dynamical mass determination, served as a benchmark for stellar evolutionary models. The exact age of the system is a current subject of discussion: 40−50 Myr for AB Dor A and 25−120 Myr for AB Dor C in Azulay *et al.* (2017), 40−60 Myr in Zuckerman *et al.* (2004) and López-Santiago *et al.* (2006), 30−100 Myr in Close *et al.* (2005), 40−100 Myr in Nielsen *et al.* (2005), 75−150 Myr in Luhman & Potter (2006), 50−100 Myr in Janson *et al.* (2007) and Boccaletti *et al.* (2008), 40−50 Myr in Guirado *et al.* (2011), >110 Myr for the AB Dor nucleus star in Barenfeld *et al.* (2013) and 130-200 Myr in Bell *et al.* (2015).

Here we present a new analysis of the 8.4 GHz VLBI data from Azulay *et al.* (2017) in order to look for more information beyond the astrometric position.

Table 1. Model components of the fit of circular Gaussians on the uv-plane for VLBI observations.

Epoch	Comp.[a]	S (mJy)	θ (mas)	T_b (K)
2007	1	3.4 ± 0.5	1.41 ± 0.18	$4.2 \cdot 10^7$
	2	1.3 ± 0.2	0.60 ± 0.1	$8.9 \cdot 10^7$
2010	1	2.31 ± 0.17	1.30 ± 0.09	$3.4 \cdot 10^7$
	2	1.20 ± 0.16	1.51 ± 0.18	$1.3 \cdot 10^7$
2013	1	2.7 ± 0.2	0.90 ± 0.08	$8.3 \cdot 10^7$
	2	0.80 ± 0.15	0.40 ± 0.07	$1.2 \cdot 10^8$
	3	1.48 ± 0.19	0.65 ± 0.08	$8.7 \cdot 10^7$

Notes. [a]We adopt the convention that the central component will be denoted by subindex 1. In case of detection, the subindex 2 will indicate the presence of a second component to the east. In 2013, subindex 2 indicates the closest component to the east while subindex 3 the furthest one. S represents the flux density, θ the FWHM diameter of the circular Gaussian component and T_b the minimum brightness temperature.

2. Observations and data reduction

We observed the binary system AB Dor A/C using the Australian Long Baseline Array during 11 November 2007, 25 October 2010 and 16 August 2013 at X band (see Table 1 in Azulay *et al.* 2017). Each observation lasted 12 hours. The system was observed in phase referencing mode using the ICRF-defining source BL Lac PKS 0516-621 (about 3.6° away) as a phase calibrator. The sequence calibrator-target lasts about 3.5 minutes. Both RCP and LCP polarizations were recorded using four 16 MHz bandwidth subbands per polarization.

We reduced and analyzed the data using the Astronomical Image Processing System (AIPS) of the National Radio Astronomy Observatory (NRAO) with standard routines. Firstly, we calibrated the ionospheric delay and corrected the instrumental phases. We then calibrated the visibility amplitudes using the nominal sensitivity for each antenna and corrected the phases for parallactic angles. Finally, we performed a fringe-search on the phase calibrator to minimize the residual contributions to the phases and applied these new corrections to our target. The phase-referenced channel-averaged images were obtained using the Caltech imaging program DIFMAP (Shepherd *et al.* 1994) with the clean algorithm while selecting the polarization of interest in each case. The AB Dor A image for each epoch and band is shown at Fig. 1. In addition to producing an image of AB Dor A/C for each LBA dataset, we also fitted circular Gaussians to the interferometric visibilities (*uv* plane) using the DIFMAP task modelfit (the use of elliptical Gaussians resulted in very small non-physical FWHM values). The fitting results can be found in Table 1.

3. Results

3.1. *VLBI Imaging and Model Fitting of AB Dor A*

All of our VLBI images of AB Dor A are presented in Fig. 1. As can be appreciate, we found a complex structure in all three epochs. In all cases, the brightest peak of emission is located at the map center, coincident (to within one beam size) with the expected position of AB Dor A according to the kinematics reported in Azulay *et al.* (2017). In 2007 and 2010, two emission peaks or components can be identified, separated by 3.1 $\pm$ 0.2 mas ($\sim$10 R_{star}) and clearly oriented east-west in both epochs. In the latter epoch, the double point-like structure is not as clearly separated as in 2007 resulting in a double core-halo morphology. Later on, in 2013 the structure becomes even more complex: the pair of brightest peaks (components 1 and 3) are separated by 5.7 $\pm$ 0.3 mas ($\sim$18 R_{star})

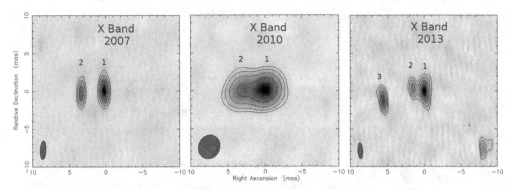

Figure 1. LBA images of all our observations of AB Dor A. Here and hereafter, north is up and east to the left. Numbers (if any) indicate the index assigned to each component. The contour levels, peak brightness and background rms noise for each image are as follows: *(2007)*: 11%, 22%, 44% and 88%, 2.35 mJy beam^{-1}, 0.06 mJy beam^{-1}; *(2010)*: 6%, 12%, 24%, 48%, and 96%, 2.16 mJy beam^{-1}, 0.05 mJy beam^{-1}; *(2013)*: 10%, 20%, 40% and 80%, 1.80 mJy beam^{-1}, 0.05 mJy beam^{-1}. All the images are centered at the expected positions of AB Dor A.

with a component between them (component 2) at 1.6 ± 0.1 mas ($\sim 5\,R_{\mathrm{star}}$) away from the center. The possible nature of this structure is discussed in Sect. 4; however it is important to notice that, given this complex morphology, the kinematics of AB Dor A might be artificially biased towards the centered peak of the images, not necessarily associated to the photosphere of the star, and not necessarily associated to the same feature for different epochs. Actually, this is likely the limiting factor of the precision of the orbit determination provided for AB Dor A (Azulay *et al.* 2017).

3.2. *Time analysis of AB Dor A images*

The images shown in Fig. 1 correspond to the structure of AB Dor A obtained with the full interferometric data set extending throughout the complete duration of each observation (typically 10 hr), which actually covers nearly one rotation period of the star (~ 12 hr). Therefore these images, and in particular, the complex structures observed at X-band, represent the emission of AB Dor A averaged over the turnaround period of the star.

If the morphology of AB Dor A happens to vary within the duration time of each observation, this will show a dependence of the observed structure with its rotation period. To further investigate this dependence we divided each X-band VLBI observation into two time intervals which allowed us to obtain two "snapshot" images for each epoch (shorter time intervals resulted in very sparse *uv* coverage and, therefore, maps of degraded quality with unreliable structures). We should emphasize that each snapshot image conserves its own astrometric information (referenced to the external quasar), that is, the snapshot images can be properly registered. As can be seen in Fig. 2, the snapshot maps corresponding to the same observational epoch (but different UT ranges) show remarkable differences. In 2007 and 2010 the snapshot images indicate that the double core-halo morphology is present at both intervals, a feature also visible in the entire data set images. However, the details of the structures change significantly on timescales of a few hours: the easternmost component seems to have rotated between snapshots with respect to the central, brightest emission an angle of $40 \pm 3°$ ($47 \pm 3°$) at epoch 2007 (2010). On the other hand, the distance between the peaks of the components increases (decreases) 0.7 ± 0.3 mas (1.5 ± 0.4 mas) between snapshots at epoch 2007 (2010). Both epochs start at similar rotational phases of the star: 0.30 for 2007 and 0.39 for 2010.

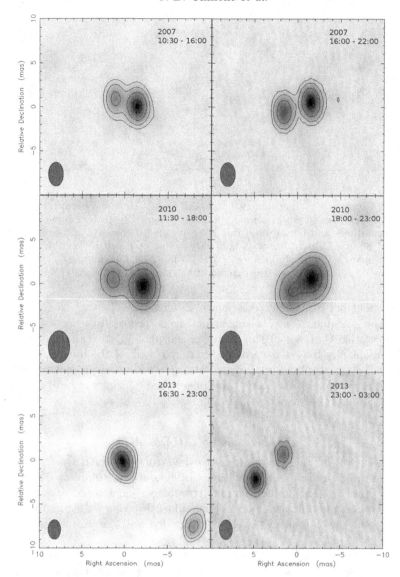

Figure 2. LBA snapshot images of AB Dor A at X band. The left column shows the first half of the observation while the right column the second half. The different epochs are shown in different rows. We have forced both the contour levels and the beam size to be exactly the same during the same epoch, choosing the average of the beam sizes of the observation halves. The FWHM beam size, its orientation, contour levels and coordinates of the map center for each epoch are as follows: *(2007)*: 1.75×2.7 mas at $0°$, 10%, 20%, 40% and 80% of 3.01 mJy beam^{-1}, $5{:}28{:}44.916066$, $-65{:}26{:}53.866084$; *(2010)*: 2.55×3.65 mas at $0°$, 15.5%, 31% and 62% of 2.46 mJy beam^{-1}, $5{:}28{:}44.943182$, $-65{:}26{:}53.454600$; *(2013)*: 1.5×2.2 mas at $0°$, 8%, 16%, 32% and 64% of 3.84 mJy beam^{-1}, $5{:}28{:}44.964236$, $-65{:}26{:}52.995000$.

Finally, the appearance of the snapshots at epoch 2013 (see Figure 2) shows a very different behaviour with a strong time dependence: The first snapshot, corresponding to the first half of the observation, shows a unique central component coincident with the brightness peak found in the maps constructed with the entire data set (named as comp. 1 in Fig. 1). However, somewhat surprisingly, this component 1 is not present in the second snapshot, where only components 2 and 3 are visible. We notice that, as seen in

Table 2. Flux density of the components
present in the snapshots (Fig. 2).

Epoch	S_1 (mJy)	S_2 (mJy)
11 Nov. 2007		
10:30-16:00	3.6 ± 0.3	1.1 ± 0.2
16:00-22:00	3.5 ± 0.6	1.5 ± 0.3
25 Oct. 2010		
11:30-18:00	1.8 ± 0.3	0.8 ± 0.3
18:00-23:00	1.7 ± 0.4	1.3 ± 0.4
16 Aug. 2013		
	S_1 (mJy)	S_4 (mJy)
16:30-23:00	5.4 ± 0.3	2.5 ± 1.1
	S_2 (mJy)	S_3 (mJy)
23:00-03:00	1.4 ± 0.2	2.1 ± 0.3

the maps, neither component 2 nor component 3 are spatially coincident with component 1, the latter corresponding to the predicted position by the known orbit of AB Dor A.

Following the same procedure described in Sect. 2, we fitted circular Gaussians to the interferometric visibilities for each time interval (Table 2). Due to the resemblance with the entire data set image and in order to make the comparison easier, we fixed the component sizes to those measured in Table 1. Both in 2007 and 2010, the brightest component flux remains constant during the entire observation while the second component might be slightly increasing in flux during the second half. The total detected flux (sum of the components) decreased a factor 2 from 2007 to 2010. Since the full observation image shows the average flux during the entire time, the 2013 components fluxes as they appear in different time intervals (Table 2) are greater than those measured in the entire data set image (Table 1).

4. Discussion

X-band detections (Fig. 1 and Fig. 2) present a challenge to the "electrons in a closed coronal loop" scenario. Possible explanations will be explored in detail in Climent *et al.* (in prep) and must account for the observed properties:

a) A complex internal structure. Not only an explanation for the emission at the expected position of AB Dor A but also for one (2007 and 2010) or two (2013) extra components is needed.

b) Extra components located at 5 R_{star}, 10 R_{star} and 18 R_{star} away from the central component in east-west direction.

c) Variability of the components position on a timescale of hours (see Fig. 2).

d) Low degree of circular polarization ($<10\%$) in all the components.

e) Brightness temperatures between 10^7 K and 10^8 K in all the components.

Previous studies of Algol (Mutel *et al.* 1998) and UV Ceti found that a strong, large-scale, dipole field could be consistent with the double-lobe structure observed in VLBI images of these objects. In these cases, one emission region would be located above one polar cap of the star while the other emission region would originate in a region above the other polar cap. Our 2007 and 2010 X-band images posses a great morphological resemblance with the double-lobe structure detected in Algol and UV Ceti. Hence, a polar cap model should be fully tested to see if it can properly explain our observations.

A flaring model where two magnetic loop structures are anchored to opposite sides of the star, similarly to what has been proposed for UX Arietis (Franciosini *et al.* 1999) and HR 1099 (Ransom *et al.* 2002), could explain our observations where the detected components would originate near or above the top of such loops where magnetic reconnection events occur. Due to the high frequency of slingshot prominences in AB Dor A, we may have detected the magnetic reconnection occuring on top of one of these prominences, that is, a helmet streamer similar to the solar ones. Although it is difficult to address this question with the limited data, one month after our 2007 observations, two big slingshot prominences were present in AB Dor A (Jardine, private communication). Although the lifespans of these phenomena are 2–3 days, this might be indicative that this scenario is, at least, plausible.

Although it may be tempting to interpret the X band images (Fig. 1) as a binary system (identifying AB Dor A as component 1 and the companion as component 2), the temporal analysis of Fig. 2 makes this scenario highly unlikely. Assuming that the axis of the orbital plane is parallel to the rotational axis of AB Dor A, at 3 mas separation, the orbital motion of component 2 (in 2007 and 2010) would imply a value of the radial velocity semi-amplitude of the stellar reflex motion much greater than the measured upper limit of ~ 1 Km·s^{-1}. Moreover, this hypothesis fails to explain why no motion is detected in component 1 in 2007 while, in 2010, component 1 clearly moves. Finally, this scenario is unable to properly explain the 2013 snapshot images since no companion would be detected during the first half of the observation while AB Dor A would disappear during the second half. For these reasons we conclude that the scenario of a companion to AB Dor A is highly unlikely and unable to reproduce our images.

As previously stated, the details of these hypothesis, how well they explain our observations, and their possible consequences will be further explored in Climent *et al.* (in prep).

References

Azulay, R., Guirado, J. C., Marcaide, J. M., *et al.* 2017, *A&A*, 607, A10

Barenfeld, S. A., Bubar, E. J., Mamajek, E. E., & Young, P. A. 2013, *ApJ*, 766, 6

Bell, C. P. M., Mamajek, E. E., & Naylor, T. 2015, *MNRAS*, 454, 593

Boccaletti, A., Chauvin, G., Baudoz, P., & Beuzit, J.-L. 2008, *A&A*, 482, 939

Close, L. M., Lenzen, R., Guirado, J. C., *et al.* 2005, *Nature*, 433, 286

Close, L. M., Thatte, N., Nielsen, E. L., *et al.* 2007, *ApJ*, 665, 736

Franciosini, E., Massi, M., Paredes, J. M., & Estalella, R. 1999, *A&A*, 341, 595

Guirado, J. C., Reynolds, J. E., Lestrade, J.-F., *et al.* 1997, *ApJ*, 490, 835

Guirado, J. C., Martí-Vidal, I., Marcaide, J. M., *et al.* 2006, *A&A*, 446, 733

Guirado, J. C., Marcaide, J. M., Martí-Vidal, I., *et al.* 2011, *A&A*, 533, A106

Herbst, W. & Shevchenko, V. S. 1999, *AJ*, 118, 1043

Janson, M., Brandner, W., Lenzen, R., *et al.* 2007, *A&A*, 462, 615

Lindegren, L. & Kovalevsky, J. 1995, *A&A*, 304, 189

Lestrade, J.-F., Jones, D. L., Preston, R. A., *et al.* 1995, *A&A*, 304, 182

López-Santiago, J., Montes, D., Crespo-Chacón, I., & Fernández-Figueroa, M. J. 2006, *ApJ*, 643, 1160

Luhman, K. L. & Potter, D. 2006, *ApJ*, 638, 887

Mutel, R. L., Molnar, L. A., Waltman, E. B., & Ghigo, F. D. 1998, *ApJ*, 507, 371

Nielsen, E. L., Close, L. M., Guirado, J. C., *et al.* 2005, *Astron. Nachr.*, 326,1033

Ransom, R. R., Bartel, N., Bietenholz, M. F., *et al.* 2002, *ApJ*, 572, 487

Shepherd, M. C., Pearson, T. J., & Taylor, G. B. 1994, *BAAS*, 26, 987

Zuckerman, B., Song, I., & Bessell, M. S. 2004, *ApJ*, 613, L65

Solar and Stellar Magnetic Fields: Origins and Manifestations
Proceedings IAU Symposium No. 354, 2019
A. Kosovichev, K. Strassmeier & M. Jardine, ed.
doi:10.1017/S1743921319010020

The impact of magnetism on tidal dynamics in the convective envelope of low-mass stars

A. Astoul[1], S. Mathis[1], C. Baruteau[2], F. Gallet[3], A. Strugarek[1],
K. C. Augustson[1], A. S. Brun[1] and E. Bolmont[4]

[1]AIM, CEA, CNRS, Université Paris-Saclay, Université Paris Diderot,
Sorbonne Paris Cité, F-91191 Gif-sur-Yvette, France
email: aurelie.astoul@cea.fr

[2]IRAP, Observatoire Midi-Pyrénées, Université de Toulouse,
14 avenue Edouard Belin, 31400 Toulouse, France

[3]Univ. Grenoble Alpes, CNRS, IPAG, 38000 Grenoble, France

[4]Observatoire de Genève, Université de Genève,
51 Chemin des Maillettes, CH-1290 Sauverny, Switzerland

Abstract. For the shortest period exoplanets, star-planet tidal interactions are likely to have played a major role in the ultimate orbital evolution of the planets and on the spin evolution of the host stars. Although low-mass stars are magnetically active objects, the question of how the star's magnetic field impacts the excitation, propagation and dissipation of tidal waves remains open. We have derived the magnetic contribution to the tidal interaction and estimated its amplitude throughout the structural and rotational evolution of low-mass stars (from K to F-type). We find that the star's magnetic field has little influence on the excitation of tidal waves in nearly circular and coplanar Hot-Jupiter systems, but that it has a major impact on the way waves are dissipated.

Keywords. MHD, waves, stars: evolution, stars: rotation, stars: magnetic fields, planetary systems

---

1. Introduction

Over the last two decades, about 4000 exoplanets have been discovered around low-mass stars (Perryman 2018). In close-in star-planet systems, tidal dissipation in the host star is known to affect the semi-major axis (and thus the orbital period) of the companion as well as the spin of the star over secular timescales (see e.g. Ogilvie 2014 for a review on this topic). In particular, the dissipation of the stellar dynamical and equilibrium tides (Zahn 1977) can vary significantly along the evolution of the star. It is highly dependent on stellar parameters like the angular velocity or the metallicity (Mathis 2015, Gallet *et al.* 2017 and Bolmont *et al.* 2017). Therefore, it is very important to identify and quantify in the most realistic way the dissipation processes that come into play. In this respect, we have examined the effect of stellar magnetism on the excitation and dissipation of dynamical tides inside the convective envelope of low-mass stars throughout their evolution. For this purpose, we have used detailed grids of rotating stellar models computed with the stellar evolution code STAREVOL, as well as databases of observed star-planet systems. We first examine (in Sect. 3) the impact of the star's magnetic field on the effective tidal forcing exciting magneto-inertial waves. The amplitude of a relatively large scale magnetic field is estimated via physical scaling laws at the base and the top of the convective envelope (Sect. 3). We then assess the ratio of the magnetic and

Table 1. Dynamo-like magnetic field derived from
simple energy or force balances.

Regime	Balance	Estimation of B_{dyn}
Equipartition	$\mathrm{ME} = \mathrm{KE}$	$\sqrt{\mu_0 2\mathrm{KE}}$
Buoyancy dynamo	$\mathrm{ME}/\mathrm{KE} = \mathrm{Ro}^{-1/2}$	$\sqrt{\mu_0 2\mathrm{KE}/\mathrm{Ro}^{1/2}}$
Magnetostrophy	$\boldsymbol{F}_{\mathrm{L}} = 2\rho_0 \boldsymbol{\Omega} \times \boldsymbol{u}$	$\sqrt{\mu_0 2\mathrm{KE}/\mathrm{Ro}}$

Notes. KE and ME are the kinetic and magnetic energy densities of
the convective flow, respectively. Ro is the fluid Rossby number.

hydrodynamic tidal forcings for several short-period exoplanets (Sect. 4) before analysing
the relative importance of viscous over Ohmic dissipation of kinetic and magnetic energies
(Sect. 5).

2. Influence of magnetism on the effective tidal forcing

In the presence of stellar magnetic fields, both the excitation and dissipation of tidal
waves are theoretically modified when compared to the hydrodynamical case because
of the Lorentz force and the magnetic diffusion. The linearised momentum equation for
tidal waves in a convective region can be written as (Lin & Ogilvie 2018):

$$\rho_0(\partial_t \boldsymbol{u} + 2\boldsymbol{\Omega} \times \boldsymbol{u}) + \boldsymbol{\nabla}p - \boldsymbol{F}_\nu - \boldsymbol{F}_{\mathrm{L}} = \boldsymbol{f}_{\mathrm{hydro}} + \boldsymbol{f}_{\mathrm{mag}}, \tag{2.1}$$

where we have introduced ρ_0 the mean density, Ω the spin of the star, $\boldsymbol{u}$, p the perturbed
flow and pressure, and $\boldsymbol{F}_\nu$, $\boldsymbol{F}_{\mathrm{L}}$ the effective viscous and Lorentz forces, respectively.
Magneto-inertial waves (left-hand side of Eq. (2.1)) are forced by an effective tidal forcing
(right-hand side), resulting mainly from the action of the Coriolis pseudo-force and the
Lorentz force on the equilibrium tide ($\boldsymbol{f}_{\mathrm{hydro}}$ and $\boldsymbol{f}_{\mathrm{mag}}$, respectively). Since $\boldsymbol{f}_{\mathrm{mag}}$ is often
neglected in studies of tidal interactions (see, e.g., Lin & Ogilvie 2018 and Wei 2016,
2018), we propose to examine its amplitude (f_{mag}) relative to f_{hydro} (the amplitude
of $\boldsymbol{f}_{\mathrm{hydro}}$) when varying the mass and age of low-mass stars. Using typical scales of
a star-planet system such as R the radius of the star and σ_{t} the tidal frequency, the
magneto-to-hydrodynamical forcing ratio can be recast as:

$$\frac{f_{\mathrm{mag}}}{f_{\mathrm{hydro}}} \sim \mathrm{Le}^2/\mathrm{Ro}_{\mathrm{t}}/\hat{\sigma}_{\mathrm{max}}, \tag{2.2}$$

with $\mathrm{Le} = B/(\sqrt{\rho\mu_0}2\Omega R)$ the Lehnert number (Lehnert 1954), $\mathrm{Ro}_{\mathrm{t}} = \sigma_{\mathrm{t}}/(2\Omega)$ the tidal
Rossby number, and $\hat{\sigma}_{\mathrm{max}} = \max\{\sigma_{\mathrm{t}}/(2\Omega), 1\}$ a dimensionless factor close to unity. We
refer to Astoul *et al.* (2019) for a detailed derivation of Eq. (2.2).

3. Scaling laws to estimate stellar magnetic fields

To evaluate the ratio of the effective tidal forces (Eq. 2.2), we made use of simple
energy and force balances to give a rough estimate of the dynamo-generated magnetic
field strength inside the convective zone of low-mass stars (see, e.g., Brun *et al.* 2015 and
Augustson *et al.* 2019). These scaling laws are listed in Table 1. From this dynamo-like
magnetic field (B_{dyn}), we also estimate a large-scale dipolar magnetic field at the top of
the convective envelope:

$$B_{\mathrm{dip}} = \gamma(r/R)^3 B_{\mathrm{dyn}}, \tag{3.1}$$

where γ can be understood as the ratio of the large-scale to small-scale magnetic fields,
or as the fraction of the total energy stored in the dipolar component of the magnetic
field (see Astoul *et al.* 2019 for more details). Unless otherwise stated, B_{dyn} is computed
at the radial interface (r) between the radiative and the convective zones, which is the
expected location for the development of a large-scale dynamo (Brun & Browning 2017).

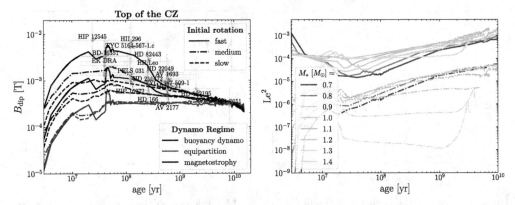

Figure 1. *Left:* surface dipolar magnetic field versus time for a $0.9 M_\odot$ star. The curves are computed with a grid of STAREVOL models, for different scaling laws and initial stellar rotation rates (see legend). The symbols ★ depict the mean dipolar magnetic field observed at the surface of $0.9 M_\odot$ stars (See *et al.* 2019). *Right:* evolution of the Lehnert number squared over time, at the base (*solid curves*) and the top (*dashed dotted curves*) of the convective envelope for various low-mass stars, using the magnetostrophic dynamo regime and the median initial rotation.

In the left panel of Fig. 1, we display the time evolution of the surface dipolar magnetic field (B_{dip}, Eq. 3.1) for a $0.9 M_\odot$ star and for the three scaling laws in Table 1. A few observational data are overplotted for comparison (See *et al.* 2019). To compute B_{dip}, parameters like the convective turnover time and velocity were obtained from grids of models computed with the 1D stellar evolution code STAREVOL (Amard *et al.* 2019). These grids were calculated for low-mass stars between $0.7 M_\odot$ and $1.4 M_\odot$, from the early pre-main sequence until the end of the main sequence. Three initial rotation rates (fast, median and slow) have been chosen (see Amard *et al.* 2019 for more details). We see that the observed and estimated surface dipolar magnetic fields are in good agreement when using the magnetostrophic regime, in particular when assuming fast initial rotation. Note that the large-scale to small-scale ratio γ has been kept constant in this analysis. The right panel in Fig. 1 shows the Lehnert number squared against time for stars of various masses for the magnetostrophic regime and median initial rotation. The panel reveals that Le^2 is higher at the base than at the top of the convective zone, similar to what can be expected from the magnetic field amplitude inside the convective envelope of a low-mass star. Moreover, Le^2 increases overall with time and with mass (except for the most massive stars).

4. Magnetic tidal forcing in observed star-planet systems

The tidal frequency σ_{t} in Eq. (2.2) depends on the orbital frequency of the planet Ω_{o} and the spin frequency Ω of the star. When the orbit is quasi-circular and coplanar, the quadrupolar component of the tidal potential dominates (Ogilvie 2014) and the tidal frequency writes $\sigma_{\mathrm{t}} = 2(\Omega_{\mathrm{o}} - \Omega)$. In the left panel of Fig. 2, the ratio of magnetic and hydrodynamical effective forcings has been calculated for observed quasi-circular and coplanar star-planet systems in the main-sequence. The masses, radius, age, and the orbital and/or spin frequencies of the planets and their host star have been extracted from the Extrasolar Planets Encyclopaedia. For each system, Le^2 is calculated via the STAREVOL grid models. The ratio $f_{\mathrm{mag}}/f_{\mathrm{hydro}}$ is higher at the base of the convective zone than near the surface, and grows with stellar mass, in line with the trends identified in the previous section. We emphasize that the forcing ratio is far from unity for all considered star-planet systems, meaning that the contribution of the Lorentz force to the effective tidal forcing is weak.

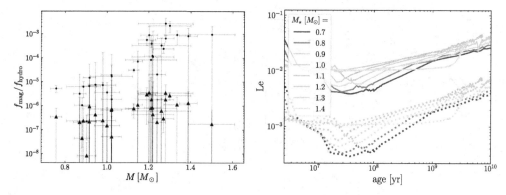

Figure 2. *Left:* ratio of the Lorentz and Coriolis tidal forcing against the mass of the star for various observed short-period exoplanetary systems. The ratio $f_{\rm mag}/f_{\rm hydro}$ (Eq. 2.2) is estimated at the base (*blue bullet*) and the top (*red triangle*) of the convective zone, and the magnetostrophic regime is used to calculate $B_{\rm dyn}$. *Right:* Lehnert number at the base of the convective zone (*solid lines*) versus age for different masses of the star. The typical Lehnert number above which Ohmic dissipation of magneto-inertial waves dominates over viscous dissipation (Lin & Ogilvie 2018) is shown by dotted curves using a turbulent magnetic Ekman number.

5. Dissipation of kinetic and magnetic energies

In the previous section, we have highlighted the negligible impact of the star's magnetic field in the effective tidal forcing. However, the star's magnetic field may still play an important role in the propagation and dissipation of magneto-inertial waves. In that regard, Lin & Ogilvie (2018) have shown that the Ohmic dissipation of magneto-inertial waves becomes comparable to viscous dissipation when the Lehnert number becomes of order $\mathrm{Em}^{2/3}$, where Em is the magnetic Ekman number. The right panel of Fig. 2 displays the time evolution of the Lehnert number at the base of the convective zone for various low-mass stars, as well as the threshold derived by Lin & Ogilvie (2018). We observe that Le is always an order of magnitude larger than this threshold. Therefore, Ohmic dissipation prevails over viscous dissipation for all low-mass stars at the base of the convective zone. At the top of the convective zone (not shown here), this statement is less clear-cut, especially for stars with $M \gtrsim 1.2 M_{\odot}$, for which both Ohmic and viscous dissipations are comparable (Astoul *et al.* 2019).

6. Conclusion

We have shown that the large-scale dynamo-generated magnetic field of a star has a limited impact on the forcing of tidal waves in the convective envelope of K, G, F-type stars all along their evolution from the pre-main sequence until the terminal age main-sequence. Nevertheless, stellar magnetism is found to have a strong influence on the dissipation mechanism of dynamical tides inside the convective envelope of these stars. Our results therefore indicate that a full magneto-hydrodynamic treatment of the propagation and dissipation of tidal waves is needed to assess the impact of star-planet tidal interactions for all low-mass stars along their evolution.

Acknowledgements

A. Astoul, K. Augustson, E. Bolmont, and S. Mathis acknowledge funding by the European Research Council through the ERC grant SPIRE 647383. The authors acknowledge the PLATO CNES funding at CEA/IRFU/DAp, IRAP and INSU/PNP. The authors further thank V. See for fruitful discussions and the use of his data. F. Gallet acknowledges financial support from a CNES fellowship. A.S.Brun acknowledges funding by ERC WHOLESUN 810218 grant, INSU/PNST, and CNES Solar Orbiter. This work has been carried out within the framework of the NCCR PlanetS supported by the Swiss National Science Foundation.

References

Amard, L., Palacios, A., Charbonnel, C., Gallet, F., Georgy, C., Lagarde, N., Siess, L., *et al.* 2019, *A&A*, 631, A77

Astoul, A., Mathis, S., Baruteau, C., Gallet, F., Strugarek, A., Augustson, K.C., Brun, A. S., Bolmont, E., *et al.* 2019, *A&A (in press)*, https://doi.org/10.1051/0004-6361/201936477

Augustson, K. C., Brun, A. S., Toomre, J., *et al.* 2019 *ApJ*, 876, 83

Bolmont, E., Gallet, F., Mathis, S., Charbonnel, C., Amard, L., *et al.* 2019, *EAS Publications Series*, 82, 71

Bolmont, E., Gallet, F., Mathis, S., Charbonnel, C., Amard, L., Alibert, Y., *et al.* 2017, *A&A*, 604, A113

Brun, A. S. & Browning, M. K. 2017, *Living Reviews in Solar Physics*, 14, 4

Brun, A. S., Garcia, R. A., Houdek, G., Nandy, D., Pinsonnneault, M., *et al.* 2015, *Space Sci. Rev.*, 196, 303

Gallet, F., Bolmont, E., Mathis, S., Charbonnel, C., Amard, L., *et al.* 2017, *A&A*, 604, A112

Lehnert, B. 1954, *ApJ*, 119, 647

Lin, Y. & Ogilvie, G. I. 2018, *MNRAS*, 474, 1644

Mathis, S. 2015, *A&A*, 580, L3

Ogilvie, G. 2014, *ARA&A*, 52, 171

Perryman, M. 2018, *The Exoplanet Handbook*

See, V., Matt, S. P., Finley, A. J., Folsom, C. P., Boro Saikia, S., Donati, J. F., Fares, R., Hébrard, É. M., Jardine, M. M., Jeffers, S. V., Marsden, S. C., Mengel, M. W., Morin, J., Petit, P., Vidotto, A. A., Waite, I. A., & The BCool Collaboration 2019, *ApJ (in press)*, arXiv:1910.02129

Wei, X. 2016, *ApJ*, 828, 30

Wei, X. 2018, *ApJ*, 854, 34

Zahn, J. P. 1977, *A&A*, 500, 121

Solar and Stellar Magnetic Fields: Origins and Manifestations
Proceedings IAU Symposium No. 354, 2019
A. Kosovichev, K. Strassmeier & M. Jardine, ed.
doi:10.1017/S1743921319009797

The rotation of low mass stars at 30 Myr in the cluster NGC 3766

Julia Roquette[1] [ID], Jerome Bouvier[2], Estelle Moraux[2],
Herve Bouy[3], Jonathan Irwin[4], Suzanne Aigrain[5]
and Régis Lachaume[6,7]

[1]Department of Physics and Astronomy, University of Exeter, Physics Building,
Stocker Road, Exeter, EX4 4QL, United Kingdom
email: jt574@exeter.ac.uk

[2]Univ. Grenoble Alpes, CNRS IPAG, F-38000 Grenoble, France

[3]Laboratoire d'astrophysique de Bordeaux, Univ. Bordeaux,
CNRS, B18N, allée Geoffroy Saint-Hilaire, 33615 Pessac, France.

[4]Smithsonian Astrophysical Observatory 60 Garden Street Cambridge, MA 02138 USA

[5]Department of Physics, University of Oxford, Keble Road, OX3 1RH, UK

[6]Instituto de Astronomía, Facultad de Física, PUC-Chile, casilla 306, Santiago 22, Chile

[7]Max-Planck-Institut für Astronomie, Königstuhl 17, D-69117 Heidelberg, Germany

Abstract. Together with the stellar rotation, the spotted surfaces of low-mass magnetically active stars produce modulations in their brightness. These modulations can be resolved by photometric variability surveys, allowing direct measurements of stellar spin rates. In this proceedings, we present results of a multisite photometric survey dedicated to the measurement of spin rates in the 30 Myr cluster NGC 3766. Inside the framework of the Monitor Project, the cluster was monitored during 2014 in the i-band by the Wide Field Imager at the MPG/ESO 2.2-m telescope. Data from Gaia-DR2 and *grizY* photometry from DECam/CTIO were used to identify cluster members. We present spin rates measured for ∼200 cluster members.

Keywords. stars: low-mass, brown dwarfs, stars: pre–main-sequence, stars: rotation, stars: spots

---

1. Introduction

Along with the mass and initial composition, angular momentum is a fundamental stellar property, and it has a direct influence in the stellar structure, close environment, magnetic field, and their evolution with time. Inside the context of the angular momentum evolution of low mass stars, it is relatively well understood, from both models and observations, that once pre-main-sequence (PMS) stars lose their disks, they follow a spin-up phase, which is later followed by a decrease of their spin rates with age. Even though tenths of thousands of rotational periods for stars of all ages have been measured in the past decades, very few data is available for stars with ages around 20−40 Myr. Filling this age gap may help us better understand the end of the PMS spin-up phase.

In this ongoing study, we are working on measuring spin rates for low mass stars in the 30 Myrs old open cluster NGC 3766. With a higher mass population of more than 200 B-type stars, NGC 3766 is at a distance of ∼2kpc (dM = 11.8 in our estimation with Gaia DR2 data) and has a moderate extinction of E(B-V) = 0.22 (Aidelman *et al.* 2012).

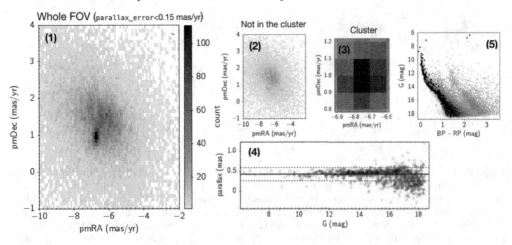

Figure 1. Panels **(1)** to **(3)** show Gaia DR2 proper motion diagrams with: **(1)** data for the whole field of view observed by WFI@ESO; **(2)** non-cluster members, and **(3)** selection of cluster members done using a squared-box. The sources selected inside the squared-box in **(3)** had their parallaxes values plotted as a function of G magnitude and are shown in panel **(4)**. Continuous and dashed black lines show median and 1σ values, which were used as a cut to clean member selection in **(3)**. Panel **(5)** shows a Gaia colour-magnitude diagram (CMD) with selected cluster members (black dots). The Gaia DR2 data used include the same cuts for high-quality astrometric data as in CG18, but we did not make any cut for the magnitude of fainter stars.

2. Data Analysis

2.1. *Dataset*

NGC 3766 was observed during 2014 inside the framework of the Monitor Project. *WFI-dataset*: It was monitored in the i-band by the Wide Field Imager at the MPG/ESO 2.2-m telescope at La Silla in Chile, with a total of 402 exposures of 300 s in the i Sloan band, carried out in 20 nights between May 6 and Jun. 29, and spanning 54 nights. Light-curves were created as in Irwin *et al.* (2007). *CTIO-dataset*: Additional single-epoch *grizY* data were obtained from DECam/CTIO archive. We also used Gaia DR2 astrometric data.

2.2. *Membership evaluation*

Brighter members down to $\sim$1 $M\odot$ were previously identified in proper motion studies by Yadav *et al.* (2013, hereafter Y13) and Cantat-Gaudin *et al.* (2018, Gaia DR2 study, hereafter CG18). Y13 gives 274 members with $P_{\mathrm{memb}} > 70\%$ and *WFI-data*, and CG18 gives 57 members with $P_{\mathrm{memb}} > 70\%$ and *WFI-data*. The two combined give a list of 327 good member candidates with valid light-curve in the *WFI-dataset*.

We used Gaia DR2 data to complement the list of members, identify likely background and foreground stars and eliminate those from the posterior analysis. Figure 1 shows the steps followed to distinguish possible members from non-members. The procedure selects 231 member candidates shown in Figure 1 **(5)**: 51 of which were also identified by CG18 or Y13, 60 were new member candidates with $G > 16$ mag, but 120 member-candidates from Y13 and CG18 that had P_{memb} in the range 10−70% were reintroduced to the list. Those flagged but kept in the member-candidate list. The combination of the selection of members in Figure 1 with previously identified members accounts for a list of 507 member candidates. In the remaining text, we refer to this sample as *Proper Motion Members*. The selection of background and foreground sources in Figure 1 **(2)** accounts for 14,402 sources eliminated from the *WFI-dataset*.

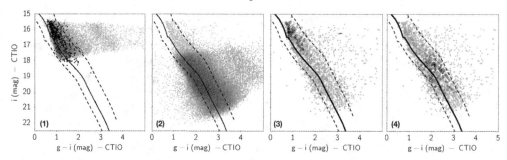

Figure 2. Example of *gi*-CMD built with the *CTIO-dataset* and used for identifying cluster members. Panel **(1)**: all sources with WFI light-curve and with good quality Gaia DR2 data are shown as grey dots, dark dots show the position of the *Proper Motion Members* including literature members and also those Figure 1. Panel **(2)**: sources with WFI light-curve and without Gaia DR2 data are shown as grey dots; Panel **(3)**: Gray dots show sources from (1) that showed periodic variability in their WFI-light-curves. *Proper Motion Members* that were also periodic are shown as black circles. Panel **(4)**: Gray dots show sources from (2) that showed periodic variability in their WFI-light-curves. Black circles show *Photometry member candidates* that were also periodic. The bold line shows a MESA 30 Myrs isochrone at cluster distance and median reddening. The dashed lines show the shifted isochrones used for selecting member-candidates. *Photometry member candidates* were selected based on their position between the dashed line in multiple colour-diagrams.

Fainter members of NGC 3766 were identified based on their position in the cluster loci in the colour-space. Figure 2 shows some of the steps followed: We built CMDs for various sets of colours (*grizY*) from our *CTIO-dataset*. We split each CMD diagram into two samples: 1- a sample of sources that had good quality data in Gaia DR2 (Figure 2 **(1)**), and 2- a sample of sources without counterpart in Gaia (Figure 2 **(2)**). In 1- we only consider as members those sources that were *Proper Motion Members*, and we ruled out the sample of stars that were not in the cluster according to their proper motion and parallax. For the data in 2-, we used a MESA 30 Myrs isochrone to identify the position of the Cluster sequence in the CMD, at the distance and median reddening of NGC3766: we then shifted this isochrone by ±0.25mag in each axis, plus 0.75 mag in the y-axis, and we selected member candidates sources that fell in between the shifted isochrones. A source was considered as a good photometric member candidate if this procedure selected it in several of the CMD for the *grizY* colours.

2.3. *Time-series analysis*

To select periodic stars, we initially used a combination of Lomb-Scargle Periodogram (Lomb 1976; Scargle 1982), String-Length statistics Clarke (2002) and Saunder-Statistics (Saunders *et al.* 2006) and visual inspection of light-curves and periodograms. The left panel in Figure 3 shows a distribution of the power of maximum peak in the Lomb-Scargle periodogram as a function of period.

3. A first glimpse of the rotational scenario in NGC 3766

Among cluster members, we measured possible spin rates for 289: 221 of which are associated with member candidates identified with photometry only and 68 are associated with member candidates identified by proper motion studies. Figure 3 shows the current period distribution and period-colour diagram for our sample. The period distributions seem to have intermediate characteristics between those in the period distribution for the 13 Myr old hPer (Moraux *et al.* 2013) and the 40 Myr old NGC 2547 (Irwin *et al.* 2008).

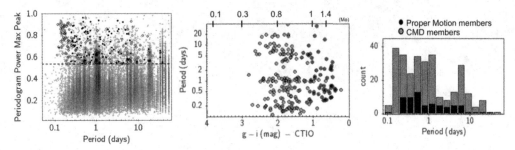

Figure 3. *Left:* Distribution of the power of the highest peak in the periodogram shown as a function of the highest peak period for each source with valid light-curve in the *WFI-dataset*. The dashed line shows the power for a False Alarm Probability of 0.01. Members from astrometry selected as periodic are shown as full black circles and member candidates selected in the CMD of Figure 2 that were periodic are shown as black circles with a grey filling. Period-colour (mass) diagram (*middle*) and Period distribution (*left*) are shown for the 68 proper motion members and for the 221 CMD members.

3.1. *Conclusions and Future work*

The period distributions in Figure 3 still show an excess of stars with periods around 1 day, which are likely aliases. Ongoing analysis of the light-curves with the fast-χ^2 method Palmer (2009) will help to improve the contamination due to the 1 day aliases. Ongoing completeness analysis and mass derivation will complement the results. A comparison with other well studied young clusters will help to contextualise the results inside the observational and theoretical scenario for the rotational evolution of young stars.

References

Aidelman, Y., Cidale, L. S., Zorec, J., *et al.* 2012, *A&A*, 544, A6
Cantat-Gaudin, T., Jordi, C., Vallenari, A., *et al.* 2018, *A&A*, 618, A93
Clarke, D. 2002, *A&A*, 386, 763
Irwin, J., Irwin, M., Aigrain, S., *et al.* 2007, *MNRAS*, 375, 1449
Irwin, J., Hodgkin, S., Aigrain, S., *et al.* 2008, *MNRAS*, 383, 1588
Lomb, N. R. 1976, *Ap&SS*, 39, 447
Moraux, E., Artemenko, S., Bouvier, J., *et al.* 2013, *A&A*, 560, A13
Palmer, D. M. 2009, *ApJ*, 695, 496
Saunders, E. S., Naylor, T., & Allan, A. 2006, *Astronomische Nachrichten*, 327, 783
Scargle, J. D. 1982, *ApJ*, 263, 835
Yadav, R. K. S., Sariya, D. P., & Sagar, R. 2013, *MNRAS*, 430, 3350

Chapter 5. Role of magnetic fields in solar and stellar variability

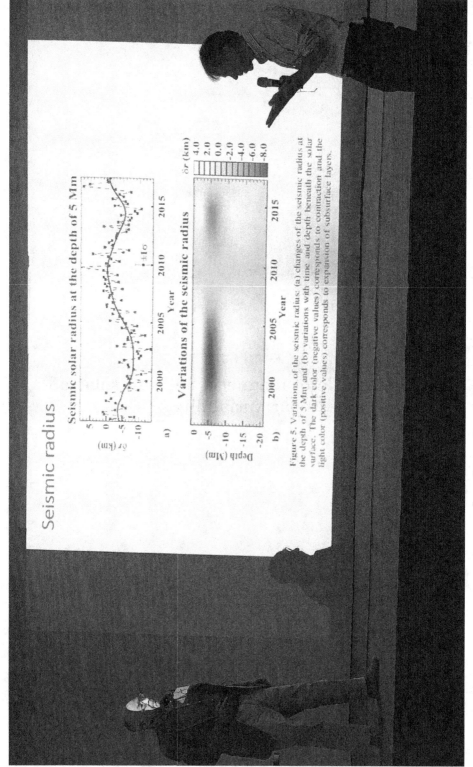

Jean-Pierre Rozelot and Axel Brandenburg

Solar and Stellar Magnetic Fields: Origins and Manifestations
Proceedings IAU Symposium No. 354, 2019
A. Kosovichev, K. Strassmeier & M. Jardine, ed.
doi:10.1017/S1743921319009852

Possible evidence for a magnetic dynamo in hot Algols

R. E. Mennickent[1]⊙, J. Garcés[1], G. Djurašević[2,3], G. Rojas[1], D. Schleicher[1] and S. Otero[4]

[1]Universidad de Concepción, Departamento de Astronomía, Casilla 160-C, Concepción, Chile
emails: rmennick@udec.cl, jgarcesletelier@gmail.com, gonzrojas@udec.cl,
dschleicher@astro-udec.cl

[2]Astronomical Observatory, Volgina 7, 11060 Belgrade 38, Serbia
email: gdjurasevic@aob.rs

[3]Issac Newton institute of Chile, Yugoslavia Branch, 11060, Belgrade, Serbia

[4]American Association of Variable Star Observers, USA
email: sebastian@aavso.org

Abstract. We present the case of hot semi-detached Algols showing photometric cycles longer than the orbital period. The evidence indicating that this long cycle might be due to a magnetic dynamo operating in the rapidly rotating donor star is examined.

Keywords. binaries: spectroscopic, close, stars: evolution.

Double Periodic Variables (DPVs) are hot semi-detached Algols with β Lyrae type light curves showing a long cycle lasting in average about 33 times the orbital period (Mennickent *et al.* 2003). The more evolved star transfers matter onto the hotter, less-evolved B-type star through Roche-lobe overflow, forming an accretion disc around it. More than 200 DPVs have been found in the Galaxy and Magellanic Clouds. Catalogs of DPVs have been published by Poleski *et al.* (2010), Pawlak *et al.* (2013) and Mennickent (2017). Recently, G. Rojas (M.Sc. thesis, University of Concepción, 2019, Fig. 1) found 34 new Galactic DPVs after a search in the catalog "Eclipsing and ellipsoidal binary systems towards the Galactic Bulge" by Soszyński *et al.* (2016). DPVs are hotter and more luminous than classical Algols. Galactic DPVs include β Lyrae (Guinan 1989) and AU Mon (Lorenzi 1980). Kalv (1979) interpreted the 516-d periodicity of RX Cas as pulsations of the Roche-lobe filling star. Pulsation is also suggested by Guinan (1989) for the B8 II secondary of β Lyrae. Peters (1984) suggests that the long cycles in AU Mon are due to cyclic pulsations of the mass transferring star. The DPVs have been reviewed by Mennickent *et al.* (2016) and Mennickent (2017).

A recent hypothesis for the long cycle involves a magnetic dynamo operating in the donor star through the Applegate mechanism (Schleicher & Mennickent 2017). A modulation of the strength of the wind generated in the stream/disc interaction region has been proposed as the cause for the long cycle. This could happen when the Applegate (1992) mechanism modifies the stellar quadrupole momentum redistributing angular momentum in the binary and producing cyclic mass loss through the inner Lagrangian point linked to changes in the equatorial radius of the donor. Actually, magneto-hydrodynamical simulations reveal changes in the stellar structure due to magnetic cycles and suggest how relevant the stellar rotation can be in rapidly rotating orbitally synchronized binary star components (Navarrete *et al.* 2019). Rapid rotation might generate stellar dynamos even in early A-type giants, as suggested by the presence of chromospheric emission

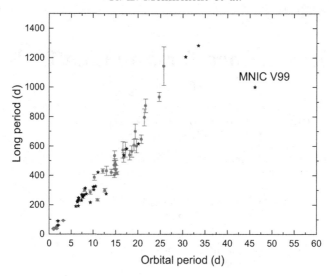

Figure 1. Long versus orbital period for Galactic DPVs. The new DPVs found by Rojas (M.Sc. thesis 2019) are shown by dots while the rest are listed in Mennickent (2017), Rosales & Mennicken (2017, 2018, 2019) and the VSX database (https://www.aavso.org/vsx/). MNIC V99, reported by Nikolay Mishevskiy in VSX, has a period ratio of 21.67, similar to β Lyrae (21.25).

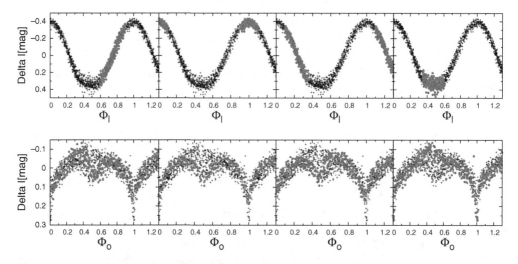

Figure 2. Disentangled long cycle (up) and orbital (down) light curves phased with the respective periods for OGLE-LMC-DPV-097. Crooses show the complete datasets, dots show segments of the data of the long cycle (Garcés *et al.* 2018). The changes in the orbital light curve might reflect structural changes in the accretion disk due to variable mass transfer.

in some DPVs and the increase of the dynamo number during mass transfer episodes (e.g. Mennickent, Schleicher & San Martin-Perez 2018, San Martín-Pérez *et al.* 2019). In cases of synchronous rotation Schleicher & Mennickent (2017) arrive to a theoretical expression that fits relatively well the observed DPV cycle lengths. The scenario of a magnetic dynamo might explain remarkable changes observed in the orbital light curve of OGLE-LMC-DPV-097, OGLE-BLG-ECL-157529 and some other DPVs during the long cycle (Garcés *et al.* 2018, Garcés M.Sc. thesis, University of Concepción, 2019, Mennickent *et al.* 2020, Fig. 2 and Fig. 3). Future studies should try to get direct evidence of magnetic fields in these systems, for example through spectropolarimetric

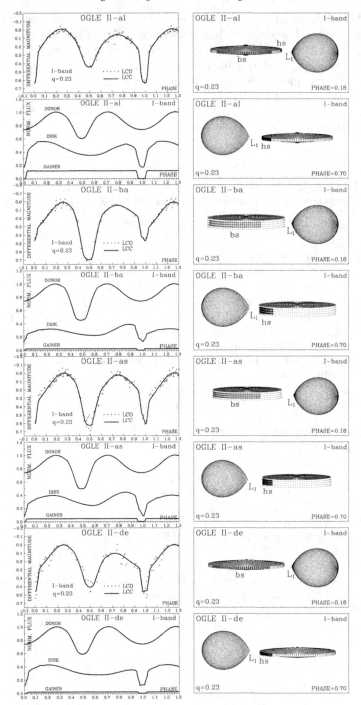

Figure 3. Orbital light curve models for OGLE-BLG-ECL-157529 at the high stage (al), low stage (ba), ascending branch (as) and descending branch (de) of the long cycle (Mennickent *et al.* 2020).

techniques. Actually, magnetic fields have been inferred in β Lyrae from the analysis of polarized light; these fields could produce magnetically driven streams onto the accretion disk (Skulskij 1982, 2018). Still is not clear if magnetic fields are present in DPVs in general, and why some semi-detached binaries containing B-type components do not show long-term photometric modulations (e.g. Koubský *et al.* 2019).

We acknowledge support by VRID-Enlace 216.016.002-1.0, the BASAL Centro de Astrofísica y Tecnologías Afines (CATA) PFB–06/2007, FONDECYT 1190621 and FONDECYT 1161247.

References

Applegate, J. H. 1992, *ApJ*, 385, 621

Garcés, L, J., Mennickent, R. E., Djurašević, G., Poleski, R., Soszyński, I., *et al.* 2018, *MNRAS*, 477, 11

Guinan, E. F. 1989, *SSRv*, 50, 35

Kalv, P. 1979, *TarOT*, 58, 3

Koubský, P. *et al.*. 2019, *A&A*, 629, A105

Mennickent R. E. 2017, *SerAJ*, 194, 1

Mennickent, R. E., Pietrzyński, G., Diaz, M., Gieren, W., *et al.* 2003, *A&A*, 399, 47

Mennickent, R. E., Otero, S., Kołaczkowski, Z., *et al.* 2016, *MNRAS*, 455, 1728

Mennickent, R. E., Schleicher, D. R. G., San Martin-Perez, R., *et al.* 2018, *PASP*, 130, 94203

Mennickent, R. E., Garcés, L. J., Djurašević, G., Poleski, R., Soszyński, I., Iwanek, P., Schleicher, D., *et al.* 2020, submitted

Navarrete, F. H., Schleicher, D. R. G., Käpylä, P. J., Schober, J., Völschow, M., Mennicken, R. E., *et al.* 2019, *MNRAS.tmp*, 2642

Pawlak, M., Graczyk, D., Soszyński, I., Pietrukowicz, P., Poleski, R.,Udalski, A., Szymański, M. K., Kubiak, M., Pietrzyński, G., Wyrzykowski, Ł., Ulaczyk, K., Kozłowski, S., Skowron, J., *et al.* 2013, *Acta Astron*, 63, 323

Peters, G. J. 1994, *ASPC*, 384, ASPC...56

Poleski, R. *et al.*, 2010, *AcA*, 60, 179

Rosales, G. J. & Mennickent, R. E. 2017, *IBVS*, 6207, 1

Rosales J. A. & Mennickent, R. E. 2018, *IBVS*, 6248, 1

Rosales J. A. & Mennickent R. E. 2019, *IBVS*, 6268, 1

San Martín-Pérez, R. I., Schleicher, D. R. G., Mennickent, R. E., *et al.* 2019, *Boletin de la Asociacion Argentina de Astronomia*, La Plata Argentina, 61, 107

Schleicher, D. R. G. & Mennickent, R. E. 2017, *A&A*, 602, 109

Skulskij M. Y. 1982, *SvAL*, 8, 126

Skulsky M. Y. 2018, *CoSka*, 48, 300

Soszyński, I., Pawlak, M., Pietrukowicz, P., Udalski, A., Szymański, M. K., Wyrzykowski, Ł., Ulaczyk, K., Poleski, R., Kozłowski, S., Skowron, D. M., Skowron, J., Mróz, P., Hamanowicz, A., *et al.* 2016, *Acta Astron*, 66, 405

Solar and Stellar Magnetic Fields: Origins and Manifestations
Proceedings IAU Symposium No. 354, 2019
A. Kosovichev, K. Strassmeier & M. Jardine, ed.
doi:10.1017/S1743921320000381

New Candidates for Chromospherically Young, Kinematically Old Stars

Eduardo Machado Pereira[ID] and Helio J. Rocha Pinto

Observatório do Valongo, Universidade Federal do Rio de Janeiro,
Ladeira do Pedro Antônio, 43, - Centro, 20080-090, Rio de Janeiro - RJ, Brazil
email: `eduardo11@astro.ufrj.br`

Abstract. Roughly speaking, young stars are associated to intense chromospheric activity (CA), whereas it decreases with stellar aging. However, some objects that show high kinematical components – in turn, associated to older stars – reveal CA similar to that of young ones; we call these stars chromospherically young and kinematically old (CYKOs). One hypothesis that could explain their occurrence is the merge of a short-period binary, from which the outcome would be a chromospherically active, kinematically evolved star. Considering that they evolved separately, we expect them to be lithium depleted, and therefore we look for CYKO stars by analyzing their lithium content (λ 6707 Å). We present a preliminary list of 48 stars matching this criteria, aiming to either confirm or discard the coalescence of a short-period pair hypothesis.

Keywords. late type stars, stars: chromospheric activity, space velocities, stars: spectroscopy

---

1. Chromospheric activity as a proxy for youth

The chromospheric activity (CA) of a star is a set of phenomena responsible for dumping mechanical energy into the so called chromosphere, a layer just above although hotter than the photosphere. Because of this activity, radiative equilibrium does not explain alone the observed heating. Information on the CA can provide insights on the time evolution of stellar magnetic fields, as well as on mass and radiative fluxes through their complex magnetic topology (Hall 2008).

One of the various methods to measure the CA of a star is through the H and K lines from the singly ionized calcium. Vaughan *et al.* (1978) introduced the now famous *S*-index, for the Mount Wilson Observatory HK project, after projecting two triangular filters in order to measure the purely chromospheric emission seen as reversals at the centers of those lines. Aiming to normalize this emission to stellar continuum, other two neighbor continuum bandpasses were also designed by these authors. These features can also be seen in more recent works, e.g. in Figure 3 by Schröder *et al.* (2009). However, due to this continuum normalization, the *S*-index is color-dependent, hence Noyes *et al.* (1984) updated it into a new index, R'_{HK}, that takes into account both the continuum normalization and an eventual photospheric contamination in H and K lines reversals, enabling the comparison of the CA of different stars.

The relation between CA and age arises from the fact that it is tightly linked to stellar rotation and magnetic activity, since these features evolve in time (Skumanich 1972). It is well established in literature (e.g. Skumanich 1972; Soderblom *et al.* 1991; Lorenzo-Oliveira *et al.* (2018) that for sun-like dwarfs stellar rotation and magnetic activity tend to decrease with age, due to angular momentum loss through magnetized winds and structural variations on evolutionary timescales, and therefore a CA–age relation can be calibrated. A relation derived by Lorenzo-Oliveira *et al.* (2018) can be seen in left panel of their Figure 7, showing that intense chromospheric activity is associated to young stars.

2. Kinematical features as a proxy for old ages

Since their birth, stars experience kinematical evolution as they age and go through their orbits around the Galactic center. This evolution results in a statistical correlation known as disk heating (Wielen 1977), a net increase in Galactic space motions with time. Almeida-Fernandes & Rocha-Pinto (2018) summarize some mechanisms that can explain the fluctuations that a star can encounter when traveling around the Galactic center: encounters with giant molecular clouds, interaction with non-axisymmetric Galactic structures, interactions with satellite galaxies, etc.

In order to quantify this, stellar space velocities can be derived from proper motions and radial velocities, and can be put in a common reference system, such as the Local Stantard of Rest (LSR), valid for stars in the solar neighborhood. In a three dimensional motion, stars move around the Galactic center with velocities u (pointing towards the center), v (pointing towards the rotation direction) and w (pointing towards Galactic north pole).

In Figure 2 from Rocha-Pinto *et al.* (2004), it is easy to qualitatively visualize this dispersion behaviour: space velocities u, v, w (with respect to LSR and corrected for accounting solar motion) for stars in the solar neighborhood show a clear spread towards older (in this case, chromospheric) ages; this is seen to be true for any of the tree components. One can derive an age–velocity dispersion relation, and we can state that high space velocities components, i.e. motions that are way faster (or slower) than the LSR, are more commonly associated with old stars.

3. Identifying CYKOS

Here we revisit the work of Rocha-Pinto *et al.* (2002) (hereafter RP+02) and follow the same formalism introduced by these authors. We used 3 samples of stars having known chromospheric activity: a list assembled by one of us (Rocha-Pinto) in the course of other various studies; the largest CA compilation we found in the literature, from Boro-Saikia *et al.* (2018); and a catalogue build by Murgas *et al.* (2013), in an investigation on comoving groups, which ended up providing space velocities components as well. For the first two samples, with no kinematical data available from the source, we used the Geneva-Copenhagen Survey (Holmberg *et al.* 2009) to retrieve u, v, w. In order to have zero-centered values, we took solar motion into account by using Almeida-Fernandes & Rocha-Pinto (2018) results.

After crossing and merging these samples into one list, we plotted the chromospherically active stars in a space velocities diagram, from which we selected objects lying outside a 3σ dispersion limit, yielding a final list of 48 CYKOs candidates, 15 more than in Rocha-Pinto *et al.* (2002). The result of applying this methdology can be seen in Figure 1, where small dots, inside the full ellipse, are stars with normal space velocities components, whereas larger dots, outside the full ellipse, represent our candidates.

In an effort to explain the ocurrence of these stars, we perform spectroscopic observations and compare them to normal objects, under the hypothesis explained in the next section.

4. CYKOS's nature: coalescence and lithium depletion

We look for objects that, at first, seem inconsistent: chromospherically active stars with high space velocities components, firstly noticed by Soderblom (1990). Rocha-Pinto *et al.* (2002) dedicated a whole paper to study these objects, and concluded that they could be single stars formed out of the coalescence of short-period binaries. The star should look chromospherically younger — since the orbital angular momentum of the former binary is transformed into rotational angular momentum of the new single star — and keep

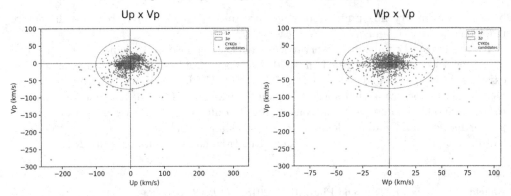

Figure 1. Space velocities diagram with chromospherically active stars: the full, outer ellipse represents the 3σ dispersion limit for space velocities components. CYKOS candidates are seen outside this line, shown by larger dots than those inside.

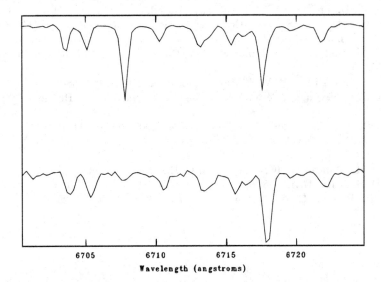

Figure 2. For comparison, two representative stars that we observed for this work. Top: a normal chromospherically active star, showing a strong lithium line (λ 6707 Å); bottom: a CYKOS candidate selected from our list, showing almost no lithium content.

the kinematical signatures of an advanced age — since the movement of the binaries' center of mass around the Galactic center is inherited by the new single star. In this way, CYKOs would be formed by stars that evolved individually until coalescence.

We can confirm this hypothesis by analyzing the lithium content in these objects, more specifically through the absorption line 6708 Å: it is well known that Sun-like stars deplete their surface Li abundances over time (e.g. Soderblom 2010, and references therein), and thus the presence of this element in stellar spectra can be used as a youth indicator. Assuming the coalescence scenario, we expect to find no or low presence of lithium, as these are individually evolved objects.

In Figure 2 we compare spectra for two representative stars that we observed up to date: at the top, a normal active star with ordinary space velocities components, showing a proeminent lithium line at λ 6707 Å; at the bottom, a CYKOS candidate, with high space velocities components, showing almost no lithium content. The vertical axis must be read as arbitrary flux units, and it was not made explicit merely because this figure is for comparison means only.

We are currently finishing observational reduction phase and awaiting for other two observing runs to be performed in 2020. We also want to take advantage of already existing databases (e.g. ESO-HARPS, Gaia-ESO, California-Kepler Survey, etc.) that have available spectra in that spectral region. Aditionally, GAIA data is part of our plans to enhance our kinematical data, considering it could provide precise measurements for proper motions and radial velocities. The results and conclusions achieved in this work will be presented as a M.Sc. thesis of one of us (Machado-Pereira) in July 2020.

References

Almeida-Fernandes, F. & Rocha-Pinto, H. J. 2018, *MNRAS*, 476, 184

Boro Saikia, S., Marvin, C. J., Jeffers, S. V., Reiners, A., Cameron, R., Marsden, S. C., Petit, P., Warnecke, J., Yadav, A. P., *et al.* 1993, *A&A*, 616, A108

Hall, J. C. 2008, *Solar Phys.* (Review), 5, article id. 2

Holmberg, J., Nordström, B., Andersen, J., *et al.* 2009, *A&A*, 501, 941

Lorenzo-Oliveira, D., Freitas, F. C., Meléndez, J., Bedell, M., Ramírez, I., Bean, J. L., Asplund, M., Spina, L., Dreizler, S., Alves-Brito, A., Casagrande, L., *et al.* 2018, *A&A*, 619, A73

Murgas, F., Jenkins, J. S., Rojo, P., Jones, H. R. A., Pinfield, D. J., *et al.* 2013, *A&A*, 552, A27

Noyes, R. W., Hartmann, L. W., Baliunas, S. L., Duncan, D. K., Vaughan, A. H., *et al.* 1984, *ApJ*, 279, 763

Rocha-Pinto, H. J., Castilho, B. V., Maciel, W. J., *et al.* 2002, *A&A*, 384, 912

Rocha-Pinto, H. J., Flynn, C., Scalo, J., Hänninen, J., Maciel, W. J., Hensler, G., *et al.* 2004, *A&A*, 423, 517

Schröder, C., Reiners, A., Schmitt, J. H. M. M., *et al.* 2009, *A&A*, 493, 1099

Skumanich, A. 1972, *ApJ*, 171, 565

Soderblom, D. R. 1990, *AJ*, 100, 204

Soderblom, D. R., Duncan, D. K., Johnson, D. R. H., *et al.* 1991, *ApJ*, 375, 722

Soderblom, D. R. 2010, *ARAA*, 48, 581

Vaughan, A. H., Preston, G. W., Wilson, O. C., *et al.* 1978, *PASP*, 90, 267

Wielen, R. 1977, *A&A*, 60(n. 2), 263

Solar and Stellar Magnetic Fields: Origins and Manifestations
Proceedings IAU Symposium No. 354, 2019
A. Kosovichev, K. Strassmeier & M. Jardine, ed.
doi:10.1017/S174392132000037X

The dynamo-wind feedback loop : Assessing their non-linear interplay

Barbara Perri[1,2]⬤, Allan Sacha Brun[1], Antoine Strugarek[1]⬤ and Victor Réville[3]

[1]AIM, CEA, CNRS, Université Paris-Saclay, Université Paris-Diderot, Sorbonne Paris Cité, F-91191 Gif-sur-Yvette, France

[2]Institut d'Astrophysique Spatiale, CNRS, Université Paris-Sud, Université Paris-Saclay, Bât. 121, 91405 Orsay Cedex, France

[3]IRAP, Université de Toulouse, CNRS, UPS, CNES, Toulouse, France, 14 Avenue Edouard Belin, F-31400 Toulouse, France

Abstract. Though generated deep inside the convection zone, the solar magnetic field has a direct impact on the Earth space environment via the Parker spiral. It strongly modulates the solar wind in the whole heliosphere, especially its latitudinal and longitudinal speed distribution over the years. However the wind also influences the topology of the coronal magnetic field by opening the magnetic field lines in the coronal holes, which can affect the inner magnetic field of the star by altering the dynamo boundary conditions. This coupling is especially difficult to model because it covers a large variety of spatio-temporal scales. Quasi-static studies have begun to help us unveil how the dynamo-generated magnetic field shapes the wind, but the full interplay between the solar dynamo and the solar wind still eludes our understanding.

We use the compressible magnetohydrodynamical (MHD) code PLUTO to compute simultaneously in 2.5D the generation and evolution of magnetic field inside the star via an α-Ω dynamo process and the corresponding evolution of a polytropic coronal wind over several activity cycles for a young Sun. A multi-layered boundary condition at the surface of the star connects the inner and outer stellar layers, allowing both to adapt dynamically. Our continuously coupled dynamo-wind model allows us to characterize how the solar wind conditions change as a function of the cycle phase, and also to quantify the evolution of integrated quantities such as the Alfvén radius. We further assess the impact of the solar wind on the dynamo itself by comparing our results with and without wind feedback.

Keywords. (magnetohydrodynamics:) MHD, Sun: activity, Sun: corona, Sun: magnetic fields, (Sun:) solar wind, (Sun:) solar-terrestrial relations

1. Introduction

The Sun exhibits a magnetic activity cycle, which has an 11-year period for amplitude. One general framework explaining such a generation of large-scale magnetic field is the interface dynamo (Parker 1993): the differential rotation profile in the convection zone of the star (Thompson *et al.* 2003) leads to the generation of strong toroidal fields at the tachocline (Spiegel & Zahn 1992), which in turn is used to regenerate poloidal fields thanks to the combination of turbulence, buoyancy and Coriolis force at the surface. This dynamo loop allows for the amplification of the initial magnetic field until saturation, thus sustaining it against ohmic dissipation (Brun & Browning 2017). A simplified yet efficient approach to model it is the mean-field dynamo framework (Moffatt 1978), focusing on large-scale fields and assuming axisymmetry. The generation of toroidal field through differential rotation is then deemed the Ω effect, and the regeneration of the poloidal or toroidal field via turbulence is deemed the α effect. This description has the advantages of being easy to implement in MHD simulations with low computational costs, and yielding realistic results (Charbonneau 2010). On the other hand, in the outer layers of the Sun, one of the main phenomena is the transsonic and transalfvénic solar wind. The first hydrodynamical description was given by Parker (1958), and magnetism was added by Weber & Davis (1967) and Sakurai (1985) to yield a better description of the corresponding torque applied to the star. Stellar wind of solar-like stars can be described using 2.5D axisymmetric MHD simulations as well (Keppens & Goedbloed 1999; Matt & Pudritz 2008).

Recent observations, by the satellite *Ulysses* for cycle 22 and the beginning of cycle 23 (McComas *et al.* 2008), or by the satellite *OMNI* for cycles 23 and 24 (Owens *et al.* 2017), have shown that there is a correlation between the 11-year dynamo cycle and the evolution of the corona. During a minimum of activity, the magnetic field is low in amplitude and its topology is mostly dipolar ; the corona is very structured with fast wind at the poles (around 800 km/s) associated with coronal holes, and slow wind at the equator (around 400 km/s) associated with streamers. During a maximum of activity, the magnetic field is high in amplitude and its topology is a mixture of high modes, dominated by the quadrupolar modes (DeRosa *et al.* 2012) ; fast and slow solar winds can be found at all latitudes. This suggests that there is a coupling operating between the interior and the exterior of the Sun, but we still don't know precisely how it is operating and on which timescales. From a theoretical and numerical point of view, it is however very difficult to study all of these layers simultaneously : the magnetic field and the wind evolve over very different scales (from hours to years and from a few solar radii to 1 AU). The physical properties of their respective environment are also very different; take for instance the rapidly changing β plasma parameter, which is the ratio of the thermal pressure over the magnetic pressure, from more than 1 inside the star to less than 1 in the chromosphere (Gary 2001). Finally, from a mathematical point of view, the MHD equations are stiff, meaning that it requires small time and grid steps to be solved. All of these disparities make the modeling of this coupling a numerical challenge.

There have been various attempts to resolve this problem with different approaches. A first attempt is to use a quasi-static approach, meaning that the coupling is modeled through a series of wind relaxed states corresponding to a sequence of magnetic field configurations evolving in time. These models can be data-driven, using series of magnetic field observations (Luhmann *et al.* 2002; Réville & Brun 2017), or rely on the numerical coupling between two codes dedicated respectively to the inner and outer layers of the Sun (Pinto *et al.* 2011; Perri *et al.* 2018b). Another approach is to zoom on the surface with a numerical box of a few tens of Megameters to capture the small time and spatial scales, which means this approach can include small-scale physical processes (for example convection or radiative transfer at the surface) but on a short period of time and only for a specific region of the Sun (Vögler *et al.* 2005; Stein & Nordlund 2006; WedemeyerBöhm *et al.* 2009; Gudiksen *et al.* 2011). Finally there have been some attempts to model a dynamical coupling on a global scale, for example in von Rekowski & Brandenburg (2006) with a simulation box including both the star and its corona, but for a T-Tauri star and with a disk interaction. Our aim is to focus on solar-like stars and on the large-scale field by using mean-field and axisymmetry assumptions. We design a 2.5D numerical model including the star and its corona, and design an interface to control the complex and diverse interactions between the two zones.

2. Numerical setup

2.1. *Wind model*

Our wind model is adapted from Réville *et al.* (2015a). We solve the set of the conservative ideal MHD equations composed of continuity equation for the density ρ, the momentum equation for the velocity field $\mathbf{v}$ with its momentum written $\mathbf{m} = \rho\mathbf{v}$, the equation for the total energy E and the induction equation for the magnetic field $\mathbf{B}$:

$$\frac{\partial}{\partial t}\rho + \nabla \cdot \rho\mathbf{v} = 0, \qquad (2.1)$$

$$\frac{\partial}{\partial t}\mathbf{m} + \nabla \cdot (\mathbf{mv} - \mathbf{BB} + \mathbf{I}p) = \rho\mathbf{a}, \qquad (2.2)$$

$$\frac{\partial}{\partial t}E + \nabla \cdot ((E + p)\mathbf{v} - \mathbf{B}(\mathbf{v} \cdot \mathbf{B})) = \mathbf{m} \cdot \mathbf{a}, \qquad (2.3)$$

$$\frac{\partial}{\partial t}\mathbf{B} + \nabla \cdot (\mathbf{vB} - \mathbf{Bv}) = 0, \qquad (2.4)$$

where p is the total pressure (thermal and magnetic), $\mathbf{I}$ is the identity matrix and $\mathbf{a}$ is a source term (gravitational acceleration in our case). We use the ideal equation of state $\rho\varepsilon = p_{th}/(\gamma - 1)$, where p_{th} is the thermal pressure, ε is the internal energy per mass and γ is the adiabatic exponent. This gives for the energy : $E = \rho\varepsilon + \mathbf{m}^2/(2\rho) + \mathbf{B}^2/2$.

PLUTO solves normalized equations, using three variables to set all the others: length, density and speed. If we note with $*$ the parameters related to the star and with 0 the parameters related to the normalization, we have $R_*/R_0 = 1$, $\rho_*/\rho_0 = 1$ and $V_K/V_0 = \sqrt{GM_*/R_*}/V_0 = 1$, where V_K is the Keplerian speed at the stellar surface and G the gravitational constant. In our set-up, we choose $R_0 = R_\odot = 6.96\ 10^{10}$ cm, $\rho_0 = \rho_\odot = 6.68\ 10^{-16}$ g/cm^3 and $V_0 = V_{K,\odot} = 4.37\ 10^2$ km/s. Our wind simulations are then controlled by three parameters : the adiabatic exponent $\gamma = 1.05$ for the polytropic wind, the rotation of the star normalized by the escape velocity $v_{rot}/v_{esc} = 2.93\ 10^{-3}$ and the speed of sound normalized also by the escape velocity $c_s/v_{esc} = 0.243$. Note that the escape velocity is defined as $v_{esc} = \sqrt{2}V_K = \sqrt{2GM_*/R_*}$. Such values correspond to a $1.3\ 10^6$ K hot isothermal corona rotating at the solar rotation rate.

We assume axisymmetry and use the spherical coordinates (r, θ, ϕ). We choose a finite-volume method using an approximate Riemann Solver (here the HLL solver, cf. Einfeldt (1988)). PLUTO uses a reconstruct-solve-average approach using a set of primitive variables $(\rho, \mathbf{v}, p, \mathbf{B})$ to solve the Riemann problem corresponding to the previous set of equations. The time evolution is then implemented via a second order Runge-Kutta method. To enforce the divergence-free property of the field, we use a hyperbolic divergence cleaning, which means that the induction equation is coupled to a generalized Lagrange multiplier in order to compensate the deviations from a divergence-free field (Dedner *et al.* 2002). We do not use the traditional approach of splitting between the curl-free background field and the fluctuation field $\delta\mathbf{B}$, because in our case the background field will be the dynamo field generated inside the star and evolving at each time-step.

The numerical domain dedicated to the wind computation is an annular meridional cut with the colatitude $\theta \in [0, \pi]$ and the radius $r \in [1.01, 20]R_\odot$. We use an uniform grid in latitude with 256 points, and a stretched grid in radius with 400 points; the grid spacing is geometrically increasing from $\Delta r/R_* = 0.002$ at the surface of the star to $\Delta r/R_* = 0.02$ at the outer boundary. At the latitudinal boundaries ($\theta = 0$ and $\theta = \pi$), we set axisymmetric boundary conditions. At the top radial boundary ($r = 20R_*$), we set an outflow boundary condition which corresponds to $\partial/\partial r = 0$ for all variables, except for the radial magnetic field where we enforce $\partial(r^2 B_r)/\partial r = 0$. Because the wind has opened the field lines and under the assumption of axisymmetry, this ensures the divergence-free property of the field. We initialize the velocity field with a polytropic wind solution and the magnetic field with a dipole. The right panel of Figure 1 shows an example of a wind simulation only.

2.2. *Dynamo model*

As we have seen in equation 2.4, PLUTO solves the full non-linear ideal induction equation. For the dynamo inside the star, we will consider that the magnetic field $\mathbf{B}$

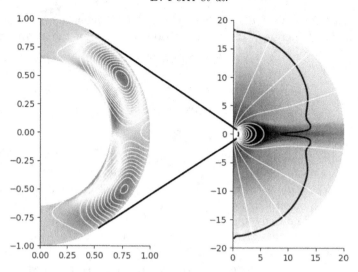

Figure 1. Examples of simulations with the dynamo (on the left) or the wind (on the right) models. For the dynamo the colorscale represents the toroidal magnetic field B_ϕ and the white lines are the poloidal magnetic field lines. For the wind the colorscale represents the quantity $\mathbf{v} \cdot \mathbf{B}/(c_s \| \mathbf{B} \|)$, the black line corresponds to the Alfvén radius and the white lines to the poloidal magnetic field lines.

is the large-scale mean field, and thus implement an alternative form of the induction equation with ohmic diffusion:

$$\frac{\partial}{\partial t}\mathbf{B} + \nabla \cdot (\mathbf{v}\mathbf{B} - \mathbf{B}\mathbf{v}) = \nabla \times (\alpha\mathbf{B}) - \nabla \times (\eta \times \nabla \times \mathbf{B}), \tag{2.5}$$

where η is the effective magnetic diffusivity and α is a coefficient for the α effect obtained by First Order Smooth Approximation (FOSA) of the electro-motive force (emf) (Pouquet *et al.* 1976). Hence the Ω effect is taken into account with the second term on the left, and the α effect with the first one on the right. This form of the induction equation is only active inside the star ($r < R_\odot$); no other equation is solved there. This new induction equation follows the same normalization as described before. However, when talking about dynamo parameters, the community usually refers to the parameters $C_\alpha = \alpha_0 R_\odot/\eta_t$, $C_\Omega = \Omega_0 R_\odot^2/\eta_t$ and $R_e = V_0 R_\odot/\eta_t$. To make it more convenient, we will use in this article the traditional control parameters of the dynamo models, just note that there is a factor $\eta_t/(R_\odot V_K)$ to switch to the PLUTO normalization (where η_t is the turbulent magnetic diffusivity, see eq. 2.8).

For the physical parameters, we got inspiration from case B of Jouve *et al.* (2008). The rotation in this zone is solar-like with a solid body rotation below $0.66 R_\odot$ and differential rotation above, with the equator rotating faster than the poles:

$$\Omega(r, \theta) = \Omega_c + \frac{1}{2}\left(1 + \mathrm{erf}\left(\frac{r - r_c}{d}\right)\right)\left(1 - \Omega_c - c_2 \cos^2\theta\right), \tag{2.6}$$

where $\Omega_c = 0.92$, $r_c = 0.7 R_\odot$, $d = 0.02$ and $c_2 = 0.2$. We recall that the physical amplitude of the Ω effect is given by the C_Ω parameter. In this model, we do not have any poloidal flows, and hence no meridional circulation.

As a first simple approximation, the α effect has no latitudinal and radial dependence in the convection zone, and is zero in the radiative zone, with a smooth transition between the two zones:

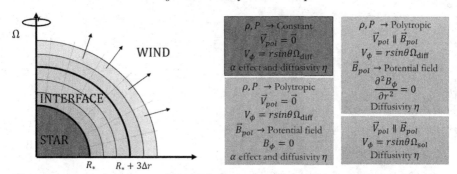

Figure 2. Schematic of our interface region. The grey box describes the conditions for inside the star, the orange one for the first layer of the interface, the green one for the last two points of the interface and the blue one for the first two points of the wind model.

$$\alpha(r, \theta) = \frac{3\sqrt{3}}{4} \sin^2\theta \cos\theta \left(1 + \mathrm{erf}\left(\frac{r - r_c}{d}\right)\right) \left(1 + \left(\frac{B_\phi(r_c, \theta, t)}{B_0}\right)^2\right)^{-1}. \tag{2.7}$$

The factor $3\sqrt{3}/4$ is used as a normalization to have a maximum amplitude of 1 in the convection zone for this profile. The quenching term (which is the last term in equation 2.7) allows for saturation of the magnetic field near the reference magnetic field value B_0.

We have a jump in diffusivity between the radiative and the convection zone of two orders of magnitudes:

$$\frac{\eta}{\eta_t}(r) = \frac{\eta_c}{\eta_t} + \frac{1}{2}\left(1 - \frac{\eta_c}{\eta_t}\right)\left(1 + \mathrm{erf}\left(\frac{r - r_c}{d}\right)\right), \tag{2.8}$$

with $\eta_c/\eta_t = 10^{-2}$ and η_t a parameter that we fix. The magnetic field is initialized with a dipole confined in the convection zone. Hence B_ϕ is initially equal to 0, but will grow through dynamo action.

The numerical domain dedicated to the dynamo computation is an annular meridional cut with the colatitude $\theta \in [0, \pi]$ and the radius $r \in [0.6, 1.01]R_\odot$. We use a uniform grid in latitude with 256 points, and a uniform grid in radius with 200 points, which yields a grid spacing of $\Delta r/R_* = 0.002$. At the latitudinal boundaries ($\theta = 0$ and $\theta = \pi$), we set axisymmetric boundary conditions. For the bottom boundary condition ($r = 0.65R_\odot$) we use a perfect conductor condition. For the top boundary condition ($r = R_\odot$), we use the two first layers of the interface, which will be described in the next section. The left panel of Figure 1 shows an example of a dynamo simulation only.

2.3. Interface principles

The interface layer between the dynamo and the wind computational zones is shown in Figure 2 by the orange and green areas. This interface is crucial to control and understand the interactions between the two zones.

The interface is divided into three layers of one radial grid point for all latitudes. This is tailored to a numerical method using a 2-point stencil linear reconstruction, but could be adapted to other methods. The first two layers constitute the boundary condition for the dynamo, and the last two layers constitute the boundary condition for the wind. We also alter the solution in the first two points of the wind computational domain for more stability. The first layer in orange is very similar to the dynamo zone, except that now the density and pressure decrease following a polytropic law (but with a continuous link with the constant value inside the star), and the magnetic field is extrapolated as a potential field using the value of the last point of the dynamo zone. We still have an

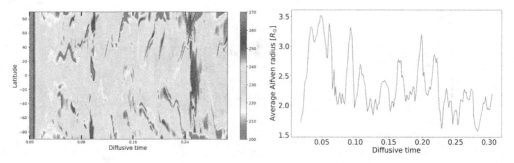

Figure 3. Evolution of the wind in the coupled model. On the left we show a time-latitude diagram of the wind speed at $20R_\odot$ in km/s. On the right we show the time evolution of the average Alfvén radius as a fraction of the diffusive time $t_\eta = R_\odot^2/\eta_t$.

α effect for continuity, but the equations are not evolved in the interface. In the second and third layers of the interface in green, we set conditions for the wind : the poloidal speed is aligned with the poloidal magnetic field extrapolated from below and $\partial_r^2 B_\phi$ is set to 0 to limit the generation of currents at the surface of the star (imposed from right to left). In the first two layers of the wind computational domain in blue, we impose the poloidal speed to remain parallel to the wind poloidal field to limit again the generation of currents. We also impose some diffusivity a bit further than the star surface using the following expression :

$$\eta = \frac{1}{2\eta_{norm}}\left(\eta_c + \frac{1}{2}(\eta_t - \eta_c)\right)\left(1 + \mathrm{erf}\left(\frac{r - r_c}{d}\right)\right)\left(1 + \tanh\left(\frac{r_\eta - r}{d_\eta}\right)\right), \qquad (2.9)$$

with $r_\eta = 1.015$ and $d_\eta = 0.003$. This allows the diffusivity to drop only after around ten grid points above the interface. The way the interface is designed, the dynamo magnetic field can influence the wind via the potential extrapolation of B_r and B_θ, and the wind can back-react on the dynamo via the B_ϕ condition that changes the dynamo boundary conditions.

3. Evolution along activity cycles

To validate the coupling from a theoretical point of view, we focus on a model whose dynamo period is shorter than the solar dynamo, thus more related to a young Sun. This helps bring closer the timescales of the dynamo and the wind. To design this model, we adapt the magnetic diffusivity to set the cycle period by adjusting the diffusive time $t_\eta = R_\odot^2/\eta_t$ using $\eta_t = 3.7\ 10^{14}$ cm^2.s^{-1}; then we set $C_\Omega = 3.4\ 10^1$ to have the solar rotation rate ; finally we set $C_\alpha = 1.44\ 10^4$ to have a dynamo number $D = C_\Omega \times C_\alpha$ above the dynamo threshold. This yields a case where C_Ω is smaller than C_α, which corresponds to a strong generation of poloidal field due to convective turbulence.

This model allows us to obtain in one simulation and in a completely self-consistent way without breaking causality the dynamical evolution of the corona in response to activity cycles. In Figure 3, we can see the time-latitude diagram for the wind speed. Extrapolated to 1 AU, we obtain speeds between 420 and 560 km/s, which correspond to the slow wind component due to our polytropic approximation. However we see faster and slower components inside our wind with a difference of up to 70 km/s between the two components. The evolution of the corona is highly dynamical with a lot of transients associated to the continuous response of the wind to the changing field. The associated streamers are very thin and evolve quickly, the wind has the time to adapt to the oscillatory dynamo field but cannot reach a stationary state.

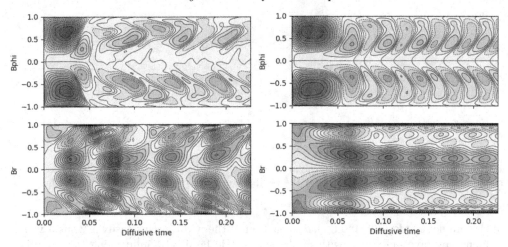

Figure 4. Butterfly diagrams with and without the feedback of the wind (respectively on the left and on the right).

We can also see the evolution of integrated quantities, as shown in the right panel of Figure 3. We will focus on the average Alfvén radius $\langle r_A \rangle$. The Alfvén radius is defined as the distance at which the wind speed equals the Alfvén speed $v_A = B/\sqrt{4\pi\rho}$, and the average Alfvén radius, as defined in Pinto *et al.* (2011), corresponds to the Alfvén radius averaged by the mass flux through the surface of a sphere. The average Alfvén radius evolves between 1.5 and 3.5 $R_\odot$; this is smaller than for the present Sun because the coupling needs a weaker field to operate with such a fast dynamo. The mass loss evolves between 3.5 and 6.5 10^{-14} $M_\odot$/yr, which is just a bit more than the solar values estimated between 2.3 and 3.1 10^{-14} $M_\odot$/yr (McComas *et al.* 2008; Réville & Brun 2017). The angular momentum loss evolves between 1.0 and 8.0 10^{29} cgs, which is the same order of magnitude as for the Sun. We see a lot of modulations in time with the evolution of the activity cycles, obtained for the first time in a completely auto-coherent way.

4. Feedback loop : influence of the wind on the dynamo solution

In this last section we will focus on the feedback loop between the dynamo and the wind. As said in the introduction, a variety of codes have shown the influence of the magnetic field generated by the dynamo on the coronal structures, but it is more difficult to evaluate the potential impact of the wind on the interior of the star. To test the impact of this feedback loop, we have run two models : in the first case, we used the interface described before, which allows the wind to back-react on the boundary conditions ; in the second case, we imposed the condition $B_\phi = 0$ in the second interface layer, thus cutting the feedback from the wind. The corresponding butterfly diagrams are shown in Figure 4. We see a clear difference, starting from $0.03t_\eta$ with $t_\eta = R_\odot^2/\eta_t$. Without the wind influence (right panel), the tachocline toroidal and surface radial magnetic fields are equatorially anti-symmetric and the cycle is regular with a period of $0.05t_\eta$. With the influence of the wind (left panel), the cycle takes a longer time to stabilize to finally reach a period of about $0.05t_\eta$. The cycle tends to become equatorially symmetric, although at most times a North-South asymmetry is still noticeable. To understand this difference, we looked at the evolution in time of the dipolar and quadrupolar modes ($\ell = 1$ and $\ell = 2$) for these two cases (cf. Figure 5). Without the influence of the wind, the dipolar mode is almost constant, slightly decreasing, while the quadrupolar mode has an amplitude 5 orders of magnitude less. Hence the symmetric family is negligible. But with the influence of the wind, the quadrupolar mode grows to an amplitude equivalent

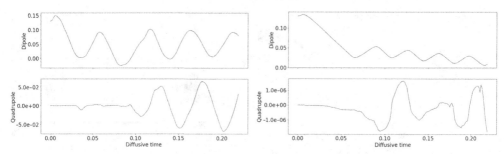

Figure 5. Time evolution of the dipolar ($\ell = 1$) and quadrupolar ($\ell = 2$) modes of the surface radial magnetic field with and without the feedback of the wind (respectively on the left and on the right).

of the dipolar mode (70%). The back-reaction of the wind in this case has thus a visible influence by favoring the growth of the symmetric family by influencing the boundary conditions of the dynamo.

It is too soon to know if this result can be generalized to any coupled system. Indeed, this case has pretty extreme values for a dynamo; Tavakol *et al.* (1995) has shown that in such parameter regimes, the dynamo is highly non-linear and can easily switch from symmetric to anti-symmetric regimes because of the non-linear quenching or asymmetry of the physical parameters. We have also performed a threshold study similar to Jouve & Brun (2007) and have determined that for this set of parameters the quadrupolar mode has a growth critical threshold lower than the dipolar mode ($C_\alpha^Q = 80$ versus $C_\alpha^D = 100$). This study is a highly sophisticated proof of concept to demonstrate that the feedback-loop between the dynamo and the wind is present in simulations, and needs to be more thoroughly investigated to understand the physical implications for stars.

References

Brun, A. S. & Browning, M. K. 2017, *Living Reviews in Solar Physics*, 14, 4
Charbonneau, P. 2010, *Living Reviews in Solar Physics*, 7, 3
Dedner, A., Kemm, F., Kröner, D., *et al.* 2002, *Journal of Computational Physics*, 175, 645
DeRosa, M. L., Brun, A. S., & Hoeksema, J. T. 2012, *ApJ*, 757, 96
Einfeldt, B. 1988, *SIAM Journal on Numerical Analysis*, 25, 294
Gary, G. A. 2001, *Sol. Phys.*, 203, 71
Gudiksen, B. V., Carlsson, M., Hansteen, V. H., *et al.* 2011, *A&A*, 531, A154
Jouve, L. & Brun, A. S. 2007, *A&A*, 474, 239
Jouve, L., Brun, A. S., Arlt, R., *et al.* 2008, *A&A*, 483, 949
Keppens, R. & Goedbloed, J. P. 1999, *A&A*, 343, 251
Luhmann, J. G., Li, Y., Arge, C. N., *et al.* 2002, *Journal of Geophysical Research (Space Physics)*, 107, 1154
Matt, S. & Pudritz, R. E. 2008, *ApJ*, 678, 1109
McComas, D. J., Ebert, R. W., Elliott, H. A., *et al.* 2008, *Geophys. Res. Lett.*, 35, L18103
Moffatt, H. K. 1978, *Cambridge Monographs on Mechanics and Applied Mathematics*
Owens, M. J., Lockwood, M., & Riley, P. 2017, *Scientific Reports*, 7, 41548
Parker, E. N. 1958, *ApJ*, 128, 664
Parker, E. N. 1993, *ApJ*, 408, 707
Perri, B., Brun, A. S., Réville, V., *et al.* 2018, *Journal of Plasma Physics*, 84, 765840501
Pinto, R. F., Brun, A. S., Jouve, L., *et al.* 2011, *ApJ*, 737, 72
Pouquet, A., Frisch, U., & Leorat, J. 1976, *Journal of Fluid Mechanics*, 77, 321
Réville, V., Brun, A. S., Matt, S. P., *et al.* 2015, *ApJ*, 798, 116
Réville, V. & Brun, A. S. 2017, *ApJ*, 850, 45
Sakurai, T. 1985, *A&A*, 152, 121

Spiegel, E. A. & Zahn, J.-P. 1992, *A&A*, 265, 106
Stein, R. F. & Nordlund, Å. 2006, *ApJ*, 642, 1246
Tavakol, R., Tworkowski, A. S., Brandenburg, A., *et al.* 1995, *A&A*, 296, 269
Thompson, M. J., Christensen-Dalsgaard, J., Miesch, M. S., *et al.* 2003, *ARA&A*, 41, 599
Vögler, A., Shelyag, S., Schüssler, M., *et al.* 2005, *A&A*, 429, 335
von Rekowski, B. & Brandenburg, A. 2006, *Astronomische Nachrichten*, 327, 53
Weber, E. J. & Davis, L. 1967, *ApJ*, 148, 217
Wedemeyer-Böhm, S., Lagg, A., & Nordlund, Å. 2009, *Space Sci. Rev.*, 144, 317

Solar and Stellar Magnetic Fields: Origins and Manifestations
Proceedings IAU Symposium No. 354, 2019
A. Kosovichev, K. Strassmeier & M. Jardine, ed.
doi:10.1017/S1743921320000903

Statistical analysis of geomagnetic storms and their relation with the solar cycle

Paula Reyes[1] , Victor A. Pinto[2] and Pablo S. Moya[1]

[1]Departamento de Física, Facultad de Ciencias, Universidad de Chile, Santiago, Chile
emails: paula.reyes.n@gmail.com, pablo.moya@uchile.cl

[2]Institute for the Study of Earth, Oceans, and Space, University of New Hampshire Durham,
New Hampshire, USA

Abstract. Geomagnetic storms can be modeled as stochastic processes with log-normal probability distribution function over their minimum D_{st} index value measured during the main phase of each event. Considering a time series of geomagnetic storm events between 1957 and 2019 we have analyzed the probability of occurrence of small, moderate, strong and extreme events. The data were separated according to solar cycle (SC) and solar cycle phases and fitted through maximum likelihood method in order to compare rates of occurrence of the last Solar Cycle (SC24) with previous ones. Our results show that for $D_{st} < -100$ nT events in SC24 are similar to those in SC20, obtaining ~42 vs 21 median rate storms per cycle with 95% confidence intervals using Bootstrap Method. As SC24 has been the least active solar cycle in over 200 years, we conclude that this method tends to overestimate geomagnetic storms occurrence rates even for small events.

Keywords. Space Weather, Geomagnetic Storms, Solar Cycle

1. Introduction

Geomagnetic storms are disturbances in the Earth's magnetic field caused by interactions with magnetized plasma ejected from the sun. These events transfer extreme amounts of energy to the magnetosphere that can result in a wide range of damages to satellites and communication systems (Wrenn *et al.* 2002), as well as pose a threat to human exploration at high altitudes, thus resulting in technological disruptions, economic losses and dangers to human life. For this reason, the study of geomagnetic storm occurrence and intensity over time as well as the relationship between geomagnetic storms rate occurrence for different solar cycles and their phases is fundamental to improve our forecasting models, and thus to prevent and reduce the risk associated with them.

Geomagnetic storms are traditionally classified according to the strength on their impact in the magnetospheric system, which is recorded from ground-base observations in a series of indices such as D_{st}, SYM-H and Kp among others. In particular, the Disturbance Storm Index (D_{st}) is an indicator of enhancement of the magnetospheric ring current near equator that results in an effective decrease of the Earth's magnetic field. Large drops are generally associated with storms produced by Coronal Mass Ejections (CME), although solar flares and high-speed streams associated with coronal holes can also produce similar magnetospheric effects. In general, geomagnetic disturbances in which the minimum $D_{st} < 50$ nT are considered as geomagnetic storms. Moderate storms correspond to minimum $D_{st} > -100$ nT, and strong storms are events with $D_{st} > -200$ nT. (Gonzalez *et al.* 1994). Stronger storms are traditionally considered extreme events and tend to occur only sporadically, generally no more than a few times per solar cycle.

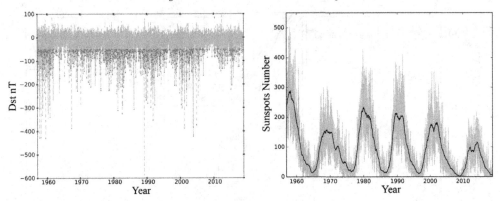

Figure 1. (left) 1367 storms found with $D_{st} < -50$ nT between 1957–2019. (right) raw sunspots number count are shown in grey with black lines to indicate yearly moving average sunspot number.

The direct relation between solar activity and the solar cycle has been known for a long time (Allen 1944). The solar activity can be measured through sunspots numbers, and its periodic variation of ~11 years is used to define the solar cycle. Sunspots are visual manifestations of the Suns magnetic activity and the presence of sunspots on the sun is related to CME (Hayakawa *et al.* 2018), CIR and sometimes flaring. Several studies have explored and quantified such relation between storms and sunspot number (see for example Riley & Love 2018). and have found that generally the number and magnitude of geomagnetic storms during a given solar cycle increase as the number of sunspots increases. To better understand the relationship between the occurrence of geomagnetic storms and the solar cycle, here we present a statistical study in which, treating storms as stochastic processes, we compare the occurrence rate of storms during the Solar Cycle 24 (SC24) with predictions based on previous solar cycles.

2. Data: geomagnetic storms and and solar indexes

Geomagnetic storms can be treated as a stochastic processes. Thus, the probability distribution function (PDF) of geomagnetic storms occurrence as a function of the D_{st} index can be fitted with a log-normal distribution. This is believed to be due to different processes (solar cycle dynamo action, the geo-effectiveness of the solar wind-magnetospheric coupling and the dynamic evolution of a geomagnetic storm) all acting together (Love *et al.* 2015). Namely;

$$F(x|\mu, \sigma^2) = \frac{1}{2}\text{erfc}\left[\frac{\ln(x) - \mu}{\sqrt{2\sigma^2}}\right] \tag{2.1}$$

gives us the occurrence probability F for an event with size exceeding $x = -D_{st}$. Here, μ and σ represent the average and standard deviation of the distribution respectively. By obtaining a good fit for the PDF it is possible to make predictions for the occurrence rate of storms and, more important, to evaluate the PDF and extrapolate the probability of occurrence for large and extreme events. To build the PDF we considered two indexes: D_{st} index to characterize storm strength, and sunspot activity index to separate the data on solar cycles (see right panel in Figure 1). D_{st} index data was obtained from the World Data Center for Geomagnetism of Kyoto's website from 1957 to 2019 at 1 hour time resolution. The dataset of sunspots number was obtained from World Data Center for the Production, Preservation and Dissemination of the International Sunspot Number (Silso's web).

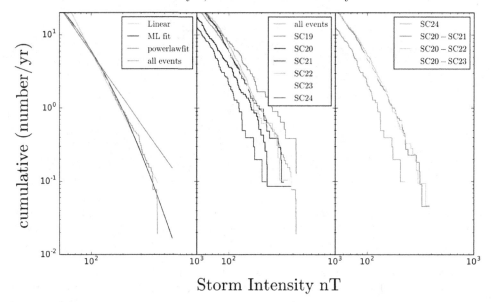

cumulative (number/yr)

Storm Intensity nT

Figure 2. (left) Three different fits to exceedance cumulative of all events, Next figures shows data separated by (center) solar cycle and (right) a combination of them.

3. Methods and results

For the statistics we represent storms by their maximum intensity during main phase with $D_{st} < -50\,\text{nT}$ as shown in Figure 1 (left panel), where data was grouped by solar cycles from 20 to 23 or different combinations of them. Then, $-D_{st}$ was fitted through Maximum Likelihood method, assuming that data corresponds to a log-normal function [Eq. (2.1)] to finally use a bootstrap method to estimate extrapolated storms median rate occurrence error with 95% confidence, in order to compare each fitted data to SC24 number of events.

Bootstrap method is a statistical technique which its main application is estimate the variation of point estimates (confidence intervals). For this purpose, $-D_{st}$ data sample is re-sampled and calculate its median x^* to compute differences $\delta^* = x^*_{median} - x_{median}$. Thus, our estimated 95% median bootstrap confidence interval is $\left[x_{median} - \delta^*_{.0025}, \right.$ $\left. x_{median} - \delta^*_{.0975} \right]$.

Figure 2 shows histograms of binned maxima storm values. Left panel includes three different fits used over exceedances cumulative. A simple look lead us to believe that ML fit is a good representation of storms rate occurrence, specially for extreme events. In the others panels storm maxima data was separated by Solar Cycle or combinations of them.

A quick comparison with SC24 shows that all cycles have more activity than SC24. A comparison of median rate occurrence with 95% confidence interval of different set of data and events from SC24, listed on Tables 1 and 2, shows better accuracy from extreme events prediction, accompanied with less uncertainties.

4. Conclusions

Our results show that ML is the most accurate method to characterize $-D_{st}$ PDF, specially for tail values that correspond to extreme events. Then, a revision of storm extrapolation with $100 < -D_{st}$ lead us to conclude that this method tends to overestimate the rate occurrence of storms in comparison of number of events occurred during SC24,

Table 1. Bootstrap and 95% Confidence Intervals for Maximum Exceedances for all events, SC20-21, SC20-22, SC20-2322 events, -Dst and comparison with number of events in SC24.

-Dst (nT)	All events	SC20-21	SC20-22	SC20-23	SC24
100	71.35	58.70	63.44	53.98	21
	[65.75, 77.01]	[50.94, 66.80]	[55.16,71.81]	[46.80,61.28]	
200	10.89	6.93	7.41	9.15	2
	[8.55,13.21]	[3.98,9.66]	[4.39,10.25]	[6.26,12.12]	
589	0.19	0.07	0.06	0.29	0
	[0.00,0.32]	[−0.11,0.12]	[−0.09,0.10]	[−0.14,0.48]	

Table 2. Bootstrap and 95% Confidence Intervals for Maximum Exceedances for all SC20-21, SC20-21, SC20-23 events, -Dst and comparison with number of events in SC24.

-Dst (nT)	SC20	SC21	SC22	SC23	SC24
100	42.63	76.98	89.05	64.38	21
	[32.99, 51.73]	[63.79,90.43]	[74.72,103.20]	[53.74,75.16]	
200	5.75	8.16	9.13	12.26	2
	[2.64,8.82]	[3.15,12.70]	[3.70,13.83]	[7.73,17.15]	
589	0.11	0.04	0.03	0.49	0
	[−0.24,0.0.21]	[−0.19,0.09]	[−0.14,0.06]	[−0.32,0.90]	

but as we move towards larger $-D_{st}$ values, this difference tends to decreased. Regardless of what combination of solar cycles were used to predict SC24, the prediction always overestimate the number of storms that actually occurred. A possible explanation is the the fact that SC24 was the least active cycle in the past 200 years. Thus, it is possible that if the trend of weak solar cycles continues, using previous solar cycles data to forecast the next cycle would most likely be unreliable as it will keep overestimating the number of storms that will actually be recorded. We expect to increase the scope of the present work in a subsequent manuscript.

Acknowledgments

The D_{st} index used in this paper was provided by the WDC for Geomagnetism, Kyoto (http://wdc.kugi.kyoto-u.ac.jp/wdc/Sec3.html). Sunspot data was obtained from the World Data Center SILSO, Royal Observatory of Belgium, Brussels. We are grateful for the support of CONICyT, Chile through FONDECyT grant No. 1191351 (P.S.M.). P.R. also thanks the travel grant handed by the IAU 354 Symposium organizing committees.

References

Allen, C. W. 1944, *Monthly Notices of the Royal Astronomical Society*, 104(1), 13–21

Gonzalez, W. D., Joselyn, J. A., Kamide, Y., Kroehl, H. W., Rostoker, G., Tsurutani, B. T., & Vasyliunas, V. M. 1994, *J. Geophys. Res.*, 99(A4), 5771–5792

Hayakawa, H., Ebihara, Y., Willis D. M., Hattori, K., Giunta, A. S., Wild, M. N., Hayakawa, S., Toriumi, S., Mitsuma, Y., Macdonald, L. T., Shibata, K., & Silverman, S. M. 2018, *Astrophys. J.*, 862, 15

Love, J. J., Rigler, E. J., Pulkkinen, A., & Riley, P. 2015, *Geophys. Res. Lett.*, 42, 6544–6553

Riley, P. & Love, J. J. 2018, *Space Weather*, 15, 53–64

Wrenn, G. L., Rodgers, D. J., & Ryden, K. A. 2002, *Ann. Geophys.*, 20, 953–956

Solar and Stellar Magnetic Fields: Origins and Manifestations
Proceedings IAU Symposium No. 354, 2019
A. Kosovichev, K. Strassmeier & M. Jardine, ed.
doi:10.1017/S1743921320000393

Examining the optical intensity and magnetic field expansion factor in the open magnetic field regions associated with coronal holes

Chia-Hsien Lin[1]⬤, Guan-Han Huang[1] and Lou-Chuang Lee[2]

[1]Graduate Institute of Space Science, National Central University, Taiwan
email: `chlin@jupiter.ss.ncu.edu.tw`

[2]Institute of Earth Sciences, Academia Sinica, Taiwan

Abstract. Coronal holes can be identified as the darkest regions in EUV or soft X-ray images with predominantly unipolar magnetic fields (LIRs) or as the regions with open magnetic fields (OMF). Our study reveals that only 12% of OMF regions are coincident with LIRs. The aim of this study is to investigate the conditions that affect the EUV intensity of OMF regions. Our results indicate that the EUV intensity and the magnetic field expansion factor of the OMF regions are weakly positively correlated when plotted in logarithmic scale, and that the bright OMF regions are likely to locate inside or next to the regions with closed field lines. We empirically determined a linear relationship between the expansion factor and the EUV intensity. The relationship is demonstrated to improve the consistency from 12% to 23%. The results have been published in Astrophysical Journal (Huang *et al.* 2019).

Keywords. magnetic fields, Sun: corona, Sun: magnetic fields, methods: statistical

---

1. Introduction

Coronal holes are major source regions of high speed solar wind streams, and their magnetic fields are the largest scale solar magnetic fields. Therefore, accurate determination of the locations and areas of the coronal holes is important for space weather forecast and the study of solar cycle variation of solar magnetic fields.

Two commonly used methods to identify the on-disk coronal holes are to identify them as (1) the darkest regions in EUV images with predominantly unipolar magnetic fields, and (2) the regions with "open" magnetic fields, which are the fields with field lines extending far away from the Sun.

While some studies reported that the coronal holes identified by the two methods are statistically associated (e.g., Levin 1982; Obridko & Shelting 1989; Mogilevsky *et al.* 1997; Neugebauer *et al.* 1998; Obridko & Shelting 1999; Hayashi *et al.* 2016), others have shown significant differences between the two (e.g., Lowder *et al.* 2014, 2017; Linker *et al.* 2017). The aim of this study is to investigate the conditions leading to different brightness of OMF regions. We use the magnetic field expansion factor f_s (Wang & Sheeley 1990) to represent the structure of open magnetic field, and apply statistical analysis to examine the relationship between the expansion factor and the EUV intensity.

2. Identification of coronal holes

The AIA 193Å images and HMI line-of-sight magnetograms are used for the identification of low EUV intensity coronal holes (LIR_{193}), and the Wilcox Solar Observatory

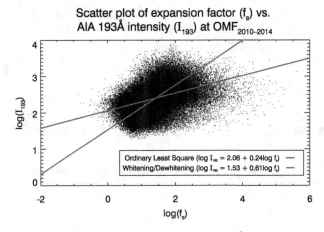

Figure 1. Scatter plot of expansion factor $\log f_s$ vs. AIA 193Å intensity I_{193}. The red and blue lines are the regression lines of using an OLS fit and WD transformation, respectively. (adapted from Huang *et al.* 2019)

radial-field synoptic maps are used for the identification of OMF coronal holes. To determine the LIR$_{193}$s, we first constructed synoptic maps of AIA images and HMI magnetograms, and identified LIR$_{193}$s from the maps based on a thresholding method developed by Krista & Gallagher (2009). To determine the OMF regions, we first applied Potential Field Source Surface (PFSS) model to construct 3D magnetic field between the solar surface and a upper boundary (source surface), and then identify the OMF regions as the footpoints of open magnetic field lines.

3. Statistical analysis

The statistical distribution profiles (histograms) of the expansion factor (f_s) and the intensity of AIA 193Å (I_{193}) show that both profiles are approximately log-normal, and that the shape of $\log f_s$ histogram changes with solar activity and latitude.

A scatter plot of $\log f_s$ vs. $\log I_{193}$, indicates a weak positive correlation between the two, with a correlation coefficient ≈ 0.39. To determine this linear relationship, we applied two linear regression methods: the ordinary least square (OLS) method and a regression method based on the concept of whitening/dewhitening (WD) transformation (Mayer *et al.* 2003). The best-fit OLS regression line is

$$\log I_{193} = 0.24 \log f_s + 2.05, \tag{3.1}$$

and the best-fit WD regression line is

$$\log I_{193} = 0.62 \log f_s + 1.51. \tag{3.2}$$

The scatter plot and the best-fit results of OLS and WD methods are presented in Figure 1 (adapted from Huang *et al.* 2019).

To evaluate the capability of using the OLS and WD regression lines to predict the brightness of an OMF region, we use Equations (3.1) and (3.2) to predict and extract the OMF regions with AIA 193Å intensities lower than the threshold used to identify the LIR$_{193}$coronal holes. The accuracy of OLS and WD predictions is quantified by computing the consistency level between the predictions and the LIR$_{193}$ coronal holes:

$$C_{\text{WD(OLS)}} \equiv \frac{\text{LIR}_{\text{WD(OLS)}} \cap \text{LIR}_{193}}{\text{LIR}_{\text{WD(OLS)}}}, \tag{3.3}$$

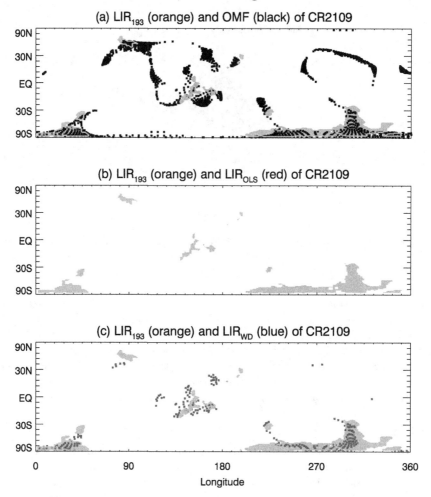

Figure 2. An example of comparison between AIA 193Å LIR_{193} (orange in all panels) and OMF, OLS-predicted low intensity regions (LIR_{OLS}, red in panel b), and WD-predicted low intensity regions (LIR_{WD}, blue in panel c). The selected synoptic map is CR2109. (adapted from Huang *et al.* 2019)

where $LIR_{WD(OLS)}$ are the WD-predicted (OLS-predicted) LIRs. The average consistency level for our data set 2010–2014 is 0.10 for OLS method and 0.23 for WD method. Comparing these with the consistency level between OMF and LIR_{193}, which is 0.12, we can see that while all three are not high, WD method clearly improved the average consistency level. Figure 2 shows an example of comparing LIR_{193} with LIR_{OLS} and LIR_{WD}. The plot shows that the WD regression line removed most of OMF regions that are inconsistent with LIR_{193} while keeping the OMF regions that conincide with LIR_{193}.

References

Hayashi, K., Yang, S., & Deng, Y. 2016, *JGRA*, 121, 1046
Huang, G.-H., Lin, C.-H., & Lee, L.-C. 2019, *ApJ*, 874, 45
Krista, L. D. & Gallagher, P. T. 2009, *Solar Physics*, 256, 87
Levine, R. H. 1982, *Solar Physics*, 79, 203
Linker, J. A., Caplan, R. M., Downs, C., *et al.* 2017, *ApJ*, 848, 70
Lowder, C., Qiu, J., & Leamon, R. 2017, *Solar Physics*, 292, 18

Lowder, C., Qiu, J., Leamon, R., & Liu, Y. 2014, *ApJ*, 783, 142

Mayer, R., Bucholtz, F., & Scribner, D. 2003, *ITGRS*, 41, 1136

Mogilevsky, E. I., Obridko, V. N., & Shilova, N. S. 1997, *Solar Physics*, 176, 107

Neugebauer, M., Forsyth, R. J., Galvin, A. B., *et al.* 1998, *JGR*, 103, 14587

Obridko, V. N. & Shelting, B. D. 1989, *Solar Physics*, 124, 73

Obridko, V. N. & Shelting, B. D. 1999, *Solar Physics*, 187, 195

Wang, Y.-M. & Sheeley, N. R., Jr. 1990, *ApJ*, 355, 726

Solar and Stellar Magnetic Fields: Origins and Manifestations
Proceedings IAU Symposium No. 354, 2019
A. Kosovichev, K. Strassmeier & M. Jardine, ed.
doi:10.1017/S1743921319009918

Solar oblateness & asphericities temporal variations: Outstanding some unsolved issues

Jean P. Rozelot[1], Alexander G. Kosovichev[2] and Ali Kilcik[3]

[1]Université Côte d'Azur, 77 Chemin des Basses Moulières, 06130 Grasse, France
email: jp.rozelot@orange.fr

[2]Center for Computational Heliophysics and Department of Physics,
New Jersey Institute of Technology, Newark, NJ 07102, USA
email: alexander.g.kosovichev@njit.edu

[3]Akdeniz University Faculty of Science, Department of Space Science and Technologies,
07058, Antalya, Turkey
email: alikilcik@akdeniz.edu.tr

Abstract. Solar oblateness has been the subject of several studies dating back to the nineteenth century. Despite difficulties, both theoretical and observational, tangible results have been achieved. However, variability of the solar oblateness with time is still poorly known. How the solar shape evolves with the solar cycle has been a challenging problem. Analysis of the helioseismic data, which are the most accurate measure of the solar structure up to now, leads to the determination of asphericity coefficients which have been found to change with time. We show here that by inverting even coefficients of f-mode oscillation frequency splitting to obtain the oblateness magnitude and its temporal dependence can be inferred. It is found that the oblateness variations lag the solar activity cycles by about 3 years. A major change occurred between solar cycles 23 and 24 is that the oblateness was greater in cycle 24 despite the lower solar activity level. Such results may help to better understand the near-subsurface layers as they strongly impacts the internal dynamics of the Sun and may induce instabilities driving the transport of angular momentum.

Keywords. Sun: heliosismology; Sun: activity; Sun: rotation; Sun: fundamental parameters

1. Introduction

The spherically symmetrical state represents a unique solution of the problem of hydrostatic equilibrium for a fluid mass at rest in the three-dimensional space. The problem complicates when the mass is rotating. For stars, the axial rotation modifies the shape of equilibrium by adding a centrifugal acceleration term to the total potential, breaking the spherical symmetry. The stellar sphere becomes an oblate figure, and we have no a priori knowledge of its stratification, the boundary shape, planes of symmetry, the angular momentum transfer, etc. Moreover, when the rotation rate is not constant in radius and latitude, the surface deviates from a simple oblate figure, and it shape becomes more complicated, particularly, in the presence of internal stresses caused by magnetic fields, for instance.

Considering the Earth as a rotating ellipsoid in uniform rotation ω, Newton gave in 1687 for the first time, an approximate formulation of its flattening f, as a function of surface gravity g_s: $f = \frac{5}{4}\omega^2 \cdot R_{eq}/g_s$, where R_{eq} is the equatorial radius. Huyghens, in 1690, reformulated the flattening in the form $f = \frac{1}{2}\omega^2 \cdot R_{eq}/g_s$, still commonly used as a first approximation.

Let us consider the case of a mass of polytropic gas of index n, rotating at a constant angular velocity ω. The equilibrium configuration and shape of such a body is known since the works of Milne (1923) and Chandrasekhar (1933). By writing the mechanical equilibrium equations and seeking a solution in the form of a perturbed case of the non-rotating configuration, and neglecting high-order effects arising from ω^4, and defining the boundary of the star by a constant null density level, the flattering is given by an equation of the type $f = v\omega^2/G\rho_c$, where G is the constant of gravitation, ρ_c the density of the core, and v is a term depending on the chosen polytropic index. Extensive computations can be found in Chandrasekhar (1933); for instance for the solar case ($n = 3$):

$$f = \left(0.5 + 0.856\frac{\rho_m}{\rho_c}\right)\frac{\omega^2 R_{eq}}{g_s} \tag{1.1}$$

where ρ_m/ρ_c, is the ratio of the mean to central density. Even if such a formalism can be now considered as outdated, it could be noticed that the approximation is still rather good for non polytropic structures with discontinuous variation of density, such as the Earth.

In the solar case, taking $\rho_c/\rho_m = 107.168$, $\omega = 2.85 \times 10^{-6}$ rad/s, $R_\odot = 6.955080 \times 10^{10}$ cm and $g = 2.74 \times 10^4$ cm/s^2 (Allen (2000)), it follows that $f = 1.04 \times 10^{-5}$, in satisfying agreement with the best up-to-date determination of 8.55×10^{-6}.

The story of the solar oblateness began in 1891 when Harzer (1891) introduced for the first time in a theory of solar rotation an oblateness of the Sun, estimating f as $\cong 6.32 \times 10^{-3}$. The history continued in 1895 when Newcomb (1895) described a rapidly rotating solar interior in "such a way that the surfaces of equal density are non spherical". He demonstrated that if the difference between the equatorial and polar radii $\Delta r = R_{eq} - R_{pol}$ reached $\cong 500$ mas, it would explain the discrepancy between the prediction of the Newtonian gravitational theory and the perihelion advance of Mercury observed by Le Verrier in 1859. However, measurements soon ruled out this hypothesis. The discrepancy between the observed advance of Mercury's perihelion and the gravitational theory of planets was explained by the formalism developed by Einstein in 1905. In recent times, even though general relativity had given a satisfactory prediction of Mercury's perihelion, the argument was once again debated after Dicke's historical measurement of $\Delta r = 41.9 \pm 3.3$ mas (Dicke 1970). We know today that such measurements were inaccurate; nevertheless they have been a source of progress. Based on theoretical premises, Dicke (1970) proposed that the magnitude of the oblateness should be 8.1×10^{-6}, without any stress generated by other constraints (magnetic fields at first). Discussion of the historical data is certainly an interesting tour through different techniques. The precision required for determination of changes of the solar oblateness at the cutting edge of modern available techniques was set up, for instance, at the Pic du Midi observatory where a number of measurements was made (Rozelot *et al.* (2011), Table 2). But, even with a deconvolution of atmospheric effects, the measurements still suffered from atmospheric disturbances. The community was attentive to further progress coming from dedicated space experiments, first on balloon flights, and then on board of spacecraft, mainly SoHO, SDO and in a lesser measure RHESSI (Fivian *et al.* (2008), Hudson & Rozelot (2010)) and Picard (Irbah *et al.* (2019)). The main conclusions from this brief review have been summarized in Damiani *et al.* (2011).

Through helioseismic measurements, considerable efforts have been made, at least since the eighties and up to now, in measuring from the odd-order frequency splitting coefficients the internal differential rotation of the Sun. Less progress has been made in analyzing the solar asphericity from the even-order frequency splitting measurements. Kuhn (1988) was the first to note that frequencies of solar oscillations vary systematically during the solar cycle, inferring the corresponding temperature change, but also noting that these variations could reflect changes in the solar structure due to variations of the Reynold's stresses or turbulent pressure. We try here to derive the global outer-limb shape temporal variations, assuming for this study that the site of the perturbation is very close to the surface.

2. Data

Thanks to the Michelson Doppler Imager (MDI) (Scherrer *et al.* 1995) on Solar and Heliospheric Observatory (SoHO) and the Helioseismic and Magnetic Imager (HMI) (Scherrer *et al.* 2012) aboard NASA's Solar Dynamics Observatory (SDO), and their capability to observe with an unprecedented accuracy the surface gravity oscillation (f) modes, it is possible to extract information concerning the coefficients of rotational frequency splitting, a_n. The odd a_n coefficients ($n = 1, 3 \ldots$) measure the differential rotation, whilst the even one ($n = 2, 4 \ldots$) measure the degree of asphericity (i.e. depar-ture from sphericity). The analysis was focused on the low-frequency medium-degree f-modes in the range of $\ell = 137-299$, using the data covering nearly two solar cycles, from April 30, 1996, to June 4, 2017. The a_n (n even) coefficients are a sensitive probe of the symmetrical (about the equator) part of distortion described by Legendre poly-nomials $P_n(\cos\theta)$. Results published by Kosovichev & Rozelot (2018a,b) showed that the asphericity of the Sun dramatically changes from the solar minimum to maximum. During the solar minimum (from 1996 to 1998) the asphericity was dominated by the P_2 and P_4 terms, while the P_6 contribution was negligible. It was shown that the ellipticity of the Sun is strongly affected during the solar cycle. We will try here to better quantify such temporal variations.

According to the von Zeipel's theorem (1924), the solar-limb contours of temperature, density, or pressure should be nearly coincident near the photosphere. Rotation, magnetic fields, and turbulent pressure are the largest local acceleration sources that violate the von Zeipel's theorem Dicke (1970). Since (geometrical) asphericities are relatively small in the Sun, we may describe the distance from the center, for instance, in terms of a constant isodensity level, (or, similarly, in terms of isotemperature or isogravity) by:

$$R(\cos\theta)|_{\rho=\text{constant}} = R_{sp}\left[1 + \sum_n c_n(R_{sp})P_n(\cos\theta)\right] \tag{2.1}$$

where R_{sp} is the mean limb contour radius, θ the angle to the symmetry axis (colatitude), and P_n the Legendre polynomial of degree n. The asphericity is described by coefficients c_n, which are called quadrupole for $n = 2$ (c_2) and hexadecapole for $n = 4$ (c_4). Terms of higher orders are conventionally named by adding "-pole" to the degree number. It is straightforward to determine f from Eq. 2.1 by means of the asphericities coefficients, c_2, c_4 and c_6, as $f = -\frac{3}{2}c_2 - \frac{5}{8}c_4 - \frac{21}{16}c_6$.

The measured splitting coefficients a_n are related to the shape coefficients c_n through a normalization factor K. An efficient method for calculating this factor was developed by Kuhn (1989) who showed that it was possible to invert the splitting data to obtain the structural asphericity; he obtained $a_n = Kc_nR_n(\ell)$. Assuming $R_n = R_{sp}$, as this analysis is conducted only very close to the surface (i.e. the seismic radius at the surface), the corresponding average factors are:

$$a_2 = -6 \times 10^{-4}c_2R_{sp}; \quad a_4 = 1 \times 10^{-4}c_4R_{sp}; \quad a_6 = -14 \times 10^{-4}c_6R_{sp}$$

where the a_n frequency splitting coefficients are measured in Hz (Kuhn (1989)).

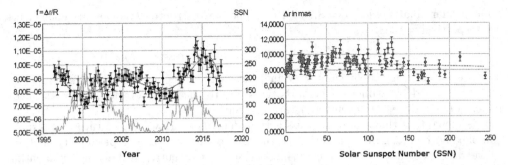

Figure 1. Left: Solar oblateness f (left scale) and the solar sunspot number SSN (red, right scale) as a function of time. A periodic oscillation appears, with two minima around 2000 and 2011 and two maxima around 2005 and 2016. Right: The difference between the equatorial and polar radius Δr (in mas) versus the solar activity described by the sunspot numbers. A slight anticorrelation is visible.

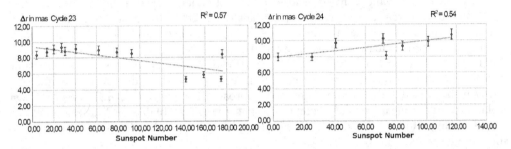

Figure 2. Annual mean difference between the equatorial and polar radius Δr (in mas) versus the solar activity during cycles 23 (left) and 24 (right). The two cycles show a different behavior: a negative trend for cycle 23 and a positive one for cycle 24. (Source of the sunspot data: WDC-SILSO, Royal Observatory of Belgium, Brussels).

3. Results

Results displayed in Fig. 1(left) show the solar oblateness f as a function of time from 1996 to 2017. A periodic oscillation appears, with two minima around the years of 2000 and 2011 and two maxima around 2005 and 2016. If these minima and maxima correspond to the minima and maxima of the solar activity cycles, then there is a shift between the asphericity and activity of around 3 years. Fig. 1(right) displays the difference between the equatorial and polar radii, Δr, in millisecond of arc (mas) versus the solar activity, the sunspot number taken as a proxy (Clette *et al.* 2016). In order to get a better view of the two cycles that are significantly different in the level of magnetic activity, we calculated variations of the Δr annual means separately for these cycles. The two cycles show a different behavior as seen in Fig. 2, left panel for cycle 23 and right panel for cycle 24. Straight lines show a linear regression fit. A negative trend for cycle 23 is noticeable, while a positive trend appears for cycle 24.

4. Conclusion

The analysis of the helioseismology data from the SoHO and SDO space missions permits to determine accurately the splitting rotational coefficients together with the structural shape parameters.

The preliminary results obtained here by averaging the f mode frequency variations over two solar cycles for the whole observed angular degree range, $\ell = 137\text{--}299$, lead to a mean solar oblateness of $f = 8.76 \times 10^{-6}$. The mean structural asphericity coefficients are respectively:

$$c_2 = -1.17 \times 10^{-6}; \quad c_4 = 2.42 \times 10^{-5} \quad \text{and} \quad c_6 = -5.03 \times 10^{-7}$$

These even splitting coefficients vary in time as they depend on latitudinal inhomogeneities caused by aspherical perturbations due to the solar rotation, magnetic fields beneath the surface, and even temperature variations.

It is shown that the solar oblateness is time dependent. However, its variation is quite complex, both in magnitude and time. If the solar oblateness shows a periodicity of about 11 years, it does not follow exactly the solar cycle. Currently, we have no explanation for the ~3-year time lag of the flattering parameter f relative to the activity cycle, bearing in mind that the a_2, a_4 and a_6 coefficients are respectively shifted from the solar cycle by around 0.1, 1.6 and -1.6 years (Kosovichev & Rozelot (2018a)). The significant variations in time and the phase shifting according to the solar cycle activity (as seen in Fig. 1) are probably two main reasons why the observational results from ground based instruments, balloon flights and satellite instruments seem to be inconsistent. An explanation has already put forward by Rozelot *et al.* (2009) by considering the temporal variation caused by a change in the relative importance of the hexadecapolar and dipolar terms. At the time of high activity, only the dipolar moment c_2 has a significant effect, but at the time of low activity, c_4 is predominant; this results in a decrease of the total value of the oblateness. Contribution of the c_6 term is less important due its low magnitude, but can be considered in a more detailed approach. Irbah *et al.* (2019) revisiting past solar oblateness measurements concluded that the solar oblateness "variations are in phase during odd cycles and anti-phase during even cycles", but the situation seems to be more complex.

Clearly, we are very close to having the required data and boundary conditions to investigate deeper the solar shape structural coefficients and their changes during the two solar cycle that are significantly different in the level of magnetic activity. Should the solar oblateness be determined accurately from space, this could help to disentangle the various contributions to the asphericity splittings of solar oscillation frequencies, and get insight into the physical processes that may be at play in the leptocline.

The work was partially supported by the NASA grants NNX14AB7CG and NNX17AE76A.

References

Allen, C. W. & Cox, A. N. 2000, *Allen's Astrophysical Quantities*, 4th edition, Springer-Verlag, New York, p.719

Chandrasekhar, S. 1933, *Mon. Not. Roy. Astron. Soc.*, 93, 390

Clette, F., Cliver, E. W., Lefèvre, L., Svalgaard, L., Vaquero, J. M., & Leibacher, J. W. 2016, *Sol. Phys.*, Vol. 291, 2479–2486. DOI: 10.1007/s11207-016-1017-8

Rozelot, J. P. & Damiani, C. 2011, *European Journal of Physics*, H., Vol. 36, 407–436 DOI: 10.1140/epjh/e2011-20017-4

Dicke, R. H. & Goldenberg, H. M. 1967, *Phys. Rev. Lett.*, 18, 313

Dicke, R. H. 1970, *ApJ*, 159, 1–23

Emilio, M., Bush, R. I., Kuhn, J., & Scherrer, P. 2007, *ApJ*, 660, L161

Fivian, M. D., Hudson, H. S., Lin, R. P., & Zahid, H. J. 2008, *Science*, 322, 560. DOI: 10.1126/science.1160863 2008

Harzer, P. 1891, "Uber die Rotationsbewegung der Sonne". Astronomische Narchrichten, 3026

Hudson, H. & Rozelot, J. P. 2010, RHESSI science nugget: http://sprg.ssl.berkeley.edu/~tohban/wiki/index.php/History_of_Solar_Oblateness

Irbah, A., Mecheri, R., Damé, L., & Djafer, D. 2019, *Astrophys. J. Lett.*, 875(2), pp.art. L26

Kuhn, J. R. 1988, *ApJ*, 331: L131–L134

Kuhn, J. R. 1989, *Sol. Phys.*, 123, 1–5

Kosovichev, A. K. & Rozelot, J. P. 2018, *ApJ*, 861, Issue 2, article id. 90, 5 pp. DOI: 10.3847/1538-4357/aac81d, arXiv:1805.09385 [astro-ph.SR]

Kosovichev, A. K. & Rozelot, J. P. 2018, *J. Atmos. Sol. Terr. Phys.*, Vol. 481, p 2981–2985. DOI: 10.1016/j.jastp.2017.08.004

Milne, E. A. 1923, *Mon. Not. Roy. Astron. Soc.*, 83, 118

Newcomb, S. 1865, *"Fundamental Constants of Astronomy"*, US GPO, Washington, D.C., p. 111

Rozelot, J. P., Damiani, C., & Pireaux, S. 2009, *ApJ*, 703(2) 1791

Damiani, C., Rozelot, J. P., Lefebvre S., Kilcik, A. & Kosovichev, A. G. 2011, *J. Atmos. Sol. Terr. Phys.*, 73, 241–250 DOI: 10.1016/j.jastp.2010.02.021

Scherrer, P. H., Bogart, R. S., Bush, R. I., Hoeksema, J. T., Kosovichev, A. G., Schou, J., Rosenberg, W., Springer, L., Tarbell, T. D., Title, A., Wolfson, C. J., Zayer, I., MDI Engineering Team, 1995. The solar oscillations investigation - Michelson Doppler Imager. *Sol. Phys.*162, 129–188

Scherrer, P. H., Schou, J., Bush, R. I., Kosovichev, A. G., Bogart, R. S., Hoeksema, J. T., Liu, Y., Duvall, T. L., Zhao, J., Title, A. M., Schrijver, C. J., Tarbell, T. D., Tomczyk, S., 2012. The Helioseismic and Magnetic Imager (HMI) investigation for the Solar Dynamics Observatory (SDO). *Sol. Phys.*275, 207–227

von Zeipel, H. 1924, *Mon. Not. Roy. Astron. Soc.*, 8, 665

Discussion

KRYSTOF HELMINIA: About the radius variations -do we expect such variations in stars with stronger magnetic fields, like late type dwarfs? Could such variations be stronger and measurable?

JEAN PIERRE ROZELOT: Within the Sun, stronger magnetic fields lead to more important radius variability, particularly just below the surface. We do expect the same for stars. Measurements are still difficult as we don't have yet accurate measurements devices. That's must be done in a next future.

Chapter 6. Star-planet relations

Observations (Vidal-Madjar et al., 2003; Ben-Jaffel, 2008) indicate th
HD 209458b the eclipse in the Ly is a few times deeper th
the planet itself (9–15% vs 1.8° his effect was also
nd Si line al-Madjar et Jaffel, 2009; Lins

Dmitry Bisikalo

Solar and Stellar Magnetic Fields: Origins and Manifestations
Proceedings IAU Symposium No. 354, 2019
A. Kosovichev, K. Strassmeier & M. Jardine, ed.
doi:10.1017/S1743921319009669

Solar activity influences on planetary atmosphere evolution: Lessons from observations at Venus, Earth, and Mars

J. G. Luhmann®

Space Sciences Laboratory, University of California, Berkeley, CA 94720, USA
email: jgluhman@ssl.berkeley.edu

Abstract. The Pioneer Venus and Venus Express missions, and the Mars Express and MAVEN missions, along with numerous Earth orbiters carrying space physics and aeronomy instruments, have utilized the increasing availability of space weather observations to provide better insight into the impacts of present-day solar activity on the atmospheres of terrestrial planets. Of most interest among these are the responses leading to escape of either ion or neutral constituents, potentially altering both the total atmospheric reservoirs and their composition. While debates continue regarding the role(s) of a planetary magnetic field in either decreasing or increasing these escape rates, observations have shown that enhancements can occur in both situations in response to solar activity-related changes. These generally involve increased energy inputs to the upper atmospheres, increases in ion production, and/or increases in escape channels, e.g. via interplanetary field penetration or planetary field 'opening'. Problems arise when extrapolations of former loss rates are needed. While it is probably safe to suggest lower limits based simply on planet age multiplied by currently measured ion and neutral escape rates, the evolution of the Sun, including its activity, must be folded into these estimations. Poor knowledge of the history of solar activity, especially in terms of coronal mass ejections and solar wind properties, greatly compounds the uncertainties in related planetary atmosphere evolution calculations. Prospects for constraining their influences will depend on our ability to do a better job of solar activity history reconstruction.

Keywords. solar wind Interactions, solar activity, space weather effects

---

1. Introduction

Planetary atmospheric loss to space, often referred to as escape, is but one element in efforts to understand what led to the present conditions of each member of our solar system. Of special interest are the terrestrial planets within (or close to) the habitable zone. These both relate to our own circumstances here on Earth, and also continue to be discovered in increasing numbers among the extrasolar planets (e.g. Tsiaras *et al.* 2019), giving impetus to the search for life elsewhere. As a result, the trio of Venus, Earth and Mars has been subject to targeted investigations in the form of space exploration toward understanding the similarities and contrasts among the three, as well as implications for their past and future, and for other worlds. In particular, Mars currently has a relatively thin CO_2 atmosphere, with pressure roughly equivalent to what would be present if the Earth's atmosphere started at stratospheric altitudes. Yet there is surface evidence suggesting an early atmosphere that had sufficient pressure to allow lakes, and perhaps even seas, to form (e.g. Villanueva *et al.* 2015). Alternatively, Venus has the atmospheric equivalent of Earth's carbonate rocks present in gaseous form, forming a thick CO_2 atmospheric blanket that has produced a 'runaway greenhouse', hostile to familiar forms

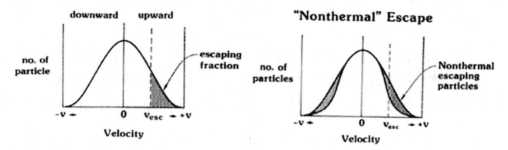

Figure 1. Illustrations of particle velocity distributions leading to Jeans or thermal escape (left), and its 'non-thermal' counterpart (right). In thermal escape, the distributions are described by Maxwell-Boltzmann functions, while in non-thermal escape, they depart from that behavior in many different ways that alter the distribution above the escape velocity, including the broadening at high energies suggested here, or even a shift to the right or left.

of life and even robotic explorers (Kasting 1988; Hall 2019). Yet this thick atmosphere is also largely devoid of water, as is the extremely hot surface. How did these planets and their atmospheres become so different, presuming they formed in roughly the same pre-planetary nebula around the Sun? Their respective distances from the Sun surely played some part, as did the smaller mass of Mars. But did their magnetic field histories also play a role? The different consequences of their photochemistry? More specifically, in light of the subject of this paper, did their atmospheric escape histories differ in ways that led to their present states?

One thing that is certain is that the Sun plays a dominating role in planetary atmosphere behavior through its effects on heating and atmospheric chemistry. Some of these effects are straightforward and concern the usual vaporization, thermalization, and photochemical processes. Others are less direct and often involve complex chains of different physical phenomena. But all essentially depend on the intensities and spectra of solar photon and particle emissions, which vary greatly with solar activity. Available observations now provide a good basic picture of some of these atmospheric energization processes, including their dependence on the planets' seasons and solar activity cycle phase. Coupled with the expectation that the more active solar conditions must have prevailed early in the planets' evolution, which is inferred from observations of Sun-like stars, these give further incentive for considering the long-term consequences of escape. On the threshold of new observations and rapidly expanding interests in terrestrial planet-star interactions, it is worth considering how well we understand the solar activity control of the current atmospheric losses to space, as well as identifying what more can be done to better constrain its impacts.

2. Escape processes

The basic physics of atmospheric escape is relatively simple: any process or chain of processes that energize some particles in a planetary atmosphere to speeds greater than the escape velocity Vesc = $(2MG/r)^{1/2}$ ($\sim$10-11 km/ for Earth and Venus, $\sim$5 km/s for Mars) can lead to loss of constituents to space. Most of these losses occur from the upper terrestrial atmospheres, where the surface boundary layer and thermalizing collisions become rare. Escape processes may be 'thermal' or 'nonthermal', 'bulk' or 'kinetic', and often involve ionized species affected by electric and magnetic fields. Figure 1 illustrates a few of these. In thermal or 'Jeans' escape of neutral gas above the 'exobase', the altitude at which Venus, Earth and Mars atmospheres transition to a more collisionless behavior can be lost if they are outward-directed and not on ballistic or orbiting trajectories. This is due to density falloff with altitude, outward-bound particles belonging to the typical Maxwell-Boltzmann velocity distribution above the escape speed. In this case, a hotter gas results in more escape. In the contrasting case of 'non-thermal' escape, some acceleration or energization process creates a non-Maxwellian distribution where the numbers of particles above the escape speed increase. Non-Maxwellian distributions can take many forms, and the processes leading to them may not affect the entire velocity distribution, as in the case illustrated here. However, as for the thermal process, the requirement for a particle's leaving the planet is the same. If the species is ionized, the basic requirement regarding escape speed and trajectory are the same, but a host of new energization mechanisms comes into play. In fact, ionization is often a key factor in enabling significant escape, as described below in the context of our three planetary examples.

3. Contrasts between Venus, Earth, and Mars Escape

The present-day compositions of the Venus, Earth and Mars atmospheres tell part of the story of long-term atmospheric escape. As mentioned above, both Venus and Mars have CO_2-dominated atmospheres that show a relatively extreme lack of water in their atmospheres and on their surfaces. In contrast, Earth's air blanket is nitrogen-rich, in large part because its CO_2 content has been removed by our liquid water, which has transformed it to carbonate rocks (e.g. Fegley 2014). The focus of interest is thus often on H and O escape, and their role in determining a planet's surface water. In particular, debate often centers on what role a planetary magnetosphere like Earth's plays in 'shielding' the atmosphere from processes that lead to escape of these constituents. In exploring this still open question, potentially important information is available in the physics of current atmospheric loss at each of these bodies, apart from the thermal escape differences that are due to their different heliocentric locations.

Earth's magnetospheric solar wind interaction converts the incident energy and momentum of the solar wind into a number of different forms. The solar wind convection electric field ($E = -Vsw \times B$) maps into the high latitude ionosphere along open magnetic field lines, where it drives large-scale, cross-polar cap motion in the partially ionized upper atmosphere. The upper atmosphere there is the main reservoir for escape. Light hydrogen atoms undergo Jeans escape, but hydrogen ions also move outward along the 'open' magnetic field lines in the polar cusp and polar cap where interconnections between Earth's magnetic field and the interplanetary field occur, as illustrated in Figure 2a. This light ion 'polar wind', which also includes helium ions, is enabled by an ambipolar electric field that develops along the field lines due to gravitational separation of the oppositely charged heavier ion and light electron populations. Other energization of ions occurs in the magnetotail, where there is internal magnetospheric magnetic reconnection occurring as part of the magnetosphere's global circulation. These dynamics lead to suprathermal electron precipitation in the auroral zone that heats and ionizes upper atmosphere neutrals, resulting in the upward acceleration of an 'auroral wind' of heavier ions, including O+. The details of the latter (see Figure 2b) appear to involve wave-particle interactions that add energy perpendicular to the locally vertical magnetospheric field, which is then converted by the magnetic mirror force to form upward moving, field-aligned 'beams' and 'conic' ion distributions with speeds greater than the escape speed. The efficiency, and thus effects, of this conversion, as well as the extent to which the outgoing O+ is contained within the closed fields of the magnetosphere (e.g. as part of the ring current), versus escape on open field lines are still open questions.

If the upward-flowing ions eventually escape Earth, estimates of the flux based on their density $\approx 10^5$ cm^{-3}, and velocity ≈ 1 km/s, give values $\approx 10^{10}$ cm^{-2}s^{-1}. When the area of the atmospheric footprint over which they are observed is considered, (e.g., 1000 km $\times$ 100 km, based on global average precipitation maps such as that in Figure 3a), the

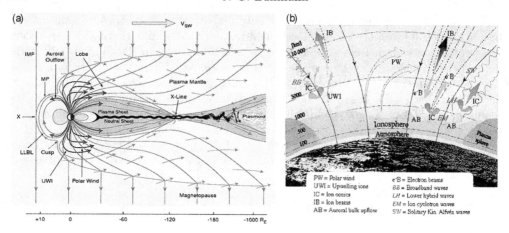

Figure 2. a) Illustration from Moore & Horwitz (2007) of the magnetic topology of the coupled solar wind-magnetosphere system (for Southward interplanetary field), showing the various ion flows within the system. Outflows from the high-latitude ionosphere can either escape to space along open magnetic field lines or enter the plasma sheet region of the magnetotail, where they either recirculate in the magnetosphere or are ejected as part of plasmoids resulting from the magnetotail reconnection processes. b) Illustration from Moore & Horwitz (2007) of the many plasma physical processes occurring in the high latitude ionosphere that can lead to ion energization and escape.

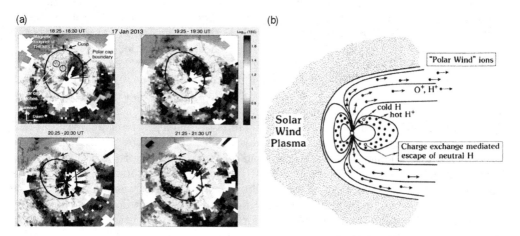

Figure 3. a) Illustration of the statistical footprints of electron precipitation inferred from high-latitude total electron content (TEC) obtained from GPS satellite transmissions (Walsh *et al.* 2014). Noon is at the top in these figures. The red tongue-like feature is the magnetospheric cusp location. b) Illustration of ion escape, including the closed field locations where charge exchange with ambient neutrals provides additional loss in the form of the produced energetic neutrals.

net outflow rate is $\approx 10^{25}$ s^{-1}. In addition, for the ions injected from below that are on closed field lines and contribute to the magnetospheric ring current (Figure 3b), charge exchange with high altitude neutral atoms of the primarily H exosphere leads to added non-thermal escape of the resulting energetic neutrals. This is especially important for H escape, with rates of up to $\sim 10^{27}$/s if limited by the rate of supply from below.

In contrast to the Earth, planetary magnetic fields do not prevent a direct solar wind interaction with the atmospheres at Venus and Mars, as can be seen by comparing Figure 3a with Figures 4a,b. The solar wind both upstream of the bow shock and in the magnetosheath around these planetary 'obstacles' penetrates into their neutral upper

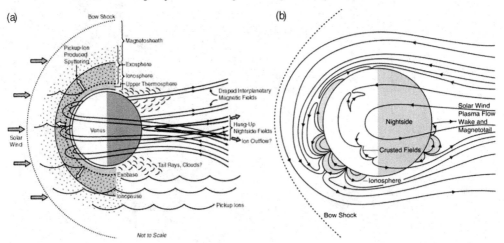

Figure 4. a) Illustration of the Venus-solar wind interaction; (b) The Mars solar wind interaction has the addition of crustal magnetic fields which add to the obstacles' and processes' complexities, but is in some basic ways, similar to the Venus case (reproduced from Russell *et al.* 2016).

atmospheres. As a result, atmospheric/ionospheric energization processes are different than at Earth. In addition, due to the scaling of these much smaller obstacles, both fluid-like and kinetic (test particle-like) processes come into play. For example, the solar wind convection electric field 'picks up' the ions produced in the region of overlap by photoionization, electron impact, or charge exchange with solar wind protons, as well as ions transported there from below. This process effectively 'mass loads', and thus further slows the solar wind plasma that is being deflected around Venus and Mars by combinations of ionospheric induced currents and (in the case of Mars) crustal magnetic fields, helping to create the highly draped fields of their comet-like induced magnetotails. But the gyroradii of the picked-up heavy (e.g. O+) ions can be comparable to, or larger than, the planetary radius, as illustrated in the left panel of Figure 4a and in Figure 5b, which show the Venus-solar wind interaction features. This leads to partial deposition of the energized planetary pickup ions back into the planets' upper atmospheres, with the possible outcome of additional neutral upper atmosphere losses by the sputtering process (e.g. see discussion in Curry *et al.* 2015). The field draping in the magnetosheath and magnetotail regions also exerts another force on the planetary ions, referred to as the magnetic tension force in Figure 5a. Also known as the magnetic 'slingshot' force, it sweeps up planetary ions where the ionosphere becomes denser and more fluid-like in its behavior. In addition, thermal ionosphere ion pressure gradients exist on the draped, penetrating magnetic fields that can accelerate upper atmosphere ions outward in a 'polar wind' like fashion (also indicated in Figure 5a).

Observations of planetary ions at both Venus and Mars are well-modeled, assuming the processes in Figure 5b are at work (see Figure 6, which show locations of energetic O+ ions detected on PVO around Venus, compared to a test particle picture of ion pickup in the region of atmosphere-solar wind overlap).

Similarly, the Mars Polar O+ Ion 'Plume', as observed on MAVEN (Figure 7a) and similarly modeled with test particles (Figure 7b), accounts for a significant fraction (∼30%) of Mars' total O+ escape (Dong *et al.* 2015). The lower-energy escaping ions in the solar wind wakes that occupy the Venus and Mars draped magnetotail 'plasma sheets' are generally considered to result from the magnetic tension force, while the thermal pressure gradient forces contribute additional outgoing ions both at and beyond the terminator

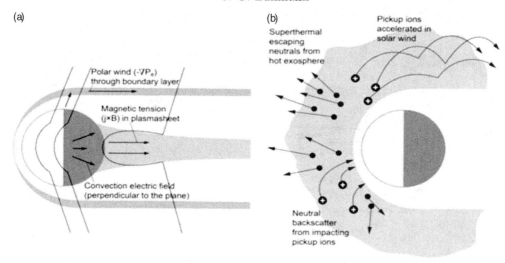

Figure 5. a) Illustration from Futaana *et al.* (2017) of the fluid-like processes involved in the escape of ions from Venus (and to some extent at Mars); (b) The other escape processes at Venus (and Mars) that involve the more kinetic aspects of their planetary ion behavior.

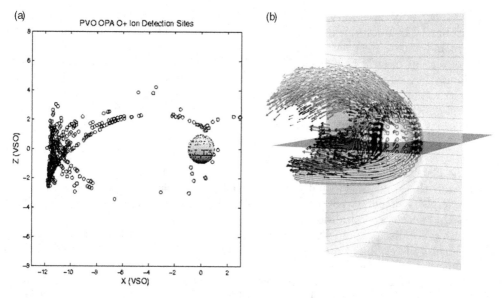

Figure 6. (a) Locations in the PVO orbit where O+ pickup ions were observed (from Luhmann *et al.* 2006), compared to (b) a model of the picked up ions (Jarvinen *et al.* 2010). The locations of H+ pickup ions are also shown here in blue.

(e.g. Dubinin *et al.* 2017). Mars' crustal fields are also thought to provide an additional loss mechanism associated with reconnection between the crustal fields and draped interplanetary fields (Brain *et al.* 2010), but the relative importance of that effect has yet to be evaluated, as does potential erosion associated with solar wind/ionosphere boundary shear-related instabilities (e.g., steepening Kelvin-Helmholtz waves (Ruhunusiri *et al.* 2016)) and the typically time-dependent boundary conditions (e.g. from heliospheric current sheet/interplanetary field sector boundary crossings (Edberg *et al.* 2011)). It has also been suggested that Venus ion loss can be affected by magnetic field reconnection

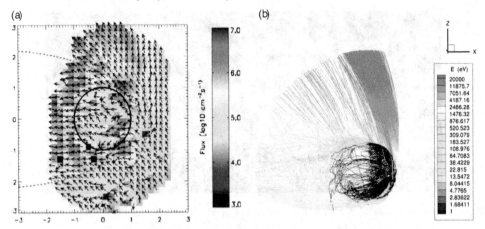

Figure 7. (a) MAVEN observations of escaping oxygen ions showing the statistical pattern of the fluxes, and the upward 'plume' extension in the coordinated system organized by the solar wind convection electric field (from Dong *et al.* 2015). (b) O+ test particles in a Mars' solar wind interaction model, for a similar geometry with ion trajectories color-coded by their energies (Fang *et al.* 2008).

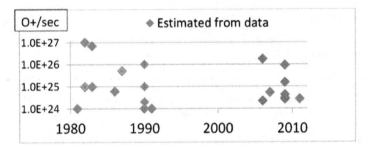

Figure 8. Figure illustrating previous estimates of Venus O+ escape rates, based on both measurements (blue) and models (red), from McEnulty (2012).

across its draped magnetotail lobes, when it occurs close to the planet (Zhang *et al.* 2012).

Ion escape rate estimates from some of these additional processes have been attempted. For example, from PVO ionospheric 'cloud' measurements (ionospheric ions seen in the magnetosheath adjacent to the main ionosphere), Brace *et al.* (1982) arrived at the number 7×10^{26} ions/sec, based on measured transit times, probability of occurrence, statistical distribution, and average electron density. Meanwhile, Russell *et al.* (1982) inferred 2×10^{25} ions/sec loss during one cloud event, assuming a similar cloud in the south. These are significant compared to estimates/measurements of escape via the other processes at Venus. In general, escape rate estimates (in Figure 8, adapted from McEnulty 2012) show that Venus O+ escape rates versus the year of published estimate vary by orders of magnitude (blue indicates those derived from observations, red indicates those derived from models).

The primarily CO_2 atmospheres also have a photochemical channel for neutral O escape, which is especially important at Mars. The reaction $CO_2^+ + O \rightarrow O_2^+ + CO$, followed by dissociative recombination $O_2^+ + e \rightarrow O^* + O^*$, proceeds rapidly in Venus' and Mars' ionospheres. This photochemical process produces 'hot O coronas' around both planets. The energies of some of these hot atoms are >2 eV, which puts them above the lower escape velocity for Mars, though not for Venus, where $\sim$10 eV is required. This

Figure 9. The upper C, O, and H atmospheres of Mars, as observed on MAVEN prior to its orbital insertion (from Schneider *et al.* 2015).

	H Jeans	O ion	O Dissoc Recomb	O sputtering
Present-day loss rate from MVN (s⁻¹)	1.6-11 x 10²⁶	5 x 10²⁴	5 x 10²⁵	3 x 10²⁴
4.2 b.y. at present rate, H₂O	3.6-25.2 m	0.2 m	2.2 m	0.14 m
4.2 b.y. at present rate, CO₂		6 mbar	68 mbar	4 mbar

Figure 10. Rates of escape and associated equivalent global amounts of water and CO_2 loss over time ($\sim$3.5 Gyr) for Mars, based on measurements by MAVEN. (Adapted from Jakosky *et al.* 2018).

process also increases O+ escape, even for Venus, by putting more O at higher altitudes, where it can be ionized and picked up in the magnetosheath and solar wind. A similar process also works for C loss at Mars (e.g. Hu *et al.* 2015). MAVEN observations of atomic C, O and H coronas, as seen in Figure 9, have allowed new escape rate estimates to be made for these neutral constituents. Also, it has been found that the H corona, with its escaping component, is probably enhanced by dust storm activity (e.g. Chaffin *et al.* 2014).

To evaluate the overall impacts of all of these processes, e.g. for water loss, one needs to add the different escape rates for H and O, as recently done for Mars by Jakosky *et al.* (2018) (see Figure 10):

The current atmospheric escape rates at all three planets are too low to explain inferred losses of evolutionary interest. For example, the estimated volume of an early ocean of Mars is $\sim$6 $\times$ 10⁷ km³ H_2O (from surface features). This amount contains about 2 $\times$ 10⁴⁵ H_2O molecules. It is relatively easy to lose the light hydrogen by extra (e.g. EUV) heating. But to remove the oxygen in this ocean over a few Gyr requires an average loss rate of at least $\sim$10²⁸ O atoms/s (over 100 times greater than present rates).

4. Escape Enhancers

The Sun produces, in addition to its varying EUV outputs, flares, enhanced solar wind flows and fields, coronal transients, and solar energetic particles (SEPs). The two kinds of solar wind structures that produce the greatest solar wind and SEP enhancements are illustrated in Figure 11. The interplanetary coronal mass ejections (ICMEs) that produce

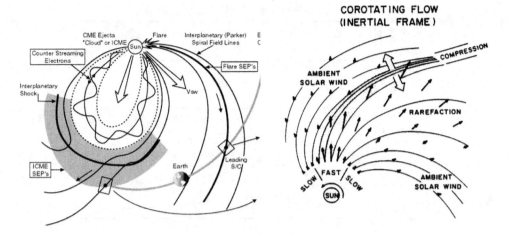

Figure 11. (Left panel) Illustration of an interplanetary coronal mass ejection's (ICME's) effects on the solar wind and interplanetary magnetic field, including solar wind compression/pileup ahead of it that is sometimes preceded by a shock (from Luhmann *et al.* 2008). (Right panel) Illustration of the effect of interacting solar wind streams, which result in spiral-shaped compressions that can appear to rotate with the Sun (from]Pizzo 1978).

the greatest effects under the present solar conditions (left panel) result from erupting coronal structures that occur most often around the solar cycle maximum phase, while solar wind stream interactions (right panel) occur throughout the cycle. The ICME produces a number of different effects on planetary space environments. First, the largest events are often initiated around the time of a flare on the Sun, which is then closely followed in some cases by the arrival of high energy particles, mainly protons and electrons, accelerated in the corona. Concurrently, the ejection of coronal material occurs, traveling at speeds up to nearly 4-5 times the typical solar wind speed. This ICME structure plows through the ambient corona and solar wind, piling up the density and magnetic field, and for fast ejections, producing a leading interplanetary shock, which provides a source of more energetic particles that speed ahead of it throughout its several-day transit times to reach these planets. Then the planetary interaction reacts to the shock, the compressed solar wind behind it, and the usually stronger than average and sometimes highly inclined magnetic fields of the coronal material, which is sometimes well-described as a large flux rope of a few tenths of an AU (in a cross section). This entire sequence from the flare to the coronal ejecta passage can last several days, as illustrated in Figure 12.

Passage of ICMEs can greatly enhance the overall magnetospheric ion energization, precipitation, and outflow processes. The Earth's auroras during and following solar activity provide a measure of the energy deposited in the atmosphere and its spatial extent. Related ion outflow rates follow suit, showing dependence on disturbance parameters such as incident solar wind pressure (see Figure 13).

The Venus electron density altitude profiles in Figure 14 illustrate its much different, more direct dayside ionosphere boundary response to enhanced solar wind pressure, which arrives with the leading compressed solar wind portion of the ICME event. In addition to the inferred erosion of the topside ionosphere, the overlying draped magnetosheath field is both present at lower altitudes, and penetrates into the ionosphere. Mars exhibits its own version of this consequence, in that its crustal fields show increasing degrees of open topologies due to enhanced reconnection with these penetrating fields (Xu *et al.* 2018).

Associated diffuse auroral emissions, examples of which are shown in Figure 15, are seen on the night sides of both Venus and Mars (Phillips *et al.* 1986; Schneider *et al.* 2015, 2018). These occur in coincidence with the local enhancements of SEPs, which,

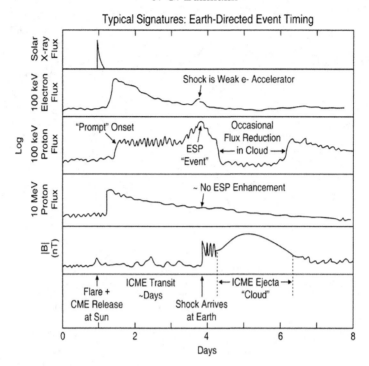

Figure 12. The different phenomena and timing of events associated with ICMEs result in a range of effects in planetary space environments. Here, a 'classic' time series of observations (solar X-rays and in-situ particles and fields) around the time of a major ICME event illustrates a particular sequence that occurs with a 'direct' impact (see Figure 11, left panel).

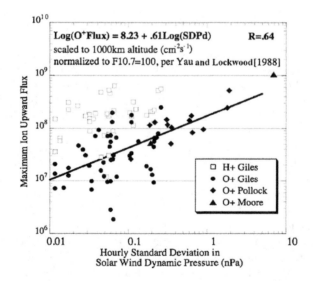

Figure 13. Figure from Moore & Horwitz (2007) showing the intensification of outflowing hydrogen and oxygen ions in the Earth's polar regions in response to increases in solar wind pressure.

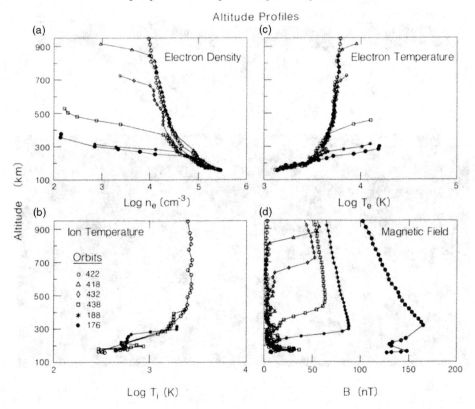

Figure 14. PVO dayside ionospheric density altitude profiles selected to illustrate their dependence on the incident solar wind pressure. In particular, the increasing solar wind erosion of the topside ionosphere is seen in the top left plot, while the corresponding lowering of the overlying magnetosheath and penetration of the magnetic field into the ionosphere is seen in the lower right plot. (From Luhmann *et al.* 1987.)

as mentioned above, can precede the ICME shock arrival by days and last throughout the event, sometimes peaking in intensity at the shock arrival (see Figure 12). Venus auroral emissions in the visible green line have also been seen from the ground in the days following coronagraph observations of Venus-directed CMEs (Gray *et al.* 2014).

Lee *et al.* (2018) summarize the details of how Mars responded to a significant ICME impact witnessed by MAVEN instruments in the form of the time series in Figure 16. Upper atmosphere heating and expansion briefly occurs in response to the flare, but the SEPs and their effects can be present for days, because the shock that travels outward ahead of the ICME is a relatively long-lasting source that populates a large swath of heliosphere in front of and around it with SEPs. The access of these ionizing particles to the lower atmosphere is enhanced by the observation that the crustal magnetic fields open up in response to solar wind compression and increased external field penetration. The enhanced magnitude of the external field also plays a role in this access. Whether atmospheric escape is significantly impacted by such events is to be determined.

Because spacecraft observations are restricted to the few orbits occurring prior to and during ICME passage, which cover only a small portion of the Mars-solar wind interaction space, MHD simulation results for the events (Ma *et al.*) are used to obtain a global picture and the related global escape rates (Ma *et al.*, 2017; Dong *et al.*, 2015). Figure 17 shows snapshots of meridional Mars planetary ion flux contours (log flux $(cm^{-2} s^{-1})$)

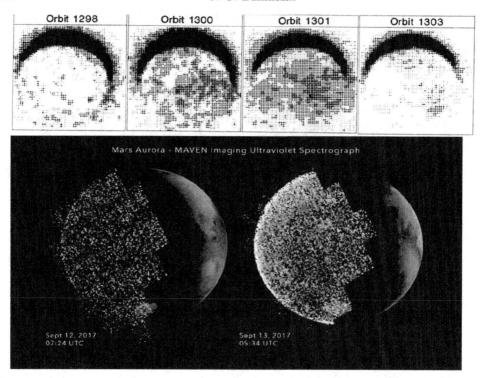

Figure 15. (Top) PVO UV images showing, in negative shading, the brightening of a diffuse nightside aurora seen in the 130.4 nm spectral line (from Phillips *et al.* 1986). (Bottom) MAVEN UV images of an auroral brightening at Mars that accompanied a SEP event and ICME impact (Schneider *et al.*, 2017).

before and during the event from the simulation, where the effect of the ICME passage is clearly seen. The estimated global ion escape in this case changed $\sim$10x from 10^{24} ions/s to 10^{25} ions/s. During the MAVEN mission, the encountered solar events have been relatively moderate. Had Mars experienced an 'extreme' ICME event such as that observed on the STEREO A spacecraft in 2012 (e.g. Liu *et al.*, 2013), similar model results suggest the global ion escape rate would have increased to 10^{27}ions/s. PVO was in orbit around Venus during a much stronger solar cycle, and experienced larger and more frequent events. PVO > 36 eV and anti-sunward ion data (see Figure 18) suggest Venus O+ escape rates increased during ICMEs by 100x or more (Luhmann *et al.*, 2007).

5. Effects over time

The challenge of reconstructing the history and consequences of these effects requires many assumptions involving poorly constrained conditions. Nonetheless, such exercises help identify specific gaps that new observables can sometimes fill. It is most straightforward to start calculations $\sim$3.5-3.8 Gyr ago, rather than at 4.5 Gyr, when solar system formation processes, including impacts, were still at work (e.g. Jakosky *et al.*, 2019), and to assume that today's planetary magnetic fields were already established. This also limits the effects of recently uncovered uncertainties in solar EUV history (Tu *et al.* 2015), based on observations of the range of EUV emitted by hundreds of G-type stars, including fast and slow rotators, as in Figure 19. It is currently unknown where the solar EUV evolution track falls, leaving up to an order of magnitude uncertainties for up to a third of the Sun's main sequence life.

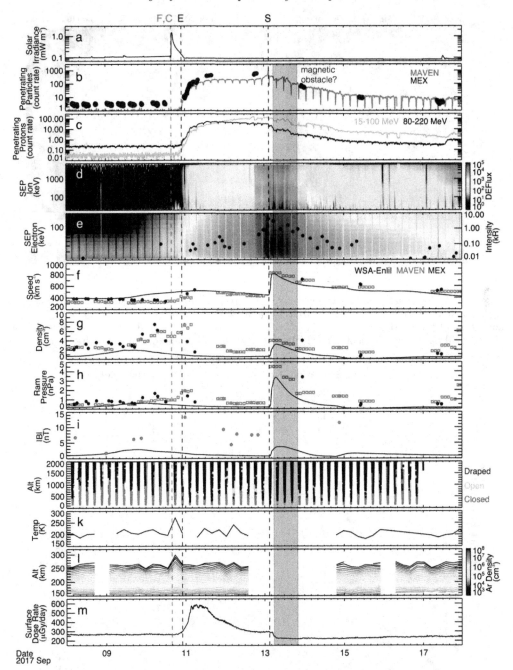

Figure 16. A time series of diverse observations from MAVEN, Mars Express and MSL, showing details of how Mars responded to a significant ICME impact in September 2017 (from Lee *et al.* 2018). The panels show (from the top), the solar flare in EUV intensity, the SEPs (including electrons and ions, the solar wind plasma parameters (density, velocity, temperature), and the magnetic field, all showing the shock arrival, a color bar indicating the change of local magnetic topology (red = closed fields, blue = open fields), and the upper atmosphere temperature and density response. The bottom panel is the highest energy SEP signature from MSL RAD on the surface. While these observations show that many different ICME responses occur at Mars, estimates of the related changes in globally escaping planetary ion flux require complementary modeling.

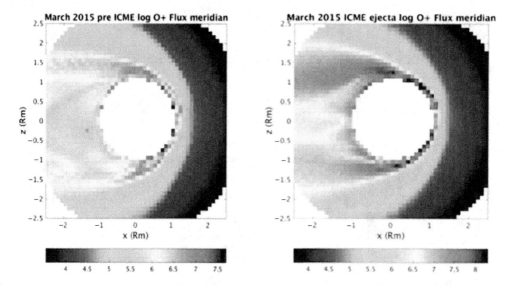

Figure 17. Meridian contours of planetary ion fluxes around Mars based on a data-validated model of an ICME passage in March 2015 (Ma *et al.* 2017; Luhmann *et al.* 2017). These represent snapshots of the conditions before the ICME arrived (left panel), and during the period when the coronal ejecta was present (right), illustrating the global enhancement of the escaping ion fluxes.

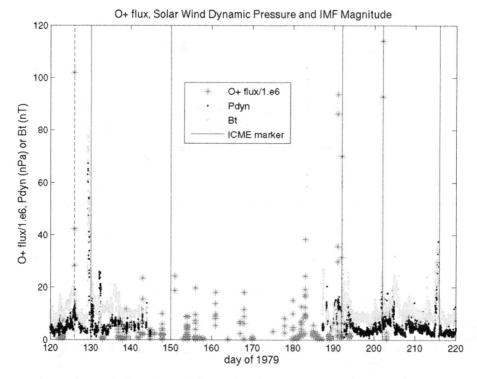

Figure 18. Extended timeline comparing escaping suprathermal planetary ion fluxes observed on PVO with the incident solar wind dynamic pressure (Pdyn) and interplanetary magnetic field (IMF) strength (Bt) at the time. The inferred escaping fluxes increase by up to ~100x during periods of high Pdyn and Bt, which are associated with solar wind inter-stream compression regions and ICMEs. (From Luhmann *et al.* 2007.)

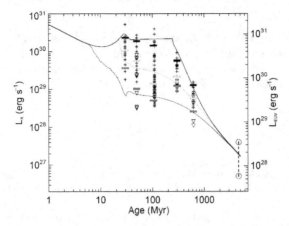

Figure 19. Figure from Tu *et al.* (2015), illustrating the range of possible EUV histories of Sun-like stars, depending on their rotational histories.

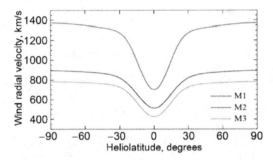

Figure 20. This figure from Airapetain & Usmanov (2016) shows models for solar wind velocities for the present-day Sun (M3), and for solar ages of 2.0 Gyr (M2) and 0.7 Gyr (M1).

Even less constrained are early solar wind models (e.g. see Wood *et al.* 2015). A recent model by Airapetain & Usmanov (2016) has conditions in its early epoch (see Figure 20, curve labeled M1), as extreme as today's observed ICME events. This early solar wind would have had major 'impacts' by itself.

6. Summary

It probably does not matter much if a planet is magnetized or not. Rather, the history of the solar and interplanetary conditions can determine atmospheric evolution in post-impact and post-hydrodynamic outflow-dominated escape epochs. We must better constrain them from times of about 1 Gyr of age. While Sun-like star observations are valuable, our Sun's own history is of utmost importance. Is the answer to be found at the moon, where samples may contain evidence of the early solar wind and solar activity? Kepler mission observations of 'superflares' on Sun-like stars of various ages (Shibayama *et al.* 2013) include flare energies of $\sim 10^{34}$ ergs or more, compared to large present-day solar flares that have up to 10^{32} ergs. The work of Aarnio *et al.* (2011) suggests that the empirical relationship between solar CME size (from coronagraphs) and solar flare intensity may apply to flaring Sun-like stars. But CMEs at these flaring early 'Suns' have been difficult to identify (e.g. Osten & Wolk 2016). Further observations are key to investigating this and other pertinent questions about both early solar wind and solar activity, and the planetary consequences they left behind.

Acknowledgments

The work described in this review paper is based largely on results from NASA's PVO, MAVEN and MSL missions and ESA's Venus and Mars Express missions. The author is also supported by a NASA grant (for related space weather observations and modeling as part of the STEREO mission science team effort).

References

Aarnio, A. N., Stassun, K.G., Hughes, W.J., McGregor, S.L. 2011. Solar Flares and Coronal Mass Ejections: A Statistically Determined Flare Flux – CME Mass Correlation. *Sol. Phys.* 268, 195–212.

Airapetain V. S., Usmanov, A.V. 2016 Reconstructing the Solar Wind from Its Early History to Current Epoch. *ApJL.* 817. doi:10.3847/2041-8205/817/2/L24.

Brace, L.H., Theis, R.F., Hoegy, W.R. 1982. Plasma clouds above the ionopause of Venus and their implications. *Planet. Space Sci.* 30, 29-37, doi:10.1016/0032-063390069-1.

Brain, D. A., Baker, A. H., Briggs, J., Eastwood, J. P., Halekas, J. S., and Phan, T.D. 2010. Episodic detachment of Martian crustal magnetic fields leading to bulk atmospheric plasma escape, *Geophys. Res. Lett.*, 37, L14108. doi:10.1029/2010GL043916.

Chaffin, M. S., Chaufray, J.Y., Stewart, I., Montmessin, F., Schneider, N. M., and Bertaux, J.L. 2014. Unexpected variability of Martian hydrogen escape. *Geophys. Res. Lett.*, 41, 314-320, doi:10.1002/2013GL058578.

Curry, S.M., Luhmann, J.G., Ma, Y.J., Liemohn, M., Dong, C., Hara, T. 2015. Comparative pick-up ion distributions at Mars and Venus: Consequences for atmospheric deposition and escape. *Planet. Space Sci.* 115, 35, 47.

Dong, Y., Fang, X., Brain, D. A., McFadden, J. P., Halekas, J. S., Connerney, J. E., Curry, S. M., Harada, Y., Luhmann, J. G., Jakosky, B. M. 2015, Strong plume fluxes at Mars observed by MAVEN: An important planetary ion escape channel. *Geophys. Res. Lett.*, 42, 8942– 8950, doi:10.1002/2015GL065346.

Dong, C., Ma, Y., Bougher, S. W., Toth, G., Nagy, A. F., Halekas, J. S., Dong, Y., Curry, S. M., Luhmann, J. G., Brain, D., *et al.* 2015, Multi-fluid MHD study of the solar wind interaction with Mars' upper atmosphere during the 2015 March 8th ICME event. *Geophys. Res. Lett.* 42, 9103– 9112, doi:10.1002/2015GL065944.

Dubinin, E., Fraenz, M., Pätzold, M., McFadden, J., Halekas, J. S., DiBraccio, G. A., Zelenyi, L. 2017. The effect of solar wind variations on the escape of oxygen ions from Mars through different channels: MAVEN observations. *J. Geophys. Res. Space Phys.*, 122, 11285-11301. doi:10.1002/2017JA024741.

Edberg, N. J. T., *et al.* 2011. Atmospheric erosion of Venus during stormy space weather. *J. Geophys. Res.* 116, A09308.

Fang, X., Liemohn, M. W., Nagy, A. F., Ma, Y., De Zeeuw, D. L., Kozyra, J. U., Zurbuchen, T. H. 2008, Pickup oxygen ion velocity space and spatial distribution around Mars. *J. Geophys. Res.* 113, A02210, doi:10.1029/2007JA012736.

Fegley, B., Jr. 2014. Venus. In H. D. Holland, & K. K. Turekian (Eds.), *Treatise on Geochemistry*, 2nd ed., vol. 2 (pp. 127–148). Amsterdam, Netherlands: Elsevier.

Futaana, Y., Stenberg Wieser, G., Barabash, S., *et al.* 2017. Solar Wind Interaction and Impact on the Venus Atmosphere, *Space Sci. Rev.* 212, 1453. doi:10.1007/s11214-017-0362-8.

Gray, C.L., Chanover, N.J., Slanger, T.G., Molaverdikhani, K. 2014. The effect of solar flares, coronal mass ejections, and solar wind streams on Venus' 5577Å oxygen green line. *Icarus.* 233, 342-347. doi:10.1016/j.icarus.2014.01.029.

Hall, S. 2019. Venus is Earth's evil twin - and space agencies can no longer resist its pull. *Nature.* 570, 20-25. doi:10.1038/d41586-019-01730-5.

Hu, R., Kass, D., Ehlmann, B. *et al.* 2015. Tracing the fate of carbon and the atmospheric evolution of Mars. *Nat. Comm.* 6, 10003. doi:10.1038/ncomms10003.

Jakosky, B. Brain, D., Chaffin, M., Curry, S., Deighan, J., Grebowsky, J., Halekas, J., Leblanc, F., Lillis, R., Luhmann, J.G. *et al.* 2018. Loss of the Martian atmosphere to space: Present-day loss rates determined from MAVEN observations and integrated loss through time. *Icarus*. 315, 146-157. doi.org/10.1016/j.icarus.2018.05.030.

Jarvinen, R., Kallio, E., Dyadechkin S., *et al.* 2010. Widely different characteristics of oxygen and hydrogen ion escape from Venus. *Geophys. Res. Lett.* 37, L16201. doi:10.1029/2010GL044062.

Kasting, J. F. 1988. Runaway and moist greenhouse atmospheres and the evolution of Earth and Venus. *Icarus*, 74, 472–494.

Lammer, H., Kasting, J. F., Chassefière, E., *et al.* 2008. Atmospheric escape and evolution of terrestrial planets and satellites. *Space Sci. Rev.* 139, 399–436. doi:10.1007/s11214-008-9413-5.

Lee, C. O., Jakosky, B. M., Luhmann, J. G., Brain, D. A., Mays, M. L., Hassler, D. M., *et al.* 2018. Observations and impacts of the 10 September 2017 solar events at Mars: An overview and synthesis of the initial results. *Geophys. Res. Lett.* 45, 8871– 8885. https://doi.org/10.1029/2018GL079162.

Liu, Y., Luhmann, J., Kajdič, P. *et al.* 2014. Observations of an extreme storm in interplanetary space caused by successive coronal mass ejections. *Nat Comm.* 5, 3481. doi:10.1038/ncomms4481.

Luhmann, J. G., *et al.* 2017, Martian magnetic storms. *J. Geophys. Res. Space Physics.* 122, 6185– 6209, doi:10.1002/2016JA023513.

Luhmann, J.G., Curtis, D.W., Schroeder, P. *et al.* 2008. STEREO IMPACT Investigation Goals, Measurements, and Data Products Overview. *Space Sci. Rev.* 136, 117. doi:10.1007/s11214-007-9170-x.

Luhmann, J. G., Kasprzak, W. T., and Russell, C. T. 2007, Space weather at Venus and its potential consequences for atmosphere evolution. *J. Geophys. Res.* 112, E04S10. doi:10.1029/2006JE002820.

Luhmann, J.G., Ledvina, S.A., Lyon, J.G., Russell, C.T. 2006. Venus O+ pickup ions: Collected PVO results and expectations for Venus Express. *Planet. Space Sci.* 54, 1457-1471, doi:10.1016/j.pss.2005.10.009.

Luhmann, J. G., Russell, C. T., Scarf, F. L., Brace, L. H., and Knudsen, W. C. 1987. Characteristics of the Marslike limit of the Venus-solar wind interaction. *J. Geophys. Res.* 92 (A8), 8545– 8557. doi:10.1029/JA092iA08p08545.

Ma, Y. J., *et al.* 2017. Variations of the Martian plasma environment during the ICME passage on 8 March 2015: A time-dependent MHD study. *J. Geophys. Res. Space Physics.* 122, 1714– 1730, doi:10.1002/2016JA023402.

McEnulty, T. 2012. Oxygen Loss from Venus and the Influence of Extreme Solar Wind Conditions, PhD thesis, University of California, Berkeley. https://www.worldcat.org/title/oxygen-loss-from-venus-and-the-influence-of-extreme-solar-wind-conditions/oclc/842823603.

Moore, T. E., Horwitz J. L. 2007. Stellar ablation of planetary atmospheres. *Rev. Geophys.* 45, RG3002. doi:10.1029/2005RG000194.

Osten, R., Wolk, S. 2016. A Framework for Finding and Interpreting Stellar CMEs. *Proceedings of the International Astronomical Union*, 12(S328), 243-251. doi:10.1017/S1743921317004252.

Phillips, J. L., Stewart, A. I. F., Luhmann, J. G. 1986. The Venus ultraviolet aurora: Observations at 130.4 nm. *Geophys. Res. Lett.*13,1047-1050, doi:10.1029/GL013i010p01047.

Pizzo, V. J. 1978. A Three-Dimensional Model of Corotating Streams in the Solar Wind - I. Theoretical Foundations. *J. Geophys. Res.* 83, 5563–5572.

Ruhunusiri, S., *et al.* 2016, MAVEN observations of partially developed Kelvin-Helmholtz vortices at Mars. *Geophys. Res. Lett.* 43, 4763– 4773. doi:10.1002/2016GL068926.

Russell, C.T., Luhmann, J.G., Strangeway, R.J. 2016. *Space Physics: An Introduction*, Cambridge University Press.

Russell, C. T., Luhmann, J. G., Elphic, R. C., Scarf, F. L. and Brace, L. H. 1982. Magnetic field and plasma wave observations in a plasma cloud at Venus. *Geophys. Res. Lett.* 9, 45-48. doi:10.1029/GL009i001p00045.

Schneider, N. M., Jain, S. K., Deighan, J., Nasr, C. R., Brain, D. A., Larson, D., *et al.* 2018. Global aurora on Mars during the September 2017 space weather event. *Geophys. Res. Lett.*, 45, 7391-7398. https://doi.org/10.1029/2018GL077772.

Schneider, N. M., Deighan, J. I., Jain, S. K., Stiepen, A., Stewart, A. I. F., Larson, D., Mitchell, D. L., Mazelle, C., Lee, C. O., Lillis, R. J., Evans, J. S., Brain, D., Stevens, M. H., McClintock, W. E., Chaffin, M. S., Crismani, M., Holsclaw, G. M., Lefevre, F., Lo, D. Y., Clarke J. T., Montmessin, F., Jakosky, B.M. 2015. Discovery of diffuse aurora on Mars. *Science.* 350, doi:10.1126/science.aad0313.

Shibayama, T,. Maehara, H., Notsu, S., Notsu, Y., Nagao, T., Honda, S., Ishii, T., Nogami, T., Daisaku, T., Kazunari, S. 2013. Superflares on Solar-type Stars Observed with Kepler. I. Statistical Properties of Superflares. *ApJ Supp.* 209.

Tsiaras, A., Waldmann, I.P., Tinetti, G. *et al.* 2019. Water vapour in the atmosphere of the habitable-zone eight-Earth-mass planet K2-18 b. *Nat Astron.* doi:10.1038/s41550-019-0878-9.

Tu, L., Johnstone, C.P., Guedel, M., Lammer, H. 2015. The extreme ultraviolet and X-ray Sun in Time: High-energy evolutionary tracks of a solar-like star, *Astron. Astrophys.* 577, doi:10.1051/0004-6361/201526146.

Villanueva G. L., Mumma, M. J., Novak, R. E., Käufl, H. U., Hartogh, P., Encrenaz, T. Tokunaga, A., Khayat, A., Smith, M. D. 2015. Strong water isotopic anomalies in the Martian atmosphere: Probing current and ancient reservoirs, *Science* 348. 218-221.doi:10.1126/science.aaa3630, 2015.

Walsh, B. M., Foster, J. C., Erickson, P. J., Sibeck, D. G. 2014. Simultaneous Ground- and Space-Based Observations of the Plasmaspheric Plume and Reconnection. *Science.* 343, 1122-1125, doi:10.1126/science.1247212. Jarvinen, R., Kallio, E., Dyadechkin S., *et al.* 2010. Widely different characteristics of oxygen and hydrogen ion escape from Venus. *Geophys. Res. Lett.* 37, L16201. doi:10.1029/2010GL044062.

Wood, B.E., Linsky, J.L., Güdel, M. 2015. Stellar Winds in Time. In: Lammer H., Khodachenko M. (eds) *Characterizing Stellar and Exoplanetary Environments.* Astrophysics and Space Science Library, 411. Springer.

Xu, S. *et al.* 2018. Investigation of Martian Magnetic Topology Response to 2017 September ICME, *Geophys. Res. Lett.*, 45, 7337–7346. doi:10.1029/2018GL077708

Zhang, T.-L. *et al.* 2012. Magnetic reconnection in the near Venusian magnetotail. *Science.* 336, 567-570.

Discussion

DMITRY BISIKALO: Could you kindly comment on the role of the planetary magnetic field in the mass loss?

JANET LUHMANN: Observations suggest that for the most extreme external conditions, the presence of the field may not matter. The planetary magnetic field may not be able to prevent energization and losses, although it may change the detailed physics of the processes involved. The escape rates may ultimately depend on the atmospheric production and delivery (e.g. by photochemistry and diffusion) of species to regions from which they can escape (e.g. the exobase).

Solar and Stellar Magnetic Fields: Origins and Manifestations
Proceedings IAU Symposium No. 354, 2019
A. Kosovichev, K. Strassmeier & M. Jardine, ed.
doi:10.1017/S1743921319009979

Different types of star-planet interactions

A. A. Vidotto[ID]

School of Physics, Trinity College Dublin, The University of Dublin, Dublin 2, Ireland
email: `aline.vidotto@tcd.ie`

Abstract. Stars and their exoplanets evolve together. Depending on the physical characteristics of these systems, such as age, orbital distance and activity of the host stars, certain types of star-exoplanet interactions can dominate during given phases of the evolution. Identifying observable signatures of such interactions can provide additional avenues for characterising exoplanetary systems. Here, I review some recent works on star-planet interactions and discuss their observability at different wavelengths across the electromagnetic spectrum.

Keywords. stars: magnetic fields, stars: winds, outflows, stars: planetary systems

---

1. Introduction

Stellar magnetic fields drive the space weather of exoplanets. In cool dwarf stars, their magnetic activity are responsible for driving stellar winds and coronal mass ejections, and also for generating high-energy irradiation in the extreme ultraviolet (UV) and X-rays. Hence, understanding the host star magnetism is a key ingredient for characterisation of exoplanetary environments. The magnetic properties of cool stars depends on several ingredients, such as their rotation, age, and internal structure (Vidotto *et al.* 2014).

Planets that orbit closer to their host stars are embedded in harsher stellar environments, as magnetism and irradiation, for example, decay with some power of the distance. The extreme architecture of most of the known exoplanetary systems, in addition to the differences in magnetic properties of host stars compared to those of our Sun, can give rise to planet-star interactions that are not present in the solar system. These interactions can generate observable signatures, thus providing additional avenues for characterising exoplanetary systems.

In this review, I present some recent works on star-planet interactions. I consider four different types of interactions, as shown in the diagram (orange boxes): magnetic, tidal, interaction with a stellar wind and with stellar irradiation. Each of these interactions can have different effects on either the star or the planet (pink boxes): causing chromospheric hotspots, migration of orbits, etc. These effects in turn can generate different observables (green boxes) throughout the electromagnetic spectrum: from radio to ultraviolet/X-rays.

In the Sections that follow, I discuss the effects of different types of star-planet interactions, based on the review talk I presented in the IAU Symposium 354 "Solar and Stellar Magnetic Fields: Origins and Manifestations". It is important to note that the signatures of star-planet interactions can occur in the star or in the planet. These signatures are not necessarily recurrent. In some cases, the signatures are not very strong, which have led to multiple interpretations of the same dataset. The potential material to be reviewed is thus very large! Instead of discussing a particular system in detail, including multiple analyses and interpretations of data, instead, here, I have chosen to present different flavours of interactions between stars and exoplanets. I hope that the reader will appreciate how vibrant this research field is and will use this paper as a starting point to deepen their interest/research in this exciting field!

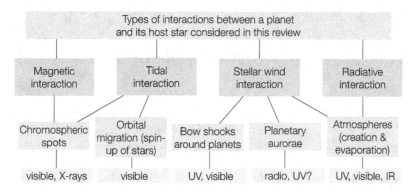

Figure 1. Diagram illustrating the different types of interactions between a planet and its host star discussed in this review.

2. Anomalous stellar activity

It did not take long after the discovery of the first exoplanet orbiting a cool dwarf for researchers to propose that close-in planets experimenting strong interactions with their host stars could excite 'hot spots' on the chromospheres of their hosts (Cuntz *et al.* 2000). Spots are seen at the surface of the Sun and indirectly detected in stars, through rotational modulation in their lightcurves. Bright hot spots, on the other hand, are seen in close binary stars, near the sub-binary point. Similarly, the idea is that close-in exoplanets would excited additional, or anomalous, activity in their hosts, in the case of strong star-planet interactions. In both cases though the orbital period and stellar rotational period must differ in order to disentangle the signature of the anomalous activity from that of normal stellar activity.

Imagine, for example, an exoplanetary system that has tidal interactions similar to the Earth-moon system. As the planet orbits around its host, it raises tidal bulges that are in constant expansion and contraction. The movement of these bulges can dissipate energy and generate heating, that would give rise to anomalous stellar activity. Because two bulges are raised in tidal interactions, this means that anomalous activity created by tidal interactions should be modulated by half the orbital period of the planet, $P_{\rm orb}/2$ (Saar & Cuntz 2001). †

Another example consists of magnetic interactions between a magnetised host star and a magnetised planet. In certain configurations, magnetic field lines of opposite polarities belonging to these two different bodies can undergo reconnection, releasing energetic particles that travel towards the star along stellar magnetic field lines (Ip *et al.* 2004). As these particles impact the stellar chromosphere at the magnetic footpoints, chromospheric hot spots can be excited. As the planet moves through its orbit, the planet interacts with different stellar field lines, so hot spots would be 'activated' through the orbit and thus be modulated by the planet's orbital period, $P_{\rm orb}$ (Saar & Cuntz 2001). ‡

Figure 4 shows that the magnetic field lines linking the planet to the star are a combination of open lines and closed loops. These closed magnetic loops will cause the SPI-related

---

† In fact, the modulation should occur at the beat period, which is a combination of stellar rotation and orbital periods (Fares *et al.* 2010).

‡ This scenario implies that the star has a simple magnetic field geometry, while, in reality, the modulation might be more complicated due to complex magnetic field topologies. For example, consider a star with a tilted dipole geometry. This star would have magnetic field lines linking to the planet that are a combination of open lines and closed loops. As the planet moves from one branch of the closed loops to another, the activated chromospheric hot spots would appear to be jumping at different positions in the stellar chromosphere, thus creating a phase lag effect or a jumping effect (e.g., McIvor *et al.* 2006, Strugarek *et al.* 2019, Folsom *et al.*, in prep).

chromospheric spots on the star to move differently from the planetary orbit, with large jumps occurring where the planet moves from one branch of the closed loop to another. This phase lag and jumping effect is evident from Fig. 5 (see also predictions by McIvor *et al.* 2006). Recently, Strugarek *et al.* (2019) showed a similar effect in the case of Kepler-78, where the magnetic topol- ogy of the host star can greatly affect the transient nature of SPI. Although their simulations did not explain the amplitude of en- hanced activity observed by Moutou *et al.* (2016), in stars with stronger magnetic fields (e.g. HD 179949, Fares *et al.* 2012) the effect may be more detectable.

Hence, the modulation of anomalous activity could tell us something about whether the interaction is of tidal or magnetic type. Shkolnik *et al.* (2005, 2008) searched anomalous activity in some close-in planets finding in some cases modulation with P_{orb}, and interpreting those as caused by magnetic interactions, which would allow us to learn about planetary magnetic moment. Recently, Cauley *et al.* (2018) investigated anomalous activity in the K dwarf HD189733, triggered by its hot-jupiter. They found that anomalous activity is present in one observational epoch, and is modulated with the orbital period of the planet. This epoch happens to be when the stellar magnetic field is observed to be stronger (Fares *et al.* 2017), indicating that the level of stellar magnetic activity would either affect when the hotspots are triggered or at the very least when they are detectable.

An important concept to keep in mind in these interactions is the orbital distance in relation to the Alfven surface of the star. The Alfven surface delineates the regime of the stellar wind that is dominated by magnetic forces (within the surface) or by inertial forces (beyond the surface). Inside the Alfven surface, the stellar wind is sub-Alfvenic, and Alfven "wing" currents, connecting star and planet, develop (Strugarek *et al.* 2015). However, in the super-Alfvenic regime, this direct connection no longer exists. This implies that, as the stellar magnetism evolves (e.g., through an activity cycle), so does the Alfven surface and the star-planet connectivity (Nicholson *et al.* 2016; Kavanagh *et al.* 2019). This evolution might, for example, affect when anomalous activity through star-planet interaction is triggered, giving rise to an 'on/off' nature (Shkolnik *et al.* 2008).

This 'on/off' regime can also happen in orbital timescales. Imagine the case of an eccentric system: at periastron, the planet interacts with a larger stellar magnetic field, hence there is a stronger star-planet interaction. At apastron, the opposite happens: the large distance means that the stellar magnetic field is weaker, hence would generate a weaker star-planet interaction. The eccentric planet HD17156b was observed at different orbital phases in X-rays by Maggio *et al.* (2015), who found enhanced stellar emission a few hours after periastron. This enhanced stellar emission could be interpreted as a magnetic reconnection event that was triggered (or became more powerful) at close orbital separation. Alternatively, it could also be interpreted as some sort of planetary material that evaporated and was accreted into the star, generating a bright X-ray event. Similar to other works on star-planet interaction, the X-ray emission was occasional, and not seen in all observed epochs.

- *In the electromagnetic spectrum: anomalous modulations are usually observed in the visible, with ground-based spectrographs. Observations in X-rays have recently been conducted with missions like XMM and Chandra and, in the future, with could be possible to conduct with Athena.*

3. Interactions with stellar winds

As the wind outflows from the star, it permeates the interplanetary medium. Stellar winds consist of particles and magnetic fields, similar to the solar wind. There are, however, important differences between the solar wind around the Earth and a stellar wind around a close-in exoplanet. Compared to the Earth, the close-in location of hot-Jupiters imply they interact with (1) higher density wind, (2) higher ambient magnetic fields, (3) in general, lower wind velocities, as close-in planets are 'parked' inside the acceleration zone of stellar winds. Even though the stellar wind is expected to have lower velocities around close-in planets, the fact that the orbital velocities of these planets are higher imply that, in the reference frame of a close-in exoplanet, the relative velocity between the planet and the host star wind can reach several km/s, going above local sonic speeds (or Alfven speeds) (Vidotto et al. 2011). Overall, the extreme conditions of local stellar winds around close-in exoplanets imply stronger planet-wind interactions than those seen in the solar system (Vidotto et al. 2015).

In super-sonic (or super-Alfvenic) interactions, ie, bow shocks can develop around the planets (Vidotto et al. 2010a; Carolan et al. 2019). In addition, depending on the characteristics of the system (ie, physical conditions of the stellar winds and the planets), evaporated planetary material may trail the planet (Villarreal D'Angelo et al. 2018; Daley-Yates & Stevens 2019), forming a comet-like tail structure. Alternatively, in some configurations, material can be funnelled towards the star, forming an inspiralling accretion stream (Matsakos et al. 2015).

Observations of these stellar-wind–planet interactions have been done mostly in ultraviolet lines, observed during transits (e.g. Fossati et al. 2010; Bourrier et al. 2016). In the case of the ultra-hot Jupiter WASP-12b, HST observations demonstrated that the planet starts transiting in the UV much before the geometric transit (Fossati et al. 2010). This asymmetric transit requires the presence of asymmetric material around the planet. One interpretation is that this asymmetry is created by shocked stellar wind material ahead of the planetary orbit. Contrary to the bow shock surrounding the Earth, which is oriented towards the Sun, the high orbital velocity of WASP-12b causes the bow shock to swing around, appearing at an angle ahead of its orbit. If the bow shock is formed around the magnetosphere of the planet (similar to Earth's bow shock), then these observations can constrain the magnetic field of the planet (Llama et al. 2011).

These observations also have the potential to reveal the physical conditions of stellar winds. In the case of the warm-Neptune GJ436b, models of the asymmetric distribution of material around the planet can constrain local stellar wind conditions (Bourrier et al. 2016), which can then be used to probe the global properties of stellar winds (Vidotto & Bourrier 2017). Winds of cool dwarfs are very rarefied and often challenging to be observed (Wood 2004; Fichtinger et al. 2017; Ó Fionnagáin et al. 2019; Jardine & Collier Cameron 2019).

In the solar system, the interaction between the solar wind and a planet's magnetosphere give rise not only to a bow shock, but also to radio (auroral) emission. The "radio Bode's law" is an empirical finding that shows that the power emitted by the magnetised solar system planet in radio is proportional to the power dissipated by the solar wind on the interaction with the planet (Zarka 1998). Extrapolations of this empirical scaling have been used to predict the amount of radio power a close-in exoplanet would emit. The predictions often result in radio powers that are several orders of magnitude larger than that of Jupiter, but so far, most searches for exoplanetary radio emission have come out empty. A hint of detection was reported in Lecavelier des Etangs et al. (2013) from the hot-Jupiter HAT-P-11b. In this work, it was not the emission of the planet that was seen, but rather the reduced radio emission of the system at the phases where the planet was occulted by the star (secondary transit). Because these are cyclotron emissions, whose frequencies are proportional to the planetary magnetic field, their detections would allow us to derive the magnetic properties of exoplanets. Magnetic fields are windows towards the interior of the planets, as they are generated by dynamo mechanism in planetary interiors.

• *In the electromagnetic spectrum: UV observations are currently being done with HST. The cubesat CUTE (PI: K. France; Fleming et al. 2017) is a dedicated mission to study spectroscopic transits of close-in exoplanets and will be launched in 2020. Radio observations of planetary radio emission have been conducted with a wide variety of radio telescopes. It is believed that low-frequency observations are more appropriate - if these exoplanets host 1 – 100G magnetic field, they would generate emission at cyclotron frequencies of 2.8 – 280 MHz. LOFAR and, in the future, SKA could be appropriate for detecting these signals.*

4. Spin-up of stellar rotation

The orbits of planets can change in time due to several mechanisms, such as tidal interactions, stellar winds that carry away stellar angular momentum (or, conversely, mass accretion that increases stellar angular momentum), friction and planet evaporation. As a consequence, orbital migration could lead a planet to be engulfed by its host star. As the planet moves to inner orbits, conservation of angular momentum of the system implies that a reduction in orbital angular momentum is accompanied by an increase in stellar angular momentum. Therefore, if a planet is engulfed, an accompanied increase in stellar rotation is expected. The timescale for this to happen, if it ever happens in a given planetary system, depends on several characteristics, from stellar properties (including its internal structure, Privitera *et al.* 2016b) to planetary properties and orbital parameters.

In studies investigating spin up of stars in the red giant phase, Privitera *et al.* (2016a,b,c) suggested that planet engulfment could be the origin of fast rotating red giants. A planet, with mass on the order of 1 to 15 Jupiter masses, initially orbiting at 0.5 au, could spin up the star to equatorial rotation velocities of up to few tens of km/s. These high velocities cannot be explained by models that only take into account the evolution of a single star. In line with what happens in the main sequence (Vidotto *et al.* 2014), rotation and magnetic fields are linked (Aurière *et al.* 2015) are also linked in the post-main sequence. Thus, it is possible that a high surface magnetic field in red giants could be a detectable signature of planet engulfment (Privitera *et al.* 2016a).

Similarly, a close-in planet engulfed during the stellar main-sequence phase could generate high stellar rotation. Benbakoura *et al.* (2019) investigated potential physical parameters in which an inward migrating planet could spin up a main-sequence solar mass star. They showed that lighter planets ($\lesssim 1 M_{\rm jupiter}$), and/or planets with orbits greater than $\sim$ 0.03au, would not spin up their hosts during the main sequence. But ultra hot Jupiters, i.e, more massive planets and at closer orbital distances, could.

Stellar rotation also plays a role in planetary migration (Lovelace *et al.* 2008): a change in sign in tidal forces occur at the corotating radius, which is given by

$$r_{\rm co} = \left(\frac{GM_\star P_{\star,\rm rot}^2}{4\pi^2} \right)^{1/3} = 0.02 \text{ au} \left(\frac{P_{\star,\rm rot}}{1 \text{ day}} \right)^{2/3} \left(\frac{M_\star}{M_\odot} \right)^{1/3} \tag{4.1}$$

where $P_{\star,\rm rot}$ is the stellar rotation period, $M_\star$ its mass and G is the gravitational constant. A fast rotating star, has a lower corotating radius. For example, at a 1-day period rotation, the corotating radius is at 0.02 au for a solar mass star. Thus, a planet orbiting outside this radius gains angular momentum from the stellar spin, causing the star to spin down and the planet to be pushed away from the star (Lovelace *et al.* 2008; Vidotto *et al.* 2009, 2010b). Thus, stars born as fast rotators could push away their inner planets at the beginning of their lives. The farther the planet is, however, means reduced tidal forces, until any orbital evolution becomes insignificant. Benbakoura *et al.* (2019) showed that as the star spins down during its main-sequence phase, the corotating radius increases and a planet, initially orbiting beyond $r_{\rm co}$, could end up orbiting below $r_{\rm co}$. In this case, even though the transfer of angular momentum would happen from the orbit to the star, the forces might be too low to change the orbital evolution significantly. They showed that, planets orbiting fast rotating hosts would take longer to be engulfed (and could never be engulfed in fact), regardless of its mass.

• *In the electromagnetic spectrum: detecting rotation rates can be done from the ground and from space. We have seen a boost in rotational data with planet detection surveys (Kepler, K2, TESS and in the future Plato, Cheops) and recently with GAIA. I have the impression that it might be difficult to attribute an 'atypical' stellar rotation (i.e., that related to a planet engulfment) in the main sequence, but this could be more easily disentangled in the red giant phase (Privitera et al. 2016a). Another possible signature of a planet engulfment could be a change in stellar metallicity ("pollution"), possibly detected using ground-based spectrographs. However, I am not aware how much metallicity change to expect nor how long the signature would remain until planetary material becomes mixed with stellar material and, effectively, disappears.*

5. Atmospheric creation or evaporation

The interaction between a stellar wind and a planetary atmosphere has long been attributed to causing planet erosion. This is, for example, what is believed to have happened to Mars, which does not have an intrinsic magnetic field, therefore lacking a protective 'umbrella' against the solar wind. Recently, there has been a debate in the literature whether this is indeed the case, with some authors arguing that magnetic fields do not affect atmospheric escape, with Mars and Earth being counter examples of unmagnetised and magnetised planets, respectively, with similar outflow rates (Strangeway et al. 2010, see also Blackman & Tarduno 2018; Egan et al. 2019). Although escape in solar system planets can help us understand exoplanet evaporation (or atmospheric survival), the different, and often very extreme, architectures of exoplanetary systems compared to the solar system does not necessary guarantee that the same evaporation mechanisms taking place in the solar system would operate (or be as strong) in the exoplanets knows to date.

For example, HST observations have shown that close-in giant planets have huge atmospheric escape rates (e.g. Vidal-Madjar et al. 2003; Fossati et al. 2010; Ehrenreich et al. 2015; Bourrier et al. 2016; Kulow et al. 2014). † The only mechanism that can generate such a large outflow rate is that of hydrodynamic escape caused by stellar extreme UV (EUV) irradiation. This mechanism is not currently important in the solar system planets. Irradiation influences exoplanetary atmospheric temperatures and affects mass loss. Due to geometrical effects, close-in planets receive large levels of irradiation from their stars. This increased heating causes their atmospheres to inflate and more likely to outflow through a hydrodynamic escape mechanism. Given that stellar EUV radiation changes through stellar evolution, this means that the survival of atmospheres depends on the EUV history of host star (Tu et al. 2015; Johnstone et al. 2015). EUV fluxes are related to stellar activity, which is observed to decrease with age. Therefore, a close-in giant planet orbiting a young star will have a higher outflow rate, which will decrease with the evolution of stellar activity. In comparative terms, a lower gravity giant cannot hold on to its atmosphere in a same way as a higher gravity giant can. The two processes combined thus indicate that young, Saturn-mass planets are more easily to evaporate

---

† More recently, escape has also been studied in the infrared triplet line of neutral Helium at 10830Å (Nortmann et al. 2018; Spake et al. 2018; Allart et al. 2019) A more indirect detection of planetary mass loss comes from *Kepler* observations of the distributions of planetary radii and orbital periods (Fig. 1, Beaugé & Nesvorný 2013; Mazeh et al. 2016; Fulton et al. 2017), which showed that planetary evaporation is necessary to explain the depletion of small planets with short orbital periods (Jin et al. 2014; Helled et al. 2016).

than older, Jupiter-mass planets orbiting at close distances to their host stars (Allan & Vidotto 2019).

In a more counter-intuitive mechanism, Vidotto *et al.* (2018) showed that planetary atmospheres can be *created* (not eroded!) in the interaction with stellar winds. The proposed mechanism is that of sputtering caused by stellar wind particles incident on the bare surface of close-in terrestrial planets. This is modelled in a similar way as to sputtering in Mercury (Pfleger *et al.* 2015), but occur with a few orders of magnitude larger incident kinetic energy of the solar wind. Vidotto *et al.* (2018) studied atmospheric creation in the close-in terrestrial planets HD219134b and HD219134c, showing that sputtering releases refractory elements from the entire dayside surfaces of the planets, with elements such as O and Mg creating an extended neutral exosphere around these planets.

- *In the electromagnetic spectrum: Hydrogen escape has been observed in the UV Ly-α line with HST and some attempts done also in H-α, with the benefit that the latter can be conducted from the ground (e.g. Cauley et al. 2017). Other UV (metal) lines have also tracked escape in observations conducted with HST (Fossati et al. 2010) and, in the future, could be done with CUTE (Fleming et al. 2017). In the infrared, detections can be made with ground-based spectrographs, such as CARMENES or SpIRou, or from space-based missions, such as HST and, in the future, JWST.*

6. Conclusions

In this (too) brief review, I discussed four different types of star-planet interactions: magnetic, tidal, with stellar wind and with stellar radiation. All these interactions can produce observable signatures that can take place in the star or in the planet. Their detections could help further characterise the physical conditions of planetary systems, such as magnetism in exoplanets, stellar wind properties at the orbit of exoplanets, etc. These signatures can take place at different wavelengths across the electromagnetic spectrum, with some new (and old) instruments with great potential for researching star-planet interactions: in the infrared (SpIRou, CARMENES, JWST), low-frequency radio (Lofar, SKA), optical (TESS, Plato, GAIA), UV (HST, CUTE), X-rays (XMM, Chandra, Athena), just to cite a few.

Acknowledgements

I would like to thank (again) the organisers of this fantastic symposium for its organisation, for the invitation to give this review and for partial financial support to attend the meeting. I also acknowledge funding received from the Irish Research Council Laureate Awards 2017/2018, and partially from the European Research Council (ERC) under the European Union's Horizon 2020 research and innovation programme (grant agreement No 817540, ASTROFLOW).

References

Allan, A. & Vidotto, A. A. 2019, *MNRAS*, 490, 3760
Allart, R., Bourrier, V., Lovis, C., *et al.* 2019, *A&A*, 623, A58
Aurière, M., Konstantinova-Antova, R., Charbonnel, C., *et al.* 2015, *A&A*, 574, A90
Beaugé, C. & Nesvorný, D. 2013, *ApJ*, 763, 12
Benbakoura, M., Réville, V., Brun, A. S., Le Poncin-Lafitte, C., & Mathis, S. 2019, *A&A*, 621, A124
Blackman, E. G. & Tarduno, J. A. 2018, *MNRAS*, 481, 5146
Bourrier, V., Lecavelier des Etangs, A., Ehrenreich, D., Tanaka, Y. A., & Vidotto, A. A. 2016, *A&A*, 591, A121
Carolan, S., Vidotto, A. A., Loesch, C., & Coogan, P. 2019, *MNRAS*, 489, 5784

Cauley, P. W., Redfield, S., & Jensen, A. G. 2017, *AJ*, 153, 217

Cauley, P. W., Shkolnik, E. L., Llama, J., Bourrier, V., & Moutou, C. 2018, *AJ*, 156, 262

Cuntz, M., Saar, S. H., & Musielak, Z. E. 2000, *ApJ Letters*, 533, L151

Daley-Yates, S. & Stevens, I. R. 2019, *MNRAS*, 483, 2600

Egan, H., Jarvinen, R., Ma, Y., & Brain, D. 2019, *MNRAS*, 488, 2108

Ehrenreich, D., Bourrier, V., Wheatley, P. J., *et al.* 2015, *Nature*, 522, 459

Fares, R., Bourrier, V., Vidotto, A. A., *et al.* 2017, *MNRAS*, 471, 1246

Fares, R., Donati, J., Moutou, C., *et al.* 2010, *MNRAS*, 406, 409

Fichtinger, B., Guedel, M., Mutel, R. L., *et al.* 2017, *A&A*, 599, A127

Fleming, B. T., France, K., Nell, N., *et al.* 2017, in Society of Photo-Optical Instrumentation Engineers (SPIE) Conference Series, Vol. 10397, Society of Photo-Optical Instrumentation Engineers (SPIE) Conference Series, 103971A

Fossati, L., Haswell, C. A., Froning, C. S., *et al.* 2010, *ApJ Letters*, 714, L222

Fulton, B. J., Petigura, E. A., Howard, A. W., *et al.* 2017, *AJ*, 154, 109

Helled, R., Lozovsky, M., & Zucker, S. 2016, *MNRAS*, 455, L96

Ip, W.-H., Kopp, A., & Hu, J.-H. 2004, *ApJ Letters*, 602, L53

Jardine, M. & Collier Cameron, A. 2019, *MNRAS*, 482, 2853

Jin, S., Mordasini, C., Parmentier, V., *et al.* 2014, *ApJ*, 795, 65

Johnstone, C. P., Guedel, M., Stökl, A., *et al.* 2015, *ApJ Letters*, 815, L12

Kavanagh, R. D., Vidotto, A. A., Ó. Fionnagáin, D., *et al.* 2019, *MNRAS*, 485, 4529

Kulow, J. R., France, K., Linsky, J., & Loyd, R. O. P. 2014, *ApJ*, 786, 132

Lecavelier des Etangs, A., Bourrier, V., Wheatley, P. J., *et al.* 2012, *A&A*, 543, L4

Lecavelier des Etangs, A., Sirothia, S. K., Gopal-Krishna, & Zarka, P. 2013, *A&A*, 552, A65

Llama, J., Wood, K., Jardine, M., *et al.* 2011, *MNRAS*, 416, L41

Lovelace, R. V. E., Romanova, M. M., & Barnard, A. W. 2008, *MNRAS*, 389, 1233

Maggio, A., Pillitteri, I., Scandariato, G., *et al.* 2015, *ApJ Letters*, 811, L2

Matsakos, T., Uribe, A., & Königl, A. 2015, *A&A*, 578, A6

Mazeh, T., Holczer, T., & Faigler, S. 2016, *A&A*, 589, A75

McIvor, T., Jardine, M., & Holzwarth, V. 2006, *MNRAS*, 367, L1

Nicholson, B. A., Vidotto, A. A., Mengel, M., *et al.* 2016, *MNRAS*, 459, 1907

Nortmann, L., Pallé, E., Salz, M., *et al.* 2018, Science, 362, 1388

Ó Fionnagáin, D., Vidotto, A. A., Petit, P., *et al.* 2019, *MNRAS*, 483, 873

Pfleger, M., Lichtenegger, H. I. M., Wurz, P., *et al.* 2015, *Planetary Space Science*, 115, 90

Privitera, G., Meynet, G., Eggenberger, P., *et al.* 2016a, *A&A*, 593, L15

Privitera, G., Meynet, G., Eggenberger, P., *et al.* 2016b, *A&A*, 591, A45

Privitera, G., Meynet, G., Eggenberger, P., *et al.* 2016c, *A&A*, 593, A128

Saar, S. H. & Cuntz, M. 2001, *MNRAS*, 325, 55

Shkolnik, E., Bohlender, D. A., Walker, G. A. H., & Collier Cameron, A. 2008, *ApJ*, 676, 628

Shkolnik, E., Walker, G. A. H., Bohlender, D. A., Gu, P.-G., & Kürster, M. 2005, *ApJ*, 622, 1075

Spake, J. J., Sing, D. K., Evans, T. M., *et al.* 2018, *Nature*, 557, 68

Strangeway, R. J., Russell, C. T., Luhmann, J. G., *et al.* 2010, in AGU Fall Meeting Abstracts, Vol. 2010, SM33B–1893

Strugarek, A., Brun, A. S., Donati, J. F., Moutou, C., & Réville, V. 2019, *ApJ*, 881, 136

Strugarek, A., Brun, A. S., Matt, S. P., & Réville, V. 2015, *ApJ*, 815, 111

Tu, L., Johnstone, C. P., Guedel, M., & Lammer, H. 2015, *A&A*, 577, L3

Vidal-Madjar, A., Lecavelier des Etangs, A., Désert, J.-M., *et al.* 2003, *Nature*, 422, 143

Vidotto, A. A. & Bourrier, V. 2017, *MNRAS*, 470, 4026

Vidotto, A. A., Fares, R., Jardine, M., Moutou, C., & Donati, J.-F. 2015, *MNRAS*, 449, 4117

Vidotto, A. A., Gregory, S. G., Jardine, M., *et al.* 2014, *MNRAS*, 441, 2361

Vidotto, A. A., Jardine, M., & Helling, C. 2010a, *ApJ Letters*, 722, L168

Vidotto, A. A., Jardine, M., & Helling, C. 2011, *MNRAS*, 411, L46

Vidotto, A. A., Lichtenegger, H., Fossati, L., *et al.* 2018, *MNRAS*, 481, 5296

Vidotto, A. A., Opher, M., Jatenco-Pereira, V., & Gombosi, T. I. 2009, *ApJ*, 703, 1734

Vidotto, A. A., Opher, M., Jatenco-Pereira, V., & Gombosi, T. I. 2010b, *ApJ*, 720, 1262
Villarreal D'Angelo, C., Esquivel, A., Schneiter, M., & Sgró, M. A. 2018, *MNRAS*, 479, 3115
Wood, B. E. 2004, Living Reviews in Solar Physics, 1, 2
Zarka, P. 1998, *JGR*, 103, 20159

Discussion

CHIA-HSIEN LIN: Could you explain the relation between the increase in planetary evaporation on HD189733b and the presence of a flare/CME?

VIDOTTO: The evaporation of the atmosphere of a close-in planet is affected by both the amount of stellar irradiation (which heats the atmosphere of the planet) and the stellar wind/CME (which interacts with the upper atmosphere of a planet). In transit observations of HD189733b performed at the Ly-α line, Lecavelier des Etangs *et al.* (2012) noticed an increase in planetary evaporation that occurred about 8 hours after a flare of the host star. Although we do not know if the two events are related, the timing is very suggestive that the flare, somehow, led to an enhance of planet evaporation. There are two possible explanations on how this enhancement took place. For example, it could be that the flare caused an increase in irradiation arriving at the planet atmosphere, which caused an enhance in the evaporation. Alternatively, it could be that, associated to this, there was an ejection of a CME, which interacted with the atmosphere and could have led to an enhance in atmospheric escape.

KUTLUAY YÜCE: What are the most-common spectral types of known planet hosting stars? What do you think about earlier type stars (B,A,F) as potential planet-hosts?

VIDOTTO: Due to biases in planet detection techniques, most of the planet hosting stars known today are cool dwarfs with masses $\lesssim 1.3 M_\odot$ and, in most of the cases, these stars tend to be inactive. It is more challenging to detect planets around earlier type stars (A and B spectral types, for instance). This happens because the transit signature is proportional to the ratio $(R_{\text{planet}}/R_\star)^2$ (i.e., the area of the planet divided by the area of the star), so the bigger the star, the smaller the signature. In the radial velocity technique, the orbit of the star about the common centre of mass (the reflex motion) is very small and the Doppler shifts caused by the planet motion are thus harder to detect. Having said that, there are now some (a few tens) detections of planets orbiting around earlier type stars, like KELT-9b and WASP-33b, whose hosts are A0 and A5 stars, respectively.

ZHANWEN HAN: What are the physical causes of the planet engulfment? Is it due to the expansion of the star in the post-main sequence or the interaction with the stellar wind that causes the loss of orbital angular momentum?

VIDOTTO: The reasons why orbital angular momentum changes over time are due to several physical processes, such as tidal interactions, stellar winds that carry away stellar angular momentum (or, conversely, mass accretion that increases stellar angular momentum), friction and planet evaporation. Changes in orbits have an impact on the rotation of the star, which also modifies the structure of the star. This, in turn, also affect the mechanisms of orbital decay.

Solar and Stellar Magnetic Fields: Origins and Manifestations
Proceedings IAU Symposium No. 354, 2019
A. Kosovichev, K. Strassmeier & M. Jardine, ed.
doi:10.1017/S1743921320000083

Influence of the magnetic field of stellar wind on hot jupiter's envelopes

Dmitry V. Bisikalo and Andrey G. Zhilkin🆔

Institute of Astronomy of the RAS, 48 Pyatnitskaya str., Moscow, Russia
email: bisikalo@inasan.ru

Abstract. Hot Jupiters have extended gaseous (ionospheric) envelopes, which extend far beyond the Roche lobe. The envelopes are loosely bound to the planet and, therefore, are strongly influenced by fluctuations of the stellar wind. We show that, since hot Jupiters are close to the parent stars, magnetic field of the stellar wind is an important factor defining the structure of their magnetospheres. For a typical hot Jupiter, velocity of the stellar wind plasma flow around the atmosphere is close to the Alfvén velocity. As a result stellar wind fluctuations, such as coronal mass ejections, can affect the conditions for the formation of a bow shock around a hot Jupiter. This effect can affect observational manifestations of hot Jupiters.

Keywords. MHD, stars: winds, stars: planetary systems

---

1. Introduction

Hot Jupiters are exoplanets which have mass of the same order as Jupiter and are located in the close proximity to the parent star (Murray-Clay *et al.* (2009)). The first hot Jupiter was discovered in 1995 (Mayor & Queloz (1995)). Due to their proximity to the parent star and relatively large dimensions of the envelopes hot Jupiters can overflow the Roche lobes. This leads to the outflows from the vicinity of the nearest Lagrange points (Li *et al.* (2010); Bisikalo *et al.* (2013)). This circumstance is indirectly indicated by the the excess absorption in the near UV, observed for some planets of this type (Vidal-Madjar *et al.* (2003, 2008); Ben-Jaffel (2007); Vidal-Madjar *et al.* (2004); Ben-Jaffel & Sona Hosseini (2010); Linsky *et al.* (2010)). Therefore, the study of the mechanisms of mass loss by hot Jupiters is one of the most urgent tasks of modern astrophysics.

The structure of the gaseous envelopes of hot Jupiters was investigated in a series of our studies, using three-dimensional numerical models (see. e.g.. (Bisikalo *et al.* (2013); Cherenkov *et al.* (2014); Bisikalo & Cherenkov (2016); Cherenkov *et al.* (2017, 2018); Bisikalo *et al.* (2018)). It is shown that gaseous envelopes of three main types may form around hot Jupiters, depending on the model parameters. In the case of close envelope, planet's atmosphere is completely located inside its Roche lobe. Open envelope is formed by outflows from the nearest Lagrange points. If dynamic pressure of the stellar wind stops the outflow beyond Roche lobe, quasi-closed envelope is formed. The rate of mass-loss substantially depends on the type of the gaseous envelope being formed.

Arakcheev *et al.* (2017), Bisikalo *et al.* (2017) presented results of the three-dimensional numerical simulation of the flow structure in the vicinity of a hot Jupiter WASP 12b that took into account the influence of the planet's proper magnetic field. Computations have shown that the presence of the planet's magnetic field can lead to the additional decline of the mass-loss rate compared to the purely gas-dynamic case. The analysis by Zhilkin & Bisikalo (2019) has shown that a very important factor is magnetic field of the stellar wind, since many hot Jupiters are located in the sub-Alfvén zone of the stellar wind, where magnetic pressure exceeds dynamic one. In this case, the flow can be shockless (Ip *et al.* (2004)).

Various perturbations of the stellar wind can lead to significant changes in the structure of the gaseous envelopes of hot Jupiters and, consequently, to the variations of the mass-loss rate. The most significant wind disturbances (coronal mass ejections, henceforth, CME) arise due to giant ejections of matter from the stellar corona. Bisikalo & Cherenkov (2016), Cherenkov *et al.* (2017), using three-dimensional numerical simulations, have shown that even in the case of a typical solar-like CME external parts of the asymmetric gaseous envelope of a hot Jupiter that are outside Roche lobe can be torn and carried away into the interplanetary medium. This leads to a sharp increase in the rate of mass loss by a hot Jupiter at the moment, when CME passes it.

In the present paper, we study the effect of stellar wind magnetic field on the structure of the ionospheric envelopes of hot Jupiters. In particular, we consider some interesting features of the interaction of a CME with the magnetosphere of a hot Jupiter due to the variation of the parameters of magnetic field of stellar wind. The paper is based on our recent studies (Zhilkin & Bisikalo (2019); Zhilkin, Bisikalo & Kaygorodov (2020)).

2. MHD model of stellar wind

To describe stellar wind in the vicinity of hot Jupiters in our numerical model, we will rely on the well-studied properties of the solar wind. As it is shown by numerous Earth-based and space studies (see, for instance, recent review by Owens & Forsyth (2013)), solar wind magnetic field has a rather complex structure. In the corona region, magnetic field structure is defined mainly, by the intrinsic magnetism of the Sun and, therefore, it is essentially non-radial. At the border of the corona, which is at a distance of several solar radii, magnetic field to a large accuracy becomes completely radial. Outside this region, the heliospheric region is located, where magnetic field is largely determined by the properties of the solar wind. In the heliospheric region, magnetic field lines, as they gradually recede from the center, twist into a spiral due to the solar rotation and, therefore (especially at large distances), magnetic field of the wind with a good accuracy can be described using the simple Parker (1958) model.

In our calculations, we do not take into account possible sectoral structure of the magnetic field of the wind, focusing on the account of the influence of its global parameters. We assume that the orbit of the hot Jupiter is located in the heliospheric region beyond the border of the corona. To describe the structure of the wind (including its magnetic field **B**) in the heliospheric region, as the first approximation, one can apply axisymmetric magnetohydrodynamic model (Weber & Davis (1967)). The model of the wind will be considered in the inertial reference frame in spherical coordinates (r, θ, φ). One can neglect the dependence of wind parameters on the angle θ, because we are interested in the flow structure close to the orbital plane. Therefore, we will assume that all quantities depend on the radial coordinate r only.

Under such assumptions, steady-state structure of the wind is defined by the continuity equation

$$\frac{1}{r^2}\frac{d}{dr}\left(r^2 \rho v_r\right) = 0, \tag{2.1}$$

equations of motion for radial v_r and azimuthal v_φ components of the velocity vector **v**

$$v_r\frac{dv_r}{dr} - \frac{v_\varphi^2}{r} = -\frac{1}{\rho}\frac{dP}{dr} - \frac{GM}{r^2} - \frac{B_\varphi}{4\pi\rho r}\frac{d}{dr}\left(rB_\varphi\right), \tag{2.2}$$

$$v_r \frac{dv_\varphi}{dr} + \frac{v_r v_\varphi}{r} = \frac{B_r}{4\pi\rho r}\frac{d}{dr}\left(rB_\varphi\right),$$ (2.3)

induction equation

$$\frac{1}{r}\frac{d}{dr}\left(rv_r B_\varphi - rv_\varphi B_r\right) = 0$$ (2.4)

and Maxwell equation $(\nabla \cdot \mathbf{B} = 0)$

$$\frac{1}{r^2}\frac{d}{dr}\left(r^2 B_r\right) = 0.$$ (2.5)

Here, ρ is density, P — pressure, G — gravity constant, M — mass of the central star. Density, pressure, and temperature satisfy the equation of state for ideal polytropic gas

$$P = P_0 \left(\frac{\rho}{\rho_0}\right)^\gamma = \frac{2k_B}{m_p}\rho T,$$ (2.6)

where k_B — Boltzmann constant, m_p — proton mass, γ — polytropic exponent. Mean molecular weight of the wind matter is assumed to be 0.5, corresponding to the completely ionized hydrogen plasma containing electrons and protons only. Index 0 defines the quantities in point r_0.

From Maxwell equation (2.7) we obtain

$$B_r = B_s \left(R_s/r\right)^2,$$ (2.7)

where R_s is stellar radius, B_s — the field strength at the stellar surface. From continuity equation (2.1) it is possible to derive the integral of motion, corresponding to the mass conservation law:

$$r^2\rho v_r = \text{const.}$$ (2.8)

Equation (2.2) allows to derive the motion integral, corresponding to the energy conservation law:

$$\frac{v_r^2}{2} + \frac{v_\varphi^2}{2} + \frac{c_s^2}{\gamma - 1} - \frac{GM}{r} - \frac{B_\varphi^2}{4\pi\rho} - \frac{B_r B_\varphi v_\varphi}{4\pi\rho v_r} = \text{const},$$ (2.9)

where sound velocity

$$c_s = \sqrt{\gamma P/\rho}.$$ (2.10)

Note, it follows from Eqs. (2.7) and (2.8) that

$$\frac{B_r}{4\pi\rho v_r} = \text{const.}$$ (2.11)

This circumstance allows to derive from Eqs. (2.3) and (2.4) two integrals of motion more:

$$rv_\varphi - \frac{B_r}{4\pi\rho v_r}rB_\varphi = L,$$ (2.12)

$$rv_r B_\varphi - rv_\varphi B_r = F.$$ (2.13)

The value of the constant F in the motion integral (2.13) could be found from the boundary conditions at the stellar surface at $r = R_s$:

$$B_\varphi = 0, \quad B_r = B_s, \quad v_\varphi = \Omega_s R_s,$$ (2.14)

where Ω_s is angular velocity of the proper stellar rotation. Therefore,

$$F = -\Omega_s R_s^2 B_s = -\Omega_s r^2 B_r.$$ (2.15)

This integral can also be interpreted in the sense that in the reference frame of the rotating star velocity vector **v** of the perfectly conducting wind plasma must be collinear to the magnetic field induction vector **B**. With account of this expression, solutions of the Eqs. (2.12) and (2.13) can be written as:

$$v_\varphi = \frac{\Omega_{\rm s} r - \lambda^2 L/r}{1 - \lambda^2}, \tag{2.16}$$

$$B_\varphi = \frac{B_r}{v_r} \lambda^2 \frac{\Omega_{\rm s} r - L/r}{1 - \lambda^2}. \tag{2.17}$$

Here, as λ we denote Alfvén Mach number for radial components of velocity and magnetic field,

$$\lambda = \frac{\sqrt{4\pi\rho} v_r}{B_r}. \tag{2.18}$$

Close to the surface of a star, radial wind velocity v_r should be lower than Alfvén velocity $u_A = B_r/\sqrt{4\pi\rho}$ and the parameter $\lambda < 1$. On the contrary, at large distances, radial velocity v_r, exceeds Alfvén velocity u_A ($\lambda > 1$). This means that at some distance from the stellar center $r = r_A$ (Alfvén point) the parameter λ becomes equal to 1. The region $r < r_A$ can be named *sub-Alfvén* zone of the stellar wind, while the region $r > r_A$, accordingly, *super-Alfvén* zone.

The values of v_φ and B_φ in the Eqs. (2.16) and (2.17) should be continuous in the Alfvén point $r = r_A$. Therefore, it is necessary to set

$$L = \Omega_{\rm s} r_A^2. \tag{2.19}$$

As a result, we find the final solution

$$v_\varphi = \Omega_{\rm s} r \frac{1 - \lambda^2 r_A^2/r^2}{1 - \lambda^2}, \tag{2.20}$$

$$B_\varphi = \frac{B_r}{v_r} \Omega_{\rm s} r \lambda^2 \frac{1 - r_A^2/r^2}{1 - \lambda^2}. \tag{2.21}$$

These relations, with the integrals of mass (2.8) and energy (2.9) taken into account, allow, using algebra, to obtain the distributions of all magnetohydrodynamic quantities describing the structure of the wind. At this, to single out a unique solution, it is necessary to use continuity conditions in three critical points at which the radial wind velocity v_r is equal, respectively, to slow magneto-sonic

$$u_S = \left\{ \frac{1}{2} \left[c_s^2 + a^2 - \sqrt{(c_s^2 + a^2)^2 - 4c_s^2 u_A^2} \right] \right\}^{1/2}, \tag{2.22}$$

Alfvén u_A, and fast magneto-sonic

$$u_F = \left\{ \frac{1}{2} \left[c_s^2 + a^2 + \sqrt{(c_s^2 + a^2)^2 - 4c_s^2 u_A^2} \right] \right\}^{1/2}, \tag{2.23}$$

velocities. Here,

$$a^2 = \frac{B_r^2 + B_\varphi^2}{4\pi\rho}. \tag{2.24}$$

It is interesting to note that solutions with accelerating wind (positive value of the radial velocity gradient, $dv_r/dr > 0$) are realized for polytropic exponent $\gamma < 3/2$ only. This indicates the presence of effective sources of heating and cooling. It is known that the solar wind at small distances from the Sun ($r < 15R_\odot$) turns out to be almost isothermal, since the effective adiabatic exponent $\gamma = 1.1$ (Steinolfson & Hundhausen (1988);

Roussev *et al.* (2003)). At large distances, $r > 25R_\odot$, the effective adiabatic exponent can be estimated as $\gamma = 1.46$ (Totten *et al.* (1995)). As the orbits of hot Jupiters are located close to the parent stars, in the calculations of the wind structure we used adiabatic exponent $\gamma = 1.1$. Model parameters corresponded to the the the distance from the star $r_0 = 10R_\odot$, which is typical for the orbits of hot Jupiters. Number density $n_0 = 1400$ cm^{-3} and temperature $T_0 = 7.3 \cdot 10^5$ K were chosen (Withbroe (1988)). Corresponding resulting distributions were used in our numerical models for description of the magnetohydrodynamic structure of stellar wind in the neighborhood of hot Jupiter. In particular, radial wind velocity at the distance $r_0 = 10R_\odot$ from the star turned out to be of the order of $v_{r,0} = 130$ km/s.

3. Magnetspheres of hot jupiters

In the solar wind, Alfvén radius $r_A = 0.11$ AU $= 24.3R_\odot$ (Weber & Davis (1967)). Since the semi-major axis of the orbit of the innermost planet, Mercury, is 0.38 AU $= 82R_\odot$, this means that all planets of the solar system are in the super-Alfvén zone of the solar wind. Sonic point, where wind velocity becomes comparable to the sound speed in the solar wind is even closer to the Sun, at a distance of approximately 0.037 AU $= 8R_\odot$. It follows then, that the magnetospheres of all planets in the solar system (if they have them) have similar structure, like that of the terrestrial magnetosphere. They are characterized by the following set of basic elements: bow shock, transition region, magnetopause, radiation belts, magnetospheric tail.

In the case of hot Jupiters, due to their proximity to the parent star, the structure of magnetospheres may be completely different. To analyze possible configurations, we processed actual data for a sample of 210 hot Jupiters from the database at www.exoplanet.eu. Selection was carried out according to the masses of the planets ($M_p > 0.5M_{jup}$, where M_{jup} is the mass of Jupiter), orbital period ($P_{orb} < 10$ day), and semi-major axis of the orbit ($A < 10R_\odot$). Additionally, only planets for which all necessary data were known were left in the sample.

As the model of stellar wind in the immediate vicinity of the Sun, at the distances $1R_\odot < r < 10R_\odot$, we applied the results of computations by Withbroe (1988). According to the obtained profiles of density $\rho(r)$ and radial velocity $v_r(r)$, for each hot Jupiter in the sample dynamic wind pressure at the orbit of the planet

$$P_{\rm dyn} = \rho(A)v_r^2(A) \tag{3.1}$$

and magnetic pressure

$$P_{\rm mag} = \frac{B_r^2(A)}{8\pi} \tag{3.2}$$

were calculated. The radial field value was computed by the formula $B_r(A) = B_s(R_\odot/A)^2$ with the parameter $B_s = 1$ G. Thus obtained distribution of hot Jupiters in the two-dimensional diagram $P_{\rm mag}$–$P_{\rm dyn}$ is presented in the left panel of Fig. 1. Positions of the planets correspond to the centers of the circles with radii determined by the mass M_p (in the logarithmic scale). Solid line shows position of the Alfvén point of the solar wind, which corresponds to a simple relation $P_{\rm dyn} = 2P_{\rm mag}$.

As can be seen from the obtained distribution, all hot Jupiters from this sample are located in the sub-Alfvén zone of the stellar wind. However, in the reference frame, associated with an orbiting planet, the nature of the flow is determined not only by the wind velocity, but also by the orbital velocity of the planet. If this is taken into consideration, dynamic pressure becomes

$$P_{\rm dyn} = \rho(A)\left[v_r^2(A) + \frac{G(M_s + M_p)}{A}\right]. \tag{3.3}$$

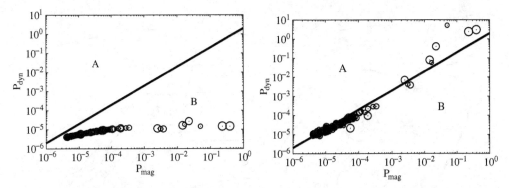

Figure 1. Distribution of hot Jupiters in the scattering diagram P_{mag}–P_{dyn}. Alfvén Mach numbers calculated with account of wind velocity only are presented in the left panel; the right panel shows orbital velocities of the planets. Parameters of the planets were taken from the database at www.exoplanet.eu. The data for 210 hot Jupiters were used. Locations of the planets are at the centers of the circles. The sizes of the circles correspond to the masses of the planets in the logarithmic scale. Solid line shows the position of Alfvén point of the solar wind. The letters mark super-Alfvén zone ("A") and sub-Alfvén zone ("B").

Note that the orbital velocity of the planet depends not only on the radius of the orbit, but also on the mass of the planet. The corresponding diagram is shown in the right panel of Fig. 1. Account of the orbital velocity shifts whole sequence significantly up toward the super-Alfvén zone of the wind. Note that most of the planets in this diagram form a regular sequence (lower left corner of the diagram). These planets are located rather far from the star, where the dependences of density and wind velocity on the radius are well described by power laws. Planets that are close to the star are scattered over the diagram rather chaotic. For these planets, the dynamic wind pressure (3.3) is determined mainly by their orbital velocity.

Because for hot Jupiters with orbits located in the sub-Alfvén zone, Alfvén Mach number $\lambda = v_r/u_A$ turns out to be less than one, the ratio v_r/u_F, where u_F is the fast magneto-sonic velocity (2.23) will also be less than one, since, obviously, $u_F > u_A$ and, therefore, the ratio $v_r/u_F < v_r/u_A$. In other words, in the neighborhood of such a hot Jupiter, stellar wind velocity will be lower than the fast magneto-sonic speed. In the pure gas dynamics, this case corresponds to the subsonic flow around the body, in which the bow shock does not form. Thus, we come to the following conclusion: the flow of stellar wind around such a hot Jupiter should be shockless (Ip *et al.* (2004)). In the structure of the magnetosphere of hot Jupiter, the bow shock should be absent.

It should be borne in mind that this distribution was obtained for the solar wind in the model of a quite Sun. At this, we assumed that the average value of the magnetic field at the surface of the Sun is 1 G. Even for the Sun, during its activity cycle, the position of hot Jupiters in the right panel of Fig. 1 can change respective to the Alfvén point in both directions. In reality, each planet in our sample is flown around not by the solar wind, but by the stellar wind of the parent star. The parameters of this wind may differ significantly from the solar one. This means that the mode of the flow of stellar wind around the atmosphere of the planet must be investigated separately in every specific case, taking into account individual characteristics of the planet and the parent star.

Let treat as the ionospheric envelope the upper layers of the atmosphere of a hot Jupiter, which are composed by almost completely ionized gas (Cherenkov *et al.* (2018)). A closed ionospheric envelope corresponds to the case when the atmosphere of a hot Jupiter is entirely located inside its Roche lobe. An open ionospheric envelope corresponds to the case when hot Jupiter overflows its Roche lobe, forming outflows from the vicinity

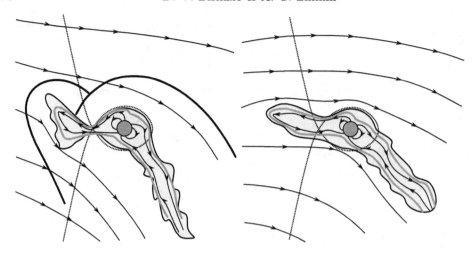

Figure 2. A sketch of the structure of an impact-induced magnetosphere (left panel) and shockless induced magnetosphere (right panel) for the case of open ionospheric envelope of a hot Jupiter. The lines with arrows correspond to the magnetic field lines. Dashed line shows the Roche lobe border. Light-gray region corresponds to the gaseous envelope of the star. Solid outer line shows location of the shock. Position of magnetopause is shown by inner gray line.

of the Lagrange points L_1 and L_2. Since proper magnetic field of hot Jupiters is rather weak, the magnetopause is located inside the ionospheric envelope. In such a situation, the most likely types of magnetospheres of hot Jupiters are an impact induced magnetosphere (see the left panel in Fig. 2) and a shockless induced magnetosphere (see the right panel in Fig. 2).

An induced magnetosphere (Russell (1993)) is formed by the currents that are excited in the upper layers of the ionosphere. These currents partially shield the magnetic field of the wind. As a result, magnetic lines of the the generated field enshroud the ionosphere of the planet, forming a peculiar magnetic barrier or ionopause. Bow shock wave sets immediately in the front of this barrier. On the night side a magnetospheric tail is formed, which can be partially filled by the plasma from the ionosphere. At difference to proper magnetosphere (similar to that of the Earth or the Jupiter), orientation of the magnetic field in the induced magnetosphere is completely determined by the field of the wind. As the result, when the planet moves in its orbit, entire structure of the magnetosphere tracks direction to the star. In the solar system, this situation for the case of a closed ionospheric envelope corresponds to the magnetosphere of the Venus and, in some sense, to that of the Mars. Induced magnetospheres with open ionospheric envelopes can form in the comets coming close to the Sun.

A curious situation may arise in the intermediate case (a "gray" zone), when the orbit of a hot Jupiter is close to the Alfvén point. In particular, in this case the planet itself may be located in sub- or super-Alfvén zone of the wind, while the outflowing ionospheric envelope, due to its rather large extent, can cross the Alfvén point and partially flow into the opposite wind zone. For hot Jupiters this case may turn out to be quite common, because the overwhelming majority of them are located just close to the Alfvén point.

4. Hot jupiters in sub-alfvenic and super-alfvenic zones

Our analysis allows to conclude that in the vicinity of almost of all currently known hot Jupiters velocity of the stellar wind turns out to be close to the Alfvén velocity. However, many of them can be located even in the sub-Alfvén zone, where magnetic pressure of the stellar wind exceeds its dynamic pressure. This means that for the study

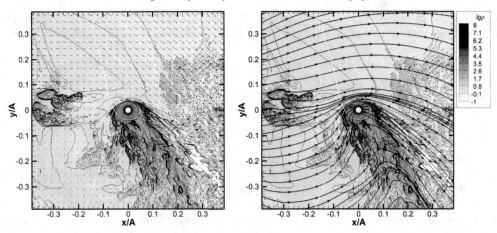

Figure 3. Distributions of the density (gray scale and level lines), velocity and magnetic field (lines with arrows) in the plane of the orbit of a hot Jupiter. Presented is solution for a model with a strong stellar wind field ($B_s = 0.5$ G) at the instant $0.5P_{orb}$ from the reference-starting point. Dashed line shows the boundary of the Roche lobe. The circle corresponds to the photometric radius of the planet.

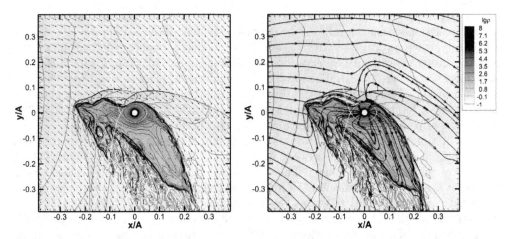

Figure 4. Distributions of the density (gray scale and level lines), velocity and magnetic field (lines with arrows) in the plane of the orbit of a hot Jupiter. Presented is solution for a model with a weak stellar wind field ($B_s = 0.5$ G) at the instant $0.26P_{orb}$ from the reference-starting point. Dashed line shows the boundary of the Roche lobe. The circle corresponds to the photometric radius of the planet.

of the stellar wind flow around the ionospheric envelope of a hot Jupiter magnetic field of the wind is an important factor, which is absolutely necessary to take into account both in theoretical modeling and in interpretation of the observational data.

In Figs. 3 and 4 we present results of the three-dimensional numerical simulation of the flow structure in the vicinity of the hot Jupiter HD 209458b (Charbonneau *et al.* (2000)). For calculations, we applied numerical model described by Zhilkin & Bisikalo (2019). The main parameters of the model were the same as in our previous studies (see, e.g., Bisikalo *et al.* (2013)). Spectral type G0 parent star HD 209458 has the mass $M_s = 1.15M_\odot$ and the radius $R_s = 1.2R_\odot$. Proper rotation of the star is characterized by the period $P_{rot} = 14.4$ day, which corresponds to the angular velocity $\Omega_s = 5.05 \cdot 10^{-6}$ s^{-1} or linear velocity at the equator $v_{rot} = 4.2$ km/s. Mass of the planet is $M_p = 0.71M_{jup}$, its photometric radius is $R_p = 1.38R_{\mathrm{jup}}$, where R_{jup} is the radius of the Jupiter. Semimajor axis of the orbit of the planet $A = 10.2R_\odot$, which corresponds to the period of orbital revolution $P_{\mathrm{orb}} = 84.6$ hr.

At the initial moment of time, a spherically symmetric isothermal atmosphere was set around the planet, the density distribution in which was determined from the condition of hydrostatic equilibrium. The radius of the atmosphere was determined from the condition of pressure equilibrium with stellar wind matter. The temperature of the atmosphere was set equal to $T_{\mathrm{atm}} = 7500$ K, while number density of the particles at the photometric radius was taken as $n_{\mathrm{atm}} = 10^{11}$ cm^{-3}.

As stellar wind parameters were taken the values appropriate for the solar wind at the distance of $10.2R_\odot$ from the center of the Sun (Withbroe (1988)): temperature $T_w = 7.3 \cdot 10^5$ K, velocity $v_w = 100$ km/s, number density $n_w = 10^4$ cm^{-3}. Magnetic field of the wind was set according to the formulas given in Zhilkin & Bisikalo (2019). Calculations were performed for two models, corresponding to the cases of weak and strong magnetic field of stellar wind. The average magnetic field at the surface of the star was set equal to $B_s = 0.5$ G for the first model (strong field) and $B_s = 0.01$ G for the second model (weak field). Given the fact that the radius of the star is slightly greater than the radius of the Sun, the magnitude of the field in the first model practically corresponds to the average magnetic field at the surface of the Sun, if corresponding magnetic moments of stars are compared.

In the numerical model, we also took into account proper magnetic field of the planet. At this, we assumed that the value of the magnetic moment of the hot Jupiter HD 209458b $\mu = 0.1\mu_{\mathrm{jup}}$, where $\mu_{\mathrm{jup}} = 1.53 \cdot 10^{30}$ G $\cdot$ cm^3 is the magnetic moment of the Jupiter. This value is consistent both with observations (Kislyakova et $al.$ (2014)) and with theoretical estimates (Stevenson (1983)). The axis of the magnetic dipole was tilted by $30°$ to the axis of rotation of the planet, in the direction opposite to the star. It was assumed in the model that rotation of the planet is synchronized with the orbital revolution and the axis of proper rotation is collinear with the axis of orbital rotation.

In the first model (strong field), the sub-Alfvén flow regime is realized. As can be seen from Fig. 3, in this case, the process of interaction of stellar wind with the ionospheric envelope of the planet is shockless. The detached shock wave is formed neither around the planet atmosphere nor around the matter ejected from L_1. This is clearly seen from the density and velocity distributions (see the left panel in Fig. 3). Magnetic field of the wind is so strong that it prevents the free motion of the plasma in the direction transverse to the field lines. Therefore, the ejected matter moves toward the star mainly along the magnetic field lines of the wind (see the right panel in Fig. 3). For that reason, we can say that in this process electromagnetic force, due to the magnetic field of the wind, plays an important role, comparable to the role of stellar gravity, centrifugal force, and Coriolis force.

In the second model (weak field), the super-Alfvén flow regime is realized. As a result of the interaction of the stellar wind with the planet's ionospheric envelope, a detached shock wave forms. This is clearly seen in Fig. 4. It can be stated that this shock wave consists of two separate shock waves, one of which occurs when the wind interacts directly with the atmosphere of the planet, and the other — when it interacts with the matter of the jet from the inner Lagrange point L_1. Inside Roche lobe of the planet magnetic field retains the dipole structure. However, in the flows forming on the day- and night-sides, the lines of the original dipole field are strongly stretched and distorted. Since in this model magnetic field of the stellar wind is weak and does not play any significant dynamic role, flow pattern corresponds to the purely gas-dynamic case described in Bisikalo et $al.$ (2013).

5. Effect of CME on the magnetospheres of hot jupiters

Various perturbations of the stellar wind, such as, for instance, CMEs, can lead to to the significant changes in the structure of the gaseous envelopes of hot Jupiters and, therefore, to variations in the rate of mass loss. Even in the case of a typical solar CME external parts of the asymmetric gaseous envelope of a hot Jupiter located outside its Roche lobe can be torn and carried away into interplanetary medium (Bisikalo & Cherenkov (2016); Cherenkov *et al.* (2017, 2019)). This leads to a sharp increase of the rate of mass loss by a hot Jupiter at the moment when CME passes along it.

As the basis of the numerical model of stellar wind at the time of the passage of CME in the vicinity of the planet, one can take the results of measurements of solar wind parameters at the orbit of the Earth, obtained by space missions ACE, WIND, SOHO in May 1998 during such an event (Farrell *et al.* (2012)). As these measurements show, the process of the passage of CME can be divided into four separate phases. The first phase corresponds to the state of the undisturbed solar wind. The second phase begins with the passage of the front of the MHD-shock and is characterized by an increase of the density n with respect to the unperturbed value n_w by a factor about 4. Velocity v in this case increases 1.3 times with respect to the unperturbed value of v_w. Beyond the front of the shock wave, magnetic field induction B increases 2.25 times with respect to the unperturbed value of B_w. Behind the shock follows the sheath of the heaped matter. The third phase (early CME) begins with the passage of the tangential MHD discontinuity, which propagates following the shock wave. The density drops about two times compared to the unperturbed value. Finally, the fourth phase (late CME) is distinguished by a sharp increase in density (about 10 times relative to the undisturbed wind). However, this phase does not have pronounced limits and, apparently, its beginning is not connected with the passage of any discontinuity. After that, wind parameters return to their original values.

It should be noted that the time profiles of variables in the CME can also have a more general form (Cherenkov *et al.* (2017)). Even for the Sun, intensities of CMEs can greatly vary. For the parent stars of hot Jupiters, these variations may manifest themselves even stronger. A more general approach allows to vary in time the profiles of magnetohydrodynamic quantities, both relative changes of the parameters during the phases and the duration of the phases themselves. In this case, one can describe CMEs of various types, corresponding to slow, medium (Möstl *et al.* (2014)), and fast (Liu *et al.* (2014)) ones.

Let consider now what features of the CME interaction process with the ionospheric envelope of a hot Jupiter may arise, if magnetic field of the stellar wind is taken into account. The influence of the magnetic field can be estimated by the value of the Alfvén Mach number. The changes of the value of the Alfvén Mach number at the different stages of the CME passage can be described by the following expression

$$\frac{\lambda}{\lambda_w} = \sqrt{\frac{n}{n_w}} \frac{v}{v_w} \frac{B_w}{B}. \tag{5.1}$$

As the measurements described by Farrell *et al.* (2012) show, the value of λ varies non-monotonously. In the first phase, λ slightly exceeds the unperturbed value λ_w. In the second phase, λ becomes lower than the unperturbed value λ_w. In the third phase, λ again sharply increases and exceeds unperturbed value λ_w more than three times.

If the planet is deep in the sub-Alfvén zone or, conversely, far in super-Alfvén zone, the nature of the flow during the passage of the CME will not change. In the first case, it will be shockless, as in our calculations presented in Fig. 3. In the second case, entire process, from the beginning to the end, will be accompanied by the formation of detached shock waves, as in our calculations, presented in Fig. 4 and as observed in purely gas-dynamic calculations (Bisikalo & Cherenkov (2016); Cherenkov *et al.* (2017); Kaigorodov *et al.* (2019)). However, if the planet's orbit is close to the Alfvén point, the process of interaction of the CME with the magnetosphere may turn out to be more complex and interesting. We recall that for hot Jupiters this should be a very common case (Zhilkin & Bisikalo (2019)).

Let imagine that such a planet is located near the Alfvén point, but at the side of sub-Alfvén zone of the wind. Then, in the second phase, the flow regime should remain shockless, because at this phase the Alfvén Mach number is lower than the unperturbed value λ_w. In the first and third phases, Alfvén Mach number, on the contrary, increases compared to the unperturbed value. Depending on the specific situation, this may be quite enough for the flow velocity to exceed the fast magneto-sonic velocity either in the third phase of the CME or immediately in the first and the third phases. In the first case, in the third phase of the CME a bow shock will emerge, which will disappear again at the end of the entire process and return of the system to the original unperturbed state. In the second case, the shock arises already in the first phase, in the second phase it disappears, then reappears in the third phase and, finally, disappears after passage of CME.

Now, let assume that the hot Jupiter is close to the Alfvén point, but at the side of the super-Alfvén zone of the wind. Then the flow regime can change in the second phase of the CME passage, when the Alfvén Mach number decreases compared to the unperturbed value. That may be quite sufficient to switch the flow into shockless mode, when the flow velocity becomes smaller than the fast magneto-sonic velocity and the bow shock already does not form any more, as in the calculation shown in Fig. 3. As a result of the change in the flow regime bow shock may for some time "switch off" and then "turn on" again after the end of the second phase of the ejection.

6. Conclusion

It is shown in the study that hot Jupiters are located in the sub-Alfvén zone of the stellar wind. However, if the planet's orbital velocity is taken into account, the nature of the flow around the planet approaches the conterminal state separating sub-Alfvén and super-Alfvén regimes. This conclusion was derived using the parameters of the solar wind. Since the parameters of the stellar wind of the parent stars of hot Jupiters can differ significantly from solar wind ones, the corresponding mode of the flow should be investigated separately in each specific case. These results lead to the conclusion that in the studies of the flow of the stellar wind around the atmosphere of a hot Jupiter magnetic field of the wind is an extremely important dynamic factor, which is absolutely necessary to take into account both in theoretical modeling and in the interpretation of observational data.

Magnetospheres of hot Jupiters in the super-Alfvén and sub-Alfvén modes of stellar wind flow around them have significantly different structure. In particular, in the case of the sub-Alfvén regime of the flow around the ionospheric envelope of a hot Jupiter a bow shock wave does not form. In other words, the flow around such a planet is shockless.

The effect of the perturbation of stellar wind parameters, caused by the passage of the coronal mass ejection, upon the nature of the flow near hot Jupiter is considered too. If the orbit of a hot Jupiter is located close to the Alfvén point, the passage of CME may cause temporary formation or disappearance of the shock wave, since the flow can switch from the sub-Alfvén regime to the super-Alfvén one and vice versa. Such phenomena, which significantly change envelope structure, can lead not to the variations in the rate of mass-loss only and, therefore, affect directly the long-term evolution of hot Jupiters, but also affect their observational manifestations.

This work was supported by the Russian Science Foundation (project 18-12-00447). This work has been carried out using computing resources of the federal collective usage center Complex for Simulation and Data Processing for Mega-science Facilities at NRC "Kurchatov Institute", http://ckp.nrcki.ru/. This work has been carried out using computational clusters of the Interdepartmental Supercomputer Center of the Russian Academy of Sciences.

References

Arakcheev, A.S., Zhilkin, A.G., Kaigorodov, P.V., *et al.* 2017, *Astronomy Reports*, 61, 932

Ben-Jaffel, L. 2007, *ApJ*, 671, L61

Ben-Jaffel, L., & Sona Hosseini, S. 2010, *ApJ*, 709, 1284

Bisikalo, D.V., Kaigorodov, P.V., Ionov, D.E., & Shematovich, V.I. 2013, *Astronomy Reports*, 57, 715

Bisikalo, D.V., Arakcheev, A.S., & Kaigorodov, P.V. 2017, *Astronomy Reports*, 61, 925

Bisikalo, D.V., Cherenkov, A.A., Shematovich, V.I., *et al.* 2018, *Astronomy Reports*, 62, 648

Bisikalo, D.V., & Cherenkov, A.A. 2016, *Astronomy Reports*, 60, 183

Charbonneau, D., Brown, T.M., Latham, D.W., & Mayor, M. 2000, *ApJ*, 529, L45

Cherenkov, A.A., Bisikalo, D.V., & Kaigorodov, P.V. 2014, *Astronomy Reports*, 58, 679

Cherenkov, A., Bisikalo, D., Fossati, L., Möstl, C. 2017, *ApJ*, 846, 31

Cherenkov, A.A., Bisikalo, D.V., & Kosovichev, A.G. 2018, *MNRAS*, 475, 605

Cherenkov, A.A., Shaikhislamov, I.F., Bisikalo, D.V., *et al.* 2019, *Astronomy Reports*, 63, 94

Farrell, W.M., Halekas, J.S., Killen, R.M., *et al.* 2012, *J. Geophys. Res. (Planets)*, 117, E00K04

Ip, W.-H., Kopp, A., & Hu, J.H. 2004, *ApJ*, 602, L53

Kaigorodov, P.V., Ilyina, E.A., & Bisikalo, D.V. 2019, *Astronomy Reports*, 63, 365

Kislyakova, K.G., Holmström, M., Lammer, H., *et al.* 2014, *Science*, 346, 981

Li, S.-L., Miller, N., Lin, D.N.C., & Fortney, J.J. 2010, *Nature*, 463, 1054

Linsky, J.L., Yang, H., France, K. *et al.* 2010, *ApJ*, 717, 1291

Liu, Y.D., Richardson, J.D., Wang, C., & Luhmann, J.G. 2014, *ApJ*, 788, L28

Mayor, M., & Queloz, D. 1995, *Nature*, 378, 355

Möstl, C., Amla, K., Hall, J.R., *et al.* 2014, *ApJ*, 787, 119

Murray-Clay, R.A., Chiang, E.I., & Murray, N. 2009, *ApJ*, 693, 23

Owens, M.J., & Forsyth, R.J. 2013, *Living Rev. Solar Phys.*, 10, 5

Parker E.N. 1958, *Astrophys. J.*, 128, 664

Roussev, I.I., Gombosi, T.I., Sokolov, I.V., *et al.* 2003, *ApJ*, 595, L57

Russell, C.T. 1993, *Rep. Prog. Phys.*, 56, 687

Steinolfson, R.S., & Hundhausen, A.J. 1988, *J. Geophys. Res.*, 93, 14269

Stevenson, D.J. 1983, *Reports on Progress in Physics*, 46, 555

Totten, T.L., Freeman, J.W., & Arya, S. 1995, *J. Geophys. Res.*, 100, 13

Vidal-Madjar, A., Lecavelier des Etangs, A., Desert, J.-M., *et al.* 2003, *Nature*, 422, 143

Vidal-Madjar, A., Desert, J.-M., Lecavelier des Etangs, *et al.* 2004, *ApJ*, 604, L69

Vidal-Madjar, A., Lecavelier des Etangs, A., Desert, J.-M., *et al.* 2008, *ApJ*, 676, L57

Weber, E.J., & Davis, L., Jr. 1967, *ApJ*, 148, 217

Withbroe, G.L. 1988, *Astrophys. J.*, 325, 442

Zhilkin, A.G., & Bisikalo, D.V. 2019, *Astronomy Reports*, 63, 550

Zhilkin, A.G., Bisikalo, D.V., & Kaygorodov, P.V. 2020, *Astronomy Reports*, in press

Solar and Stellar Magnetic Fields: Origins and Manifestations
Proceedings IAU Symposium No. 354, 2019
A. Kosovichev, K. Strassmeier & M. Jardine, ed.
doi:10.1017/S1743921319009815

Star-planet interaction through spectral lines

C. Villarreal D'Angelo[1]⊙, A. A. Vidotto[1]⊙, A. Esquivel[2,3],
M. A. Sgró[3], T. Koskinen[4] and L. Fossati[5]

[1]School of Physics, Trinity College Dublin, The University of Dublin College Green,
Dublin 2, Dublin, Ireland
email: villarrc@tcd.ie

[2]Instituto de Ciencias Nucleares, UNAM, Ciudad de México, México

[3]Instituto de Astronomía Teórica y Experimental, Conicet-UNC, Córdoba, Argentina

[4]University of Arizona, Lunar and Planetary Laboratory, Tucson, Arizona, United States

[5]Space Research Institute, Austrian Academy of Sciences, Graz, Austria

Abstract. The growth of spectroscopic observations of exoplanetary systems allows the possibility of testing theoretical models and studying the interaction that exoplanetary atmospheres have with the wind and the energetic photons from the star. In this work, we present a set of numerical 3D simulations of HD 209458b for which spectral lines observations of their evaporative atmosphere are available. The different simulations aim to reproduce different scenarios for the star-planet interaction. With our models, we reconstruct the Lyα line during transit and compare with observations. The results allows us to analyse the shape of the line profile under these different scenarios and the comparison with the observations suggest that HD209458b may have a magnetic field off less than 1 G. We also explore the behaviour of the magnesium lines for models with and without magnetic fields.

Keywords. stars: individual, HD209458, stars: planetary systems, stars: winds, outflows, line: profiles, magnetic fields, methods: numerical

1. Introduction

The Lyman α line has been broadly used to detect escaping atmospheres of close-in exoplanets. The first example was the case of HD209458b, a hot-Jupiter orbiting a solar-like star. While observed in transit, this exoplanet revealed an absorption of 15%, ten times larger than the one produced at optical wavelengths (Vidal-Madjar *et al.* 2003). This implied that the obstacle obscuring the disc of the star was bigger than the size of the planet Roche lobe. The Lyman α observation also revealed that the duration of the transit in this line was longer than the one produced by the opaque planetary disc, probing the existence of a cometary tail of neutral hydrogen trailing the planet. Finally, the absorption at high blue-shifted velocities ($\sim$100 km s^{-1}) indicated that neutral hydrogen atoms were accelerated away from the star.

In transit observations in other lines like O_I, C_{II}, Si_{III} and Mg_I (Vidal-Madjar *et al.* 2004; Linsky *et al.* 2010; Vidal-Madjar *et al.* 2013; Bourrier *et al.* 2014) also presented an excess of absorption compared to the optic, confirming the blow-off state of this atmosphere and showing that lighter elements were dragging heavier elements towards the upper atmosphere. But only lighter atoms like hydrogen have been found so far above the Roche lobe.

Many theoretical works since then have been developed in order to reproduce the transit observations of HD209458b in Lyman α. In particular, 3D numerical simulations that include the stellar wind and the escaping planetary atmosphere with or without magnetic

Table 1. Stellar and planetary winds parameters employed in the simulations for the system HD 2019458. System parameters taken from Torres *et al.* (2008). (1)From Sanz-Forcada *et al.* (2011).

Stellar parameters	Sym.	HD 2019458
Radius [$R_\odot$]	$R_\star$	1.2
Mass [$M_\odot$]	$M_\star$	1.1
Wind temperature [MK]	$T_\star$	1.5
Mass loss rate [$M_\odot$ yr^{-1}]	$\dot{M}_\star$	2.0×10^{-14}
Photon rate [s^{-1}]	S_0	2.5×10^{38} [(1)]

Planetary parameters	Sym.	HD 209458b
Radius [R_J]	R_p	1.38
Mass [M_J]	M_p	0.67
Orbital period [d]	τ_p	3.52
Inclination [deg]	i	86.71
Wind launch radius [R_p]	$R_{w,p}$	3
Wind velocity at $R_{w,p}$ [km s^{-1}]	v_p	10
Wind temperature at $R_{w,p}$ [K]	T_p	1×10^4
Ionisation fraction at $R_{w,p}$	χ_p	0.8
Mass loss rate [g s^{-1}]	$\dot{M}_p$	2×10^{10}

fields (Villarreal D'Angelo *et al.* 2014, 2018), and with or without charge exchange (Esquivel *et al.* 2019), have been able to reproduce the observations. 3D simulations can capture the asymmetric nature of the interaction between the stellar and planetary wind which may leave an imprint in the line profile observed during transit.

2. Numerical simulations

A number of numerical simulations have been run using the MHD code Guacho†. The physical domain of the simulations comprises the star at the centre and the planet orbiting around it. The models focused on the system HD209458 and assume that the planet has a wind as a consequence of the heating of the planetary atmosphere. We only model hydrogen (ionised for the stellar wind and partially ionised for the planetary wind) but other elements can be accounted assuming an abundance relative to hydrogen (like magnesium). The simulations account for the gravity of the star and the planet, the radiation pressure and the ionising flux from the star. In addition, a radiative transfer module takes care of the change in the density of neutral hydrogen due to collisional ionisation, photoionisation and recombination. Heating due to photoionisation and cooling due to the atomic transitions of hydrogen are included. More detailed information of the model setup and the code itself can be found in Villarreal D'Angelo *et al.* (2018).

The stellar wind is launched from the stellar surface with the initial conditions set by a thermally driven wind with a coronal temperature of 1.5 MK. The planetary wind is launched from $3R_p$ with initial conditions set by the 1D atmospheric model of Murray-Clay *et al.* (2009). System parameters used in the models together with the winds values are presented in Table 1. We consider that both objects, star and planet, posses a dipolar magnetic field perpendicular to the orbital plane and both aligned. We ran a total of five simulations varying the strength of the magnetic field both in the star and the planet. Table 2 summarises the model names and the corresponding values of B at the stellar and planetary poles.

† Available for free at https://github.com/esquivas/guacho

Table 2. Models characteristics and the estimated values for the stand-off distance at the sub-stellar point. Last column shows the integrated Lyα absorption in the velocity range of ±300 km s^{-1}.

Models	$B_\star$ [G]	B_p [G]	R_0 [R_p]	$(1 - I/I_\star)$ [%]
B1.0	1	0	7	12.1
B1.1	1	1	7	12.1
B1.5	1	5	9	10.7
B5.1	5	1	4	12.7
B5.5	5	5	4	13.4

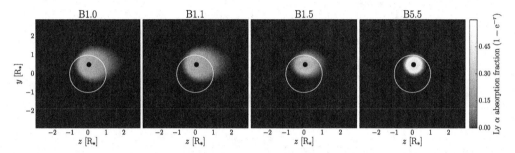

Figure 1. Lyman α absorption fraction integrated between [-300, 300] km s^{-1}, for four of the models presented in Table 2. The plot is a zoom at the position of the star (white circle) at the moment of mid-transit. The planet is represented with the black filled circle.

2.1. *Lyman α line under different magnetic field values*

For each model we calculated the Lyα absorption fraction $(I/I_\star = (1 - e^{-\tau}))$ in the line of sight (LOS) direction integrated over the velocity range [-300, 300] km s^{-1} and taking into account the orbital inclination. The optical depth is computed assuming a Voigt profile with an absorption cross section $\sigma = 0.01105$ cm^2. The results are shown in Figure 1. From the figure we can see that models with a stellar magnetic of 1 G (B1.0, B1.1 and B1.5) presents a more extended area of neutral hydrogen around the planet than models with a higher stellar magnetic field (B5.1 and B5.5). Since a higher $B_\star$ increase the velocity and temperature of the stellar wind, it confines the escaping planetary material to a smaller and spherical region. Figure 1 only shows model B5.5 as model B5.1 gave similar results.

For the models with $B_\star = 1$ G, the planetary material stop the stellar wind further from the planet and the neutral hydrogen expands producing a more extended region. In this scenario, models with $B_p = 0$ or 1 G behave similarly, indicating that the gas pressure in the atmosphere dominates over the magnetic pressure. Model B1.5 however, does show the influence of the planetary magnetic field as a higher B_p prevents the escape of material from the equatorial region (neutral are coupled to the ions) and ionise the polar one, and so the region of neutral material around the planet have a more oval distribution.

Figure 2 shows the normalised transmission spectra in the Lyα line as a function of the LOS velocity integrated over the stellar disc. Models with $B_\star = 1$ and 5 G are represented with solid and dashed lines respectively and the different B_p values with colours.

As mentioned before, models with a higher stellar or planetary magnetic field differentiate from the ones with a lower B (stellar or planetary) due to the shape of the neutral region around the planet, producing different depths and shapes of the line profile.

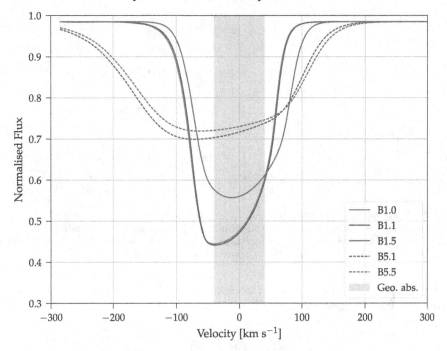

Figure 2. Normalised stellar transmission as a function of the LOS velocity in the Lyα line averaged over the stellar disk as seen by an observer. The yellow stripe corresponds to part of the line contaminated with the geo-coronal glow. Solid lines represent models with $B_\star$=1 G and dashed lines for $B_\star$=5 G. Line colours represent different values for the planetary magnetic field.

Whit a stronger stellar wind ($B_\star$=5 G) most of the neutral material is confined very close to the planet, having the smaller standoff distance at the sub-stellar point (see Table 2). However, a small amount still escape away from the planet forming a small tail due to the interaction. The line shape is then asymmetric but with a lower depth.

Another asymmetric line shape is obtained for models B1.0 and B1.1. In these cases, the stellar wind is not as strong ($B_\star$=1 G) and the neutral material is able to expand more, forming a larger cometary tail. In these cases, the interaction with the stellar wind doesn't accelerate the neutrals to a very high velocity but the column density is larger producing a larger absorption depth.

A symmetric line shape is possible when $B_\star$=1 G and B_p=5 G (model B1.5). In this case, the planetary magnetic field constrain the expansion of the neutral material into a more spherical region at the equator. The velocities of the neutral material seems to be more constrained by the planetary magnetic field than the stellar wind.

Overall, the line shape for each model reveal that the interaction with the stellar wind can accelerate the neutral material to high blue-shifted velocities, and that these velocities are related to the type of interaction produced.

Comparison with observations: Among the Lyα lines produced with the magnetic models presented above we can see that only models with B_p=0 or 1 G agreed with the 10% of absorption spotted at velocities of -100 km s^{-1} in the observed line profile.

The total absorption value is calculated integrating the line in Fig. 2 over [-300, 300] km s^{-1} excluding the geocoronal emission range from [-40,40] km s^{-1}. The computed values are shown in Table 2 for all the models together with the model's stand-off distance. Most of them agree within the errors with the 15% of absorption found in the same velocity

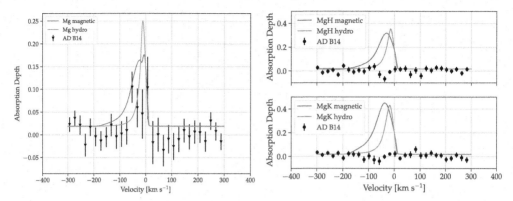

Figure 3. Magnesium absorption fraction as a function of the LOS velocity for two models: hydrodynamic (green) and magnetohydrodynamic (red) with the similar initial conditions for planetary and stellar wind. Left: Mg$_\mathrm{I}$. Right top: Mg$_\mathrm{II}$ H. Right bottom: Mg$_\mathrm{II}$ K. Observational values presented in Bourrier *et al.* (2014) are shown with black dots.

range by Vidal-Madjar *et al.* (2003), although the lowest total absorption corresponds to the model with the largest stand-off distance (model B1.5).

2.2. *The Mg lines*

To explore different type of interactions in other spectroscopic lines we have used one of the magnetic models presented above (model B1.1) together with a pure hydrodynamical model, with similar initial conditions for the stellar and planetary wind, from the work of Esquivel *et al.* (2019) (model M5a). For these two models we compute the abundance and calculate the absorption profile during transit for the Mg$_\mathrm{I}$ and Mg$_\mathrm{II}$ H & K lines. This was done as a post-processing from the output of those simulations. The results shown in this section are preliminary and so, they should be taken as such. The final purpose of these calculations is to compare the shape of the magnesium lines with two different sets of simulations (with and without magnetic fields).

To calculate the abundance of magnesium within the simulations we used a 1D hydrodynamic escape model from T. Koskinen (private comm.) that includes several species (hydrogen, magnesium, helium, sodium). As these models predict a higher mass loss rate for HD209458b than the one we used in our models, we scaled the total hydrogen in our simulations to match the total hydrogen in the 1D model at the position of our boundary for the planetary wind. The relative abundance of magnesium (neutral and ionised) to total hydrogen computed in for the 1D model is fairly constant above 1.5R$_p$ and so we used this ratio as a scale factor to convert from hydrogen to magnesium in our 3D models. Below 3R$_p$ the magnesium abundance follows the one predicted in the 1D model. An important thing to have in mind with these calculations is that we are assuming that magnesium will follow hydrogen above the 3R$_p$ and have the same photoionisation rate which may not be the case.

In the same way as for Lyα, we calculated the absorption fraction of Mg$_\mathrm{I}$ and Mg$_\mathrm{II}$ H & K in both models with cross-section coefficients of $\sigma_{\mathrm{Mg_I}} = 0.04869$ and $\sigma_{\mathrm{Mg_{II}}} = 0.00814$ cm^2 respectively. The lines profiles are shown in Figure 3 together with the observational results from Bourrier *et al.* (2014). It is clear that the model including magnetic fields produce a broader line shape compared with the hydrodynamic one with absorption present a high blue-shifted velocities. However, both of them shown a sharp jump in the red part of the line. Also, both models predict absorption in the magnesium II H & K lines which is not found in the observations.

3. Conclusion

We have shown how the spectral lines such as Lyα and the Mg lines are shaped trough the interaction of the stellar and planetary winds. We have used the output of numerical simulations based on the HD209458 system to compute the transmission spectra in these lines during transit. We have shown that the Lyman α profile is shaped by magnetic fields and that a strong stellar magnetic field or a strong planetary magnetic field (5 G) with a weak stellar magnetic field (1 G) gives symmetric profiles. From these magnetic models we have calculated the total Lyman α absorption between [10-13]% matching the amount computed from the observations.

The future launch of the Colorado Ultraviolet Transit Experiment (CUTE) satellite will produce a number of in-transit spectra in the NUV. Most prominent lines like magnesium will then be sampled for a number of extrasolar close-in gaseous planets. Our attempt was to study the behaviour of these lines in a very simplistic way assuming that magnesium will follow hydrogen above the $3R_p$. The lines shape found in these cases show lines that are less broad for the hydrodynamic models than magnetic models.

Spectral lines observed during a planetary transit can probe the escaping atmosphere of the planet and give an insight of the type of interaction that occurs between the planet and the star. We have shown that the presence of absorption in higher blue-shifted velocities indicates a strong interaction with the stellar wind. With the possibility to observe simultaneously several spectral lines in the future, for example with CUTE, and with the use of numerical models capable to reproduce such observations, we will be able to constrain the physical parameters of a planetary system.

Acknowledgements

I want to thank the organisers to allow me to give this talk in this very rich symposium and for such a good organisation and predisposition to help. I also want to acknowledge the funding from the Irish Research Council through the postdoctoral fellowship. Project ID: GOIPD/2018/659.

References

Bourrier, V., Lecavelier des Etangs, A., & Vidal-Madjar, A. 2014, *A&A*, 565, A105
Esquivel, A., Schneiter, M., Villarreal D'Angelo, C., Sgró, M. A., *et al.* 2019, *MNRAS*, 487, 5788
Murray-Clay, R. A., Chiang, E. I., & Murray, N. 2009, *ApJ*, 693, 23
Linsky, J. L., Yang, H., France, K., *et al.* 2010, *ApJ*, 717, 1291
Torres, G., Winn, J. N., & Holman, M. J. 2008, *ApJ*, 677, 1324
Sanz-Forcada, J., Micela, G., Ribas, I., *et al.* 2011, *A&A*, 532, A6
Vidal-Madjar, A., Lecavelier des Etangs, A., Désert, J.-M., *et al.* 2003, *Nature*, 422, 143
Vidal-Madjar, A., Désert, J.-M., Lecavelier des Etangs, A., *et al.* 2004, *ApJ*, 604, L69
Vidal-Madjar, A., Huitson, C. M., Bourrier, V., *et al.* 2013, *A&A*, 560, A54
Villarreal D'Angelo, C., Schneiter, M., Costa, A., *et al.* 2014, *MNRAS*, 438, 1654
Villarreal D'Angelo, C., Esquivel, A., Schneiter, M., M., Sgró, M. A. 2018, *MNRAS*, 479, 3115

Discussion

L. CAMPUSANO: Hydrodynamic vs magnetohydrodynamic models. Can one of them be discarded by observations?

C. VILLARREAL D'ANGELO: Not really as there is not too many observations and all of them have large errors too. Also models are dependent on initial conditions.

W. CAULEY: Can you guess what is causing the Mg_{II} models to be off?

C. VILLARREAL D'ANGELO: Not at this point.

Solar and Stellar Magnetic Fields: Origins and Manifestations
Proceedings IAU Symposium No. 354, 2019
A. Kosovichev, K. Strassmeier & M. Jardine, ed.
doi:10.1017/S1743921320000095

From the Sun to solar-type stars: radial velocity, photometry, astrometry and $\log R'_{HK}$ time series for late-F to early-K old stars

Nadège Meunier⬤ and Anne-Marie Lagrange

Univ. Grenoble Alpes, CNRS, IPAG, F-38000 Grenoble, France
email: nadege.meunier@univ-grenoble-alpes.fr

Abstract. Solar simulations and observations showed that the detection of Earth twins around Sun-like stars is difficult in radial velocities with current methods techniques. The Sun has proved to be very useful to test processes, models, and analysis methods. The convective blueshift effect, dominating for the Sun, decreases towards lower mass stars, providing more suitable conditions to detect low mass planets. We describe the basic processes at work and how we extended a realistic solar model of radial velocity, photometry, astrometry and LogR'HK variability, using a coherent grid of stellar parameters covering a large range in mass and average activity levels. We present selected results concerning the impact of magnetic activity on Earth-mass planet detectability as a function of stellar type. We show how such realistic simulations can help characterizing the effect of stellar activity on RV and astrometric exoplanet detection.

Keywords. Sun: activity, Sun: faculae, plages, (Sun:) sunspots, stars: spots, stars: activity, (stars:) planetary systems, techniques: radial velocities, astrometry, techniques: photometric

---

1. Introduction

Our primary objective in this work is to estimate exoplanet detection limits due to stellar activity and to find new methods to correct for the stellar signal, due to magnetic activity and flows at various scales. We focus on radial velocity (RV), and compare with high-precision astrometry. This approach is necessary to be able to reach very low mass planets around solar type stars, especially in their habitable zone. The measurement of the Doppler shift is perturbed by stellar activity, for example by spots and plages which are distorting the shape of the lines (for the same reason it allows Doppler imaging or Zeeman Doppler imaging, see contributions by Solanki & Petit, this volume). To study their impact for many stellar configurations, we built synthetic time series of various observables, and in particular RV and chromospheric emission. We also generated astrometric time series to study the performance of this technique to detect low mass exoplanets, and photometric time series to evaluate the performance of various diagnosis of stellar activity for such complex activity patterns. The outline of the paper is as follows. In Sect. 1, we describe our approach, which combines our knowledge of solar activity (our reference for such stars and for which many properties are well characterized) with stellar activity and exoplanets, and our model. Results on detectability in RV and astrometry are presented in Sect. 3. We conclude in Sect. 4.

2. Approach and model

General approach. Over the last ten years, we have implemented a dedicated approach, in three steps of increasing complexity. Our original question was: could we detect the

Earth if we were observing the Sun from the outside, with current or future instruments and methods. Therefore our first step was to reconstruct the solar integrated RVs. Secondly, we developed a solar model to generate realistic structures to simulate the Sun seen from different points of view. Finally, this model was extended to the detectability of Earth mass planets in the habitable zone around solar-type stars since we were able to reproduce complex activity patterns as observed on the Sun. We describe these steps below, together with the processes taken into account when building these time series.

Step 1: Solar RV, from observed structures and MDI/SOHO Dopplergrams. We reconstructed the solar RV using observed structures (spots and plages) as well as the inhibition of the convective blueshift and a model to compute the integrated RV from these structures (Meunier *et al.* 2010a). When considering only the distorsion of the lines due to their contrasts we obtained time series with a rms (root mean square) of ∼0.3 m/s, which is very similar to what we found for spots alone (Lagrange *et al.* 2010). We found that the dominant contribution, in particular on long timescales, is the inhibition of the convective blueshift in plages, also identified early on by Saar & Donahue (1997). This process is responsible for a net redshift with a peak-to-peak amplitude of the long-term variation of 8 m/s. This is two orders of magnitude higher than the Earth signal. In addition, we confirmed this long-term variability by reconstructing the solar RV signal from MDI Dopplergrams (Scherrer *et al.* 1995), which involves no model (Meunier *et al.* 2010b). The importance of this effect was later found using other approaches, for example by Lanza *et al.* (2016) and Haywood *et al.* (2016).

Step 2: A solar model to generate spots, plages, and network. In a second step, we built a model to generate synthetic spots and plages to be able to look at different solar inclinations, and to validate this model on the Sun, before applying it to other stars. At each time step, each spot or plage is characterized by its size and its position. A contrast is then associated to each of them (it varies with the position on the disk for plages), and their lifetime is determined by a decay time following the solar distribution. Magnetic network structures are also generated from a fraction of the plage decay. The model uses several empirical laws which are well determined for the Sun. It allowed us to generate structures over complete cycles, following a given butterfly diagram and a given cycle shape, with certain size distributions, decay time distributions, and the proper dynamics. All laws are described in Borgniet *et al.* (2015). From these, we computed the RV and other observables. This allowed us to show the importance of stellar inclinations on the time series properties (Borgniet *et al.* 2015), an example is shown in Fig. 1.

Step 3: Extension of the model to solar-type stars. We built a large number of synthetic time series for stars over the range of parameters shown in Fig. 2, from F6 to K4 and for old main sequence stars (Meunier *et al.* 2019a). We adapted some of the parameters from the solar ones, but not all of them because many properties which are well known in the solar case are not well characterized for other stars, such as the size distributions for example. We also needed to keep the number of parameters reasonnable. We list in Fig. 3 the parameters which depend on spectral type, on average activity level or on both in this model. Two examples of laws are the rotation period depending on the activity level and spectral type (Mamajek *et al.* 2007), and the inhibition of the convective blueshift depending on the spectral type, which we have measured using a large sample of HARPS spectra (Meunier *et al.* 2017a, 2017b). In addition to RV time series, we also produce chromospheric emission, astrometry, and photometry time series. The latter is not presented in detail here (Meunier & Lagrange 2019c): the brightness variability, although correlated with the $\log R'_{HK}$, exhibit a large dispersion; it increases towards low-mass stars, as observed with Kepler; The differences between brightness variability regimes dominated by spots or by plages is strongly affected by stellar inclination.

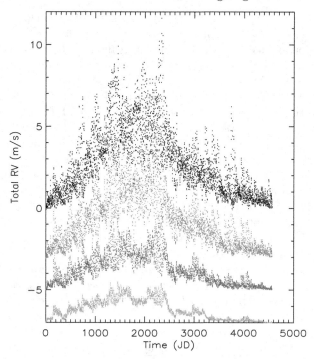

Figure 1. Solar RV from the generation of stellar spots and plages, showing the impact of inclination on short-term and long-term variability, from Borgniet *et al.* (2015).

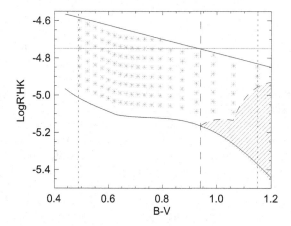

Figure 2. Range of B-V and $\log R'_{HK}$ values covered by our stellar simulations (orange stars). The upper solid line represents the transition to younger spot-dominated stars which are not modeled here (Lockwood *et al.* 2007). The lower solid line represents the basal flux (corresponding to no magnetic activity), from Meunier *et al.* (2019a).

Final sets of time series. We obtained more than 10000 independant time series of stellar structures, which are then declined in 10 inclinations and 2 spot contrasts. We also add to the RV time series granulation and supergranulation signals, using the results of Meunier *et al.* (2015). More recently, we showed that supergranulation was also affecting even long orbital periods (Meunier & Lagrange 2019a).

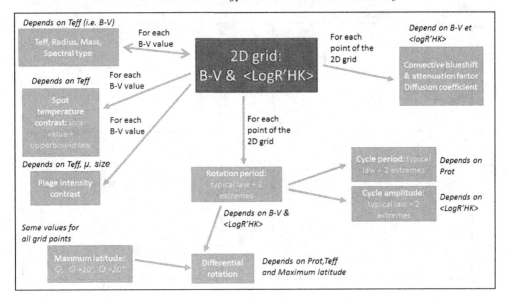

Figure 3. Parameters which are adapted in the stellar simulation and depend on B-V, on log R'_{HK}, or on both, from Meunier *et al.* (2019a).

3. Results

In this section, we analyze these stellar synthetic time series. We first compute detection limits due to activity, after correction, using a simple approach. Such simulations also allowed us to better understand how stellar activity affects RV, because there are a lot of degeneracies in the final signal and many contributions from the star. We focus here on the relationship with chromospheric emission, which allows us to improve correction techniques using activity indicators. Finally, we compare our results in RV with the performance high-precision astrometry would allow to reach.

Radial velocity jitter and exoplanet detectability. As a first simple approach to study exoplanet detectability, we have estimated the mass which could be detected given the amplitude of the RV jitter after a simple correction technique, and the number of observations. For that purpose, we used a criteria established in the fitting challenge organized by Xavier Dumusque (Dumusque *et al.* 2017), in which the criterion C depending on the RV jitter, the number of observations and the planet RV amplitude was related to what we could detect given current techniques, typically ~ 7.5, was a limit between poor and good performance. The planet RV amplitude which can be detected using current correction techniques is then deduced from each time series using this criterion. Associated to a given planet orbital period (here in the habitable zone), we estimate the typical lower masses which can be detected. This is shown in Fig. 4 as a function of spectral type and log R'_{HK}. With a very low number of points (100), the minimum mass is close to 10 M_{Earth}. For several 1000 points, covering full stellar cycles and an excellent sampling, values below 1 M_{Earth} can be reached, but only in a few cases: for K stars, edge-on configurations and excellent samplings. This illustrates the challenge we are facing to detect or even characterize (for example for a transit follow-up in RV) the mass of an Earth mass planet in RV in the habitable zone around such stars (Meunier & Lagrange 2019).

Limits to the usual correlation using a linear relationship between RV and log R'_{HK}.
A linear relationship between the long-term RV variability and log R'_{HK} is often assumed in correction techniques based on this indicator, because both are very well correlated with the plage coverage. However, this type of correction is not perfect

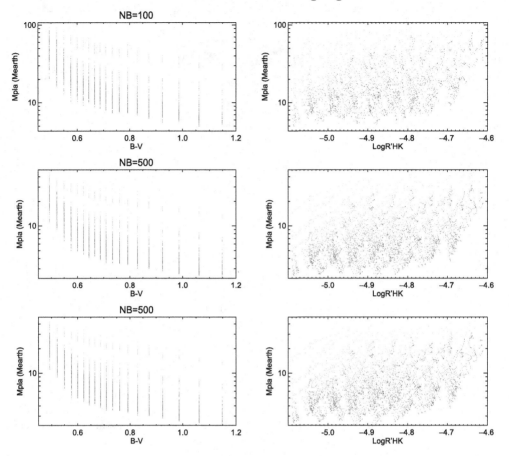

Figure 4. Lower mass which can be detected according to the criterion defined in Sect. 3 vs. B-V (left-hand side panel) and $\log R'_{HK}$ (right-hand side panel). The color code corresponds to stellar inclination, from pole-on in yellow to edge-on in green/blue, from Meunier & Lagrange (2019). Only one point out of 5 is shown for clarity.

(e.g. Meunier & Lagrange 2013). These new simulations allowed us to find out at least one important reason for this (Meunier *et al.* 2019b). Fig. 5 shows that although the chromospheric emission is roughly correlated with long-term RV as expected, there is a significant and systematic departure from a linear relationship, with difference up to 2 m/s in that example for the pole-on configuration. This example is for a star with similar activity compared to the Sun, except for the larger extent in latitude of the activity patter. Similar patterns are seen for the other simulations. The amplitude of the residuals is therefore still significantly larger than an Earth mass signal. The RV versus $\log R'_{HK}$ curves show that there is a kind of hysteresis pattern, with a different behavior between the ascending phase and the descending phase of the cycle. The sign of this pattern reverses between the two extreme configurations, with a reversal for a stellar inclination around 60°. We also observed this for the Sun and for some stars observed with HARPS. This is due to the combination of two effects: 1/ due to the butterfly diagram, the structures are not at the same distance to disk center on average over time, and the trend is reversed between pole-on and edge-on configurations; 2/ chromospheric emission and RV do not have the same dependence on the position on the disk (i.e. different projection effects). As a consequence, the ratio between the two changes over time, which leads to a departure from a linear relationship. We have

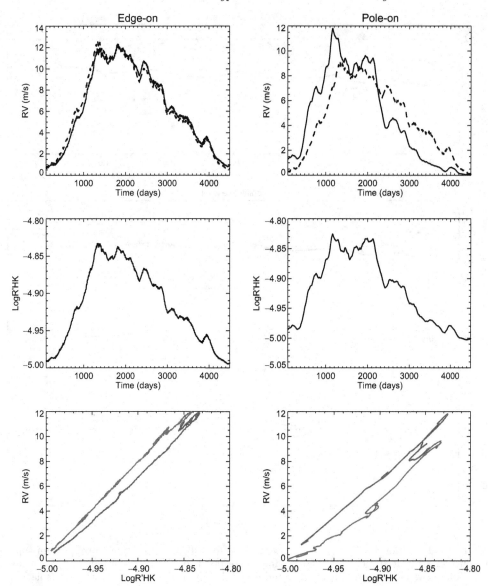

Figure 5. Example of hysteresis pattern between RV and log R'_{HK}, for the edge-on configuration (left-hand side panels) and pole-on configuration (right-hand side panels). The time series have been smoothed over 1 year to focus on long-term variability. From Meunier *et al.* (2019b).

therefore proposed a new method to take this effect into account: the correcting trend is described with functions modelling this effect for many different parameters; we choose the best function to minimize the residuals. The gains with respect to the usual linear correction are very interesting, and can reach values up to 4 or 5 in some cases (Meunier *et al.* 2019b) on the smoothed time series.

High precision astrometry. We also produce astrometric time series, corresponding to the same activity patterns. Since it is very difficult to reach very low masses with RV, it is interesting to consider the performance in astrometry, even if a mission with the necessary precision does not exist yet. For the Sun, we know this is the case: this was shown from the reconstruction of its astrometric signal over cycle 23 (Makarov *et al.* 2010;

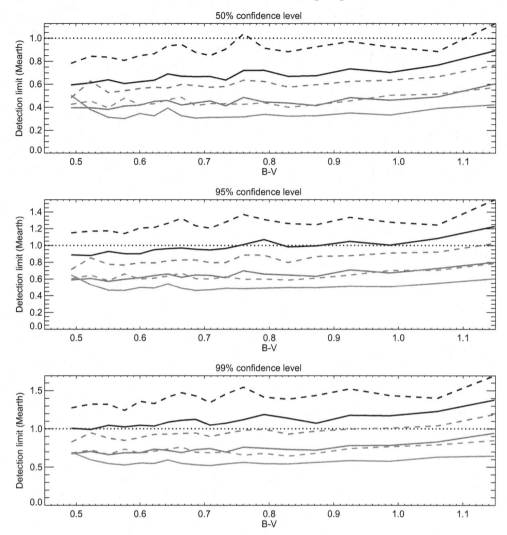

Figure 6. Detection limits in astrometry vs. B-V for different confidence levels (from top to bottom) for stars at 10 pc. The different lines correspond to different positions in the habitable zone (inner side in black, middle in red, and outer side in green) and spot contrasts (solar ones in solid lines, and upper limit in dashed lines).

Lagrange *et al.* 2011), with a typical rms of 0.07 μ as (at 10 pc) while the Earth signal at this distance would have an amplitude of 0.33 μ as so that stellar activity is not a limitation. We used the strategy proposed for the high precision Theia mission (The Theia Collaboration *et al.* 2017), i.e. 50 observations per star over 3.5 year, and 0.2 μarcsec of noise per measurement to compute detection limits on our time series (Meunier & Lagrange 2019, in prep). The detection limits for different confidence levels are shown in Fig. 6, for different positions in the habitable zone of the considered stars. We find that many of them are below 1 M_{earth}, especially for F and G stars (which are in this case slightly easier than K stars, while it was the opposite for RV). These computations were made for a star at 10 pc. The ratio between the planet signal and the stellar signal is independent on the distance. The instrumental noise is proportionnally lower for stars closer than 10 pc however (so we expect these detection limit to be improved) and worse for stars further away.

4. Conclusion

We showed that such realistic time series (in terms of typical amplitude and complexity of the activity pattern) are very useful to better understand the relationship between observables. Such studies have a strong impact for the exoplanet community, for both detectability and mass characterization, but also in stellar physics: for example, the hysteresis in the relationship between RV and $\log R'_{HK}$ is indicative of the sign of the dynamo wave, provided the inclination is known. This is complementary to other approaches, for example MHD simulation of granulation (Cegla *et al.* 2018), or the direct observation of solar integrated RV: many groups have indeed recently implemented dedicated instruments to be able to observe the integrated RV of the Sun as a star, for example with HARPS-N in La Palma (Dumusque *et al.* 2015; Collier Cameron *et al.* 2019), HARPS at the 3.60m in La Silla, Espress at Lowell Observatory. Based on the results obtained in the fitting challenge presented in Dumusque *et al.* (2017) and our simulations, we found that given current mitigating technique of stellar activity, Earth mass planet in the habitable zone around solar type stars is not possible, except for K stars observed in very good conditions. On the other hand, high-precision astrometry, although technically very challenging, would not be limited by stellar activity for most solar type stars (we consider old main sequence stars only in this work), at least for stars in the solar neighbourhood. Future work will include a precise determination of the detection limits we can expect from the frequential analysis of the synthetic RV time series, taking their frequential behavior into account as well as for granulation and supergranulation signals. More observables will also be added to the simulations and new correcting methods will be developed.

References

Borgniet, S., Meunier, N., & Lagrange, A.-M. 2015, *AA*, 581, A133

Cegla, H. M., Watson, C. A., Shelyag, S., Chaplin, W. J., Davies, G. R., Mathioudakis, M., Palumbo, III, M. L., Saar, S. H., & Haywood, R. D. 2018, *ApJ*, 866, 55

Collier Cameron, A., Mortier, A., Phillips, D., Dumusque, X., et al. 2019, *MNRAS*, 487, 1082

Dumusque, X., Udry, S., Lovis, C., Santos, N. C., & Monteiro, M. J. P. F. G. 2011, *AA*, 525, A140

Dumusque, X., Glenday, A., Phillips, D. F., Buchschacher, N., Collier Cameron, A., Cecconi, M., Charbonneau, D., Cosentino, R., Ghedina, A., Latham, D. W., Li, C.-H., Lodi, M., Lovis, C., Molinari, E., Pepe, F., Udry, S., Sasselov, D., Szentgyorgyi, A., Walsworth, R. 2015, *ApJL*, 814, L21

Dumusque, X. 2016, *AA*, 593, A5

Dumusque, X., Borsa, F., Damasso, M., Díaz, R. F., Gregory, P. C., et al. 2017, *AA*, 598, A133

Haywood, R. D., Collier Cameron, A., Unruh, Y. C., Lovis, C., Lanza, A. F., Llama, J., Deleuil, M., Fares, R., Gillon, M., Moutou, C., Pepe, F., Pollacco, D., Queloz, D., & Ségransan, D. 2016, *MNRAS*, 457, 3637

Lagrange, A.-M., Desort, M., & Meunier, N. 2010, *AA*, 512, A38

Lagrange, A.-M., Meunier, N., Desort, M., & Malbet, F. 2011, *AA*, 528, L9

Lanza, A. F., Molaro, P., Monaco, L., & Haywood, R. D. 2016, *AA*, 587, A103

Lockwood, G. W., Skiff, B. A., Henry, G. W., Henry, S., Radick, R. R., Baliunas, S. L., Donahue, R. A., Soon, W. 2007, *ApJS*, 171, 260

Makarov, V. V., Parker, D., & Ulrich R. K. 2010, *ApJ*, 717, 1202

Mamajek, E. A., & Hillenbrand, L. A. 2008, *ApJ*, 687, 1264

Meunier, N., Desort, M., & Lagrange, A.-M. 2010a, *AA*, 512, A39

Meunier, N., Lagrange, A.-M., & Desort, M. 2010b, *AA*, 519, A66

Meunier, N., & Lagrange, A.-M. 2013, *AA*, 551, A101

Meunier, N., Lagrange, A.-M., Borgniet S., & Rieutord, M. 2015, *AA*, 583, A118

Meunier, N., & Lagrange, A.-M. 2019a, *AA*, 625, L6

Meunier, N., & Lagrange, A.-M. 2019, *AA*, 628, A125

Meunier, N., & Lagrange, A.-M. 2019c, *AA*, 629, A42

Meunier, N., & Lagrange, A.-M. 2019d, *In preparation*

Meunier, N., Lagrange, A.-M., Boulet, T. & Borgniet, S. 2019a, *AA*, 627, A56

Meunier, N., Lagrange, A.-M., & Cuzacq, S. 2019b, *AA*, in press

Saar, S. H., & Donahue, R. A. 1997, *ApJ*, 485, 319

Scherrer, P. H., Bogart, R. S., Bush, R. I., Hoeksema, J. T., Kosovichev, A. G., Schou, J., Rosenberg, W., Springer, L., Tarbell, T. D., Title, A., Wolfson, C. J., Zayer, I., & MDI Engineering Team 1995, *Sol. Phys.*, 162, 129

The Theia Collaboration, Boehm, C., et al. , 2017 *arXiv e-prints*, 1707.01348

Discussion

VIDOTTO: You showed a correlation between RV and $\log R'_{HK}$ when the Sun is seen edge-on, but the correlation does not hold for pole-on view. What would be the best strategy to use this correlation to other stars?

MEUNIER: The example shown in the figure was already for a star that is not the Sun, although we do see this behavior for the Sun. We have proposed a strategy to mitigate this effect by using functions describing this geometricl effect, to correct for the difference in shape between RV and $\log R'_{HK}$. In our simulations, the sign of this effect is related to the fact that we chose a solar-like butterfly diagram (equatorward dynamo wave). For a poleward pattern, the signs would be reversed. This would not change the principle of the correction method. However, if the stellar inclination of the star, it would be a novel method to determine the sign of the dynamo wave, provided it is close to pole-on or edge-on configurations to allow for a large amplitude of the hysteresis pattern.

Solar and Stellar Magnetic Fields: Origins and Manifestations
Proceedings IAU Symposium No. 354, 2019
A. Kosovichev, K. Strassmeier & M. Jardine, ed.
doi:10.1017/S1743921319009992

Could star-planet magnetic interactions lead to planet migration and influence stellar rotation?

Jérémy Ahuir[1]📷, Antoine Strugarek[1], Allan Sacha Brun[1],
Stéphane Mathis[1], Emeline Bolmont[2,1], Mansour Benbakoura[1],
Victor Réville[3,1] and Christophe Le Poncin-Lafitte[4]

[1]Département d'Astrophysique-AIM, CEA/IRFU, CNRS/INSU, Université Paris-Saclay,
Université Paris Diderot, Université de Paris, 91191 Gif-sur-Yvette, France
email: jeremy.ahuir@cea.fr

[2]Observatoire Astronomique de l'Université de Genève, Université de Genève,
CH-1290 Versoix, Switzerland

[3]IRAP, Université Toulouse III - Paul Sabatier, CNRS, CNES, Toulouse, France

[4]SYRTE, Observatoire de Paris, Université PSL, CNRS, Sorbonne Université, LNE, 61 avenue
de l'Observatoire, 75014 Paris, France

Abstract. The distribution of hot Jupiters, for which star-planet interactions can be significant, questions the evolution of exosystems. We aim to follow the orbital evolution of a planet along the rotational and structural evolution of the host star by taking into account the coupled effects of tidal and magnetic torques from *ab initio* prescriptions. It allows us to better understand the evolution of star-planet systems and to explain some properties of the distribution of observed close-in planets. To this end we use a numerical model of a coplanar circular star-planet system taking into account stellar structural changes, wind braking and star-planet interactions, called ESPEM (Benbakoura *et al.* (2019)). We find that depending on the initial configuration of the system, magnetic effects can dominate tidal effects during the various phases of the evolution, leading to an important migration of the planet and to significant changes on the rotational evolution of the star. Both kinds of interactions thus have to be taken into account to predict the evolution of compact star-planet systems.

Keywords. stellar evolution, solar-type stars, stellar rotation, magnetism, star-planet interactions

---

1. Introduction

The discovery of more than 4000 exoplanets during the last two decades has shed light on the importance of characterizing star-planet interactions. Indeed, a large fraction of these planets have short orbital periods and are consequently strongly interacting with their host star. In particular, several planetary systems, like 55 Cancri, WASP-18 or HD 189733 (cf. Figure 1 to see the architecture of those systems), are likely to host exoplanets undergoing a migration due to tidal and magnetic torques. We consider here the joint influence of stellar wind, tidal and magnetic star-planet interactions on the star's rotation rate and planetary orbital evolution. We focus our study on the relative influence of tidal and magnetic torques on the system evolution. Our objective is to take into account simultaneously *ab initio* prescriptions of tidal and magnetic torques in exosystems, so as to improve our understanding of close-in star-planet systems and their long-term evolution.

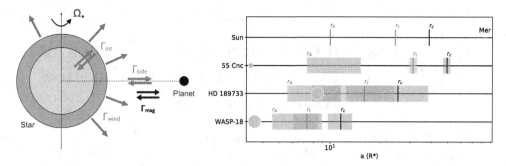

Figure 1. On the left: schematic view of the system and its interactions (adapted from Benbakoura *et al.* 2019). The various Γ quantities (int, wind, mag, tide) illustrate the various angular momentum transfer mechanisms taken into account in our study. On the right: architecture of the inner Solar System, 55 Cancri, HD 189733 and WASP 18. The semi-major axis a is expressed in stellar radii $R_\star$. Red dots: Sub-Earths. Blue dots: Super-Earths. Orange dots: jovian planets. In black: co-rotation radius r_c, with error bars. In red: transition equilibrium-dynamical tide r_t, with error bars. In blue: Estimates of the sub-alfvénic region transition r_A for realistic stellar magnetic fields.

2. Stellar rotation and planetary migration

ESPEM (French acronym for Planetary Systems Evolution and Magnetism; Benbakoura *et al.* 2019) is a 1D numerical model of a coplanar circular star-planet system allowing us to follow the joint secular evolution of the semi-major axis of the orbit and the stellar rotation. We consider in this model a solar-type star with a bi-layer internal structure following MacGregor & Brenner 1991, whose changes are monitored along the evolution of the system by relying on grids provided by the 1D stellar evolution code STAREVOL (Amard *et al.* 2016). The companion is considered a punctual mass.

A global schematic view of the different interactions involved in the system is given in Figure 1. The yellow disk depicts the stellar radiative core and the orange shell the convective envelope. The red arrows account for the angular momentum extraction by the stellar wind (Schatzman 1962, Weber & Davis 1967; see Ahuir *et al.* 2019 for the interdependencies between magnetism, wind and stellar rotation). The green arrows correspond to internal angular momentum exchanges between the core and the envelope of the star, which in absence of external disruptions evolve towards synchronization of their rotation (MacGregor & Brenner 1991). Such a coupling also takes into account the growth of the radiative zone during the Pre-Main Sequence, resulting in an angular momentum transfer from the envelope to the core (MacGregor 1992). The blue arrows account for exchanges between the stellar envelope and the planetary orbit due to both equilibrium and dynamical tidal effects (Hansen 2012, Ogilvie 2013, Mathis 2015, Bolmont & Mathis 2016).

3. Two-body magnetic interaction

Along with wind braking and tidal effects, magnetic interactions (black arrows in Figure 1) occur because of the relative motion between the planet and the ambient wind at the planetary orbit. If the planet is in a sub-alfvenic region (where the wind velocity is smaller than the local Alfvén speed), a magnetic torque applied to the planet can be associated to an efficient transport of angular momentum between the planet and the star through the so-called *Alfvén wings* (Neubauer 1998). Several regimes then appear according to the magnetic properties of the planet, at least as a first approximation. If the planet is able to sustain a magnetosphere, Alfvén waves generally do not have enough

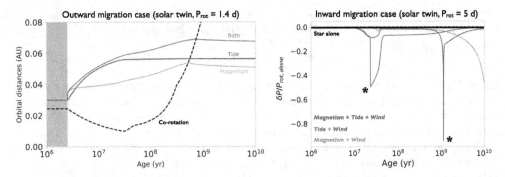

Figure 2. On the left: orbital evolution of a Jupiter-like planet, initially at $a = 0.03$ AU, orbiting a fast rotating solar twin ($P_{\rm rot} = 1.4$ d) in the open-field configuration. The black dashed line corresponds to the co-rotation radius of the star without any planet. On the right: evolution of the relative difference of rotation periods between a midly rotating solar twin ($P_{\rm rot} = 5$ d) hosting a Jupiter-like planet initially at $a = 0.03$ AU and a similar star without planet. Mark * corresponds to the planetary destruction. In black: case of a star without planet.

time to go back and forth between the star and the planet. In this case, two uniques Alfvén wings form around the planet, leading to the so-called *dipolar* interaction (Saur *et al.* 2013, Strugarek *et al.* 2015, Strugarek 2016). In the opposite case, when the ambient wind pressure is too strong or if the planetary dynamo sustains a too weak magnetic field, propagating Alfvén waves can overlap and the interaction becomes *unipolar* (Laine *et al.* 2008, Laine & Lin 2012). (Strugarek *et al.* 2017) have performed a first study on planetary migration taking into account simultaneously tidal and magnetic torques, by computing the migration timescale of the planet for both contributions. Their study reveals that both effects could play a key role depending on the characteristics of the star-planet system considered. A self-consistent secular evolution of star-planet systems under the influence of magnetic and tidal torques is thus needed to properly disentangle the importance of the two physical mechanisms.

4. Influence of the tidal and MHD torques on the orbital evolution and the stellar rotation

To assess the relative contributions of the tidal and magnetic torques on the star-planet secular evolution, we will assume in the following section that the planet is located in an open-field region to maximize magnetic interactions (for example in the equatorial plane of a star with a quadrupolar magnetic field). The stellar magnetic field is then assumed to be radial ($B_\star \sim 1/r^2$).

The left panel of Figure 2 shows the evolution of the semi-major axis of a Jupiter-like planet, initially at a distance $a = 0.03$ AU, orbiting around a fast rotating solar twin ($P_{\rm rot} = 1.4$ d). The coupled action of the tidal and magnetic torques (in dark blue in Figure 2) shows an outward migration acting at the beginning of the evolution and an evolution of the semi-major axis deviating significantly from the case where tidal effects alone are acting on the system (in red in Figure 2). The orbital evolution of the planet taking into account star-planet magnetic interactions without tidal effects (in light blue in Figure 2) also presents an important outward migration at the beginning of the simulation, then a less pronounced evolution than in the purely tidal case.

The magnetic and tidal torques lead to angular momentum exchange between the orbital motion of the planet and the rotating star. Because of the conservation of the total angular momentum of the system, stellar rotation is also affected by those interactions. The right panel of Figure 2 shows a case of planet destruction, for which a Jupiter-like planet located initially at $a = 0.03$ AU orbits a midly rotating solar twin ($P_{rot} = 5$ d). The influence of planetary migration on stellar rotation is quantified by

$$\delta P = P_{rot} - P_{rot,alone}, \qquad (4.1)$$

where $P_{rot,alone}$ is the rotation period of the star without a planet. In the case of a planetary destruction (mark * in Figure 2), the star presents a sharp spin-up when the planet falls down in the central body. This corresponds to the transfer of its orbital angular momentum to the host-star. The presence of magnetic torques (in dark blue in Figure 2) changes the survival time of the planet, making its destruction happen earlier than in the purely tidal-case. In the case of Figure 2 the planet falls during the PMS, which is at the origin of a briefer and less intense spin-up than for the tide-only case (in red in Figure 2). An evolution only driven by magnetic effects (in light blue in Figure 2) leads to a potential destruction of the planet after the end of the simulation, which is at the origin of a monotonous decrease of the stellar rotation rate. In those three configurations, an overestimated stellar rotation rate due to planetary migration, compared to what it is expected from gyrochronology (Skumanich 1972, Barnes 2003), can last several billions of years.

5. Conclusions

We presented some preliminary results obtained with the ESPEM model comparing the relative importance of tidal and magnetic torques for exosystems' evolution. Depending on the initial configuration of the system, magnetic effects can dominate tidal effects along the evolution of the system, which is in agreement with Strugarek *et al.* (2017). Furthermore, the planet has a significant influence on stellar rotation, especially in the case of an efficient inward migration and a collision during the MS. First statistical studies performed with ESPEM tend to show that initial fast stellar rotation result in the excitation of tidal inertial waves and of a stronger magnetic field, that leads to intense tidal-magnetic effects and therefore to more efficient planetary migration (Teitler & Königl 2014). Studying the evolution of a synthetic population of exoplanets may allow us to explain some features of the distribution of close-in planets (McQuillan *et al.* 2014). We intend to explore such statistical approaches in the near future. With a similar wind prescription as the one used in ESPEM we find that 20% of the exoplanets detected so far are likely to lie in a sub-alfvenic region. The dissipation of tidal gravity waves (Zahn 1975, Goodman & Dickson 1998, Terquem *et al.* 1998) will be also taken into account in future work.

The authors acknowledge funding from the European Union's Horizon-2020 research and innovation programme (Grant Agreements no. 776403 ExoplANETS-A and no. 647383 ERC CoG SPIRE), INSU/PNST, INSU/PNP and the CNES-PLATO grant at CEA.

References

Ahuir, J., Brun, A.-S. & Strugarek, A. 2019, submitted to *A&A*
Amard, L., Palacios, A., Charbonnel, C. *et al.* 2016, *A&A*, 587, A105
Barnes, S.A. 2003, *ApJ*, 586, 464
Benbakoura, M., Réville, V., Brun, A. S. *et al.* 2019, *A&A*, 621, A124
Bolmont, E. & Mathis, S. 2016, CeMDA, 126, 275.
Goodman, J., Dickson, E. S. 1998, *ApJ*, 507, 938
Hansen, B. M. S. 2012, *ApJ*, 757, 6
Laine, R. O. & Lin, D. N. C. 2012, *ApJ*, 745, 2
Laine, R. O., Lin, D. N. C. & Dong, S. 2008, *ApJ*, 685, 521

MacGregor, K. B. 1991, in Angular Momentum Evolution of Young Stars, eds. S. Catalano, & J. R. Stauffer (Dordrecht: Kluwer), 315

MacGregor, K. B. & Brenner, M. 1991, *ApJ*, 376, 204

McQuillan, A., Mazeh, T. & Aigrain, S. 2013, *ApJ*, 775, L11

Mathis, S. 2015, *A&A*, 580, L3

Neubauer, F. M. 1998, *JGR*, 103, 19843

Ogilvie, G. I. 2013, *MNRAS*, 429, 613

Saur, J., Grambusch, T., Duling, S. *et al.* 2013, *A&A*, 552, 119

Schatzman, E. 1962, *AnAp*, 25, 18

Skumanich, A. 1972, *ApJ*, 171, 565

Strugarek, A. 2016, *ApJ*, 833, 140

Strugarek, A., Bolmont, E., Mathis, S. *et al.* 2017, *ApJL*, 847, 2

Strugarek, A., Brun, A. S., Matt, S. P., & Réville, V. 2015, *ApJ*, 815, 111

Teitler, S. & Königl, A. 2014, *ApJ*, 786, 139

Terquem, C., Papaloizou, J. C., Nelson, R. P. *et al.* 1998, *ApJ*, 502, 788

Weber, E. J. & Davis, L. Jr. 1967, *ApJ*, 148, 217

Zahn, J. P. 1975, *A&A*, 41, 329

Solar and Stellar Magnetic Fields: Origins and Manifestations
Proceedings IAU Symposium No. 354, 2019
A. Kosovichev, K. Strassmeier & M. Jardine, ed.
doi:10.1017/S1743921319009761

TESS light curves of low-mass detached eclipsing binaries

Krzysztof G. Hełminiak[1]⬤, Andrés Jordán[2,3], Nestor Espinoza[4,2] and Rafael Brahm[2]

[1]N. Copernicus Astronomical Center, Polish Academy of Sciences,
ul. Rabiańska 8, 87-100, Toruń, Poland
email: xysiek@ncac.torun.pl

[2]Instituto de Astrofísica, Pontificia Universidad Católica de Chile
Av. Vicuña Mackenna 4860, 7820436 Macul, Santiago, Chile

[3]Milleium Institute of Astrophysics
Av. Vicuña Mackenna 4860, 7820436 Macul, Santiago, Chile

[4]Space Telescope Science Institute
3700 San Martin Dr., Baltimore, MD 21218, USA

Abstract. We present high-precision light curves of several M- and K-type, active detached eclipsing binaries (DEBs), recorded with 2-minute cadence by the *Transiting Exoplanet Survey Satellite* (TESS). Analysis of these curves, combined with new and literature radial velocity (RV) data, allows to vastly improve the accuracy and precision of stellar parameters with respect to previous studies of these systems. Results for one previously unpublished DEB are also presented.

Keywords. binaries: eclipsing, binaries: spectroscopic, stars: activity, stars: chromospheres, stars: fundamental parameters, stars: low-mass, stars: spots

1. Introduction

Magnetic fields in low-mass ($<$0.8 $M_\odot$) stars affect the fundamental stellar properties. In short-period, tidally locked binaries fast rotation of components strengthens the magnetic field through a form of a dynamo mechanism, enhances activity, and affects the observed radii and effective temperatures, which has been observed in low-mass detached eclipsing binaries (LMDEB) for decades. Several descriptions of this phenomenon have been proposed, but we lack good quality observational data and models of LMDEBs in order to validate or falsify them. In this paper we present improved, precise results for four already studied and one unpublished LMDEB.

2. Targets

The presented sample consists of one system that was not studied to date, ASAS J125516-3156.7 (A-125), and four LMDEBs already described in the literature. These are: ASAS J011328-3821.1 (A-011; Hełminiak *et al.* 2012), ASAS J030807-2445.6 (AE For; Różyczka *et al.* 2013), ASAS J032923-2406.1 (AK For; Hełminiak *et al.* 2014), and ASAS J093814-0104.4 (A-093; Hełminiak *et al.* 2011). All systems have component masses below 0.8 $M_\odot$, and orbital periods shorter than 4 days. All are very active, with prominent spots, occasional flares, and Hα emission lines. The four targets from literature usually have radii known with precision of ~1.5-2% at best. Masses of A-011 and A-093 were poorly constrained ($>$4%).

3. TESS photometry

In order to obtain high accuracy and precision in stellar parameters, especially radii, one needs a very precise photometry, and the best-quality data come from space borne instruments. The high-precision, 2-minute-cadence time-series photometry of our targets comes from the *Transiting Exoplanet Survey Satellite* (TESS), and was obtained through the Guest investigator program No. G011083 (PI: Hełminiak) during the first year of TESS operations. Detrended light curves were downloaded from the Mikulski Archive for Space Telescopes (MAST). Our targets were mostly observed in one sector, except for A-011 (two sectors). TESS light curves are presented in Figure 1.

4. Spectroscopy and radial velocities

Direct determination of masses of DEBs requires radial velocity (RV) measurements, which are obtained from a series of high-resolution spectra. Our targets were initially included into a large spectroscopic survey of DEBs, identified by the All-Sky Automated Survey (Pojmański 2002).

RVs and orbital solutions of AK For and A-011 remain unchanged with respect to the literature (Hełminiak *et al.* 2012, 2014). In three other cases we used our own new spectroscopy from CHIRON and CORALIE spectrographs, and calculated the RVs with the TODCOR method Zucker & Mazeh (1994). The CHIRON data for A-093 were supplemented with measurements from Hełminiak *et al.* (2011). AE For was already described in Różyczka *et al.* (2013), but we did not use their data. The RVs for A-125 were not published to date. The orbital parameters were found with the code V2FIT (Konacki *et al.* 2010). The observed and model RV curves of AE For, A-093, and A-125 are shown in Figure 2.

5. Light curve modelling

The TESS light curves were modelled with the JKTEBOP code v34 (Southworth *et al.* 2004). To account for the out-of-eclipse modulation coming from spots we applied (in JKTEBOP) a series of sine functions (up to four) and polynomials (up to fifth degree). Because the spot-originated variation may change in time quite rapidly, data were split into several (between 2 and 6) pieces, which were analyzed separately. Parameter errors for each piece were evaluated with a Monte-Carlo procedure. As the final values we adopted weighted averages, and to get final parameter uncertainties, we added in quadrature a median of individual piece errors and the *rms* of individual results. The JKTEBOP models are sjhown as blue lines in Figure 1.

6. Results

In Table 1 we present the most important results of our analysis, including RV semi-amplitudes K, orbital period P and inclination i, and absolute values of masses M and radii R.

The variations in spot pattern, in time scales of single weeks, is the main difficulty in reaching good precision in radii. The behavior of residuals during the eclipses reflect the asymmetries and deviations of the shape of an eclipse from a "clean photosphere" case, and originate from spots on a surface of the eclipsed component. However, thanks to the TESS data, we were able to successfully model the influence of spots, and the uncertainties in radii are few times better than reported in literature. The exception is AK For, where the ratio of radii R_2/R_1 is strongly correlated with the level of third light contamination.

Spots also hamper the RV measurements, introducing additional jitter to the data. Also, components of the shortest-period ($P < 1$ d) systems rotate rapidly. Nevertheless,

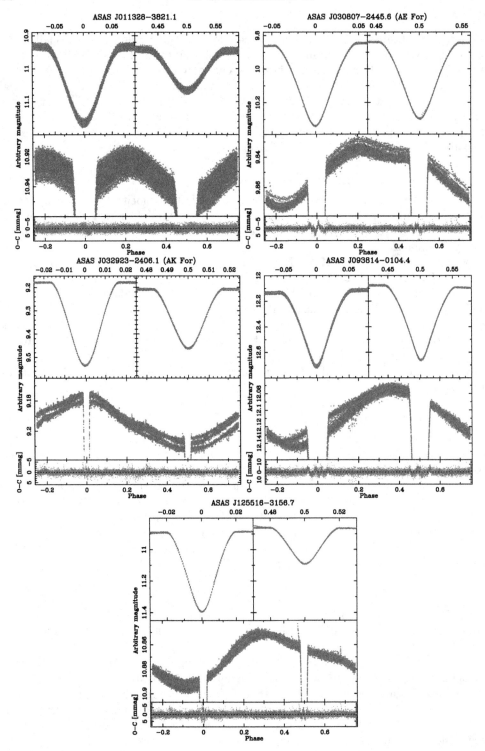

Figure 1. TESS data (red) and JKTEBOP models (blue) of the studied systems, phase-folded with orbital periods. Top rows are zooms on primary (left) and secondary (right) eclipses. Below are zooms on the out-of-eclipse modulations. Bottom panels depict the residuals. One can clearly see the evolution of spots in time, as well as flares on AE For and A-125.

Table 1. Basic orbital and stellar parameters of the studied systems.

| ASAS ID | 011328-3832.1 | 030807-2445.6 | 030807-2445.6 | 032923-2406.1 | 125516-3156.7 |
TIC	183596242	88479623	144539611	14307980	103683084
TESS Sector	2,3	4	4	8	10
P [d]	0.44559604(18)	0.918207(7)	3.9809620(45)	0.897420(2)	3.0570393(44)
K_1 [km/s]	118.4(2.0)	118.3(5)	70.47(3)	127.55(68)	73.34(5)
K_2 [km/s]	162.9(3.3)	119.5(5)	77.16(5)	127.62(97)	87.00(15)
i [°]	87.5(1.5)	87.8(1)	87.37(3)	86.87(6)	87.56(7)
M_1 [$M_\odot$]	0.597(28)	0.644(6)	0.696(1)	0.775(12)	0.7104(25)
M_2 [$M_\odot$]	0.434(17)	0.638(6)	0.6356(7)	0.774(10)	0.5989(13)
R_1 [$R_\odot$]	0.607(12)	0.674(7)	0.684(18)	0.774(6)	0.669(4)
R_2 [$R_\odot$]	0.445(12)	0.617(10)	0.628(20)	0.771(6)	0.557(8)

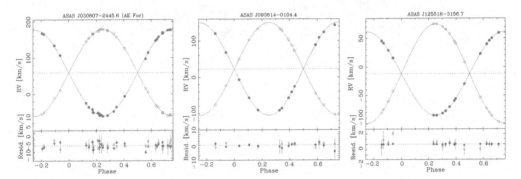

Figure 2. RV measurements (red) and model curves (blue) of three systems with our CORALIE and CHIRON observations, phase-folded with their respective orbital periods. Corresponding residuals are shown in lower panels. Solid symbols represent data for primaries, while open for secondaries. Four points with largest error bars in A-093 are data taken from Hełminiak *et al.* (2011).

new mass determination is also quite good. Errors in masses for AE For are quite low (<1%), yet larger than those from Różyczka *et al.* (2013), which is probably due to larger amount of their RV measurements. The new mass uncertainties of A-093 are 2-3 times better than in Hełminiak *et al.* (2011), at the level of 1.5%. For the new system A-125 all properties are derived with high precision.

Introduction of high-precision TESS photometry allows to improve our knowledge on the smallest, most active stars, where magnetic fields and rotation strongly influence the observed properties. The five binary systems presented here are only a sample of ~40 DEBs with K- and M-type components observed by TESS in our GI programs. Publications of the first set of final solutions is scheduled for mid-2020.

Acknowledgements

K. G. H. acknowledges support provided by the Polish National Science Center through grant 2016/21/B/ST9/01613. N. E. would like to thank the Gruber Foundation for its generous support to this research.

References

Hełminiak, K. G., Konacki, M., Złoczewski, K. *et al.* 2011, *A&A*, 527, A14
Hełminiak, K. G., Konacki, M., Różyczka, K. *et al.* 2012, *MNRAS*, 425, 1245
Hełminiak, K. G., Brahm, R., Ratajczak, M. *et al.* 2014, *A&A*, 567, A64
Konacki, M., Mutterspaugh, M. W., Kulkarni, S., & Hełminiak, K. G. 2010, *ApJ*, 719, 1293

Pojmański, G. 2002 *AcA*, 52, 397

Różyczka, K., Pietrukowicz, P., Kałużny, J., Pych, W., Angeloni, R., & Dékány, I. 2013, *MNRAS*, 429, 1840

Southworth, J., Maxted, P. F. L., & Smalley, B. 2004, *MNRAS*, 351, 1227

Zucker, S., & Mazeh, T. 1994, *ApJ*, 420, 806

Solar and Stellar Magnetic Fields: Origins and Manifestations
Proceedings IAU Symposium No. 354, 2019
A. Kosovichev, K. Strassmeier & M. Jardine, ed.
doi:10.1017/S1743921319009773

Tuning in to the radio environment of HD189733b

R. D. Kavanagh[1], A. A. Vidotto[1], D. Ó Fionnagáin[1], V. Bourrier[2], R. Fares[3,4], M. Jardine[5], Ch. Helling[6], C. Moutou[7], J. Llama[8] and P. J. Wheatley[9]

[1]School of Physics, Trinity College Dublin, The University of Dublin, Dublin 2, Ireland
email: `kavanar5@tcd.ie`

[2]Observatoire de l'Université de Genéve, Chemin des Maillettes 51, Versoix, CH-1290, Switzerland

[3]Physics Department, United Arab Emirates University, P.O. Box 15551, Al-Ain, United Arab Emirates

[4]University of Southern Queensland, Centre for Astrophysics, Toowoomba, Queensland, 4350, Australia

[5]SUPA, School of Physics and Astronomy, University of St Andrews, North Haugh, St Andrews, Fife, Scotland, KY16 9SS

[6]Centre for Exoplanet Science, University of St Andrews, St Andrews KY16 9SS, UK

[7]CNRS/CFHT, 65-1238 Mamalahoa Highway, Kamuela HI 96743, USA

[8]Lowell Observatory, 1400 W. Mars Hill Rd, Flagstaff. AZ 86001. USA

[9]Department of Physics, University of Warwick, Coventry CV4 7AL, UK

Abstract. The hot Jupiter HD189733b is expected to be a source of strong radio emission, due to its close proximity to its magnetically active host star. Here, we model the stellar wind of its host star, based on reconstructed surface stellar magnetic field maps. We use the local stellar wind properties at the planetary orbit obtained from our models to compute the expected radio emission from the planet. Our findings show that the planet emits with a peak flux density within the detection capabilities of LOFAR. However, due to absorption by the stellar wind itself, this emission may be attenuated significantly. We show that the best time to observe the system is when the planet is near primary transit of the host star, as the attenuation from the stellar wind is lowest in this region.

Keywords. stars: individual (HD189733), stars: magnetic fields, stars: winds, outflows, stars: planetary systems

---

1. Introduction

Close-in hot Jupiters are expected to be sources of strong auroral radio emission, analogous of what is observed for the magnetised solar system planets (Zarka *et al.* 2001). This is thought to occur due to magnetic interactions between the stellar wind of the host star and the intrinsic magnetic field of the orbiting planet. However, despite the large number of hot Jupiters detected to date, along with numerous radio surveys, no sources of exoplanetary radio emission have been detected (Smith *et al.* 2009; Lazio *et al.* 2010; Lecavelier des Etangs *et al.* 2013; Sirothia *et al.* 2014; O'Gorman *et al.* 2018).

HD189733b is one such exoplanet that is expected to emit strong low frequency radio emission. The planet orbits its host star just 0.03 au. The host star is magnetically active,

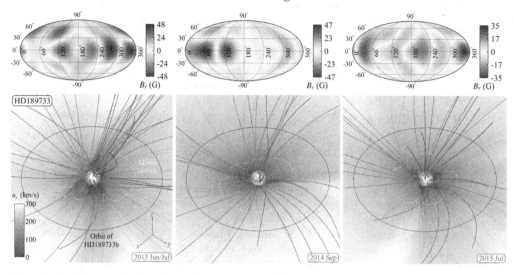

Figure 1. *Top panels:* Radial surface magnetic field maps of the host star reconstructed by Fares *et al.* (2017), at the epochs 2013 Jun/Jul, 2014 Sep, and 2015 Jul (left to right). These maps are used as boundary conditions in our stellar wind simulations. *Bottom panels:* Simulated stellar wind of the host star at 2013 Jun/Jul, 2014 Sep, and 2015 Jul (left to right). Grey lines show the large-scale structure of the magnetic field of the star, which is embedded in the stellar wind. Profiles of the radial velocity of the stellar wind in the orbital plane of the planet are shown. The planetary orbit is shown with a black circle, and Alfvén surfaces are shown in white.

with its unsigned field strength observed to vary from 18 to 42 G over a 9 year period (Fares *et al.* 2017). Here, we model the stellar wind of the host star, and use the stellar wind properties obtained from the models to predict the flux density and frequency emitted by the planet. This emission is found to be within the detection limit of LOFAR. However, we also find that the emission may be attenuated significantly by the stellar wind of the host star. A complete description of our work is published in Kavanagh *et al.* (2019).

2. Modelling the stellar wind of the host star

To model the wind of the host star, we perform 3D magnetohydrodynamic simulations using the BATSRUS code developed by Powell *et al.* (1999), modified by Vidotto *et al.* (2012). We use surface stellar magnetic field maps reconstructed from observations by Fares *et al.* (2017) as boundary conditions in our simulations, at the epochs 2013 Jun/Jul, 2014 Sep, and 2015 Jul. In our models, we adopt a coronal base density of 2 MK and number density of 10^{10} cm^{-3}. These values produce a stellar wind with a mass-loss rate of 3×10^{-12} $M_\odot$ yr^{-1}, which is within the range of inferred values for other active K-stars (see Wood 2004; Jardine & Collier Cameron 2019; Rodríguez *et al.* 2019).

Figure 1 shows the radial component of each stellar surface magnetic field map, with the corresponding wind simulation shown below each map. We see that over the modelled timescale, the stellar wind varies in response to the varying surface magnetic field.

3. Predicting radio emission from HD189733b

Using the local stellar wind properties obtained from our simulations, we use the radiometric Bode's law to compute the expected flux density and frequency of emission from the planet, for an assumed planetary magnetic field strength (see Vidotto & Donati 2017). In the model, 0.2% of the incident stellar wind's magnetic power is converted into

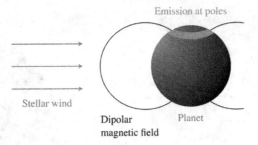

Figure 2. Sketch illustrating the stellar wind incident on the magnetic field of the planet. The interaction results in radio emission from polar cap regions near the surface.

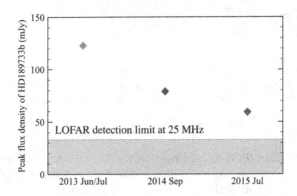

Figure 3. Peak radio flux densities emitted by the planet at each modelled epoch, for a field strength of 10 G.

radio power from the planet (Zarka 2010). The planet's magnetic field is assumed to be dipolar. This is illustrated in Figure 2.

Figure 3 shows the predicted peak flux densities received at Earth from HD189733b at 2013 Jun/Jul, 2014 Sep, and 2015 Jul, computed using the stellar wind properties at the planet's orbit obtained from our models. For an assumed planetary magnetic field strength of 10 G, we find that this emission occurs at a frequency of 25 MHz. Due to the variability of the stellar wind over the three modelled epochs, the peak flux densities from the planet also vary. At 25 MHz, the emission predicted from HD189733b place it above the detection limit of LOFAR for a 1 hour integration time (Grießmeier *et al.* 2011).

4. Absorption of the planetary radio emission in the stellar wind of the host star

While we predict that radio emission from HD189733b could be detected with LOFAR, the stellar wind itself can absorb low frequency radio emission (Panagia & Felli 1975). Here we solve the equations of radiative transfer for the stellar wind, using the numerical code developed by Ó Fionnagáin *et al.* (2019). We find that the planet orbits through regions of the stellar wind that are optically thick to the predicted frequency emitted from the planet. This is illustrated in Figure 4. As a result, emission from HD189733b may only be observable as the planet approaches and leave primary transit of the host star. This could be useful information for timing future radio observing campaigns in search of exoplanetary radio emission from systems similar to HD189733b.

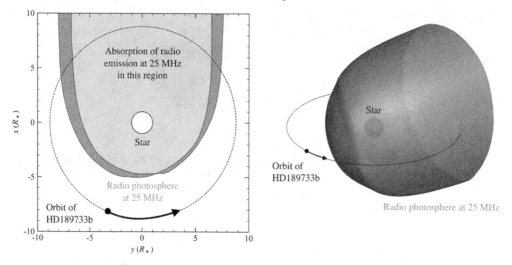

Figure 4. The planet orbits through the radio photosphere of the stellar wind, the region optically thick to the emitted planetary frequency of 25 MHz. The left panel shows the shape of the radio photosphere at 25 MHz in the orbital plane, and the right shows its shape in 3D.

5. Conclusions

The hot Jupiter HD189733b indeed may be a good target for detecting exoplanetary radio emission. However, as we have shown, the stellar wind can in fact absorb this emission for a large fraction of the planet's orbit. The best time to observe the system is when the planet is near primary transit of the host star. This is also applicable to other exoplanetary systems similar to HD189733b.

Acknowledgements

RDK acknowledges funding received from the Irish Research Council through the Government of Ireland Postgraduate Scholarship Programme. RDK and AAV also acknowledge funding received from the Irish Research Council Laureate Awards 2017/2018. VB acknowledges support by the Swiss National Science Foundation (SNSF) in the frame of the National Centre for Competence in Research PlanetS, and has received funding from the European Research Council (ERC) under the European Union's Horizon 2020 research and innovation programme (project Four Aces; grant agreement No 724427). This work was carried out using the BATSRUS tools developed at The University of Michigan Center for Space Environment Modeling (CSEM) and made available through the NASA Community Coordinated Modeling Center (CCMC). The authors also wish to acknowledge the SFI/HEA Irish Centre for High-End Computing (ICHEC) for the provision of computational facilities and support.

References

Fares, R., Bourrier, V., Vidotto, A. A., *et al.* 2017, *MNRAS*, 471, 1246
Grießmeier, J. M., Zarka, P., & Girard, J. N. 2011, *Radio Science*, 46, RS0F09
Jardine, M. & Collier Cameron, A. 2019, *MNRAS*, 482, 2853
Kavanagh, R. D., Vidotto, A. A., Ó. Fionnagáin, D., *et al.* 2019, *MNRAS*, 485, 4529
Lazio, T. J. W., Shankland, P. D., Farrell, W. M., & Blank, D. L. 2010, *AJ*, 140, 1929
Lecavelier des Etangs, A., Sirothia, S. K., Gopal-Krishna, & Zarka, P. 2013, *A&A*, 552, A65
Ó Fionnagáin, D., Vidotto, A. A., Petit, P., *et al.* 2019, *MNRAS*, 483, 873
O'Gorman, E., Coughlan, C. P., Vlemmings, W., *et al.* 2018, *A&A*, 612, A52

Panagia, N. & Felli, M. 1975, *A&A*, 39, 1

Powell, K. G., Roe, P. L., Linde, T. J., Gombosi, T. I., & De Zeeuw, D. L. 1999, *Journal of Computational Physics*, 154, 284

Rodríguez, L. F., Lizano, S., Loinard, L., *et al.* 2019, *ApJ*, 871, 172

Sirothia, S. K., Lecavelier des Etangs, A., Gopal-Krishna, Kantharia, N. G., & Ishwar-Chandra, C. H. 2014, *A&A*, 562, A108

Smith, A. M. S., Collier Cameron, A., Greaves, J., et al. 2009, *MNRAS*, 395, 335

Vidotto, A. A. & Donati, J. F. 2017, *A&A*, 602, A39

Vidotto, A. A., Fares, R., Jardine, M., *et al.* 2012, *MNRAS*, 423, 3285

Wood, B. E. 2004, Living Reviews in Solar Physics, 1, 2

Zarka, P. 2010, in Astronomical Society of the Pacific Conference Series, Vol. 430, Pathways Towards Habitable Planets, ed. V. Coudé du Foresto, D. M. Gelino, & I. Ribas, 175

Zarka, P., Treumann, R. A., Ryabov, B. P., & Ryabov, V. B. 2001, *AP&SS*, 277, 293

Chapter 7. Formation, structure and dynamics of solar and stellar coronae and winds

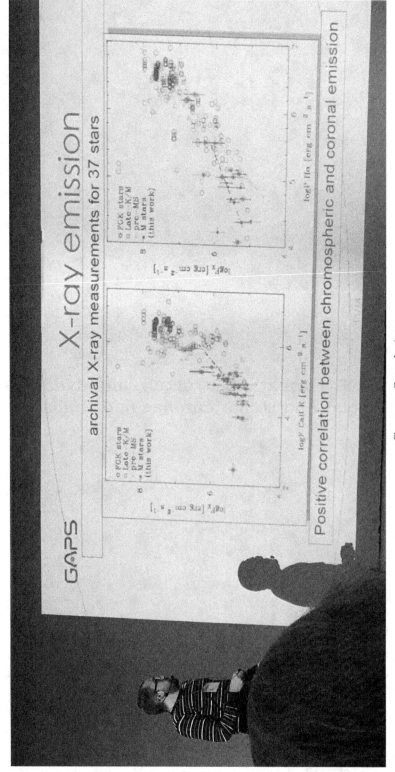

Gaetano Scandariato

Solar and Stellar Magnetic Fields: Origins and Manifestations
Proceedings IAU Symposium No. 354, 2019
A. Kosovichev, K. Strassmeier & M. Jardine, ed.
doi:10.1017/S1743921320001441

Observational constraints for solar-type Stellar winds

Manuel Güdel [ORCID]

University of Vienna, Department of Astrophysics
Türkenschanzstr. 17, 1180 Vienna, Austria
email: `manuel.guedel@univie.ac.at`

Abstract. Ionized winds from late-type main-sequence stars are important for stellar spin-down and therefore the evolution of stellar activity; winds blow an "astrosphere" into the interstellar medium that absorbs a large part of galactic cosmic rays; and the winds play a key role in shaping planetary environments, in particular their upper atmospheres. These issues have been well studied for the solar wind but little is known about winds escaping from other solar-type stars. Several methods have been devised to either detect winds directly or to infer the presence of such winds from features that are shaped by the winds. This paper summarizes these methods and discusses exemplary findings. There is need for more studies using multiple methods for the same stars.

Keywords. stars: winds, outflows; stars: rotation; radio continuum: stars; ultraviolet: stars

---

1. Introduction

The solar wind and by analogy ionized winds from cool main-sequence stars matter for many aspects of stellar and planetary evolution. Some of the most important roles of stellar winds are:

- Stellar braking and spin-down due to angular momentum removal by magnetized winds;
- Erosion of planetary atmospheres due to their interactions with stellar winds;
- Chemical processing of planetary atmospheres due to high-energy particles transported in the wind;
- Potential influence on the stellar luminosity evolution as the stellar mass may significantly decrease with time;
- Protection from a large part of galactic cosmic ray flux.

The solar wind, although directly accessible only since the beginning of the space age, is relatively well studied out to its limits at around 120 au where it is terminated in the region of interaction with the interstellar medium. Nevertheless, many aspects of the solar wind require further study, in particular,

- the heating and wind acceleration mechanism(s);
- the heating/cooling behavior across the heliosphere;
- the long-term evolution of the wind mass-loss rate which is possibly related to the magnetic dynamo operation and therefore rotation.

The solar wind is studied both *in situ* and remotely, e.g. using optical observations. In contrast, observing winds from cool main-sequence stars has turned out to be extremely difficult. In essentially all wavelength ranges, the emission from the star outshines the wind emission by far. Wind emission from far outside the stellar surface or absorption effects induced by the wind may be alternative ways to detect stellar mass loss.

On the other hand, *indirect* methods have been developed that observe easily accessible features around the star that are controlled by the wind, but that are not themselves part of the wind. Such methods have been more successful in estimating wind mass-loss rates although they often depend on assumptions and complex models linking the unobserved wind to the observed features.

This paper summarizes a variety of stellar-wind detection methods and presents results from corresponding observations. Only detection methods that can presently be applied to observations and that have provided useful results are discussed here. I do not aim to discuss wind physics beyond what is of immediate relevance to the models; nor do I intend to be comprehensive in presenting a review of all published results. For some methods, in particular those inferring wind mass loss from stellar spin-down, separate comprehensive reviews have been presented elsewhere.

2. Evidence for Winds from Stellar Spin-Down

The most fundamental evidence for the presence of ionized stellar winds is stellar spin-down. Spin-down is well documented by large surveys of rotation periods as a function of age (e.g., Irwin *et al.* 2011). Stars spin down because magnetic fields immersed in the wind plasma act as a lever arm on the stellar surface. A packet of mass leaving the stellar surface and moving out along a magnetic field line gains angular momentum as long as the field lines drag the gas mass and build up magnetic stresses, which is until it reaches the Alfvén surface (or Alfvén distance) where the Alfvén velocity drops below the wind speed and the gas drags the magnetic fields. Effectively, the angular momentum gained by reaching the Alfvén distance is extracted at the stellar surface, thus spinning down the star. This is a simplified sketch for a more gradual change from the magnetic-field dominated subalfvénic inner region to the superalfvénically expanding wind region (for an early theoretical model of angular momentum transport in a rotating solar wind, see Weber & Davis 1967).

A vast amount of literature has accumulated over the past several decades about many aspects of spin-down, angular momentum transfer from the star to the wind, and surveys of rotational evolution and rotation period distributions at given ages. I do not intend to review any of these topics here but summarize a few key findings before focusing on the spin-down evolution of a cool main-sequence star, exemplified by one model that focuses on this specific aspect rather than the reconstruction of all the underlying physical mechanisms (e.g., core-envelope decoupling, wind acceleration, torque formulae for magnetized flows, etc.).

The reality of a solar wind was recognized after extensive theoretical and predictive work by Parker (1958) and evidence provided by cometary tails (Biermann 1951), and finally *in situ* observations by space probes around 1960. The theoretical and observational basis to generalize winds to other stars was rapidly developed in (among others) three key papers I mention here briefly. In 1967, Kraft found that the average rotational velocities of stars with strong Ca II emission are higher than for weak Ca emitters. Because strong Ca II emitters were known to be younger, Kraft suggested that stars spin down with age, making magnetically coupled winds like the solar wind responsible for the angular momentum loss (Kraft 1967). He also recognized that this picture should apply to stars with outer convection zones. In the same year, Weber & Davis (1967) developed a theoretical model for the angular momentum transfer via magnetic wind torques. And in 1972, Skumanich summarized the observational aspects in a key paper (Skumanich 1972, 1.5 pages of text, 1 figure, 10 references) that laid the foundations of subsequent stellar statistical studies of age, evolution of activity, and rotation. He claimed that Ca II emission declines with the inverse square root of the age, and the same should hold for the rotation velocity. Given the proportionality between surface magnetic field strength

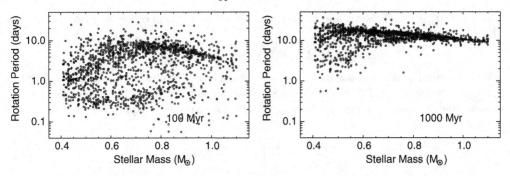

Figure 1. Snapshots of the rotation period distribution evolving in time. Rotation periods of a sample of ~1500 observed stars from clusters with known ages have been used to develop a spin-down/wind/rotation evolution model. These same stars can then be evolved forward and backward in time. Shown here are two snapshots for ages of 100 Myr (left) and 1 Gyr (right). (From Johnstone *et al.* 2015.)

and Ca II emission, the magnetic fields are also supposed to decay with the inverse square root of the age. While we now know that corrections are needed in this rough picture, age-rotation-activity relations have been a focus in open cluster studies ever since.

Models of stellar rotational evolution have been developed by many groups, for the purpose of studying basic physical concepts and test them with observations (e.g., Gallet & Bouvier 2013), or for the purpose of reproducing phenomenologically the long-term evolution of stellar rotation. To conceptually describe the relevant model steps, I confine myself to one example model that comprises many relevant aspects, published by Johnstone *et al.* (2015).

Large observational samples from open clusters have shown a wide spread of rotation periods P for any given mass back to very early evolutionary stages, essentially starting in the T Tauri phase already. While the stars spin up during the pre-main sequence phase due to contraction, they spin down during their main-sequence life but still keep their wide spread in P until, at around 0.5–1 Gyr, most of them converge to a unique rotation period only depending on age and stellar mass (Soderblom *et al.* 1993; Fig. 1). This *convergence* is ascribed to a feedback between angular momentum removal by a magnetized coronal wind and the rotationally induced operation of an internal dynamo that generates the magnetic fields. Given that magnetized and ionized winds appear to be related to stellar coronae, one suspects that they, like other magnetic activity indicators, also decline with time. The goal then is to use the observed rotational evolution to derive the evolution of the wind mass-loss rate $\dot{M}$, even if the evolutionary tracks $P(t)$ are non-unique given the widely dispersed initial conditions.

The basic ingredients of a rotational evolution model are the initial rotation rate, the internal structure of the star, and the rate at which angular momentum is removed from the star by the wind. Rotational models also include core-envelope decoupling. This is relevant when the time scale for angular momentum transport within the star is significant compared to the time scale over which the moment of inertia of the star changes and the time scale over which angular momentum is removed from the stellar surface by the wind. Due to this mechanism, stars arriving on the Zero-Age Main-Sequence (ZAMS) have cores that rotate more rapidly than the surfaces.

We now need a set of equations to solve for the angular rotation rate $\Omega_*(t)$, the wind mass-loss rate $\dot{M}(t)$, and the surface magnetic field $B(t)$. Spin-down occurs because the rotating magnetic field transfers angular momentum to the wind flow that therefore removes it from the stellar surface. A numerical recipe therefore needs a formula for the torques that act on the star, and these torques τ are a function of stellar mass M_*, radius R_*, $\dot{M}$, B, and Ω_*,

$$\tau = f(B, \dot{M}, M_*, R_*, \Omega_*) \approx B^{0.87} \dot{M}^{0.56} R_*^{2.87} \Omega_* \tag{2.1}$$

where the numerical example on the right-hand side is simplified from a torque formula given by Matt *et al.* (2012). Such torque formulae are derived from simulations using some assumptions about the magnetic-field structure. The torque is required in the spin-down formula,

$$\frac{d\Omega_*}{dt} = \frac{1}{I_*} \left(\tau - \frac{dI_*}{dt} \right) \Omega_*, \tag{2.2}$$

where I_* is the star's moment of inertia derived from stellar structure models; we assume a dependency of $\dot{M}$ on R_*, Ω_*, and M_*,

$$\dot{M} = \dot{M}_\odot \left(\frac{R_*}{R_\odot} \right)^2 \left(\frac{\Omega_*}{\Omega_\odot} \right)^a \left(\frac{M_*}{M_\odot} \right)^b \tag{2.3}$$

where the exponents a and b will be fitted to observational constraints. We further need an equation for, in the simplest case, the equatorial magnetic field strength of the dominant dipole component,

$$B = B_\odot \left(\frac{\Omega_* \tau_*}{\Omega_\odot \tau_\odot} \right)^{1.32} \tag{2.4}$$

(Vidotto *et al.* 2014); τ_* is the convective turnover time of the star. We need to also consider that B "saturates" for rapidly rotating stars, i.e., is no longer a function of Ω. By implication, $\dot{M}$ should also saturate, as do many magnetic activity indicators such as the X-ray luminosity. The onset of saturation should be described by an equation like

$$\Omega_{\text{sat}}(M_*) = \Omega_{\text{sat}}(M_\odot) \left(\frac{M_*}{M_\odot} \right)^c \tag{2.5}$$

where c is a further fit parameter, and $\Omega_{\text{sat}}(M_\odot) \approx 15 \Omega_\odot$. Equations (2.1)–(2.5) can be solved with fits to observations of rotation period distributions for a given mass at different ages. The evolutionary tracks $\Omega(t)$ and $\dot{M}(t)$ will depend on the initial condition $\Omega(t_0)$ taken from observed Ω distributions.

The results of this wind treatment are rotational evolution tracks and wind mass-loss tracks. The constants a, b, and c are 1.33, –3.36, and 2.3 in Johnstone *et al.* (2015). Examples of results are shown in Fig. 2. The rotational evolution of solar-mass stars converges to a unique age-dependent value after $\sim$700 Myr, while at 100 Myrs of age rotation periods are distributed over more than an oder of magnitude (between the 10th and the 90th percentiles). For stars older than ~ 700 Myr, the mass-loss rate declines roughly as

$$\dot{M} \propto t^{-0.75}, \tag{2.6}$$

t being the age. The mass-loss rate is also related to the rotation period as

$$\dot{M} \propto R_*^2 \, M_*^{-3.36} \, \Omega_*^{1.33} \tag{2.7}$$

showing that the mass-loss rate decreases strongly with stellar mass but increases with rotation velocity; the saturation mass-loss rate is

$$\dot{M}_{\text{sat}} = 37 \dot{M}_\odot \left(\frac{M_*}{M_\odot} \right)^{1.3}, \tag{2.8}$$

which means that higher-mass stars can achieve higher mass-loss rates. Saturation of $\dot{M}$ limits the mass-loss rates of young, very rapid rotators, as shown in Fig. 2.

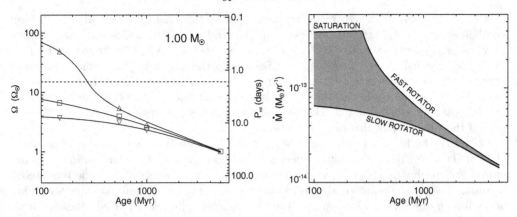

Figure 2. *Left:* Modeled rotational evolution of stars at the 10th, 50th, and 90th percentiles of the rotational distribution from 100 Myr to 5 Gyr. The blue symbols mark the same percentiles from observational distributions. The dashed line indicates the limiting rotation rate above which the wind mass-loss rate and the surface magnetic field saturate. – *Right:* The evolution of the solar wind mass loss rate with age on the main sequence. The green area includes all possible $\dot{M}$ evolutionary tracks for different initial rotation periods between the 10th percentile (bottom edge) and the 90th percentile (top edge). The fastest rotators remain saturated at a maximum $\dot{M}$ level during the first 300 Myr. (From Johnstone *et al.* 2015.)

For solar-type stars younger than ~ 700 Myr, the $\dot{M}$ history is non-unique and depends on the initial rotation period of the star. The result in Fig. 2-right shows that $\dot{M}$ of solar analogs is distributed over nearly an order of magnitude (10th to 90th percentile, green area) near the Zero-Age Main-Sequence (ZAMS) age, due to a broad distribution of initial rotation periods. Similar results were reported for lower-mass stars by Johnstone *et al.* (2015).

Specific wind models can be developed for individual stars if enough information is available about the stellar magnetic field and the stellar fundamental properties such as the rotation period. Vidotto *et al.* (2011b) developed a 3-D magnetohydrodynamic coronal-wind model using an observed stellar surface magnetic-field map for the M dwarf V374 Peg, and also applying magnetocentrifugal forces. The winds in this simulation reach final velocities of $(1500 - 2300)n_{12}^{-1/2}$ km s^{-1}, n_{12} being the coronal base density in units of 10^{12} cm^{-3}. The mass-loss rate then is $\dot{M} = 4 \times 10^{-10}\, n_{12}^{-1/2} M_\odot$ yr^{-1} where the corresponding angular momentum loss suggests that $n_{12} \lesssim 0.1$.

3. Hydrogen Walls and Lyα Absorption

The solar wind blows an ionized "bubble" into the surrounding interstellar medium (ISM), known as the heliosphere. The heliosphere interacts with the interstellar medium across three boundaries. At the termination shock, the supersonically expanding stellar wind decelerates to a subsonic flow. The (variable) location of the termination shock is known from measurements made by Voyager 1 and 2 that found it to lie at 94 au and 84 au, respectively (Stone *et al.* 2008), from the Sun, in different but roughly upstream directions as seen from the Sun. Further out, at approximately 120 au for both Voyagers (Stone *et al.* 2019), the heliopause separates the solar wind from the flow of the interstellar medium. Still further away from the Sun, a bow shock could develop in the interstellar medium where the flow changes from supersonic to subsonic. This depends on the relative velocity between the Sun and the ISM, and recent measurements show that the motion is subsonic and therefore no bow shock should form (McComas *et al.* 2012).

If the ISM contains a neutral component – as is the case in the local ISM around the heliosphere – the neutral particles can penetrate into the heliosphere where they may undergo charge exchange (CX) reactions (see Sect. 4). A hot hydrogen wall builds up outside the heliopause absorbing significantly in the Lyα line. The same mechanism supposedly also occurs around nearby stars, and it is the observation of Lyα absorption from the *stellar* hydrogen wall that is the basis of an indirect method to quantify *stellar winds*, as largely worked out by B. Wood and colleagues (see review by Wood 2004 that guided the summary of this present section).

The method relies on a very accurate analysis of the line profile of the Lyα lines from H I and D I at $\sim$1216 Å. The usually very strong Lyα emission line is formed in the stellar chromosphere, but is also subject to strong absorption by neutral H in the interstellar medium, leading to complete absorption in the line center while only the line wings are accessible to observation. In contrast, the D I absorption line is displaced blueward from the H I line by 0.33 Å and is easily detected on the H I line wings. Observations of these profiles for the α Cen system showed a discrepancy in that the H I absorption revealed excess broadening and redshift by 2.2 km s^{-1} relative to D I (Linsky & Wood 1996). The excess was attributed to the presence of a relatively small column of hot neutral hydrogen (temperature of 30,000 K) around the equivalent of a heliosphere around α Cen, i.e., a neutral hydrogen wall around α Cen's "astrosphere", but the excess should also be influenced by the heliospheric H wall.

The heliospheric H wall produces excess absorption on the red side of the Lyα ISM absorption profile because the H wall material moves slower relative to the Sun than the ISM that imprints the ISM Lyα absorption profile. Conversely, the astrospheric excess absorption acts on the blue side of the Lyα absorption.

To make progress, the neutral H walls need to be related to the the helio-/astrospheres and therefore to solar and stellar winds by means of hydrodynamic simulations. Gayley *et al.* (1997) showed that the red excess absorption is consistent with heliospheric models. Further sightlines provide more absorption data that are reasonably well explained by a heliospheric model, opening up the possibility to infer ISM properties from the observations. However, results for the ISM are strongly model dependent considering the various regions of charge exchange in a multi-fluid or kinetic approach.

Analogously, the Lyα excess absorption from the astrospheric H wall is due to the presence of a stellar wind, and fluid or kinetic models can be used to quantify the stellar mass loss. To do so, the ISM flow must be determined in the rest frame of the star. Then, the wind mass-loss rates of the star are varied in the simulations until the excess absorption is well fitted. This then leads to an estimate of the stellar mass-loss rate, $\dot{M}$.

The Lyα absorption method has so far provided an appreciable number of indirect wind mass-loss estimates for a variety of stars, as summarized recently by Wood (2018). Before summarizing the systematics in the measurements of $\dot{M}$, I will mention some caveats of the method (see Wood 2018):

- Results depend on accurate numerical models of the highly complex interaction between the partly neutral ISM and the stellar (and solar) wind. However, models can be calibrated with the well studied *solar* wind.
- The local ISM properties around the target stars need to be sufficiently well understood; variations between nearby stars are deemed to be modest, however.
- The wind speeds are assumed to be solar in all models. This may be related to similar escape speeds for the observed stars, values that are similar to wind speeds, although rapidly rotating stars may eject much faster winds due to magneto-centrifugal acceleration (Holzwarth & Jardine 2007).
- The method applies only if the ISM around the target star is partially neutral. Absence of excess absorption may therefore not signify a weak wind but could be due to

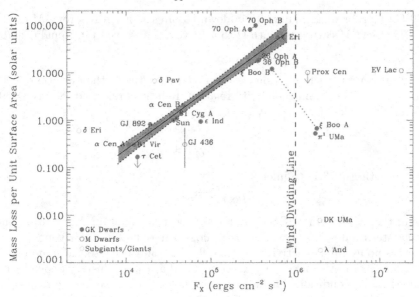

Figure 3. Stellar mass-loss rates per unit surface area as a function of the measured surface X-ray flux. The red filled, red open, and and green open circles mark main-sequence G&K stars, M dwarfs, and evolved stars, respectively. The gray band is a power-law fit with the uncertainty range for the main-sequence stars up to a maximum surface X-ray flux of 8×10^5 erg cm^{-2}s^{-1}. (From Wood 2018.)

the absence of a astrospheric hydrogen wall. Wood (2018) reports that within 7 pc of the Sun, most stars show excess absorption, while the percentage of detections drops rapidly beyond ~ 10 pc, indicating that the ISM at such distances is fully ionized.

- High ISM absorption may obscure the astrospheric excess absorption completely.
- Low ISM velocity relative to the star leads to cooler and less decelerated astrospheric H I, making the absorption narrower and difficult to separate from the ISM absorption.
- Very high ISM velocities, instead, may compress the astrosphere and heat the H wall sufficiently to make Lyα optically thin, complicating the interpretation further.

Despite some of the above caveats, the results do show important systematics. It is reasonable to correlate $\dot{M}$ with a stellar coronal activity parameter because the winds originate in the coronal region of a star, where magnetic fields are open rather than closed as in active regions. Plotting inferred stellar mass loss rates per stellar surface unit area against the average X-ray surface flux F_X (Fig. 3) finds the best-fit relation

$$\dot{M} \propto F_X^{1.34\pm0.18} \tag{3.1}$$

(Wood *et al.* 2005). This relation applies to G and K-type main-sequence stars with F_X up to about 20 times the solar F_X; for higher activity levels, $\dot{M}$ seems to be much lower than the extrapolated trend. There are some caveats here, however. Two of the detections among these very active objects are giants, and two more are small M dwarfs. The two active G/K targets that support very low $\dot{M}$ are ξ Boo A+B, a binary for which a precise separation of the contribution of A or B is not possible.

If the breakdown of $\dot{M}$ toward very active stars is real, an explanation may be sought in the magnetic field topology of the corona, e.g., dipolar magnetic fields anchored in large polar spots, inhibiting stellar winds (Wood 2018). This is problematic, however, in the light of strong spin-down exactly in the age range where active stars abound, i.e., for ages of 0.1–1 Gyr for solar analogs. Spin-down requires sufficiently strong $\dot{M}$ as discussed in Sect. 2; in fact, $\dot{M}$ correlates with Ω to achieve the observed rotational distributions,

$$\dot{M} \propto \Omega^{1.33}. \tag{3.2}$$

(see Eq. 2.7). Knowing the relation between $\dot{M}$ and activity, the evolution of $\dot{M}$ could be inferred if there were an age-activity relation. Relations reported by Ayres (1997) for the equatorial rotational velocity v_{rot},

$$v_{\text{rot}} \quad \propto \quad t^{-0.6\pm0.1} \tag{3.3}$$
$$F_{\text{X}} \quad \propto \quad v_{\text{rot}}^{2.9\pm0.3} \tag{3.4}$$

can be used with Eq. (3.1) to infer

$$\dot{M} \propto t^{-2.33\pm0.55}. \tag{3.5}$$

Because of the breakdown of Eq. (3.1) for the most active stars, Eq. (3.5) applies only for ages greater than ~700 Myr for solar analogs. I note here that for younger solar-type stars there is no useful age-activity relation like Eq. (3.3) because the spin-down history and therefore the rotational velocity history as well as the F_{X} history depend on (widely distributed) initial conditions, making gyrochronology and activity-age relations invalid at such ages (Johnstone *et al.* 2015; Tu *et al.* 2015; Sect. 2). The results above imply that at ~700 Myr of age for a solar analog, the wind mass-loss rate is about a factor of 100 higher than in the present-day Sun.

Note also the anomalously low data point for GJ 436, an M3 dwarf for which, however, absorption by an evaporating planetary atmosphere was used (see Sect. 9 below). Together with the M dwarfs above the "Wind Dividing Line" at $F_{\text{X}} = 10^6$ erg cm^{-2} s^{-1}, this may be an indication for generally weaker winds from M dwarfs, perhaps again related to the large-scale structure of the stellar magnetic field.

4. Astrospheric Charge Exchange

The solar wind can be subject to charge exchange (CX) as has been observed around comets (Lisse *et al.* 1996; Cravens 1997). In this process, a highly charged solar wind ion interacts with a neutral atom or molecule from the comet. Charge exchange can also occur between ions although this process is inefficient. Charge exchange populates levels at high n in He-like ions, followed by a cascade of decays in which, at X-ray wavelengths, most emission is from $2 \to 1$ transitions. A detailed analysis was given in Wargelin & Drake (2001).

Charge exchange should equally take place between solar wind ions and neutrals from the interstellar medium streaming into the heliosphere (Cox 1998). By analogy, a similar mechanism is expected for all cool stars with an ionized wind *provided* that the ISM contains a neutral component. Such extended emission could be identified with X-ray spectro-imaging devices (e.g., CCD cameras). Wargelin & Drake (2001) used a parameterized model for the depletion of neutral H near the Sun, depending on the wind mass loss rate, and analyzed (multiple) charge exchange of O^{7+} and O^{8+} to O^{6+} and emission of O VII in the X-ray range. They assumed a spherical, radially expanding wind with a velocity of $v_{\text{ion}} = 400$ km s^{-1}. Parameters required in this model are,

• the neutral hydrogen density profile with depletion in the solar vicinity, given by $n_{\text{H}}(r) = n_{\text{H}_0}\exp^{-\lambda_{\text{H}}/r}$ where $\lambda_{\text{H}} = 5$ au or 50 au, depending on the mass-loss rate (the higher value being appropriate for a stronger wind). The density at infinity is ~0.15 cm^{-3}. This defines the path length for CX, $\lambda_{\text{CX}} = (n_{\text{H}}\sigma_{\text{CX}})^{-1}$, where σ_{CX} is the CX cross section between $O^{7+,8+}$ and H, He, or H_2.

• The flux of solar wind O^{7+} and O^{8+} ions is estimated from measurements (3.6×10^{31} O^{8+} and 9.6×10^{31} O^{7+} ions injected into the solar wind per second). The ion

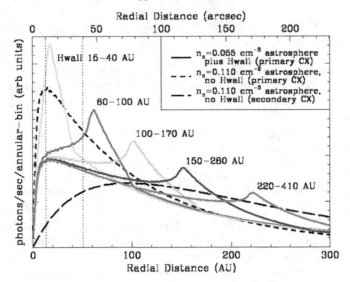

Figure 4. Photon rates in annuli of given radius modeled for CX radiation from the astrosphere around Proxima Centauri, for the *Chandra* ACIS CCD detector. Solid lines are for models with a hemispherical hydrogen wall with twice the ISM neutral density and half the ISM density inside the astrosphere. The short-dashed line refers to models with no hydrogen walls, and a neutral H density equal to the ISM. The long-dashed line is for models also without H walls but for secondary CX emission. (From Wargelin & Drake 2002.)

density profiles with radius follow from two differential equations considering depletion or addition of ions due to CX (at 1 au from the Sun, the *total* ion density is ~ 7 cm^{-3}). The emissivity of CX then is,

$$\epsilon = n_{\mathrm{H}} n_{\mathrm{ion}} v_{\mathrm{ion}} \sigma_{\mathrm{CX}}. \tag{4.1}$$

The same authors (Wargelin & Drake 2002) subsequently applied their method to X-ray observations of Proxima Centauri. For this work, they also added a realistic neutral "hydrogen wall" (as discussed in Sect. 3; Fig. 4) to their model of the astrosphere around Prox. Cen. Furthermore, they filtered counts in energy to capture only the CX-line rich region of 453–701 eV. A search for an extended structure at a projected distance of 13–51 au around the star delivered an upper limit translating into an upper limit of $\dot{M} < 3 \times 10^{-13}$ $M_\odot$ yr^{1} (3σ) or $\sim$ten times the present-day solar mass-loss rate.

5. Radio Bremsstrahlung from Ionized Stellar Winds

Because winds from cool stars are ionized, they emit bremsstrahlung as a consequence of accelerating/decelerating interactions between electrons and ions. The bremsstrahlung emissivity is a function of electron temperature T and number density n_{e} (absorption coefficient at frequency ν is roughly $\kappa_\nu \propto n_{\mathrm{e}}^2 T^{-3/2}$, therefore emissivity $\eta_\nu \propto n_{\mathrm{e}}^2 T^{-1/2}$). Measuring stellar emission of this type would constitute a direct detection of the winds. This method has indeed been successfully used to quantify wind mass loss rates from massive stars (O, B, and Wolf-Rayet; e.g., Bieging *et al.* 1982; Scuderi *et al.* 1998; Leitherer *et al.* 1995).

The radio free-free flux spectrum for an optically thick, constant-velocity, fully ionized isothermal spherical (isotropic) wind is predicted to be of the form (Panagia & Felli 1975; Wright & Barlow 1975; Olnon 1975)

$$S_\nu = 9 \times 10^{10} \left(\frac{\dot{M}}{v}\right)^{4/3} T^{0.1} \nu^{0.6} d^{-2} \text{ mJy},\tag{5.1}$$

where $\dot{M}$ is the mass loss rate in $M_\odot$ yr^{-1}, T the temperature of the plasma in K, ν the frequency in Hz, v the wind velocity in km s^{-1}, and d the stellar distance in pc. At any frequency one essentially sees emission from gas down to a level where the gas becomes optically thick.

For a wind that is completely optically thin down to the stellar surface, the following spectral flux is observed at Earth:

$$S_\nu = 5 \times 10^{39} \left(\frac{\dot{M}}{v}\right)^2 T^{-0.35} R_*^{-1} \nu^{-0.1} d^{-2} \text{ mJy},\tag{5.2}$$

where R_* is the stellar radius in cm. For non-isothermal winds, the equation of radiative transport must be iterated along all parallel sightlines across the wind region, using the temperature-dependent absorption coefficient.

As for non-isotropic winds, Reynolds (1986) expanded the wind emission theory to a conical pair of collimated polar outflows (analogous to jets). Principally, for the same mass loss rate, wind temperature, and wind velocity as for an isotropic wind, the outflows produce stronger radio emission because of the higher required density.

Searches for wind bremsstrahlung emission were conducted early on, see, e.g., Doyle & Mathioudakis (1991), or Mullan *et al.* (1992). A critical upper limit was obtained by Drake *et al.* (1993) for the wind mass-loss rate of the nearby F5 IV-V star Procyon, namely $\dot{M} < 2 \times 10^{-11}$ $M_\odot$ yr^{-1}, an estimate that considered also the *detection* of bremsstrahlung from the chromosphere and optically thick surfaces of active regions, as well as optically thin coronal bremsstrahlung emission. Sensitive millimeter measurements and theoretical arguments have constrained ionized-wind mass loss rates for M dwarfs to $\dot{M} \lesssim$ a few times $10^{-10} M_\odot$ yr^{-1} or $\dot{M} \lesssim 10^{-12} M_\odot$ yr^{-1} (Lim & White 1996; van den Oord & Doyle 1997). Specifically for the nearby Proxima Centauri, Lim *et al.* (1996) derived an upper limit to its mass loss rate of $7 \times 10^{-12} M_\odot$ yr^{-1} for a wind velocity of 300 km s^{-1}, as determined at a wavelength of 3.5 cm.

Gaidos *et al.* (2000) studied wind mass loss evolution of solar analogs to test the hypothesis that the zero-age main-sequence Sun may have been more massive than the present Sun by several percent. This would provide an interesting solution for the "Faint Young Sun Paradox" (FYSP) that confronts evidence of mild climates on early Earth and Mars from geological evidence with the significantly fainter Sun at those times (by 25–30%). Observations of three solar analogs in the age range of 0.3–1.5 Gyr again revealed only flux upper limits down to 12μJy (2σ) at 3.6 cm wavelength, corresponding to mass loss rates of $(4-5) \times 10^{-11} M_\odot$ yr^{-1}. The integrated mass loss rate would be no more than 6% of the solar mass, which however still left the possibility open to explain the FYSP.

Later, deeper observations with the VLA and ALMA of four solar analogs with ages of $\sim 100 - 700$ Myr by Fichtinger *et al.* (2017) again resulted in upper limits down to 9μJy (3σ) at 6–14 GHz wavelengths or detections that were justifiably identified with other emission processes (Fig. 5). These limits, however, were sufficiently low to exclude a higher initial solar mass required to explain the FYSP, with upper limits to a ZAMS mass excess of $\sim 2\%$, based on isotropic winds or polar outflows.

The upper limits for $\dot{M}$ as measured by Gaidos *et al.* (2000) and Fichtinger *et al.* (2017) exceed the expected $\dot{M}$ by about 2 orders of magnitude if a reasonable stellar spin-down/mass-loss evolutionary scenario adapted to the present solar mass-loss rate of $2 \times 10^{-14} M_\odot$ yr^{-1} is assumed (see Sect. 2 above, and Fig. 1 in Gaidos *et al.* 2000).

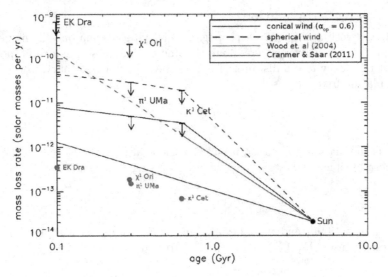

Figure 5. Mass-loss evolution for solar-type stars. The two black solid lines are upper limit estimates for the $\dot{M}$ evolution based on non-detections of stellar wind radio bremsstrahlung (Sect. 5; arrows indicate corresponding $\dot{M}$ upper limits). The upper black dashed line refers to a spherical wind, the lower solid black line to a conical wind with an opening angle of 40 degrees. The red circles are mass-loss estimates from the spin-down model by Johnstone *et al.* (2015) (Sect. 2), while the red line shows the fit from the Lyα absorption model of Wood (2018) (Sect. 3). The blue solid line relates to the theoretical model of Cranmer & Saar (2011). (From Fichtinger *et al.* 2017.)

6. Radio Free-Free Optical Depth of Stellar Winds

The wind bremsstrahlung theory discussed in Sect. 5 can also be used to put limits to the optical depth required by observations of radio emission different from the wind emission. Magnetically active, young stars from spectral type G to M occasionally produce radio flares detected due to their radio gyrosynchrotron emission or a variety of coherent radio emission types. Realistically, these flares take place, like on the Sun, in closed, somewhat compact magnetic loops or loop arcades with heights of a fraction of the stellar radius. Very Long Baseline Interferometry has explicitly supported this picture (e.g., Benz *et al.* 1995; Benz *et al.* 1998), and so has the detection of radio rotational modulation in active stars (Lim *et al.* 1992; Güdel *et al.* 1995).

If the stellar wind became optically thick to radio radiation only a few tenths of a stellar radius above the surface, then radio flares would occur in the optically thick region and therefore not be detectable. Essentially, therefore, the optically thick wind radius should be no larger than the stellar radius, making the wind optically thin (Eq. 5.2). The optically thick radius can be obtained from the (isotropic) wind theory described above, namely

$$R_{\text{thick}} = 8 \times 10^{28} \left(\frac{\dot{M}}{v} \right)^{2/3} T^{-0.45} \nu^{-0.7} \quad \text{cm.} \tag{6.1}$$

(e.g., Lim & White 1996). Solving this equation for $\dot{M}$ again provides stringent limits to the mass-loss rates of winds given wind temperatures and velocities. Lim & White (1996) obtain $\dot{M} \lesssim 5 \times 10^{-14} - 10^{-12} M_\odot$ yr^{-1} for $v = 300 - 600$ km s^{-1} and wind temperatures of $10^4 - 10^6$ K based on low-frequency observations of flare radio emission from the dMe star YZ CMi.

7. Radio Wave Propagation

Radio waves with a frequency ν propagate through a plasma only if they exceed the local plasma frequency everywhere along the line of sight. Assume a radio point source emitting at frequency ν at a distance of r from the star in a stationary, isothermal wind. The local plasma frequency is

$$\nu_p = \left(\frac{4\pi n_e e^2}{m_e}\right)^{1/2} \tag{7.1}$$

where m_e is the mass of the electron and e is the charge of the electron. The condition for a static, isotropic wind is

$$n_i(r) = \frac{\dot{M}}{4\pi r^2 m_i v} . \tag{7.2}$$

Noting that $n_e \approx 1.09 n_i$ and $m_i \approx 1.25 m_p$ (m_p being the mass of a proton) for cosmic abundances, we insert Eq. (7.2) into Eq. (7.1) to find

$$\dot{M} < \frac{r^2 m_e m_i v \nu_p^2}{1.09 e^2} \approx 1.2 \times 10^{-58} r^2 v \nu_p^2 \ M_\odot \ \text{yr}^{-1}. \tag{7.3}$$

For a wind with $v = 400$ km s^{-1} at $r = 10 R_\odot$ and $\nu = 0.1$ GHz, we find a limit of 1.5×10^{15} g s$^{-1} \approx 2.3 \times 10^{-11} M_\odot$ yr^{-1}. Vidotto & Donati (2017) applied this method to potential radio emission of a young Jupiter-mass planet around the T Tauri star V830 Tau (orbital radius 0.057 au). They assumed that the frequency of the planetary radio emission is given by the gyrofrequency determined by the magnetic field strength B in the planetary emission source; they found that $B = 10$ G (corresponding to the gyrofrequency $\Omega_c = 28$ MHz) requires a stellar mass-loss rate of $\dot{M} \lesssim 10^{-10} M_\odot$ yr^{-1}.

The condition $\nu > \nu_p$ applies to the entire line of sight. For a planet located in the front half of the isotropic stellar wind (e.g., during transit, Fig. 6), the above condition is sufficient because the wind density decreases outward. For locations in the more distant hemisphere, the maximum density along the line of sight must be determined.

8. X-Ray Limits to Stellar Winds

Essentially all cool main-sequence stars host magnetically confined coronae that are detected as variable soft X-rays sources. However, part of the X-ray emission could be due to the hot plasma of a wind that is optically thin to X-rays (like the corona itself).

For a stationary, constant-velocity, spherically symmetric wind, the particle density profile for ions follows Eq. (7.1). The total emission measure of the entire wind is

$$\text{EM} = \int_{R_*}^{\infty} n_e n_i dV = 1.09 \frac{\dot{M}^2}{4\pi R_* m_i^2 v^2} \approx 2 \times 10^{46} \frac{\dot{M}^2}{R_* v^2} \tag{8.1}$$

where the factor 1.09 is again the ratio between the electron and ion number densities for a fully ionized plasma with cosmic abundances, and $m_i \approx 1.25 m_H \approx 2 \times 10^{-24}$ g.

For a temperature of order ≈ 1 MK, the luminosity of the isothermal wind would be

$$L_X \approx 10^{-22} \ \text{EM} \tag{8.2}$$

for solar abundances (see, e.g., Audard et al. 2004, their Fig. 10). Therefore,

$$\dot{M} \approx 7.1 \times 10^{-13} \left(L_X R_*\right)^{1/2} v$$

$$\approx 3.75 \times 10^{-12} \left(\frac{L_X}{10^{27} \ \text{erg/s}}\right)^{1/2} \left(\frac{v}{400 \ \text{km/s}}\right) \left(\frac{R_*}{7 \times 10^{10} \ \text{cm}}\right)^{1/2} M_\odot/\text{yr} \tag{8.3}$$

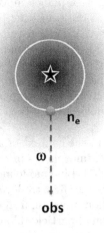
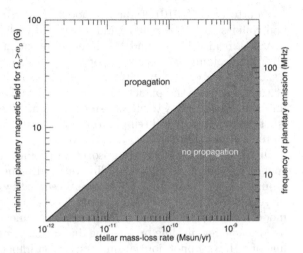

Figure 6. *Left:* Sketch illustrating a stellar wind with decreasing density outwards, and a planet orbiting in the wind where the wind density is n_e. The angular frequency ω must exceed the local plasma frequency ω_p along the entire line of sight. – *Right:* Minimum planetary magnetic field intensity required for the propagation of planetary radio emission through the wind of the host star as a function of stellar wind mass-loss rate, assuming that the planetary emission takes place at the gyrofrequency in the planetary magnetic field emission source (see right y axis). The shaded area is for parameter combinations that do not allow planetary radio emission to propagate through the wind. (Right figure from Vidotto & Donati 2017.)

For $L_X = 10^{27} - 10^{30}$ erg s^{-1} for a solar analog and $v = 400$ km s^{-1}, the maximum mass-loss rate (assuming that all observed X-ray emission is due to a wind) would be $3.75 \times 10^{-12} - 1.2 \times 10^{-10}$ $M_\odot$ yr^{-1}. This obviously does not apply to the Sun where almost all X-ray emission is from closed active regions and the measured wind mass-loss rate is $\sim 2 - 3$ orders of magnitude lower than from the above estimate, but these limits provide conservative constraints to the total mass loss rate from hot ionized winds (note that most of the wind radiation originates from close to the stellar surface where the temperatures are still high, because of the r^{-2} dependence of the density).

An estimate for the maximum possible wind mass loss rate for Proxima Centauri based on the X-ray luminosity was briefly mentioned by Lim & White (1996), with $\dot{M} \lesssim 6 \times 10^{-11} M_\odot$ yr^{-1} for the active mid-M dwarf YZ CMi. In general for mid M dwarfs with $L_X = 10^{26} - 10^{29}$ erg s^{-1} and a radius of $\approx 0.4 R_\odot$, one finds a maximum of $\dot{M} = 7.5 \times 10^{-13} - 2.4 \times 10^{-11}$ $M_\odot$ yr^{-1}.

9. Wind-Planet Interactions

Planets with atmospheres interact in various ways with stellar winds. For example, stellar winds can drag ions formed in the upper atmospheres of planets. Stellar winds also transport shock fronts from coronal mass ejections and high-energy particles produced in those shocks or in stellar flares. Such particles can interact with planetary atmospheres where they induce sputtering, ionization, and drive chemical reactions (see, for example, Lammer *et al.* 2003; Airapetian *et al.* 2016). The collision between the wind and a planetary magnetosphere may also form a detectable shock in front of the planet (Vidotto *et al.* 2011a). There is a rapidly increasing amount of literature modeling stellar wind parameters based on exoplanetary observations. I briefly discuss four exemplary cases in this section.

Vidotto *et al.* (2010) studied a model of a bow shock forming as a planet moves in, and collides with, the stellar wind. Specifically, the early ingress of the transiting planet WASP-12b in orbit around the late-F-type main-sequence star WASP-12 requires an absorbing column density for Mg of $> 1.4 \times 10^{13}$ cm^{-2} (Lai *et al.* 2010), corresponding to a shock density of $n_{\rm H} \approx 6 \times 10^6$ cm^{-3} or a pre-shock wind density of $n_{\rm obs} \approx 1.5 \times 10^6$ cm^{-3} at the location of the planet with an orbital radius of $3.15 R_*$. The coronal temperature must be $< 4.2 \times 10^6$ K to allow a shock to form. The wind velocity at the planet's position is a function of the wind temperature, which is unknown, but if known, would then allow us to calculate $\dot{M}$ from $n_{\rm obs}$ and the velocity.

Observations of Lyα excess absorption profiles (also discussed in the context of astrospheric H walls in Sect. 3) including inflated planetary hydrogen atmospheres in transit around the host star HD 209458 were presented by Ben-Jaffel (2007). Line modeling was presented in various publications, but Kislyakova *et al.* (2014) was the first to develop a model including all relevant physical effects, namely, i) Lyα radiation pressure, ii) natural line broadening, iii) line broadening due to the velocity distribution of particles along the line of sight, iv) photoionization, v) electron impact ionization, and vi) exospheric energetic charge exchange to form energetic neutral atoms. These authors find an extended H corona around the planet arranged in the shape of a cometary tail in the stellar wind and best-fit parameters for the Lyα profile of 400 km s^{-1} for the wind velocity, a wind particle density of 5×10^3 cm^{-3}, and therefore, using the orbital distance of 0.047 au, a mass-loss rate of $\dot{M} = 4 \times 10^{-14}$ $M_\odot$ yr^{-1} for this F9 V star.

Lecavelier des Etangs *et al.* (2012) estimated wind density and velocity for the planet host HD 189733 (K2 V) from the transit Lyα profile, using an N-body simulation of the upper atmospheric H atoms that interact with the wind and are subject to radiation pressure and charge exchange. The best-fit model parameters include the wind temperature $T \approx 10^5$ K, the wind density $n = 3 \times 10^3$ cm^{-3}, and the wind velocity $v = 190$ km s^{-1}. Given the orbital radius of the planet of 0.031 au, one finds $\dot{M} \approx 5 \times 10^{-15}$ $M_\odot$ yr^{-1}, which is 4 times less than the present-day solar wind.

Along similar lines, Vidotto & Bourrier (2017) modeled the interaction between the wind of the M2.5 dwarf GJ 436 and the warm Neptune GJ 436b. The model is a spherically symmetric, steady-state isothermal wind. The best-fit model indicates a wind temperature of 0.41 MK, a terminal velocity of 370 km s^{-1}, a local wind velocity of 85 km s^{-1} as the wind is still accelerating at the position of the close-in planet (orbital radius 0.029 au), and a local proton density of 2×10^3 cm^{-3}. This implies $\dot{M} \approx 1.2^{+1.3}_{-0.75} \times 10^{-15}$ $M_\odot$ yr^{-1}, or 0.059 times the present solar mass-loss rate.

10. Slingshot Prominences

Apart from a steady wind, mass may also be lost in episodic ejections, in the case discussed here in prominences. The presence of co-rotating, relatively cool material trapped in coronal magnetic fields has been demonstrated spectroscopically in Hα absorption transients in active, rapidly rotating stars (Collier Cameron & Robinson 1989). Cool prominences can form in the apex region of coronal magnetic loops when a thermal instability occurs. The drop in pressure will attract more material from below.

To analyze the situation further, Jardine & Collier Cameron (2019) distinguish three coronal regions: i) the region from the stellar surface up to the sonic radius R_s where the wind speed is equal to the sound speed. ii) The region between R_s and the Alfvén radius R_A where the wind speed reaches the Alfvén speed, and iii) the region beyond R_A. Coronal magnetic loops can be closed in regions (i) and (ii), but are open in region (iii). Assuming that the condensations form easiest if the loop apex is at the co-rotation radius R_K, what then matters is the location of R_K relative to the above three regions. Note that R_K increases with increasing Ω, like $R_K = (GM_*/\Omega^2)^{1/3}$.

If a loop-top coronal condensation (i.e., at R_K) lies below R_s, then the loop can remain hydrostatic and a hydrostatic equilibrium can always be established (hydrostatic regime). If the condensation (at R_K) forms in a loop apex above R_s, then cooling material accumulates because the flow arrives supersonically in that region and hydrostatic equilibrium in the entire loop cannot be established. The increasing pressure will eventually release the prominence, and depending on whether it is below or above the co-rotation radius, it will fall back to the star or will be episodically ejected (limit-cycle regime). If the coronal condensation (at R_K) forms beyond the Alfvén radius, then it will not accumulate further mass but will flow out along the open field lines (open regime).

The analysis by Jardine & Collier Cameron (2019) relies on the assumption that the coronal filling acts like a wind, and for this the plasma electron temperature matters because R_s is a function of T, $R_s = GM_*/2c_s^2$. The temperature required such that $R_s = R_K$ is,

$$T_{\rm crit} = 1.6 \times 10^6 \text{ K} \left(\frac{M_*}{M_\odot}\right)^{2/3} \left(\frac{P}{1 \text{ d}}\right)^{-2/3}. \tag{10.1}$$

The expected average wind mass-loss rate in particular from the episodic ejections from the limit-cycle regime where $R_s < R_K < R_A$ follows from observations. The observed masses m_p and lifetimes τ of Hα prominences provide an average mass-upflow rate into the prominence,

$$\dot{m}_{\rm p} \approx \frac{m_{\rm p}}{\tau}. \tag{10.2}$$

The footpoint area of the respective loop can be estimated from the dipole approximation,

$$A_0 = A_{\rm p} \left(\frac{R_*}{R_{\rm p}}\right)^3 \tag{10.3}$$

where $R_{\rm p}$ is the radial height of the prominence, and $A_{\rm p}$ is the cross-section area of the prominence in the loop. Then, assuming full coverage of the star with this type of wind, the wind mass-loss rate will be

$$\dot{M} = \frac{4\pi R_*^2}{2A_0} \dot{m}_{\rm p} \approx 100 \frac{m_{\rm p}}{\tau}. \tag{10.4}$$

Some $\dot{M}$ estimates for very rapidly rotating, active stars are shown in Fig. 7. The highest mass-loss rates reach up to 3000 times the solar rate per unit area and suggest that the correlation found by Wood *et al.* (2005), Eq. (3.1), continues to hold approximately toward much more active stars, and there may be no breakdown of $\dot{M}$ at some specific activity level.

11. Accretion Contamination in White Dwarf Atmospheres

In white dwarfs (WD), heavy elements are supposed to settle below the upper atmosphere on time scales $\lesssim 10^6$ yr, giving rise to pure H or He photospheres. Nevertheless, several white dwarfs with long cooling times show spectroscopic evidence of heavy metals in their atmospheres. The common explanation for this metal contamination is accretion of solar-abundance material from the interstellar medium. However, as discussed in Debes (2006), close WD + M dwarf binaries show heavy metals anomalously often. The idea is that the contaminating material in the WD originates from the stellar wind of the companion red dwarf through Bondi-Hoyle accretion. In this case, the accretion rate is,

$$\dot{M}_{\rm acc} = \frac{4\pi G^2 M_{\rm WD}^2 \rho(R)}{v_{\rm rel}^3} \tag{11.1}$$

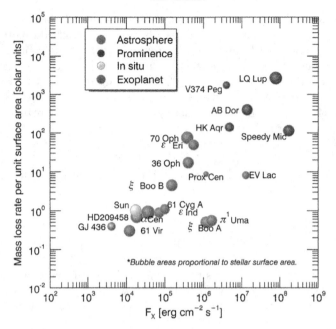

Figure 7. Stellar mass-loss rate per unit surface area plotted against the X-ray surface flux. The blue dots are for stars interpreted with the slingshot prominence method (Sect. 10). Red dots are from the astrospheric Lyα method (Sect. 3; see also Fig. 3). The green dots use exoplanetary atmospheres (Sect. 9). (From Jardine & Collier Cameron 2019.)

(Bondi & Hoyle 1944; see Debes 2006) where G is the gravitational constant, M_{WD} is the mass of the white dwarf, $\rho(R)$ is the density of the M dwarf wind at the position of the accreting WD (distance of R), and v_{rel} is the relative velocity between the WD and the wind flow, composed of the radial wind velocity and the WD orbital velocity. In an assessment presented by Debes (2006), the wind speed is assumed to be the escape speed of the M dwarf, which is approximately true also for the Sun, although physical drivers for the wind velocity (coronal heating) should make this only a rough approximation.

The convective atmosphere collects heavy elements by accretion while these elements subsequently diffuse below the convection zone. These two competing mechanisms should be in a steady-state equilibrium. Using Ca as an easily observable element, the accretion rate of Ca is (see Debes 2006)

$$\dot{M}_{acc,Ca} = \dot{M}_{acc}[Ca/H]_{\odot} \qquad (11.2)$$

where the solar abundance of Ca has been assumed for the accreting material from the M dwarf wind. The rate of diffusion of Ca out of the convection zone is,

$$\dot{M}_{diff,Ca} = \frac{qM_{WD}}{\tau}[Ca/H] \qquad (11.3)$$

where M_{WD} is the WD mass, q is the fractional mass in the convection zone and the photosphere, and τ is th diffusion time scale. The Ca abundance is $[Ca/H]$, as observed. We assume a spherical, stationary wind for the M dwarf with a rate $\dot{M}$, so that the density at a distance R is

$$\rho(R) = \frac{\dot{M}}{4\pi R^2 v}. \qquad (11.4)$$

Table 1. Summary of methods for wind mass-loss rate measurements

Method	Type	Caveats
Spin-down	indirect	torque formula
H walls, Lyα	indirect	hydrodynamic model, presence of neutral ISM
Astrospheric CX	direct	presence of neutral ISM
Radio bremsstrahlung	direct	other radio emission; wind temperature and velocity
Radio free-free absorption	direct	presence coronal emission; wind temperature and velocity
Radio wave propagation	indirect	planetary magnetic field strength and radio emission
X-ray emission	direct	assumed wind temperature; stellar X-rays
Planetary Lyα absorption	indirect	model, planetary atmosphere
White dwarf metal contamination	indirect	orbital separation, wind velocity, diffusion parameters

Requiring $\dot{M}_{\mathrm{acc,Ca}} = \dot{M}_{\mathrm{diff,Ca}}$ leads to

$$\dot{M}_{\mathrm{acc}} = \frac{qM_{\mathrm{WD}}}{\tau}\frac{[\mathrm{Ca/H}]}{[\mathrm{Ca/H}]_{\odot}} \qquad (11.5)$$

(Debes 2006). Since this must be the same as Eq. (11.1) (assuming v for v_{rel}), we find

$$\dot{M} = \frac{q}{\tau}\frac{[\mathrm{Ca/H}]}{[\mathrm{Ca/H}]_{\odot}}\frac{R^2 v^4}{G^2 M_{\mathrm{WD}}} \, . \qquad (11.6)$$

This method for $\dot{M}$ has a number of caveats, most importantly the poorly known R (if the orbit orientation and orbit eccentricity are not known), the assumption of the wind velocity, and the uncertainties in the calculated parameters q and τ. Debes (2006) suggests that the estimates are good to about one order of magnitude for close binaries. He presents three such cases, all of which show very low $\dot{M}$ in the range $\dot{M} \approx 10^{-16} - 6 \times 10^{-15}$ $M_{\odot}$ yr^{-1} for M dwarf masses of $0.095 - 0.36$ $M_{\odot}$ (spectral types ~M4–6.5) and orbital separations ≤ 0.015 au. These $\dot{M}$ values are much lower than $\dot{M}_{\odot}$. Even lower values are reported by Parsons *et al.* (2012) and references therein.

12. Discussion and Conclusions

The previous sections have summarized a variety of methods presently available to infer the presence of stellar mass loss, and in many cases provide estimates or upper limits for $\dot{M}$. The methods can be grouped into *direct* or *indirect* methods. Table 1 summarizes the methodologies and caveats. The latter refer to those issues that are poorly known or that are difficult to isolate from the wind signatures.

Direct wind measurements rely on electromagnetic emission of the wind plasma itself, or optical-depth related absorption or attenuation processes in the wind plasma. This category comprises measurements of radio bremsstrahlung, radio optical depth of the wind plasma to other stellar radio emission, radio wave propagation effects near the plasma frequency, X-ray thermal emission from the wind, and X-ray emission due to charge exchange between wind ions and neutrals penetrating into the astrosphere from the ISM. While these measurements are based on straightforward interpretation of observational data, emission and absorption effects in ionized stellar winds have turned out to be weak so that essentially all observations have provided upper limits to $\dot{M}$. This is not by itself a weakness. Stringent upper limits help constrain stellar mass-loss models; they have also rejected a model in which a significantly more massive young Sun helps solve the Faint Young Sun Paradox. Most direct methods suffer, however, from competing radiation from the star itself. This is particularly true for coronal X-rays outshining the feeble astrospheric charge exchange radiation, or for wind radio bremsstrahlung in the presence of chromospheric radio emission if the wind is optically thin.

Table 2. Selected published estimates of $\dot{M}$ using different methodologies

Method	$\dot{M}$ (F/G/K Dwarfs) $[M_\odot \ \mathrm{yr}^{-1}]$	$\dot{M}$ (M Dwarfs) $[M_\odot \ \mathrm{yr}^{-1}]$	References[1]
Spin-down	$10^{-14} - 4 \times 10^{-13}$	$10^{-14} - 10^{-13}$	1
H walls, Lyα	$< 10^{-15} - 2 \times 10^{-12}$	$< 2 \times 10^{-15}$	2
Astrospheric CX	...	$< 3 \times 10^{-13}$	3
Radio bremsstrahlung	$< (2-3) \times 10^{-11}$ [2]	$< (6-9) \times 10^{-10}, < 7 \times 10^{-12}$	4
Radio free-free absorption	$< 7 \times 10^{-10}$	$< 5 \times 10^{-14} - < 10^{-11}$	5
Radio wave propagation	$< 10^{-10}$	...	6
X-ray emission	$< 3.5 \times 10^{-12} - < 10^{-10}$	$< 7 \times 10^{-13} - < 2 \times 10^{-11}$	7
Planetary Lyα absorption	$5 \times 10^{-15} - 4 \times 10^{-14}$	1.2×10^{-15}	8
White dwarf metal contamination	...	$10^{-16} - 6 \times 10^{-15}$	9

Notes: [1]References: 1: Johnstone *et al.* (2015); 2: Wood (2004); 3: Wargelin & Drake (2002); 4: Fichtinger *et al.* (2017) (G dwarfs), Lim & White (1996) (M dwarfs AD Leo, YZ CMi), Lim *et al.* (1996) Prox. Cen. ($< 7 \times 10^{-12}$); 5: Fichtinger *et al.* (2017) (G dwarfs), Lim & White (1996) (M dwarf YZ CMi); 6: Vidotto & Donati (2017) (V830 Tau, T Tauri star); 7: this paper, Sect. 8. 8: Lecavelier des Etangs *et al.* (2012) for HD 189733 (K2 V), Kislyakova *et al.* (2014) for HD 209458 (F9 V), Vidotto & Bourrier (2017) for GJ 436 (M2.5); 9: Debes (2006), also lower values in Parsons *et al.* (2012).
[2]Only for spherical winds; lower upper limits possible for conical winds.

Indirect wind measurements use features for which the wind is responsible but that themselves are distinct from the wind. This category comprises the inference of stellar winds from stellar rotation and evolutionary spin-down, interpretation of Lyα line absorption in the hydrogen walls around astrospheres embedded in a partially neutral ISM, inferences for the stellar wind from observations of planetary atmospheres, the interpretation of slingshot prominences in very active stars, and white dwarf metal contamination in WD + M dwarf close binaries. These methods have the advantage of using features that often "amplify" the evidence of a wind. The difficulty with these methods comes from the need of complex models relating the observed features to the putative stellar winds. Nevertheless, indirect measurements have been more successful in providing estimates for $\dot{M}$ rather than upper limits even though model limitations need to be kept in mind (e.g., the breakdown of the $\dot{M}$ relation from Lyα absorption toward very active stars, or the relation between $\dot{M}$ and the angular momentum loss rate based on a "torque formula" for stellar spin-down).

Table 2 summarizes some select results from various methods. Of course, these results by now means cover a representative parameter range of stars as the observed targets are biased by target selection criteria. There is need to infer $\dot{M}$ from different methods in the hope to cross-calibrate them and verify their applicability. This has succeeded only very rarely. I mention the measurements of $\dot{M}$ using Lyα absorption, astrospheric CX, X-ray thermal emission, and radio bremsstrahlung for Proxima Centauri or Lyα absorption and stellar spin-down estimates for the solar analog π^1 UMa. In the former case, the measurements are in agreement with each other although three of them (X-ray thermal emission, radio bremsstrahlung and astrospheric CX) provided only upper limits. For π^1 UMa, in contrast, the two estimates contradict each other, as spin-down requires strong winds while the Lyα method delivered an estimate of $\dot{M}$ even below the present-day solar value. This clearly illustrates the need for further multiple $\dot{M}$ measurements.

These reservations and caveats aside, observations in the past two decades have started confining wind mass loss for cool main-sequence stars to levels that look reasonable in the context of the solar $\dot{M}$ and spin-down evolution, with results or upper limits that straddle around the solar value up to values 100–1000 times the solar $\dot{M}$ for very active stars. This can be compared to the stellar X-ray luminosities that vary in a similar range.

Acknowledgments

It is pleasure to thank the organizers for a wonderful conference and for giving me the opportunity to present this review. This research has been supported by the Austrian Science Fund FWF through project S116 *Pathways to Habitability: From Disks to Active Stars, Planets and Life* and the related subproject S11604 *Radiation & Wind Evolution from the T Tauri Phase to ZAMS and Beyond.*

References

Airapetian, V. S., Glocer, A., Gronoff, G., *et al.* 2016, *Nature Geosci.*, 9, 452
Audard, M., Telleschi, A., Güdel, M., *et al.* 2004, *ApJ*, 617, 531
Ayres, T. R. 1997, *J. Geophys. Res.*, 102, 1641
Ben-Jaffel, L. 2007, *ApJ*, 671, L61
Benz, A. O., Alef, W., Güdel, M. 1995, *A&A*, 298, 187
Benz, A. O., Conway, J., Güdel, M. 1998, *A&A*, 331, 596
Bieging, J. H., Abbott, D. C., Churchwell E. B. 1982, *ApJ*, 263, 207
Biermann, L., 1951, *ZfA*, 29, 274
Bondi, H., Hoyle, F. 1944, *MNRAS*, 104, 273
Collier Cameron, A., Robinson, R. D., 1989, *MNRAS*, 238, 657
Cox, D. P. 1998, in The Local Bubble and Beyond, ed. D. Breitschwerdt, M. J. Freyberg, & J.
    Trümper (Berlin: Springer), 121
Cravens, T. E. 1997, *Geophys. Res. Lett.*, 24, 105
Cranmer, S. R., Saar, S. H. 2011, *ApJ*, 741, 54
Debes J. H., 2006, *ApJ*, 652, 636
Doyle, J. G., Mathioudakis, M. 1991, *A&A* 241, L41
Drake, S. A., Simon, T., & Brown, A. 1993, *ApJ*, 406, 247
Fichtinger, B., Güdel, M., Mutel, R. L., *et al.* 2017, *A&A*, 599, A127
Gaidos, E. J., Güdel, M., Blake, G. A. 2000, *Geophys. Res. Lett.*, 27, 501
Gallet, F., Bouvier, J. 2013, *A&A*, 556, A36
Gayley, K. G., Zank, G. P., Pauls, H. L., Frisch, P. C., Welty, D. E. 1997, *ApJ*, 487, 259
Güdel, M., Schmitt, J. H. M. M., Benz, A. O., Elias, II, N. M. 1995, *A&A*, 301, 201
Holzwarth, V., Jardine, M. 2007, *A&A*, 463, 11
Irwin, J., Berta, Z. K., Burke, C. J., *et al.* 2011, *ApJ*, 727, 56
Jardine, M., Collier Cameron, A. 2019, *MNRAS*, 482, 2853
Johnstone, C. P., Güdel, M., Brott, I., Lüftinger, T. 2015, *A&A*, 577, A28
Kislyakova, K. G., Holmström, M., Lammer, H., *et al.* 2014, *Science*, 346, 981
Kraft, R. P. 1967, *ApJ*, 150, 551
Lai, D., Helling, C., van den Heuvel, E. P. J. 2010, *ApJ*, 721, 923
Lammer, H., Lichtenegger, H. I. M., Kolb, C., *et al.* 2003, *Icarus*, 165, 9
Lecavelier des Etangs, A., Bourrier, V., Wheatley, P. J., *et al.* 2012, *A&A*, 543, L4
Leitherer, C., Chapman, J. M., Koribalski, B. 1995, *ApJ*, 450, 289
Lim, J., White, S. M. 1996, *ApJ*, 462, L91
Lim, J., Nelson, G. J., Castro, C., *et al.* 1992, *ApJ*, 388, L27
Lim, J., White, S. M., Slee, O. B. 1996, *ApJ*, 460, 976
Linsky, J. L., Wood, B. E. 1996, *ApJ*, 463, 254
Lisse, C. M., Dennerl, K., Englhauser, J., *et al.* 1996, *Science*, 274, 205
Matt, S. P., MacGregor, K. B., Pinsonneault, M. H., Greene, T. P. 2012, *ApJ*, 754, L26
Matt, S. P., Brun, A. S., Baraffe, I., *et al.* 2015, *ApJ*, 799, L23
McComas, D. J., Alexashov, D., Bzowski, M., *et al.* 2012, Science, *336*, 1291
Mullan, D. J., Doyle, J. G., Redman, R. O., Mathioudakis M. 1992, *ApJ*, 397, 225
Olnon, F. M. 1975, *A&A*, 39, 217
Parker, E. N. 1958, *ApJ*, 128, 664
Panagia, N. & Felli, M. 1975, *A&A*, 39, 1
Parsons S. G. Marsh, T. R., Gänsicke, P. T. *et al.* 2012, *MNRAS*, 420, 3281

Reynolds, S. P. 1986, *ApJ*, 304, 713

Scuderi, S., Panagia, N., Stanghellini, C., *et al.* 1998, *A&A*, 332, 251

Skumanich, A. 1972, *ApJ*, 171, 565

Soderblom, D. R., Stauffer, J. R., MacGregor, K. B., Jones, B. F. 1993, *ApJ*, 409, 624

Stone, E. C., Cummings, A. C., McDonald, F. B., *et al.* 2008, *Nature*, 454, 71

Stone, E. C., Cummings, A. C., Heikkila, B. C., Lal, N. 2019, *Nature Astron.*, 3, 1013

Tu, L., Johnstone, C. P., Güdel, M., Lammer, H. 2015, *A&A*, 577, L3

van den Oord, G. H. J., Doyle, J. G. 1997, *A&A* 319, 578

Vidotto, A. A., Donati, J.-F. 2017, *A&A*, 602, A39

Vidotto, A. A., Bourrier, V. 2017, *MNRAS*, 470, 4026

Vidotto, A. A., Jardine, M., Helling, Ch. 2011a, *MNRAS*, 414, 1573

Vidotto, A. A., Jardine, M., Opher, M. *et al.* 2011b, *MNRAS*, 412, 351

Vidotto, A. A., Jardine, M., Helling, Ch. 2010, *ApJ*, 722, L168

Vidotto, A. A., Gregory, S. G., Jardine, M., *et al.* 2014, *MNRAS*, 441, 2361

Wargelin, B. J., Drake, J. J. 2001, *ApJ*, 546, L57

Wargelin, B. J., Drake, J. J. 2002, *ApJ*, 578, 503

Weber, E. J., Davis, Jr., L. 1967, *ApJ*, 148, 217

Wood, B. E. 2004, *Liv. Rev. Solar Phys.*, 1, id. 2

Wood, B. E., Müller, H.-R., Zank, G. P., *et al.* 2005, *ApJ*, 628, L143

Wood, B. E. 2018, in Dissipative and Heating Processes in Collisionless Plasma: The Solar Corona, the Solar Wind, and the Interstellar Medium, IOP J. Phys. Conf. Ser. 1100, 012028

Wright, A. E., Barlow, M. J. 1975, *MNRAS*, 170, 41

Discussion

HAN: In order to get the mass-loss rate, we need to know the torque in the spin-down method. The torque is inferred from a hydro simulation. How reliable is this simulation? What are the uncertainties in the simulation?

GÜDEL: Johnstone *et al.* (2015) used a specific torque formula from simulations performed by Matt *et al.* (2012). This formula (simplified above in Eq. 2.1) derives from 2D axisymmetric MHD simulations using a dipolar stellar magnetic field. The latter assumption is a first approximation because the largest-scale structure of the field is dominated by the dipolar component. As the authors mention, the formula is invalid in the limit of very weak magnetic fields, because other, e.g., viscous, effects begin to be more important. Then, the wind-driving physics, such as the coronal temperature and the heating and cooling physics, is poorly understood. The authors estimate a factor of 2 uncertainty coming from these sources. Furthermore, these simulations assumed a dipole magnetic field that is aligned with the stellar rotation axis. For a pure quadrupolar field with the same surface magnetic field strength, however, the torque would be reduced by a factor of ten! Newer torque calculations for example by Matt *et al.* (2015) still show disagreements with observations of low-mass slow rotators. Johnstone *et al.* (2015) aimed at fitting observed rotation distributions and therefore used a "fudge" factor in Matt et al.'s torque formula to include all ignored unknown physics; this factor is derived from requiring the torque to lead to the known spin-down rate for old solar-type stars (age $\sim 1 - 5$ Gyr) assuming the present-day average magnetic field and the present-day solar mass-loss rate; these assumptions are again somewhat uncertain as the state of activity of the Sun may significantly vary on time scales shorter than the general evolutionary time scales (but still on time scales $\gg$ the activity cycle time scale). But introducing such an empirical fit parameter should at least reduce the systematic – and potentially large – uncertainties due to ignored physics in the models.

Solar and Stellar Magnetic Fields: Origins and Manifestations
Proceedings IAU Symposium No. 354, 2019
A. Kosovichev, K. Strassmeier & M. Jardine, ed.
doi:10.1017/S1743921320000551

Semi-empirical 2D model of the solar corona and solar wind using solar eclipse images: Progress report

Edward C. Sittler Jr.[iD] **and Linda M. Sittler**

email: edwardsittler@gmail.com

Abstract. We present white light images of the Sun's corona acquired during the total Solar Eclipses on August 21, 2017 in mountains north of Boise Idaho USA and on July 2, 2019 south of Copiapo Chile. In both cases the viewing was excellent, altitudes ~ 1200 m and relative humidity $\sim 10\%$. We used an Orion equatorial reflecting telescope with 203 mm diameter aperture and 1000 mm focal length for f4.9 optics. A computer-controlled Canon EOS Rebel T3i digital camera was used. We plan to use our 2019 eclipse images for analysis since the Sun is near solar minimum so 2D steady state MHD equations can be used. We present a plan to process the images and convert them into a 2D empirical model of electron density and magnetic field in radial distance and co-latitude, from which 2D maps of flow velocity, effective temperature and effective heat flux can be computed.

1. Introduction

As discussed in Habbal *et al.* (2011) eclipse observations provide unique information about the solar corona just above the limb from the solar surface out to a few solar radii where the thermodynamic and magnetic properties of the coronal plasma is most complex, usually not provided by coronagraphs in space, and is a region of most importance where the expansion of the solar magnetic field and solar wind acceleration is occurring. This inner most region 3D MHD models have the most difficulty and the underlying physics is not well understood. One can model the photospheric magnetic field quite well below the corona and at higher altitudes where the flow is supersonic beyond the sonic critical point. But this region of transition cannot be modeled well theoretically, so a semi-empirical approach as first attempted by Sittler & Guhathakurta (1999, 2002) or SG model was performed using satellite and ground-based coronagraph data. Here we note in a recent paper by Woo (2019) who discusses past solar eclipse images, including the August 21, 2017 solar eclipse, some using photographic film, others based on drawings using naked eye observations and those more recently using cameras with CCD detectors and what can be learned from the various solar eclipse images. In this paper we will be just presenting the observations made for the August 21, 2017 solar eclipse in the state of Idaho and the July 2, 2019 solar eclipse south of Copiapo Chile. Regarding the analysis of the observations we will outline our approach to processing the data which is still in progress and not ready for this report. We will then present the 2D semi-empirical model we used in the Sittler & Guhathakurta (1999, 2002) papers and will mention how it could be generalized to 3D, which when combined with 3D MHD models that include the continuity, momentum and energy equations, could provide the quickest path to modeling the coronal expansion and further our understanding of the solar corona's heating, acceleration and energy transport.

Figure 1. Shows telescope used during the August 21, 2017 solar eclipse up in the mountains north of Boise Idaho at an altitude 1219 m. One can see the clear blue skies for the eclipse observations. See text and Tables 1-3 for details.

2. Telescope Description

In Figure 1 we show the telescope as used for the August 21, 2017 solar eclipse in the mountains north of Boise Idaho. The figure shows the Orion Equatorial Reflecting Telescope with 203 mm diameter aperture with focal length of 1 m for f4.9 optics, secondary mirror with 58 mm minor axis, SkyView Pro, German equatorial mount, steel tripod, and RA/Dec motor drives controlled by Orion SynScan V5 GoTo Hand Controller. The telescope mount weighs 30 lbs, optical tube 32 lbs and the two counterweights weigh 10.9 lbs each. Orion Solar Glass Filter with $\sim 100,000$ light rejection can be seen covering the telescopes aperture, and a computer-controlled Canon EOS Rebel T3i single reflex camera with 5184 x 3456 pixels (17.9 MP, 50 mm OD) CCD mounted where telescopes eye piece would normally be located. The camera can be manually moved up/down for focusing. With filter on during partial eclipse periods we used shutter speed 1/30 s for 2017 solar eclipse, therefore over-exposed, while for 2019 solar eclipse we used 1/2000 s shutter speed, images not over-exposed. USB serial interface cable from Dell Inspiron 13 laptop computer goes to metal converter box (i.e., acts as switch for shutter, circuit used came from website by Paul Beskeen 2009) which can be seen with its output red cable going to the camera's shutter control terminal which triggers when the shutter opens so images spaced ~ 1 s can be achieved. The other USB serial cable from the computer goes to the camera's digital control terminal so one can configure camera settings to ISO 100, set shutter speed, store raw images and have images stored directly to the cameras memory card so images ~ 1 s apart can be taken (i.e., speed to transfer images to the computer takes too much time). Batteries not shown were used to provide power for the equatorial mount motors, and laptop computer (not so for July 2, 2019 but computer battery lasted well after total eclipse). The camera has its own lithium batteries which were replaced just before total solar eclipse as was done for the camera's memory card. We also fine-tuned the focusing by eye using the computer screen just before total eclipse. For both eclipses the tracking was not perfect. For the 2017 and 2019 total eclipses we used two shutter times of 0.3 s and 1/30 s, and three shutter times of 0.3 s, 1/30 s and

Table 1. Canon EOS Rebel T3i Camera Parameters

Parameter	Eclipse 08/21/2017	Eclipse 07/02/2019
ISO	100	100
Shutter Trigger	External via USB	External via USB
Image	Raw	Raw
Pause	500 ms	0 ms
Store Images	Camera Memory Card	Camera Memory Card
Shutter Speed ΔT	0.3 s, 1/30 s, 1/30 s	0.3 s, 1/30 s, 1/500 s

1/500 s, respectively. In order to control the camera we used a software package called Astro Photography Tool (APT) "Eclipse Beta v3.36" (v3.36 was a delta version made by the providers for the 2017 solar eclipse so images could be recorded quickly; normally for astrophotography they usually integrate for long periods of time for good signal to noise S/N measurements). Using this method, we recorded excellent images of the Sun's corona.

3. Observational Method Used

The use of the wide aperture reflecting telescope was to achieve the highest spatial resolution allowed by atmospheric effects and to image the polar corona with altitude out beyond one solar radius. Near the Sun the E-corona from electrons should dominate the F-corona from interplanetary dust so polarized light observations would not be required. The 2019 eclipse occurred near solar minimum and as the images show the corona displays a single equatorial streamer belt which is expected when at solar minimum. We used a computer-controlled camera so one could take images as quickly as possible since for the two total eclipses used here lasted only about 1-2 minutes. The providers of the APT software v3.36 allows one to reduce the delay between images ≤ 500 ms and have control of the shutter via a different USB serial port rather through the camera itself which evidently is much slower. For the 2017 eclipse we set the APT pause (i.e., delay) parameter equal to 500 ms for delay between images, while for 2019 eclipse we used no delay. With v3.36 one can run up to 20 images for a sequence which take ∼ 1.7 s per image or ∼ 34.29 s and then restarted almost immediately by the operator when that sequence ends. For the 2017 eclipse we recorded ∼64 images when total, while for the 2019 eclipse we recorded ∼ 60 images when total. For both eclipses we had three shutter speeds (in case of 2017 eclipse two were identical at 1/30 s) so there were 21 sets of images with the 3 shutter speeds for each set for 2017 eclipse and 20 sets for the 2019 eclipse. Empirically, using the APT clock, we estimated the 2017 eclipse began on August 21, 2017 at 1:26:04 and ended at 1:28:32 for 2017 (i.e., 2 min 28 s long), while for the 2019 the start and stop times were July 2, 2019 04:39:15 and 04:40:57 (i.e., 1 min 42 s long), respectively. In Tables 1, 2 and 3 we list all relevant parameters for the two eclipses, respectively.

For both solar eclipses we arrived at the observation sites several hours before the partial eclipse phase would begin. For the 2017 eclipse we were north of Boise Idaho in the mountains at elevation ∼ 1219 m and not a cloud in the sky. Using the National Geophysical Data Center, we recorded the coordinates of the Sun/Moon at the time of the eclipse RA ∼ 10 hr 4 min and declination 11° 52 min and its magnetic declination for our location ∼ 13.5^{o} E. We then setup the telescope with tripod north equatorial axis point to the north magnetic pole using a compass, string and nails on each end and nailed to the ground. We then aligned the tripod using plumb bob, so its central axis intersected the string. We then put another string from the other two legs, so it crossed the magnetic axis string at 90°. This was achieved when the distance of the two legs and the magnetic axis string were the same. In addition, we used a circular level, so tripod was level with respect to the local ground. Around this axis we rotated the tripod by

Table 2. Orion Reflecting Telescope Parameters

Aperture	203 mm
Focal Length	1000 mm
Focal Ratio	F4.9
Shadow Parameter	0.92
2nd Mirror Focal Length	58 mm
Mount	SkyView Pro, German Equatorial
RA/Dec	Motorized Hand Controller
Focusing	Manual
Tripod	Steel
Orion Solar Filter Transmission	10^{-5}
Compass	Align tripod with magnetic north or south
360° Level	Level Telescope with local ground
Plumb Bob	Used to align tripod with geographic poles
Telescope Weight	62 lbs
Counterweights Weight	10.9 lbs each

Table 3. Solar Eclipse Site

Eclipse	Altitude	Sun RA	Sun Dec	Mag Dec	LMT[1]	Start[1,2]	Stop[1,2]
21/08/17	1219 m	10 h 4 m	11° 52 m	13.5°E	12:47:49	1:26:04[3]	1:28:32[3]
02/07/19	1200 m	6 h 46 m	23° 0 m	2.25°W	12:44:15	4:39:15[2]	4:40:57[2]

[1]LT is given.

[2]In July Chile LT is same as EDT which is -4 hours relative to GMT.

[3]In August Boise Idaho LT is -6 hours relative to GMT.

13.5^o W so it was now aligned with the Earth's geographic spin axis. Then the telescope was mounted with counter balancing weights installed, solar filter installed and then rechecked with level to make sure tripod still level. We then installed the camera with its view finder on so we could see the Sun once properly aligned. We also had to correct for precise time when Sun would intersect the local meridian which would then give the RA/Dec of the local meridian at that time since we knew the RA/Dec of the Sun. This was important since the field-of-view (FOV) of the telescope was only $\sim 1^o$ x 1°. We then tried to find the Sun and failed for the first attempt, we then realized a mistake was made and once repeated we found the Sun nearly immediately. In the case of the 2019 eclipse the magnetic declination was $\sim 2.2^o$ W so we did not attempt to make this correction. For the 2019 eclipse we did not realize the secondary mirror was miss-aligned during shipping with its four-vaned spider holder (i.e., vanes only 0.4 mm thick) were badly twisted out of shape. Fortunately, we were able to get the vanes straightened out properly and using the manual with allen wrench to make the needed fine adjustments, so the secondary mirror was approximately aligned (note, Dr. Christian Hummel helped us fix this unexpected problem). Within minutes we found the Sun and then using the equatorial motors put the telescope into solar tracking. Since the alignment was not perfect every so often, we would have to use the hand motor controller to put image back into the center of the camera FOV. During total eclipse this was not needed. Once solar tracking achieved, we completed the connections of the camera with the laptop computer and the metal converter box for the shutter control. Now the image would appear in the APT viewing window. Then as described above, we used the APT software to control the camera during both partial and total eclipse phases. During partial eclipse the images were spaced about 5 s apart. Just before total we put APT software into the total eclipse mode and based on the darkening as we approach total, we removed the solar filter and similar but reverse approach was used when total eclipse was ending. While in total eclipse mode the images were going to the camera memory card and one cannot see the

Figure 2. August 21, 2017 total solar eclipse image 7799 with 1/30 s shutter speed used. At this time the Sun is in transition from solar max to solar min so reason for several overlapping equatorial and mid-latitude streamers. With a few tenths of solar radius altitude, the equatorial streamers are over exposed, while the polar streamers are not except very near the lunar edge.

Figure 3. August 21, 2017 total solar eclipse image 7800 with 0.3 s shutter speed used. Near the Sun most of the image is over exposed as expected but when combined with image 7799 in Fig. 2 can allow extension of the observations to the outer boundaries of the image FOV.

Sun's image until after total. By inspecting images later some were over exposed but got it approximately right. Note, some of the details were recorded and some were based on memory.

4. Results: August 21, 2017 Solar Eclipse

We show in Figures 2 and 3 representative images of the solar corona taken during the 2017 solar eclipse using the telescope shown in Figure 1. These images are what some refer to as thumb nail low resolution images of the raw images which are ~ 17.9 MP in size with each pixel 16 bits deep (note, not all 16 bits are usable). We have made preliminary

Figure 4. Diamond ring effect in this image 7824 taken just after the August 21, 2017 total eclipse ended and shows that the moon is moving along a line 45° from top right to lower left which is approximately where the ecliptic plane is located.

processed images using the raw images of those shown in Figure 2. One can combine these two images most simply by just renormalizing say image 7799 by the ratio 0.3∗ 30 = 9 and then adding the two images with regions of over-exposer removed. As discussed below the processing will be more complex. Using the raw images 5184 x 3456 pixels the Sun/moon are ∼ 2195 pixels in diameter and since 0.5° wide in the sky, each pixel is ∼ 0.5 degrees x 3600 arcsec/degree / 2195 pixels ∼ 0.8 arcsec/pixel. When looking at the images under high magnification they became blurred, so the actual spatial resolution was likely several arcsec to be determined. Since the Sun is in transition from solar max to solar min the equatorial region is more complex and composed of multiple streamers and within a few tenths of solar radius above the limb the image in Figure 2 is over exposed (1/30 s exposure time). In the polar regions where polar plumes can be seen and evidently showing the dipolar curved nature of the Sun's polar magnetic field the images are not over exposed except very close to the limb this may occur. Note, these are inverted images and we note the moon's motion is tilted ∼ 45° in the image with the moon moving from upper right to lower left which is shown by Figure 4 from image 7824 which shows diamond ring effect near end of the solar eclipse. In Figure 3 out to a few tenths of a solar radius even the polar regions, as expected, are over-exposed. For our semi-empirical modeling which is presently 2D we will not use the 2017 images except for the polar corona.

5. Results: July 2, 2019 Solar Eclipse

Like the 2017 solar eclipse, we here show in Figures 5, 6 and 7 representative images of the solar corona taken during the July 2, 2019 solar eclipse using the same telescope shown in Figure 1 for the 2017 eclipse. The main difference is now we also have images with 1/500 s shutter time intervals which allows us to have good images all the way down to the solar limb. This is shown in Figure 5 and since inverted images and moon slightly offset to the north, one can see evidence of the chromosphere within the south polar region. This figure also shows evidence of active region arches on both sides near the south polar boundary and likely extends all around the Sun. These coronal arches are ∼ 45° from vertical and the symmetry axis of these arches appear to be slightly tilted

Figure 5. July 2, 2019 solar eclipse image taken south of Copiapo Chile using the telescope, camera and computer described in the text. This image is a highly compressed pdf version of raw images 18 MB in size. Shutter time 1/500 s.

Figure 6. The same as in Figure 5 for shutter time 1/30 s.

in counterclockwise direction opposite of what appears to be the case when looking at the polar plumes in Figure 6 (i.e., indicates slight tilt in clockwise direction). We do not see similar coronal arches to the north or bottom of figure for which the moon is slightly tilted to the north and thus covering more of the solar limb. For this eclipse, as shown in Figure 8 for image 9744 near beginning of the 2019 total solar eclipse, the moon is moving from lower right to upper left at angle $\sim 45^o$. Again, as for the 2017 eclipse these images are thumb nail low resolution versions of the raw images which are ~ 17.9 MP in size. We have not yet made preliminary processed images of the raw images for this eclipse. In this case one can combine all three images weighting 5 by $500/30 \sim 16.67$, replace those pixels in Figure 6 that maybe over-exposed with those in Figure 5, and reduce the amplitude of those in Figure 7 by 1/9. One can then renormalize so min count

Figure 7. The same as in Figure 5 for shutter time 0.3 s.

Figure 8. Diamond ring effect in this image taken just before total eclipse for the July 2, 2019 solar eclipse. This image indicates the moon is moving from lower right to upper left in the image. Figures 5-7 indicate the equatorial streamers are from left to right so the polar regions will be nearly vertical in the figure.

is ~ 1 at the outer boundaries of the images. Again, the processing discussed below will be more complex. For this eclipse the Sun is near solar min and as expected only a single equatorial streamer is evident in the images so will work best with our 2D semi-empirical model of the solar corona and solar wind.

6. Planned Data Analysis Method

6.1. *Processing of Images*

When it comes to processing of raw images there is good article on the internet by Sumner (2014) which describes raw sensor data structure, and how to incorporate the raw Canon CR2 images into MATLAB. The CR2 files contain uncompressed image containing pixel values plus meta-information or header information about the image referred to as tags. The first step is to convert the images into Digital Negative or DNG files using an Adobe DNG Converter. The DNG files are similar to TIFF files. Within the MATLAB environment one can perform a series of operations such as Linearization, White Balance, Demosaicing, Color Space Correction and Brightness and Contrast Control. All these steps are used to make good quality images for display purposes on one's computer monitor but for scientific purposes some we may not use, but all can be done in MATLAB. We have performed some of these steps to our 2017 images with some success for display purposes.

In order to process these images, one must perform several steps which includes the calibration of our camera + telescope as discussed in article by Clark (2013). This will involve the observation of the 0-magnitude star Alpha Lyra (RA 18h 36m 56s & Dec +38° 47' 1") as our candle (see Schild *et al.* 1971) and using filters at specific wavelengths. One can then take these measurements and compare with known intensities for Alpha Lyra $\sim 10^{11}$ photons/m^2/s/micron, then using green filter at 0.53 microns, with its FWHM $\sim$0.077 in microns, $\sim 7.7 \times 10^9$ photons/m^2/s will pass through the filter, then correcting for atmospheric transmission for clear atmosphere $\sim 85\%$ to give $\sim 6.5 \times 10^9$ photons/m^2/s near sea level. From the recorded data at known shutter exposure times, camera at ISO 100, aperture of telescope 203 mm diameter, shadow effect for telescope, and CCD counts in DN (integer Data Numbers) and typical camera 12-bit DNs are mapped into the same scale as the computer 16-bit DNs one can then estimate the number of photons/DN for the camera. In the case of the Sun we used a solar filter with transmission $\sim 10^{-5}$, while for Alpha Lyra it will be ~ 1. In the example given by Clark, 2013 the Canon camera 10D digital camera it was estimated ~ 28 photons/DN at ISO 400. In our case the lens elements are removed. Other parameters are fill factors $\sim 25\%$ to 80%, typical quantum efficiency range between 19% to 26%, green filter transmission $\sim 90\%$, and if needed how to convert lux into photon intensities when comparing different stellar sources. The Sun is $\sim 4 \times 10^{10}$ more intense than Alpha Lyra. In the case of Alpha Lyra only a few nearby pixels will receive signal from the star while in case of the Sun ~ 15 MP (occulted part of image by moon ~ 3 MP not counted) will be *illuminated*, so will need to scan the star across the CCD since the response may not be uniform and not sure how the response may have been effected when over-exposed. I would expect the performance of our Canon EOS Rebel T3i to be better than the Canon 10D camera used in Clark's observations back in 2013. In addition, our calculations of the data should be close to the typical intensities of the Sun.

6.2. *Combine Images*

It may also be necessary to combine images as discussed in the paper by Druckmüller (2009) (also see Reddy & Chatterji 1996) who developed a modified phase correlation method based on Fourier transforms which enables alignment of coronal images taken during solar eclipses, and can measure translation, rotation and scaling factors between images. This will be very important for our measurements for which the alignment was clearly not perfect. Even though the tracking needed manual updates this was not done during the total eclipse phase which was < a few minutes. Once the above is done we would be able to convert the images into electron density radial and latitudinal profiles.

6.3. *Convert Images to 2D Electron Density Maps*

The conversion of the images into electron densities would follow the approach described in Guhathakurta *et al.* (1996) and references within such as the classic paper by Hulst & van de (1950) and the review article about the interpretation of total eclipse images in the early days by Hulst & van de (1953). In our case we expect our images to be dominated by the K-corona due to Thompson scattering of light by free electrons. The F-corona is more important further out radially, but we will investigate its importance for the eclipse in 2019. The Guhathakurta *et al.* (1996) results were based on polarized brightness pB data from *Skylab* HAO white-light coronagraph (WLC) images and ground based K-coronameter images of the corona (i.e., High Altitude Observatory (HAO) Mark II K-coronameter at Mauna Loa (1.12 to 2.0 R_{SUN}) and the HAO coronal eclipse camera). One would expect such ground-based coronagraphs would be taking images of the corona during both eclipses and provide information of the K-corona and F-corona so corrections could be made if necessary.

7. Electron Density Model

We will then use the electron density model developed by Sittler & Guhathakurta (1999, 2002) and fit it to our derived electron density profiles. The model used is briefly:

$$N(r, \theta) = N_P(r) + [N_{cs}(r) - N_P(r)] e^{-\lambda^2/w^2(r)} \tag{7.1}$$

for which latitude $\lambda = \pi/2 - \theta$ and θ is the colatitude. $N_P = a_{P1} e^{a_{P2}z} z^2 P_P(z)$ and $P_P(z) = 1 + a_{P3}z + a_{P4}z^2 + a_{P5}z^3$ with a_n as free parameters to be fit to the data. The subscript P stands for polar regions and CS the current sheet region and $z = R_{Sun}/r$. The variable w(r) gives the radial variation of the equatorial helmet streamer angular width which increases with r. A similar set of equations apply for $N_{CS}(r)$. Then using Eqs. 21-23 in Sittler & Guhathakurta (1999) we solve for the field line equation and by using the lower boundary of the polar coronal hole we fit the parameters for monopole $\eta_M = 2B_M/B_0$, dipole $\eta_D = B_D/B_0$ and quadrupole $\eta_Q = B_Q/B_0$ terms. The magnetic constant B_0 is set by the radial component of the magnetic field which is ~ 3 nT at one astronomical unit or AU from the Sun. We could also use the radial component of the magnetic field measured by the Parker Solar Probe spacecraft. Once B_0 is known one can then estimate the magnetic field strength at the base of the corona and can be compared with corresponding values based on solar magnetogram observations. Once available we will also use electron density measurements made by the Parker Solar Probe instruments to get a better constraint on the equatorial radial profiles. The same can be said for the radial magnetic field measurements.

8. Theoretical Description

8.1. *Mass, Momentum and Energy Equations*

We plan to convert our empirical electron density radial and latitudinal profiles into 2D maps of solar wind velocity, effective temperature and effective heat flux using the method and equations described in the Sittler & Guhathakurta (1999, 2002). Starting from the basic MHD equations of the coronal expansion in steady state one can then transform the equations into the rotating frame of the Sun which allows one to include the Parker spiral pattern of the interplanetary magnetic field in the solution and solve for various field line constants. One then ends up with the following equations:

$$\rho \vec{V} = \alpha \vec{B} \tag{8.1}$$

$$T_{eff} = \frac{\mu m_H}{\kappa} \left(\frac{1}{\rho} \int_l^\infty W' \frac{d\rho}{ds} ds \right) \tag{8.2}$$

$$W' = \frac{1}{2} (V_{P\infty}^2 - V_P^2) + \frac{1}{2} \left(\Omega^2 R^2 - \frac{\alpha^2 B_t^2}{\rho^2} \right) + \frac{1}{2} v_{esc}^2 z \tag{8.3}$$

$$v_{esc}^2 = \frac{2GM_{Sun}}{R_{Sun}} \quad \text{and} \quad z = R_{Sun}/r \tag{8.4}$$

$$q_{eff} = \rho V \left(W' - \frac{5}{2} \frac{P_{eff}}{\rho} \right) \tag{8.5}$$

$$P_{eff} = \rho \kappa T_{eff}/\mu m_H \tag{8.6}$$

for which μ is mean molecular weight with 5% helium abundance assumed, m_H is proton mass, κ is Boltzmann constant, G is gravitational constant, M_{Sun} is solar mass, R_{Sun} is solar radius, Ω is Sun's angular velocity, $\rho = \mu\, m_H N_e$ is the mass density, N_e is electron number density, V_P is the poloidal velocity of the solar wind flow, B_t is the toroidal component of the magnetic field **B** and subscript *eff* indicates wave terms could be present. In the case of q_{eff} we assume the heat flux is field aligned. In a paper by Sittler & Ofman (2006) we made initial attempt to combine our solutions for T_{eff} and q_{eff} within the confines of a 2D MHD model. In Airapetian et al. (2011) they used 3D tomography techniques developed by Kramar *et al.* (2009) for STEREO COR1 3D observations of solar corona reducing it to 2D. These same techniques can be used for our full 3D modeling of the corona and solar wind. For 3D semi-empirical MHD model one will need a more realistic version of the empirical electron density model in Eq. 7.1 such as shown in Eq. 8.7 which will continue to display the desirable asymptotic solutions far from the Sun:

$$N_e(r, \theta, \phi) = N_0 z^2 P_1(z, \theta, \phi) \exp[-\alpha z P_2(z, \theta, \phi)] \tag{8.7}$$

For which now the functions $P_1(z, \theta, \phi)$ and $P_2(z, \theta, \phi)$ with $z = 1/r$ are Legendre Polynomials in co-latitude θ and azimuthal angle ϕ.

8.2. Set Boundary Conditions

In order to solve for the above equations certain field line constants and boundary conditions must be established. The constant $\alpha = \rho V/B$ at the equator can be set using the mass flux equation at 1 AU by knowing the solar wind speed $V_{P\infty}$, the electron density $N_e(R_{AU})$ and the radial component of the magnetic field B_r which can all be measured by near Earth spacecraft. The field line equation can be used for **B**(r) and using the density and velocity variations with latitude using the *Ulysses* plasma observations one can solve for $\alpha(r, \theta)$ for all field lines. This will then allow us to solve for the integral for T_{eff} which is integrated along an open field line to 1 AU. Once this is done, one can the use the equation for q_{eff} to solve for the heat flux for all (r, θ). An example of such 2D maps of flow velocity, effective temperature and effective heat flux are given in Sittler & Guhathakurta (1999, 2002).

9. Summary and Future Work

We have presented our solar eclipse observations for the August 21, 2017 eclipse and the July 2, 2019 eclipse. The seeing was excellent in both cases. Both sets of images have used raw data files which will allow quantitative analysis of these images and convert them into electron density profiles as outlined in Guhathakurta *et al.* (1996) and references therein. For the future we plan to process these images as outlined in this paper and then convert them into 2D maps of the solar wind velocity, effective temperature and effective heat flux for which the 2019 eclipse observations will be used when the Sun is near solar minimum and the 2D axisymmetric assumption is most applicable. Hopefully this effort will be successful and can be compared with other publications based on similar eclipse observations. Within the equatorial regions it will be interesting to see how they compare to the Parker Solar Probe observations.

Even further into the future, using a more general approach as previously discussed, 3D MHD semi-empirical model could be developed but instead of setting our boundary conditions at 1 AU or far from the Sun, one needs to set those constraints near the Sun. A key to this is to use as a local condition to solve for the field-line constant on individual flux tubes with $\alpha(r, \theta, \phi) = \rho V_{||}/B_{||}$ using the local field-aligned components of $\mathbf{V}$ and $\mathbf{B}$. Also, note locally 1/B gives the area of a unit flux tube. Once $\alpha(r, \theta, \phi)$ is known then the other equations can be integrated for effective temperature and effective heat flux. These can then be used as constraints on the 3D MHD solutions which will give the 3D magnetic field topology which can then be used for a next iteration of the semi-empirical solution until convergence occurs. A more advanced version of STEREO may be required for 3D solutions of the lower to middle corona as discussed here.

Acknowledgments

We would like to acknowledge the assistance provided by friends in Boise Idaho, Debbie Lombard and Jerry Bloom, for our August 21, 2017 Solar Eclipse observations in the Idaho mountains as shown in Figure 1 of our paper. In addition, we would like to thank the organizers of the IAU 354 conference in Copiapo Chile, especially Dr. Alexander Kosovichev who helped us throughout the process of attending the meeting from beginning to end dating back to January 2019 and Dr. Giovanni Leone for all his help and encouragement during the conference and to Dr. Christian Hummel who assisted us in getting our telescope set up and last minute repairs and realignment of the telescope at the observation site due to its damage incurred from the long trip from Maryland, USA to Copiapo, Chile.

References

Airapetian, V., Ofman, L., Sittler, E. C., Jr., & Kramar, M. 2011. Probing the thermodynamics and kinematics of solar coronal streamers, *Astrophys. J.*, 728:67, 1.

Beskeen, P. 2009, http://www.beskeen.com/projects/dsir_serial/dsir_serial.shtml version 2.12, March 2009.

Clark, R. N. 2013, http://www.clarkvision.com/articles/digital.photons.and.qe/, last updated January 4, 2013.

Druckmüller, M. 2009, Phase correlation method for the alignment of total solar eclipse images, *Astrophys. J.*, 706, 1605–1608, doi:10.1088/0004-637X/706/2/1605.

Habbal, S. R., Cooper, J. F., Daw, A., Ding, A., Druckmüller, M., Esser, R., Johnson, J., & Morgan, H. 2011, Exploring the Physics of the corona with total solar eclipse observations, https://arxiv.org/abs/1108.2323 arXiv:1108.2323 [astro-ph.SR].

Guhathakurta, M., Holzer, T. E., & MacQueen, R. M. 1996, The large-scale density structure of the solar corona and the heliospheric current sheet, *Astrophys. J.*, 458, 817–831.

Hulst, H. C. van de, 1950, The electron density of the solar corona, Bull. Astron. Inst. Netherlands, 11, 135–150.

Hulst, H. C. van de, 1953, Chapter 5: The chromosphere and the corona, in *The Sun* (Univ. of Chicago Press).

Kramar, M., Jones, S., Davila, J., Inhester, B., & Mierla, M. 2009, On the tomographic reconstruction of the 3D electron density for the solar corona from STEREO COR1 data, *Solar Physics*, 259, 109–121.

Reddy, S. B. & Chatterji, B. N., 1996, An FFT-Based technique for translation, rotation, and scale-invariant image registration, IEEE Trans. Image Process., 5, 1266–1271.

Schild, R. E., Peterson, D. M., & Oke, J. B. 1971, Effective temperatures of B- and A- Type stars, *Astrophys. J.*, 166, 95.

Sittler, E. C., Jr. & Guhathakurta, M. 1999, Semiempirical two-dimensional magnetohydrodynamic model of the solar corona and interplanetary medium, *Astrophys. J.*, 523, 812–826.

Sittler, E. C., Jr. & Guhathakurta, M. 2002, Erratum: "Semiempirical two-dimensional mag-netohydrodynamic model of the solar corona and interplanetary medium" (ApJ, 523, 812 [1999]), *Astrophys. J.*, 564, 1062–1065.

Sittler, E. C., Jr. & Ofman, L. 2006, 2D MHD model of the solar corona and solar wind: Recent results, ILWS Workshop 2006, Goa, February 19–20, 2006.

Sumner, R. 2014, Processing RAW Images in MATLAB, https://rcsumner.net/raw_guide/RAWguide.pdf, May 19, 2014.

Woo, R. 2019, Naked eye observations of the 2017 total solar eclipse: a more complete understanding of the white-light corona, *MNRAS* 485, 4122–4127.

Solar and Stellar Magnetic Fields: Origins and Manifestations
Proceedings IAU Symposium No. 354, 2019
A. Kosovichev, K. Strassmeier & M. Jardine, ed.
doi:10.1017/S1743921320001532

Realistic 3D MHD modeling
of self-organized magnetic structuring
of the solar corona

Irina N. Kitiashvili[1]![ORCID], Alan A. Wray[1], Viacheslav Sadykov[1],
Alexander G. Kosovichev[1,2] and Nagi N. Mansour[1]

[1]NASA Ames Research Center Moffett Field, MS 258-6, Mountain View, USA
email: `irina.n.kitiashvili@nasa.gov`
[2]New Jersey Institute of Technology, Newark, NJ 07102, USA
email: alexander.g.kosovichev@njit.edu

Abstract. The dynamics of solar magnetoconvection spans a wide range of spatial and temporal scales and extends from the interior to the corona. Using 3D radiative MHD simulations, we investigate the complex interactions that drive various phenomena observed on the solar surface, in the low atmosphere, and in the corona. We present results of our recent simulations of coronal dynamics driven by underlying magnetoconvection and atmospheric processes, using the 3D radiative MHD code StellarBox (Wray *et al.* 2018). In particular, we focus on the evolution of thermodynamic properties and energy exchange across the different layers from the solar interior to the corona.

Keywords. convection; plasmas; shock waves; turbulence; waves; (magnetohydrodynamics:) MHD; Sun: corona, magnetic fields; methods: numerical, Sun: transition region

1. Introduction

High interest in solar coronal structure and dynamics is primarily driven by interest in the episodic massive energy releases that can cause significant impacts on the Earth's space environment and in fundamental physical problems such as coronal heating and energy transport.

Recent achievements in realistic 3D radiative MHD modeling, in combination with multiwavelength observations, have provided a solid basis for investigating complex dynamical interactions in the solar atmosphere, from the upper layers of the convection zone to the corona. Realistic modeling, which takes into account the nonlinear coupling of turbulence, magnetic fields, and radiation, provides a physics-based interpretation of the observed phenomena and allows us to determine their primary physical mechanisms. An important feature of this approach is building the models from first physical principles, such that the physical processes develop spontaneously driven by dynamical energy flow from the solar interior. These models can reproduce a wide range of phenomena observed on different spatial scales, such as heating events in the chromosphere and transition region, small-scale dynamos, generation of different types of waves, jets, pore formation, sunspot-like and tornado-like structures, loops, etc. (e.g. Rempel *et al.* 2009; Cheung *et al.* 2010, 2019; Kitiashvili *et al.* 2010, 2013, 2015, 2019b; Stein & Nordlund 2012; Carlsson *et al.* 2016; Chen *et al.* 2017; Iijima & Yokoyama 2017; Snow *et al.* 2018).

Extension of the computational domain into the solar corona provides a unique opportunity to trace disturbances from subsurface layers through many layers of the atmosphere and allows investigation of the realistic structure and dynamics of the transition region and corona, which are critically important for understanding variations in space weather. Currently only a few realistic models of the solar corona have been developed (Gudiksen & Nordlund 2005; Rempel 2017; Cheung *et al.* 2019; Kitiashvili *et al.* 2019a, 2020) that allow detailed comparisons with observed phenomena. In this paper, we present 3D MHD simulations of the solar corona that reveal the spontaneous formation of a funnel-like magnetic structure. Such structures can play a significant role in the fine structuring of the corona and the formation of the solar wind.

2. Structure and dynamics of the solar corona

Generation of the 3D radiative MHD model, which covers dynamics of the solar plasma from the subsurface to the corona, is performed using the StellarBox code (Wray *et al.* 2018). The StellarBox code simulates the fully compressible MHD equations with radiative transfer and includes a large-eddy simulation (LES) treatment of subgrid turbulent transport. The StellarBox code includes subgrid turbulence models for flows (Smagorinsky 1963; Moin *et al.* 1991) and magnetic fields (Theobald *et al.* 1994) that are critical to obtain a more accurate description of small-scale energy dissipation and transport.

The computational domain of $12.8 \times 12.8 \times 15.2$ Mm includes a 10-Mm high layer from the photosphere to the corona. The grid spacing is 25 km in the horizontal directions, with a variable vertical grid-spacing of similar size in the photosphere. The lateral boundary conditions are periodic. The top boundary is open to mass, momentum, and energy fluxes, and also to radiation flux. The bottom boundary in these simulations is open only for radiation, to simulate the energy input from the interior of the Sun. The simulation is initialized from a standard solar model of the interior structure and the lower atmosphere (Christensen-Dalsgaard *et al.* 1996). For the initial conditions of the chromosphere and corona, the model by Vernazza *et al.* (1981) is used.

In these simulations, we introduced an initially uniform vertical magnetic field of $B_{z0} = 10$ G. The boundary conditions conserve the total flux but the field freely evolves inside the computational domain. In the near-surface layers, the magnetic field amplifies and concentrates into magnetic patches due to a small-scale dynamo and collapsing field in the intergranular lanes. Magnetic field concentrations with a field strength of about 1 kG are strong enough to hold their magnetized structure extending into the corona. The complexity of the structure and its dynamics increases with hight as it often splits into the substructures. In particular, there are spontaneously-formed helical patterns of different scales, which appear and disappear with time. The vertical velocity distribution is highly inhomogeneous (Figure 1a) and reveals strong upflows, exceeding 100 km/s, and downflows with speeds of tens kilometers per second. Notably, complex flows and numerous strong shock waves are excited during the formation and 'active' evolution of the structure. During the structure's decay phase, the amplitude of fluctuations associated with shock waves decreases.

The self-organized magnetic structure in the solar corona is primarily associated with cold plasma (Figure 1b). However, the numerous current sheets that are in the coronal structure are able to heat the plasma to a few million degrees. The magnetic field strength drops significantly with height. However, inside the magnetized region the field is still relatively strong, $24 - 27$ G (Fig. 1c) at a height 10 Mm above the photosphere.

In addition to the ubiquitous shock waves in the corona, the simulations reveal high-frequency oscillations initiated in the transition zone. Figure 1d shows a vertical slice through a fraction of the computational domain. The color-scale corresponds to divergence of the velocity field. It illustrates small-scale perturbations associated with the propagation of the high-frequency perturbations from the transition region (the location of which is indicated by two constant temperature curves for 0.1MK (black) and

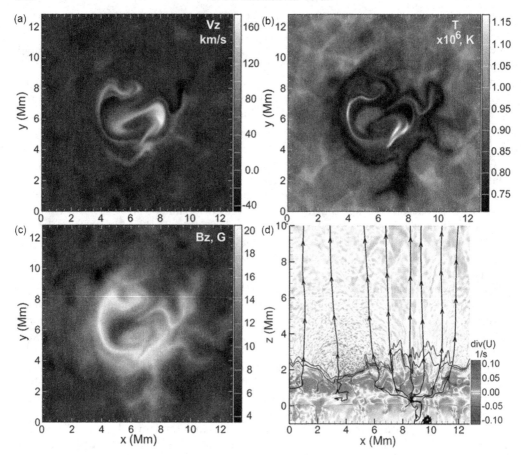

Figure 1. Distributions of (a) vertical velocity, b) temperature, and c) magnetic field in a quiet Sun region at height 10 Mm above the photosphere. The vertical slice of divergence of the velocity in panel d) shows propagation of high-frequency perturbations from the transition zone to upper layers of the corona. Two thin color curves (black and red) in panel d) show the temperature levels of 0.1MK and 0.5MK, respectively, in the transition region. Black lines with arrows indicate magnetic field lines.

0.5MK (red curve)) into the high corona. To characterize the gas pressure perturbations, we selected several areas inside and outside the magnetic structure at a height of 9 Mm above the photosphere and plotted their variations as a function of time in Figure 2. Outside the magnetic structure (warm colors) the plot reveals the high-frequency oscillations of low amplitude superimposed on a high mean, whereas in the area inside where there is a stronger magnetic field (cold colors), the variations have a lower frequency higher amplitude with low mean values. At times inside the magnetized structure, high-frequency oscillations are also present and in such cases reflect a local decrease of the magnetic field strength.

The model reveals frequent eruptive activity on small scales. Therefore there is particular interest in the associated helicity and vorticity transport. Figure 3a shows a time-height diagram (averaged over horizontal plains) of the flow enstrophy, which shows quasi-periodic enstrophy perturbations with a period about of 3 min generated near the transition region (panel b). Near the transition zone, the enstrophy transport speed is in the range of 16 − 35 km/s for different events, whereas above 4 Mm height, the speed reaches 76 − 115 km/s.

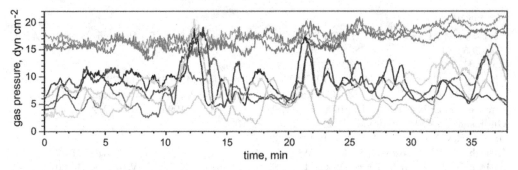

Figure 2. Temporal variations of gas pressure inside (cold colors) and outside (warm colors) the funnel structure at a height of 6Mm above the photosphere.

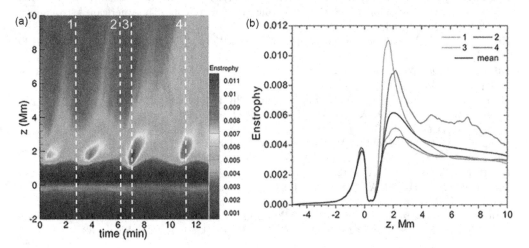

Figure 3. Enstrophy transport from the transition zone to the corona. Panel (a) Time-height diagram of the enstrophy shows propagation of the helical flows from the transition region into the corona. The numbered white dashed lines corresponds to the vertical enstrophy profiles shown in the panel b) for selected moments of time.

3. Conclusions

Understanding the fine thermodynamical and magnetic structure of the solar corona is an essential basis for interpreting observational data. The performed 3D radiative MHD simulations performed, which cover layers from the convection zone to the corona, reveal complex multi-scale dynamics, which causes formation of self-organized magnetic coronal structures originating from kG magnetic patches at the photosphere. In the solar corona, the magnetic field strength drops to several tens of Gauss, and the plasma inside the magnetized structure is colder. However, there are episodic strong plasma heating events, where the temperature reaches a few million degrees, produced by small-scale current sheets. Numerous shock waves are excited due to the dynamics of this magnetic funnel-like structure, and these also contribute to coronal heating. In addition, the simulation results predict high-frequency oscillations in weak magnetic field regions, originating in the transition zone. The mechanism of these oscillations is currently under investigation.

Acknowledgments

The research was supported by the NASA Grants NNX12AD05A, NNX14AB68G, NNX16AP05H, 80NSSC19K0630, and NSF grants: 1835958, 1916509.

References

Carlsson, M., Hansteen, V. H., Gudiksen, B. V., Leenaarts, J., & De Pontieu, B. 2016, *A&A*, 585, A4

Chen, F., Rempel, M., & Fan, Y. 2017, *ApJ*, 846, 149

Cheung, M. C. M., Rempel, M., Title, A. M., & Schüssler, M. 2010, *ApJ*, 720, 233

Cheung, M. C. M., Rempel, M., Chintzoglou, G., *et al.* 2019, Nature Astronomy, 3, 160

Christensen-Dalsgaard, J., Dappen, W., Ajukov, S. V., *et al.* 1996, Science, 272, 1286

Gudiksen, B. V. & Nordlund, Å. 2005, *ApJ*, 618, 1031

Iijima, H. & Yokoyama, T. 2017, *ApJ*, 848, 38

Kitiashvili, I. N., Kosovichev, A. G., Wray, A. A., & Mansour, N. N. 2010, *ApJ*, 719, 307

Kitiashvili, I. N., Kosovichev, A. G., Lele, S. K., Mansour, N. N., & Wray, A. A. 2013, *ApJ*, 770, 37

Kitiashvili, I. N., Couvidat, S., & Lagg, A. 2015, *ApJ*, 808, 59

Kitiashvili, I., Wray, A. A., Kosovichev, A. G., Sadykov, V. M., & Mansour, N. N. 2019a, in American Astronomical Society Meeting Abstracts, Vol. 234, American Astronomical Society Meeting Abstracts #234, 106.15

Kitiashvili, I. N., Kosovichev, A. G., Mansour, N. N., Wray, A. A., & Sandstrom, T. A. 2019b, *ApJ*, 872, 34

Kitiashvili, I. N., Wray, A. A., Kosovichev, A. G., Sadykov, V. M., & Mansour, N. N. 2020, in preparation

Moin, P., Squires, K., Cabot, W., & Lee, S. 1991, Physics of Fluids A, 3, 2746

Rempel, M., Schüssler, M., & Knölker, M. 2009, *ApJ*, 691, 640

Rempel, M. 2017, *ApJ*, 834, 10

Smagorinsky, J. 1963, Monthly Weather Review, 91, 99

Snow, B., Fedun, V., Gent, F. A., Verth, G., & Erdélyi, R. 2018, *ApJ*, 857, 125

Stein, R. F. & Nordlund, Å. 2012, *ApJ*, 753, L13

Theobald, M. L., Fox, P. A., & Sofia, S. 1994, Physics of Plasmas, 1, 3016

Vernazza, J. E., Avrett, E. H., & Loeser, R. 1981, *ApJS*, 45, 635

Wray, A. A., Bensassi, K., Kitiashvili, I. N., Mansour, N. N., & Kosovichev, A. G. 2018, Realistic Simulations of Stellar Radiative MHD, ed. J. P. Rozelot & E. S. Babayev, 39

Solar and Stellar Magnetic Fields: Origins and Manifestations
Proceedings IAU Symposium No. 354, 2019
A. Kosovichev, K. Strassmeier & M. Jardine, ed.
doi:10.1017/S1743921320000113

Coherent structures and magnetic reconnection in photospheric and interplanetary magnetic field turbulence

Rodrigo A. Miranda[1]ⓘ, **Abraham C.-L. Chian**[2,3,4,5]ⓘ, **Erico L. Rempel**[4,5]ⓘ and **Suzana S. A. Silva**[5]ⓘ

[1]UnB-Gama Campus, and Institute of Physics, University of Brasília (UnB), Brasília DF 70910-900, Brazil. email: rmiracer@unb.br

[2]School of Mathematical Sciences, University of Adelaide, Adelaide SA 5005, Australia.

[3]Institute for Space-Earth Environmental Research, Nagoya University, Nagoya 464-8601, Japan.

[4]National Institute for Space Research (INPE), São José dos Campos SP 12227-010, Brazil.

[5]Institute of Aeronautical Technology (ITA), São José dos Campos SP 12228-900, Brazil.

Abstract. In this paper it is shown that rope-rope magnetic reconnection in the solar wind can enhance multifractality in the inertial subrange and drive intermittent magnetic field turbulence. Additionally, it is shown that Lagrangian coherent structures can unveil the transport barriers of magnetic elements in the quiet Sun.

Keywords. turbulence, photosphere, intermittency, solar wind plasma

1. Introduction

Magnetic reconnection in plasmas refers to the conversion of magnetic energy into kinetic and thermal energy, resulting in a change of the topology of magnetic field lines (Yamada *et al.* 2010; Treumann & Baumjohann 2013; Lazarian *et al.* 2015). The study of magnetic reconnection is key to understand the dynamical manifestations of solar and stellar magnetic fields such as solar and stellar flares, coronal mass ejections, and their effects on star-planet interactions. Several models have been proposed to understand magnetic field reconnection (e.g., Parker 1957; Sweet 1958; Petschek 1964; Sonnerup *et al.* 1981; Lazarian & Vishniac 1999).

The interplanetary medium is permeated by the solar wind and provides a natural laboratory in which theoretical models of magnetic reconnection can be validated. The solar wind can be regarded as a network of entangled magnetic flux ropes and Alfvénic fluctuations propagating within each flux rope (Bruno et al., 2001; Borovsky, 2008). Flux ropes can emerge locally in the turbulent solar wind (Mattheaus and Montgomery, 1980, Greco et al., 2009; Telloni et al., 2016) or can be advected from the solar surface to the interplanetary medium by the solar wind (Bruno et al., 2001; Borovsky, 2008). The interaction between flux ropes can lead to magnetic reconnection and the generation of intermittent magnetic field turbulence.

The quiet Sun is the region of the solar photosphere outside of sunspots, plages and active regions (Bellot Rubio & Orozco Suárez 2019). It holds a significant fraction of the

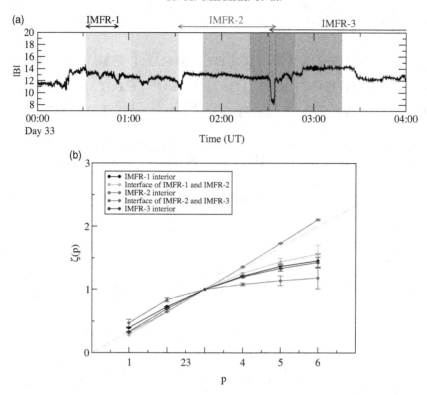

Figure 1. Upper panel: the modulus of the magnetic field detected by Cluster-1 during a rope-rope magnetic reconnection event. Horizontal lines indicate the duration of three ropes, and the violet vertical dashed lines indicate the reconnection site between the second and third IMFRs. Lower panel: the scaling exponents as a function of the order of the structure function. The grey dashed line corresponds to the K41 monofractal scaling.

photospheric magnetic flux, which emerges and disappears through cancellation processes in short time scales. For this reason, the magnetic flux observed on the quiet Sun can contribute effectively to the heating of the chromosphere and the corona. The quiet Sun displays patterns of intense magnetic fields known as the magnetic network that coexist with small-scale weaker magnetic fields called the solar internetwork. The magnetic network outlines the boundaries of supergranule cells within which the internetwork is found.

In this paper, we focus on the role of coherent structures in solar and interplanetary magnetic field turbulence. First, we show that rope-rope reconnection in the solar wind is the source of coherent structures and multifractality in the inertial subrange and drives intermittent magnetic field turbulence. Next, we detect Lagrangian coherent structures (LCS) using surface velocity data in the quiet Sun and discuss the role of LCS on the dynamics of magnetic elements in supergranular cells.

2. Magnetic reconnection in the solar wind

Figure 1(a) shows the time series of the modulus of the magnetic field $|\mathbf{B}|$ (black line) detected by Cluster-1 on 2 February 2002. During this period, Cluster collected data from the solar wind upstream of the Earth's bow shock (Chian & Miranda 2009). In this event, three interplanetary magnetic flux ropes (IMFRs) were identified, as well as magnetic reconnection and a bifurcated current sheet in the interface region between two IMFRs (Chian *et al.* 2016). The duration of each IMFR is indicated in Fig. 1(a) by

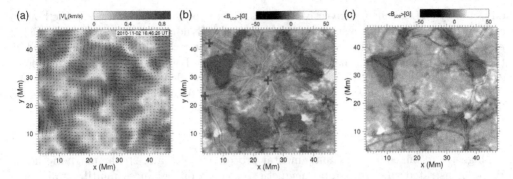

Figure 2. (a) The 2-hourly time-averaged horizontal velocity field (black arrows) deduced by applying LCT to the intensity maps and the horizontal velocity modulus (the background image) at 16:46:26 UT. (b) The forward-time FTLE field (orange), and (c) the backward-time FTLE field (green) computed for $\tau = 7$ h. The background grey-scale images represent the line-of-sight magnetic field averaged over 7 h. The value of the FTLE field is proportional to the colour intensity. Ridges of the FLTE fields are located in regions in which the colour intensity is higher. Blue crosses in (b) mark the Lagrangian centre of supergranular cells.

horizontal arrows. Five intervals with a duration of 30 minutes were selected during this interval, indicated by different background colors. These intervals represent the interior region of IMFR-1 (grey), the interface between IMFR-1 and IMFR-2 (green), the interior of IMFR-2 (red), the interface between IMFR-2 and IMFR-3 (violet) and the interior of IMFR-3 (blue). Magnetic reconnection occurs at the interface between the IMFR-2 and IMFR-3, bounded by the two dashed vertical lines.

In order to characterize the multifractality of each interval, we compute the scaling exponents ζ of structure functions within the inertial subrange. Figure 1(b) shows ζ as a function of the p-th order structure function. The Kolmogorov (K41) monofractal scaling is represented by a dashed line. Departure from the K41 scaling indicates multifractality due to the presence of coherent structures at scales within the inertial subrange. The interval with the strongest departure corresponds to the interface between the IMFR-2 and the IMFR-3, during which rope-rope magnetic reconnection occurs. This suggests that magnetic reconnection acts as the source of the intermittent magnetic field turbulence.

3. Lagrangian coherent structures in the quiet Sun

We detect LCS using data of photospheric horizontal velocity fields derived from continuum intensity images of the quiet Sun taken at the disc centre. The image data was captured by Hinode on 2 November 2010. The data have a cadence of 90 s, a field of view of 80 arcsec x 74 arcsec, a pixel size of 0.16 arcsec and spatial resolution of 0.3 arcsec (Gošić *et al.* 2014). The horizontal velocity fields shown in Fig. 2(a) for t =16:46:26 UT were extracted using the local correlation tracking method (November & Simon 1988; Molowny-Horas 1994; Requerey *et al.* 2018).

Figure 2 (b) and (c) show the forward and backward finite-time Lyapunov exponents (FTLE), superposed by the line-of-sight component of the magnetic field on a quiet sun observed by Hinode. The value of the FTLE field is proportional to the colour intensity. Ridges of the forward-FTLE and backward-FTLE indicate the locations of repelling and attracting LCS, respectively. These ridges are visualized in Fig. 2 as regions in which the corresponding colour (orange for the forward-FTLE, green for the backward-FTLE) is more intense. The blue crosses mark the centre of supergranular cells obtained by the Lagrangian method. Magnetic elements located in the interior of supergranular cells are advected by the turbulent convective flow along the repelling LCS and reach the boundaries of the supergranular cell where they are advected along the attracting LCS (Chian *et al.* 2019). From this figure it is clear that the internetwork dynamics are dominated by the repelling LCS, whereas the attracting LCS mark the Lagrangian boundaries of supergranular cells. The forward-FTLE has been shown to be closely related to the squashing Q-factor (Yeates, Hornig & Welsch 2012; Chian *et al.* 2014), therefore the repelling LCS indicate regions where magnetic elements are most likely to interact and reconnect.

4. Conclusions

Magnetic reconnection is a fundamental process of solar and stellar magnetic fields. We demonstrated that rope-rope magnetic reconnection enhances multifractality and intermittent turbulence in the solar wind. We also showed that LCS can unveil the paths of magnetic elements advected by photospheric flows, and can detect the regions where magnetic elements are most likely to interact and reconnect in the quiet Sun. These results can contribute to the understanding of magnetic reconnection and the heating processes of the solar and stellar chromospheres and coronae.

Acknowledgements

The authors thank CAPES, CNPq, FAPESP and FAPDF for support. R.A.M. thanks Juan A. Valdivia, Pablo R. Muñoz, Elías M. Ovalle and Félix A. Borotto for fruitful discussions. R.A.M. is also gratefully indebted to Norma Cerda and Rodolfo Sucre for their kind hospitality.

References

Bellot Rubio, L & Orozco Suarez, D. 2019, *Liv. Rev. Solar Phys.*, 16, 1.
Chian, A. C.-L. & Miranda, R. A. 2009, *AnGeo*, 27, 1789.
Chian, A. C.-L., Rempel, E. L., Aulanier, G., Schmieder, B., Shadden, S. C., Welsch, B. T. & Yeates, A. R. 2014, *ApJ*, 786, 51.
Chian, A, C.-L., Feng, H. Q., Hu, Q., Loew, H. M., Miranda, R. A., Muñoz, P. R., Sibeck, D, G. and Wu, D. J. 2016, *ApJ*, 832, 179.
Chian, A. C.-L., Silva, S. S. A., Rempel, E. L., Gošić, M., Bellot Rubio, R. L., Kusano K., Miranda, R. A. & Requerey, I. S. 2019, *MNRAS*, 488, 3076.
Gošić, M., L. R. Bellot Rubio, D. Orozco Suárez, Y. Katsukawa & J. C. del Toro Iniesta 2014, *ApJ*, 797, 49.
Lazarian, A. & Vishniac, E. T., 1999 *ApJ*, 517, 700.
Lazarian, A., Eyink, G., Vishniac, E. & Kowal, G. 2015, *Phil. Trans. R. Soc. A*, 373, 20140144.
November, L.-J. & Simon, G.-W. 1988, *ApJ*, 333, 427.
Molowny-Horas, R. 1994, *Solar Phys.*, 154, 29.
Parker, E. N. 1957, *J. Geophys, Res.* 62, 509.
Petschek, H. E. 1964, *AAS-NASA Symp.*, ed. W. H. Hess, 425
Requerey, I. S., Cobo, B. R. Gošić, M. & Bellot Rubio, L. R. 2018, *A&A*, 610, A84.
Sonnerup, B. U. Ö, Paschmann, G. Papamastorakis, I., Sckopke, N, Haerendel, G., Bame, S. J., Asbridge, J. R., Gosling, J. T. and Russell, C. T. 1981, *J. Geophys. Res.* 86, 10049.
Sweet, P. A. 1958 *The Observatory* 78, 30.
Treumann, R. A. & Baumjohann, W. 2013, *Front. Phys.* 1, 31.
Yamada, M., Kulsrud, R. & Ji, H. 2010, *Rev. Modern Phys.*, 82, 603.
Yeates, A. R., Hornig, G., & Welsch, B. T. 2012, *A&A*, 539, A1.

Solar and Stellar Magnetic Fields: Origins and Manifestations
Proceedings IAU Symposium No. 354, 2019
A. Kosovichev, K. Strassmeier & M. Jardine, ed.
doi:10.1017/S1743921319009931

Analysis of the chromosphere and corona of low-activity early-M dwarfs.

Gaetano Scandariato[1], E. González Álvarez[2,3], J. Maldonado[2],
A. Suárez Mascareño[4,5], M. Perger[6,7] and the HADES collaboration

[1]INAF - Osservatorio Astrofisico di Catania, via S. Sofia 78, 95123 Catania, Italy
email: gaetano.scandariato@inaf.it

[2]INAF – Osservatorio Astronomico di Palermo, Piazza del Parlamento, 1, I-90134,
Palermo, Italy,

[3]Dipartimento di Fisica e Chimica, Università di Palermo, Piazza del Parlamento 1, I-90134
Palermo, Italy,

[4]Instituto de Astrofísica de Canarias, 38205 La Laguna, Tenerife, Spain

[5]Universidad de La Laguna, Dpto. Astrofsica, 38206 La Laguna, Tenerife, Spain

[6]Institut de Cincies de lEspai (ICE, CSIC), Campus UAB, C/ Can Magrans, s/n, 08193
Bellaterra, Spain

[7]Institut d'Estudis Espacials de Catalunya (IEEC), 08034 Barcelona, Spain

Abstract. While most of the exoplanets have been found orbiting around solar-type stars, low-mass stars have recently been recognized as ideal exo-life laboratory. Currently, stellar activity is one of the limiting factors for the characterization of Earth-twins and for assessing their habitability: understanding the activity of M dwarfs is thus crucial. In this contribution I present the spectroscopic analysis of the quiet early-M dwarfs monitored within the HADES (HArps-n red Dwarf Exoplanet Survey) radial velocity survey. The spectra allow us to analyze simultaneously the Ca II H&K doublet and the Hydrogen Balmer series, while the intensive follow up gives us a large number of spectra ($\lesssim 100$) for each target. We complement this dataset with ground-based follow-up photometry and archival X-ray data. I present our results on the activity-rotation-stellar parameters and flux-flux relationships, and discuss the correlation of emission fluxes at low activity levels and the evolution timescales of active regions.

Keywords. stars: activity, stars: atmospheres, stars: chromospheres, stars: coronae,stars: late-type, stars: low-mass, brown dwarfs, stars: rotation,stars: spots

1. Introduction

Recent planet search programs have started to monitor samples of M dwarfs, as they are are extremely interesting for the discovery of Earth-like planets. As a matter of fact, M dwarfs represent $\sim 75\%$ of the stars in the solar neighborhood (Reid *et al.* 2002; Henry *et al.* 2006). Moreover, for a number of observational constraints, the chances of finding a temperate rocky planet (i.e. a planet at a convenient density and distance from its host star such to have liquid water on its surface) increase as the stellar mass decreases. Such planets are said to be inside the habitable zone of their host stars. Still, habitability is not guaranteed simply by an assessment of the distance from the star; several other factors, such as stellar activity and the strength of the stellar magnetic field, may move and/or shrink the habitability zone of a star (see Vidotto *et al.* 2013, and references therein). Thus, it is crucial to better understand the activity of M dwarfs and how it can affect the circumstellar environment.

The term "stellar activity" refers to a large class of phenomena that, in stars with convective envelopes, are triggered by the reconfiguration of the surface magnetic field. These phenomena are commonly classified as spots, plages, flares, or coronal holes, and they take place throughout the stellar atmosphere. The places where these phenomena take place are called "active regions" (ARs). In this contribution I will discuss the results of an extensive analysis of the stellar activity of a sample of low-activity M dwarfs, which have been selected and monitored for planet-search purposes and for which we have collected measurements of photospheric, chromospheric and coronal activity indicators. My discussion is based on a group of four papers which we have recently published (Maldonado *et al.* 2017; Scandariato *et al.* 2017; Suárez Mascareño *et al.* 2018; González-Álvarez *et al.* 2019) and which the reader is encouraged to check in order to obtain further details on our analysis.

2. The project, the stellar sample and the data

The HArps-N red Dwarf Exoplanet Survey project (HADES, Affer *et al.* 2016; Perger *et al.* 2016) is a collaborative program between the Global Architecture of Planetary Systems project† (GAPS, Covino *et al.* 2013), the Institut de Ciències de l'Espai (ICE/CSIC, IEEC), and the Instituto de Astrofísica de Canarias (IAC). The aim of the project is to monitor the radial velocities of a sample of low-activity M-type dwarfs and discover new exoplanets.

Within the HADES framework we have monitored 71 stars with spectral types ranging between K7.5V and M3V (corresponding to the ~3400–3900 K temperature range). These stars were observed with the HARPS-N spectrograph (Cosentino *et al.* 2012) mounted at the Telescopio Nazionale Galileo. The spectrograph covers the 383–693 nm wavelength range with a spectral resolution of ~115,000. HARPS-N spectra were reduced using the most recent version of the Data Reduction Software (DRS) pipeline (Lovis & Pepe 2007).

The targeted stars have been selected as favorable for planet search, by consequence the sample is biased towards low activity levels with some exceptions. The data analyzed in this paper have been collected over seven semesters, from September 2012 to February 2016, and for each star we collected up to ~100 spectra. Affer *et al.* (2016) give further details on the target selection and Maldonado *et al.* (2017) describe the determination of the stellar parameters.

HARPS-N spectra have been flux calibrated by comparison with a low-resolution synthetic spectral library. The stars showing the minimum emission in the cores of the Ca II H&K lines (commonly used as a tracer of chromospheric activity) are selected as spectral templates with minimum contribution by chromospheric emission, accounting that they span the full range of effective temperature covered by our stellar sample. For each star, the corresponding "quiet" spectral template is selected as the one with the closest effective temperature. Finally the spectra are subtracted by the selected template, such to obtain the "chromospheric" spectrum. Fluxes of the Ca II H&K and the hydrogen Balmer line (from Hα to Hϵ) are extracted by integrating the residual spectrum over convenient wavelength ranges.

Optical UBVRI photometry has been collected simultaneously to the HARPS-N campaign in the framework of the EXORAP project. This project consists in the nightly follow up of our sample of M dwarfs with an 80 cm f/8 Ritchey-Chretien robotic telescope (APT2) located at Serra la Nave (+14.973 E, +37.692N, 1725 m a.s.l.) on Mt. Etna and operated by INAF-Catania Astrophysical Observatory. The main goal of this project is to provide optical differential light curves by which we can characterize the photometric variability and measure the stellar rotation period.

† http://www.oact.inaf.it/exoit/EXO-IT/Projects/Entries/2011/12/27_GAPS.html

With the purpose of complementing our analysis with the coronal activity, we searched the catalogs of the X-ray missions XMM-Newton, Chandra and ROSAT, following this order of preference. Even if ROSAT is older and less sensitive than the most recent X-ray observatories, it is still the main source of X-ray measurements since it covered all the sky. In total, we find X-ray counterparts for 37 stars in our selected sample.

3. Stellar rotation, activity cycles and ARs evolution timescale

We look for periodic variability in the time-series of measurements of indicators compatible with both stellar rotation and long term magnetic cycles. We compute the power spectrum using a Generalised Lomb Scargle (GLS) periodogram (Zechmeister & Kurster 2009) and, if there is any significant periodicity, we fit the data using a sine function with the detected period. Then we repeat the same process in the residuals of the fit. Typically this allowed us to determine the stellar rotation, and in some cases to unveil the presence of an activity cycle. Following this procedure we measure the rotation periods of 32 of the stars of our sample and identify magnetic cycles in 14 stars.

Stars in our sample are typically slow rotators, as we find an average rotation period of ~30 days for early M-dwarfs and of ~80 days for mid M-dwarfs. This is consistent with the fact that our sample is biased towards low activity levels, which correspond to slow rotation rates. As a matter of fact, in Fig. 1 we show that the rotation period of stars in our sample strongly anti-correlates with the R'_{HK} activity index. This trend is consistent with other samples of M dwarfs previously analyzed.

Cycle lengths are typically of a few years, with the longest cycles detected at almost 14 years. For the cycle measurements we are limited by the time-span of the survey. Cycles longer than ~3–4 years are not measurable by our campaign on its own, they always require support data (previous HARPS spectra or ASAS photometry). There is a hint of a lower envelope of the distribution of cycle lengths against the color B-V (hence, the stellar mass) which might indicate that lower mass stars have longer magnetic cycles (Fig. 1).

We analyze the same data using the Pooled Variance (PV) approach (Donahue *et al.* 1997a). The PV measures the average variance in the data over a given time bin τ running along the series of measurements. When τ increases, the PV remains constant until the effects from processes with longer timescales become more noticeable, in which case the PV increases with τ.

We apply this algorithm to the time series of measurements of Ca II H&K and Hα indices. We complement this analysis using also the V-band photometry collected in the EXORAP survey. In general, for each star we find that the PV diagrams of the three proxies look alike (see Fig. 1 for an example): they generally increase at small timescales τ; they reach a plateau at roughly 10–40 days, and then increase to level off again at $\tau \gtrsim 50$ days. Our interpretation of these diagrams is that the first plateau corresponds to the stellar rotation period, consistently to the GLS analysis. The second flattening is likely related to the growth and decay of chromospheric ARs, as also suggested by Donahue *et al.* (1997a) and Donahue *et al.* (1997b). Our results are consistent with other analysis of M dwarfs, which indicate that ARs of main sequence M dwarfs evolve with timescales of ~100 days (Davenport *et al.* 2015; Robertson *et al.* 2015; Newton *et al.* 2016).

In Fig. 2 we plot the relationship between the stellar rotation and the X-ray luminosity for M stars. In particular we compare the sample of M dwarfs analyzed by Pizzolato *et al.* (2003) with our sample of stars, divided into two subsamples depending on stellar effective temperature. The former are in the saturated regime, i.e. the L_X/L_{bol} ratio is maximum and does not depend on the rotation period. These stars are typically more active than ours, for which we find that the X-ray luminosity decreases with increasing

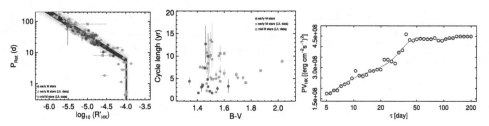

Figure 1. *Left panel* - Rotation periods in our stellar samples (red dots) as a function of the R'_{HK} activity proxy. Gray symbols mark other published samples of main sequence M dwarfs. *Middle panel* - Cycle length as a function of B-V color. Symbols are the same as in the left panel. *Central panel* - Activity cycle length vs. B-V colors for different samples of M dwarfs. *Right panel* - Example of the PV analysis showing the stellar rotation period (the flattening at ∼15 days) and the ARs lifetime (the flattening at ∼50 days).

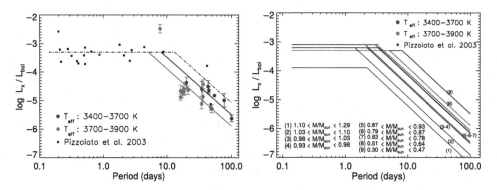

Figure 2. *Left panel* - L_X/L_{bol} vs. stellar rotation for different samples of M dwarfs. *Right panel* - L_X/L_{bol} vs. stellar rotation loci for different samples of main sequence stars.

rotation period (i.e. stars in our sample are in the non-saturated regime). Moreover, we find that the locus occupied in the diagram moves to the right side as the effective temperature (and the stellar mass) decreases. If we assume that the X-ray saturation level (the horizontal dashed-dotted line in the plot) does not depend on stellar mass, then this suggests that the saturation level is reached at increasing stellar rotation as stellar mass decreases. This behavior, across all mass bins, is shown in the right panel of Fig. 2, where we show the collection of all best-fit relations found by Pizzolato *et al.* (2003) together with the new relations found in our analysis.

4. Flux-flux relationships

For each star in our sample we compute the average excess flux for each chromospheric line. Fig. 3 shows the comparison between pairs of fluxes. Several samples from literature are overplotted for comparison: a sample of F, G, and K stars from López-Santiago *et al.* (2010), Martínez-Arnáiz *et al.* (2010), Martínez-Arnáiz *et al.* (2011), a sample of late-K and M dwarfs (from the same authors), and a sample of pre-MS M stars from Stelzer *et al.* (2013).

Our sample of M dwarfs seems to follow the same trend as FGK stars and other late-K/early-M dwarfs for the Ca II H&K doublet, for pairs of Balmer lines and for the Ca II K vs. Hα relationship without any obvious deviation between samples.

In Fig. 4 we show the averaged Ca II K vs. Hα relationship in linear scale, where for each star the error bars represent the scatter of the time series of measurements around the median value. We find that the trend is not monotonic: the flux radiated in the

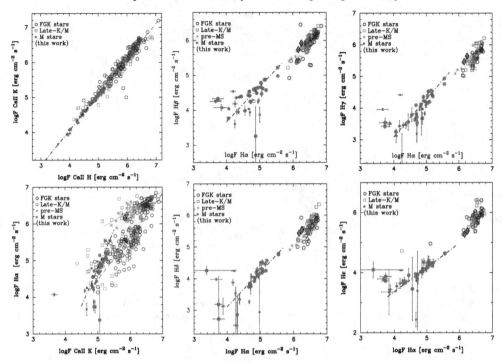

Figure 3. *Left column* - Correlations between Ca II H&K and Hα fluxes. *Central and right columns* - Correlations between the Balmer lines in the analyzed spectra. For the sake of comparison, in all plots we show other stellar samples from the studies discussed in the text.

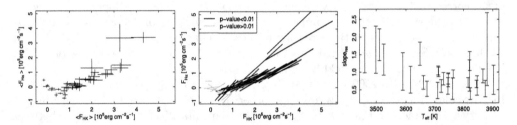

Figure 4. *Left panel* - Correlations between Ca II H&K doublet and Hα fluxes in linear scale. Error bars represent the scatter of the measurements in the time series. *Central panel* - Same as in the left panel, but representing each star as the linear best fit of the time series of measurements. The *p-values* of the correlation tests are also reported. *Right panel* - The slope of the linear best fits in black in the central panel as a function of stellar effective temperature.

Hα line initially decreases with increasing Ca II H&K excess flux, reaches a minimum, and finally increases with the Ca II H&K flux.

Previous observational studies performed on similar samples of stars (K- and M-type main sequence stars, e.g., Stauffer & Hartmann 1986, Giampapa *et al.* 1989, Robinson *et al.* 1990, Rauscher & Marcy 2006, Walkowicz & Hawley 2009) have found comparable results. This evidence has been also supported by chromospheric models (Cram & Mullan 1979, Cram & Giampapa 1987, Rutten *et al.* 1989, Houdebine *et al.* 1995, Houdebine & Stempels 1997), according to which the chromospheric Ca II H&K emission lines are collisionally dominated while the Hα line is radiation-dominated. With these assumptions, the Ca II H&K flux steadily increases with pressure, while the increase of the optical depth initially leads to a deeper absorption profile in the Hα line. This trend is reverted

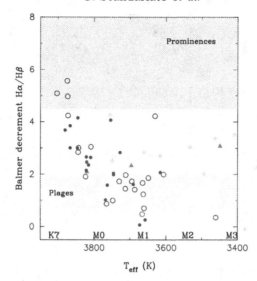

Figure 5. Balmer decrement $F_{H\alpha}/F_{H\beta}$ vs. the effective temperature. Stars in our sample are marked as circles. Green stars and red triangles mark the samples of Stelzer *et al.* (2013) and Bochanski *et al.* (2007) respectively. Typical ranges of solar plages and prominences are shown as hatched areas.

as soon as the electron density is high enough to take the Hα line into the collisionally dominated formation regime, leading to the fill-in of the line.

In Fig. 4 we also show the Ca II K vs. Hα relationship for each star using the bestfit line of the corresponding data, and we show the correlation between the slope of these bestfit lines with the stellar effective temperature. We find evidence of decreasing steepness of the flux-flux relationship with increasing effective stellar temperature with a confidence level of $\sim$4%. Stelzer *et al.* (2013) found a similar trend by analyzing a sample of pre-MS low-mass stars in the 2500–4500 K temperature range.

Meunier & Delfosse 2009 already analyzed the F_{HK} vs. $F_{H\alpha}$ relationship for the Sun. In particular, they use spatially resolved observations of the solar surface covering 1.5 activity cycles, and find that the presence of dark filaments affects the correlation between F_{HK} and $F_{H\alpha}$ on daily-to-weekly timescales, eventually leading to negative correlations. Moreover, for a fixed contrast of Hα plages on the solar disk, the slope of the F_{HK} vs. $F_{H\alpha}$ relationship tends to decrease with increasing contrast of the filaments. This is due to the fact that solar filaments are optically thicker in the Hα than in the Ca II H&K lines. Thus, the increase in slope with decreasing effective temperature that we find may indicate that the Hα absorption in equivalent active regions decreases towards later M types.

We draw the same conclusion analyzing the ratio between Hα and Hβ fluxes, which is an indicator of the physical conditions of the emitting regions (e.g. Landman, & Mongillo 1979; Chester 1991). Fig. 5 shows the Balmer decrement $F_{H\alpha}/F_{H\beta}$ as a function of the effective temperature, together with solar plages and prominences for comparison. The stars in our sample show a decreasing trend in Balmer decrement towards later spectral types. Once again, stars in the spectral range between M0 and M1.5 are compatible with a filament-dominated scenario, while chromospheres of later spectral type stars are dominated by plages. This statement is also consistent with pre-MS M stars (Stelzer *et al.* 2013) and the active M dwarf templates from the Sloan Digital Sky Survey (Bochanski *et al.* 2007).

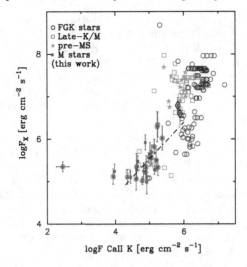

Figure 6. X-ray flux vs. Ca II K. Symbols are the same as in Fig. 3.

In addition to the flux-flux relationships between different chromospheric activity indicators, we also studied the chromospheric-coronal relation. Our sample of M dwarfs seems to follow the same trend as other literature stellar samples and extend it towards lower activity levels (Fig. 6).

5. Summary

In the context of the HADES project, we collected optical spectroscopy and photometry for a sample of 71 low-activity early-M dwarfs, and complemented our data with the X-ray fluxes reported in public data archives. In this contribution I have summarized the main results of a series of papers reporting our analysis (Maldonado *et al.* 2017; Scandariato *et al.* 2017; Suárez Mascareño *et al.* 2018; González-Álvarez *et al.* 2019).

The M dwarfs in our sample have rotation periods of the order of 20–100 days which are well correlated with the R'_{HK} activity indicators. Comparing the rotation periods with the X-ray fluxes, we argue that early-M dwarfs with rotation periods of 7–10 days are already saturated and the fluxes increase with decreasing stellar mass. We also detect activity cycles of a few years, but this result is biased by the fact that our time series are 6.5 years long, hence we fail to detect longer cycles. Finally, we find evidence that ARs evolve in a few stellar rotations. In this perspective, we find that low-activity M dwarfs are similar to other stellar samples in literature.

Our analysis of the activity proxies extends the chromospheric and coronal flux-flux relationships of samples of active/young stars towards lower activity levels. We also find evidence that at very low activity levels, and at earlier M spectral types, the chromosphere is dominated by cold filaments absorbing in the Hα, while at higher activity levels, and in general at later M types, the emission by plages dominates.

References

Affer, L., Micela, G., Damasso, M., *et al.* 2016, *A&A*, 593, A117

Bochanski, J. J., West, A. A., Hawley, S. L., *et al.* 2007, *AJ*, 133, 531

Cosentino, R., Lovis, C., Pepe, F., *et al.* 2012, *Proc. SPIE*, 8446, 84461V

Chester, M. M. 1991, Ph.D. Thesis

Covino, E., Esposito, M., Barbieri, M., *et al.* 2013, *A&A*, 554, A28

Cram, L. E., & Mullan, D. J. 1979, *ApJ*, 234, 579

Cram, L. E., & Giampapa, M. S. 1987, *ApJ*, 323, 316

Davenport, J. R. A., Hebb, L., & Hawley, S. L. 2015, *ApJ*, 806, 212

Donahue, R. A., Dobson, A. K., & Baliunas, S. L. 1997, *Solar Phys.*, 171, 211

Donahue, R. A., Dobson, A. K., & Baliunas, S. L. 1997, *Solar Phys.*, 171, 191

Giampapa, M. S., Cram, L. E., & Wild, W. J. 1989, *ApJ*, 345, 536

González-Álvarez, E., Micela, G., Maldonado, J., *et al.* 2019, *A&A*, 624, A27

Henry, T. J., Jao, W.-C., Subasavage, J. P., *et al.* 2006, *AJ*, 132, 2360

Houdebine, E. R., Doyle, J. G., & Koscielecki, M. 1995, *A&A*, 294, 773

Houdebine, E. R., & Stempels, H. C. 1997, *A&A*, 326, 1143

Landman, D. A., & Mongillo, M. 1979, *ApJ*, 230, 581

López-Santiago, J., Montes, D., Gálvez-Ortiz, M. C., *et al.* 2010, *A&A*, 514, A97

Lovis, C., & Pepe, F. 2007, *A&A*, 468, 1115

Maldonado, J., Scandariato, G., Stelzer, B., *et al.* 2017, *A&A*, 598, A27

Martínez-Arnáiz, R., Maldonado, J., Montes, D., *et al.* 2010, *A&A*, 520, A79

Martínez-Arnáiz, R., López-Santiago, J., Crespo-Chacón, I., *et al.* 2011, *MNRAS*, 414, 2629

Newton, E. R., Irwin, J., Charbonneau, D., *et al.* 2016, *ApJ*, 821, 93

Perger, M., García-Piquer, A., Ribas, I., *et al.* 2017, *A&A*, 598, A26

Pizzolato, N., Maggio, A., Micela, G., *et al.* 2003, *A&A*, 397, 147

Rauscher, E., & Marcy, G. W. 2006, *PASP*, 118, 617

Reid, I. N., Gizis, J. E., & Hawley, S. L. 2002, *AJ*, 124, 2721

Robertson, P., Endl, M., Henry, G. W., *et al.* 2015, *ApJ*, 801, 79

Robinson, R. D., Cram, L. E., & Giampapa, M. S. 1990, *ApJS*, 74, 891

Rutten, R. G. M., Zwaan, C., Schrijver, C. J., Duncan, D. K., & Mewe, R. 1989, *A&A*, 219, 239

Scandariato, G., Maldonado, J., Affer, L., *et al.* 2017, *A&A*, 598, A28

Stauffer, J. R., & Hartmann, L. W. 1986, *Cool Stars, Stellar Systems and the Sun*, 254, 58

Stelzer, B., Frasca, A., Alcalá, J. M., *et al.* 2013, *A&A*, 558, A141

Suárez Mascareño, A., Rebolo, R., González Hernández, J. I., *et al.* 2018, *A&A*, 612, A89

Vidotto, A. A., Jardine, M., Morin, J., *et al.* 2013, *A&A*, 557, A67

Walkowicz, L. M., & Hawley, S. L. 2009, *AJ*, 137, 3297

Zechmeister, M., & Kürster, M. 2009, *A&A*, 496, 577

Discussion

KOSOVICHEV: Is Hα always the emission line?

SCANDARIATO: No, it isn't. As our sample of stars is biased towards low activity levels, the Hα line is most often in absorption. This line seems to have a tricky behavior, as it is not filled in as the chromospheric activity increases. As a matter of fact, when we subtract the reference template, the residuals show an extra-absorption in the Hα, before emission comes into play leading the residuals to show a flux excess in the core of the line.

HAN: Maybe I have missed something, but how did you get the rotational period? You are using L-S approach from variations of radial velocities?

SCANDARIATO: We use the time series of the broad-band photometry and the chromospheric proxies, and we apply the L-S approach to them. Radial velocities are not used to measure the stellar rotation in the works discussed in my talk.

Solar and Stellar Magnetic Fields: Origins and Manifestations
Proceedings IAU Symposium No. 354, 2019
A. Kosovichev, K. Strassmeier & M. Jardine, ed.
doi:10.1017/S1743921320000137

Reversibility of Turbulent
and Non-Collisional Plasmas: Solar Wind

Belén Acosta⬤, Denisse Pastén⬤ and Pablo S. Moya⬤

Departamento de Física, Facultad de Ciencias, Universidad de Chile,
Las Palmeras 3425, Ñuñoa, Santiago
emails: bacostaazocar@gmail.com, denisse.pasten.g@gmail.com, pablo.moya@uchile.cl

Abstract. We have studied turbulent plasma as a complex system applying the method known as *Horizontal Visibility Graph* (HVG) to obtain the *Kullback-Leibler Divergence* (KLD) as a first approach to characterize the reversibility of the time series of the magnetic fluctuations. For this, we have developed the method on *Particle In Cell* (PIC) simulations for a magnetized plasma and on solar wind magnetic time series, considering slow and fast wind. Our numerical results show that low irreversibility values are verified for magnetic field time series associated with Maxwellian distributions. In addition, considering the solar wind plasma, our preliminary results seem to indicate that greater irreversibility degrees are reached by the magnetic field associated with slow solar wind.

Keywords. Plasmas, turbulence, methods: statistical.

1. Introduction

A particular case of turbulent and non-collisional plasma is solar wind. The study of the dissipation of these processes allows characterizing the behavior of fast or slow solar wind. To understand its dynamics from another perspective, this plasma has been modeled as a complex system, studying the information delivered in time series, since the analysis of these results is a useful tool to characterize the behavior of any system.

In statistical mechanics, the Horizontal Visibility Graph (HVG) technique has been used to measure the temporal irreversibility of data sets produced by processes that are not in equilibrium, getting information on the production of entropy generated by the physical system (Lacasa *et al.* 2012). The HVG method has been applied to obtain the Kulback Leibler Divergence (KLD) or relative entropy value and thus analyze the degree of irreversibility of magnetic fluctuations generated by the system, understanding a high degree of irreversibility as a chaotic and dissipative system (Suyal *et al.* 2014). Here we show preliminary numerical results on the use of HVGs to two plasma systems Namely, numerical Particle in Cell simulations, and the solar wind. In both cases we computed the KLD based on magnetic field time series, as a first approach to the study of the reversibility of magnetized collisionless plasma systems in astrophysical environments.

In this sense, we have used the PIC simulation method to generate different types of plasmas in order to understand the degree of the irreversibility KLD as a parameter that can be related to the shape of the particles velocity distributions considering the well-known κ distributions (Viñas *et al.* 2014). In addition, we also applied the algorithm to study the reversibility of magnetic field time series associated obtained by the Wind mission in the solar wind between 01-01-1995 and 30-06-1995, focusing on the possible relation between the KLD value and solar wind speed. This article is organized as follows: in section 2 we present the HVG graph and the calculation of the KLD. Section 3 presents our main results, and section 4 is dedicated to conclusions and discussion.

2. The Horizontal Visibility Graph

The complex network is constructed by assigning a node to each data in the series, and a height according to its value. Then, two nodes are connected if a horizontal line can be drawn on the time series graph, such that it links them and does not intersect any intermediate height.

To characterize this network, a stationary process $X(t)$ is said to be statistically time reversible if for every N the series $\{X(t_1), ..., X(t_N)\}$ and $\{X(t_N), ..., X(t_1)\}$ have the same joint probability distributions $P_{out}(k = k_{out})$ and $P_{in}(k = k_{out})$, where k_{in} and k_{out} correspond to the input and output connections in each data, respectively (Lacasa *et al.* 2012).

The KL Divergence is constructed with the degree distributions, understood as a distance between P_{in} and P_{out} to assess the difference between both distributions. It is defined by

$$D[P_{out}(k)||P_{in}(k)] = \sum_k P_{out}(k) \log \frac{P_{out}(k)}{P_{in}(k)} \tag{2.1}$$

In eq. 2.1 we have used the convention $0 \log \frac{0}{0} = 0$ and the convention $0 \log \frac{0}{q} = 0$ and $p \log \frac{p}{0} = \infty$. Thus, if there is any symbol $x \in X$ such that $p(x) > 0$ and $q(x) = 0$, then $D(p||q) = \infty$. Otherwise, if $D \to 0$, the system has a low degree of irreversibility (Thomas & Cover 2006).

3. Results

3.1. *Particle-In-Cell Simulations*

First, we apply the HVG method to study time series of magnetic fluctuations obtained from PIC simulations, considering thermal (Maxwellian) and non-thermal (kappa) distributions (please see Viñas *et al.* (2014) for details about the simulations).

Considering the Maxwellian distribution and kappa distributions with different κ values we run the simulations to generate a magnetic energy density (B^2) time series (see Fig. 1 left panel). Then, the KLD, designated by D, is calculated for each case.

Fig. 1 (Maxwellian distributions correspond to velocity distributions when $\kappa \to \infty$) is shown that the dissipative degree of the system increases as the value of κ decreases. The HVG method is able to detect and measure the different behaviors of each distribution, even when magnetic time series seem similar among themselves. These results suggest a relation between the KLD and the κ value.

3.2. *Solar Wind*

We also used the HVG and computed the KLD considering magnetic field time series measured by the Wind spacecraft in the solar wind. The fast solar wind originates from coronal holes (Feldman *et al.* 1976) although the slow wind originates from above active regions on the Sun. (Krieger *et al.* 1973; Woo & Martin 1997). In order to study possible correlations between the degree of irreversibility of magnetic fluctuations (see Fig. 3 left panel) with the solar wind speed, we applied the HVG on time intervals associated with slow or fast solar wind.

Left and right panels in Fig. 2 show electron and proton bulk velocity measured by Wind during 1995, respectively. To separate between slow and fast streams we considered a threshold value of 500 km/s. In addition, we also discarded data between 450 km/s and 550 km/s so we can be sure that each interval contains only data from slow or fast solar wind exclusively. With this procedure, 46 time windows (24 with slow wind data, and 22 with fast wind data) were created, and the KLD value was computed for each of them.

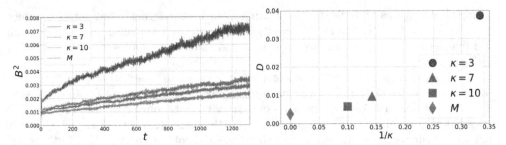

Figure 1. (left) Magnetic field obtained from PIC simulations for Maxwellian and kappa distributions considering different values of the κ parameter. Magnetic field and time values are expressed in arbitrary units. (right) KL-Divergence of magnetic field for different kappa distributions. The values for D are 0.0033 (Maxwellian), 0.0060 ($\kappa = 10$), 0.0094 ($\kappa = 7$) and 0.0381 ($\kappa = 3$).

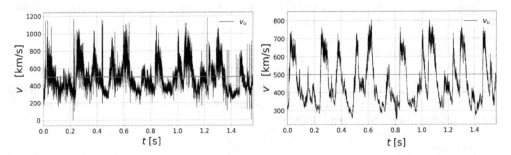

Figure 2. (left) electron and (right) proton bulk velocity magnitude. The red line separates slow and fast wind intervals.

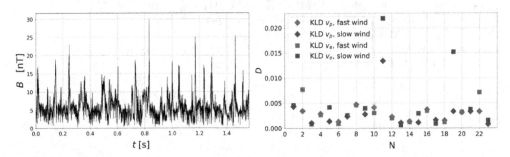

Figure 3. (left) Magnetic field magnitude. Measurements made by the Satellite Wind, data obtained from NASA CDAWeb database from 01-01-1995 until 30-06-1995. (right) KL-Divergence of magnetic field in matching time windows of slow and fast wind, protons and electrons.

Fig. 3 (right) show the results separating the 46 time windows between slow (blue) and fast wind (purple) intervals, and according the electron (squares) and proton (rhombuses) bulk velocity. It is observed that for high values of divergence, or a high degree of irreversibility, magnetic field is associated with slow solar wind, which tends to present a more chaotic behavior. However, for fast stream intervals the relation is not as clear.

4. Discussion and conclusions

In this study we have applied the method known as Horizontal Visibility Graph (HVG) to obtain the Kullback-Leibler Divergence as a first approach to study the reversibility on magnetic fluctuations, modeling turbulent plasma as a complex system. We have developed algorithms to build HVGs starting from magnetic field time series obtained from Particle In Cell simulations of collisionles magnetized plasmas, and also on solar wind magnetic time series measured by the Wind spacecraft, considering slow and fast wind streams.

Considering PIC simulations, low irreversibility values are verified for magnetic field simulations associated with Maxwellian distributions, and that the KLD value increases for decreasing κ value. Also, for the case of solar wind data, our preliminary resuls show that greater irreversibility degrees are reached by the magnetic field associated with slow solar wind. Our results seem to indicate that the shape of the particle distributions and macroscopic plasma parameters like the solar wind speed are related with the KLD value, and therefore the reversibility of the magnetic field time series, suggesting that complex networks may be a valuable alternative tool to study and characterize turbulent plasma systems.

We expect that understanding and characterizing these complex systems properties on the basis of in situ measurements could be helpful to study the characteristics of stellar winds only reachable through distant observations.

Acknowledgments

Wind data were obtained from the SPDF Web site http://spdf.gsfc.nasa.gov. We are grateful for the support of CONICyT, Chile through FONDECyT grants No. 11160452 (D.P.) and No. 1191351 (P.S.M.). B.A. also thanks the travel grant handed by the IAU 354 Symposium organizing committees.

References

Feldman, W.C., Asbridge, J.R., Bame, S.J., Gosling, J.T., 1976, *J. Geophys. Res.* 81, 5054.
Krieger, A.S., Timothy, A.F., Roelof, E.C., 1973, *Solar Phys.* 29, 505.
Lacasa, L., A. Nuñez, É. Roldán, J.M.R. Parrondo & B. Luque, 2012, *Eur. Phys. J. B*, 85: 217.
Suyal V., A. Prasad & H.P. Singh, 2014, *Solar Phys*, 289, 379–389.
Thomas, J.A., & T.M. Cover, 2006, *Elements of Information Theory*, 2 ed., 19.
Viñas, A.F., P.S. Moya, R. Navarro, & J. Araneda, 2014, *Phys. Plasmas*, 21, 012902.
Woo, R., Martin, J.M., 1997, *Geophys. Res. Lett.* 20, 2535.

Solar and Stellar Magnetic Fields: Origins and Manifestations
Proceedings IAU Symposium No. 354, 2019
A. Kosovichev, K. Strassmeier & M. Jardine, ed.
doi:10.1017/S1743921320000678

Temporal evolution of the velocity distribution in systems described by the Vlasov equation; Radiation Belts: Analytical and computational results

Abiam Tamburrini C[1]![ORCID], Iván Gallo-Méndez[1]![ORCID], Sergio Davis[2,3]![ORCID] and Pablo S. Moya[1]![ORCID]

[1]Departamento de Física, Facultad de Ciencias, Universidad de Chile, Las Palmeras 3425, Santiago.
emails: `abiam.tamburrini@uchile.cl`, `ivan.gallo@uchile.cl`, `pablo.moya@uchile.cl`

[2]Comisión Chilena de Energía Nuclear, Casilla 188-D, Santiago.
email: `sergdavis@gmail.com`

[3]Departamento de Física, Facultad de Ciencias Exactas, Universidad Andres Bello.

Abstract. An interesting problem in plasma physics, when approached from the point of view of Statistical Mechanics is to obtain properties of collisionless plasmas, which are described by the Vlasov equation. Through what we call the Ehrenfest procedure, which uses statistical mechanical relations we obtain expectation value relations for arbitrary observables, which allows us to study the dynamics of the Earth's Outer Radiation Belt. Focusing on the velocity fluctuations, the width of the distribution function and the pitch angle, a computer simulation was performed to describe the system in order to compare and test the Ehrenfest approach. Our results show that the change in the average width of the distribution follows the analytical relation. However, for the velocity fluctuation results are not conclusive yet and require more exploration. It remains as future work to verify the relation for the pitch angle.

Keywords. Statistical Mechanics, Space Plasma Physics, Ehrenfest Procedure, Radiation Belts.

1. Introduction

Throughout our lives we have witnessed the interaction between the planet we inhabit and our star. The Earth's magnetosphere is one of its main consequences, originated by the interaction of the Earth's magnetic field and the solar wind. It is well known that the magnetosphere is highly sensitive to the activity of the Sun, which gives rise to many natural phenomena that intervene in our daily lives, such as geomagnetic storms, responsible for affecting navigation instruments, an. Many effects of this interaction are permanent, for example the radiation belts, composed of charged particles that were trapped in the magnetic field, whose variability in the outer belt is intimately related to solar activity and solar wind. Most of these phenomena are mediated by space plasmas, in particular the solar wind and regions of the magnetosphere, among others, are important examples of *non-collisional plasmas*, in which the presence of long-range interactions gives rise to stationary states (but not thermodynamic equilibrium) described by non-Maxwellian distributions such as the Kappa distribution (Viñas *et al.* 2015).

The dynamics of this system can be described from the point of view of nonequilibrium Statistical Mechanics, through the Vlasov equation (Bellan 2006).

$$\frac{\partial f}{\partial t} + \boldsymbol{v} \cdot \nabla_x f + \frac{q}{m}\left(\boldsymbol{E} + \frac{1}{c}\boldsymbol{v} \times \boldsymbol{B}\right) \cdot \nabla_v f = 0. \qquad (1.1)$$

This equation is the Liuoville theorem when the Hamiltonian describes an electromagnetic interaction, and the force on each particle is given by the Lorentz force. Whatever the nature and form of the fields $\boldsymbol{E}$ and $\boldsymbol{B}$, it is possible to derive relations for the expectation values of time-dependent observables from the Vlasov equation, using what we will call the Ehrenfest procedure (as classical analog of Ehrenfest theorem in Quantum Mechanics) (Davis & González 2015), (Davis & Gutiérrez 2012).

$$\frac{\partial}{\partial t}\langle w \rangle_t = \left\langle \frac{\partial w}{\partial t} \right\rangle_t + \left\langle \boldsymbol{v} \cdot \partial_x w \right\rangle_t + \left\langle \frac{\boldsymbol{F}}{m} \cdot \partial_v w \right\rangle_t, \tag{1.2}$$

where $w = w(\boldsymbol{x}, \boldsymbol{v})$ is an arbitrary, differentiable function of position and velocity.

An example of the consequences of this differential equation is an expression for the temporal evolution of the velocity fluctuations, which in our case of interest to model, has the form (González *et al.* 2018)

$$\frac{\partial}{\partial t}\langle (\delta \mathbf{v})^2 \rangle_t = \frac{-2q}{mc} \left\{ \langle \mathbf{v} \times \mathbf{B} \rangle_t \cdot \langle \mathbf{v} \rangle_t \right\}. \tag{1.3}$$

Here if we make the observable $w = \delta(\boldsymbol{v} - \boldsymbol{v_0})$ we then get an expression for the temporal evolution of the logarithm of the velocity distribution function in a component,

$$\frac{\partial}{\partial t} \ln P(v_i|t) = \frac{-2q}{mc} \left\{ (\mathbf{v} \times \langle \mathbf{B} \rangle)_i \cdot \frac{\partial}{\partial \mathbf{v}} \ln P(v_i|t) \right\}. \tag{1.4}$$

Finally as the last analytical result, if we consider $w = \frac{\mathbf{B} \cdot \mathbf{v}}{|B||v|} = \cos \theta$ searching information about the collective behavior of the pitch angle θ, we obtain an expression for the temporal evolution of the average value of $\cos \theta$,

$$\frac{\partial}{\partial t}\langle \cos \theta \rangle = \langle \mathbf{v} \cdot (\hat{\mathbf{v}} \cdot \nabla_{\mathbf{r}}) \hat{\mathbf{B}} \rangle. \tag{1.5}$$

In this work, we seek to verify, by means of a computer simulation, the analytic relations previously shown in the outer radiation belt.

2. About the simulation

Considering the application of our theoretical results for the study of energetic particles in the Earth's outer radiation belt, we performed a test particle simulation of non interacting electrons trapped in a dipole magnetic field. As initial conditions we considered a Maxwellian distribution with a thermal velocity corresponding to an energy in the order of a few keV, and the particles were injected at a the magnetic Equator at a radial distance of 4 R_E, being R_E the Earth radius. The simulation was run for 80000 time steps in units of R_E/c, where c is the speed of light. To obtain the velocity and position of the particle, the Boris algorithm was used, which by construction based in Liouville's theorem, preserves energy without being simplectic (Hong Qin *et al.* 2013).

3. Results

The first result to show corresponds to the trajectory of a particle and its corresponding energy in the time of flight (See Fig. 1).

In order to test Eq. (1.3), the velocity and magnetic field data were retrieved at each time step and were treated in order to plot both sides of the equation.

To analyze the evolution of the velocity distribution by component, we built histograms for each time step in each component, noting the most significant differences in the start, middle and end of the dynamics. The expectation (average) values of one of the components of velocity in time was also plotted.

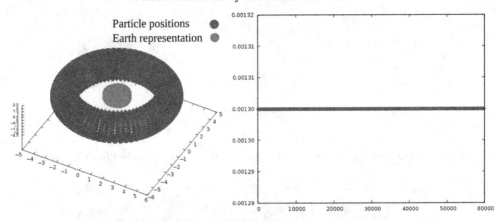

Figure 1. Left panel, trajectory of a given particle. Right panel, energy evolution during the dynamics.

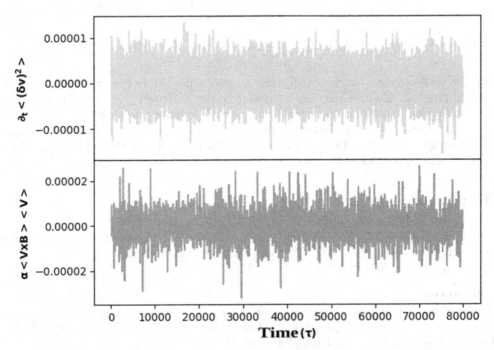

Figure 2. Temporal evolution of the left and right-hand sides of the relation 1.3, using data extracted from the simulation.

4. Summary

From Fig. 2, effects such as noise in the curve and the small number of particles, which generates a statistically limited sample, do not allow us yet to affirm that there is a statistically significant correlation for relation (1.3). We believe that, since the movement is not ergodic, following periodic orbits, it may be necessary to consider more particles to sweep away the effect of initial conditions. From relation (1.4) it is clear that we expected a variation in the average width of the distribution, which is confirmed by the results of the simulation represent in Fig. 3. It remains as a future work to calculate the variation in the average for each time step and see its correlation with Fig. 4. About relation (1.5), our results suggest a good agreement between theory and numerical simulations, but in

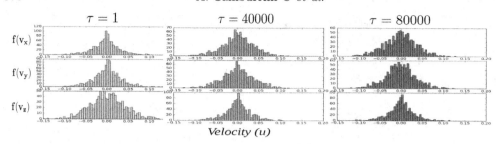

Figure 3. Evolution of the velocity distribution by component extracted from the simulation.

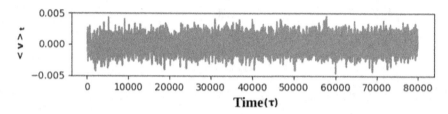

Figure 4. Temporal evolution of the expectation value of the velocity during dynamics.

order to make stronger conclusions further analysis is needed. In general, despite being a work in progress, from the relations derived from the Ehrenfest procedure to the Vlasov equation we can, as a first approximation, rescue information from the system and see from the results of the simulation, as these show expected behaviors of the dynamics, which is a good indication to consider the Ehrenfest procedure as a useful tool to address widely studied plasma systems as is the case of the outer radiation belt.

Acknowledgments

We are grateful for the support of CONICyT, Chile through Anillo ACT-172101 and FONDECyT grant 1171127(S.D) and 1191351 (P.S.M). A.T.C. and I.G.M also thank the travel grant handed by the IAU 354 Symposium organizing committees.

References

Bellan, P. *Fundamentals of Plasma Physics, Cambridge University Press*, 2006
Davis, S. & D. González 2015, *J. Phys. A: Math. Theor.*, 48, 425003
Davis, S. & G. Gutiérrez 2012, *Phys. Rev.*, 86, 051136
González, D., A. Tamburrini, S. Davis, & J. Jain 2018, *Journal of Physics: Conference Series*, 1043, 012008
Hong Qin, Shuangxi Zhang, Jianyuan Xiao, Jian Liu & Yajuan Sun 2013, *Phys. Plasmas*, 20, 084503
Viñas, A. F., P. S. Moya, R. E. Navarro, J. A. Valdivia, J. A. Araneda, & V. Muñoz 2015, *J. Geophys. Res. Space Physics*, 120, 3307–3317

Solar and Stellar Magnetic Fields: Origins and Manifestations
Proceedings IAU Symposium No. 354, 2019
A. Kosovichev, K. Strassmeier & M. Jardine, ed.
doi:10.1017/S1743921320000514

On the multifractality of plasma turbulence in the solar wind

Sebastián Echeverría[iD], Pablo S. Moya[iD] and Denisse Pastén[iD]

Departamento de Física, Facultad de Ciencias, Universidad de Chile,
Las Palmeras 3452, Ñuñoa, Santiago, Chile
emails: `s.echeverria@ug.uchile.cl`, `pablo.moya@uchile.cl`, `denisse.pasten.g@gmail.com`

Abstract. In this work we have analyzed turbulent plasma in the kinetic scale by the characterization of magnetic fluctuations time series. Considering numerical Particle-In-Cell (PIC) simulations we apply a method known as MultiFractal Detrended Fluctuation Analysis (MFDFA) to study the fluctuations of solar-wind-like plasmas in thermodynamic equilibrium (represented by Maxwellian velocity distribution functions), and out of equilibrium plasma represented by Tsallis velocity distribution functions, characterized by the *kappa* (κ) parameter, to stablish relations between the fractality of magnetic fluctuation and the kappa parameter.

Keywords. Plasma, turbulence, multifractality

1. Introduction

The upper atmosphere of the Sun is continuously releasing a stream of charged particles which constitutes the solar wind. This ejected plasma gives an extent of interesting phenomena in plasma physics. One of the fundamental problems in this area is the understanding of the relaxation process in a collisionless plasma and the resultant state of the electromagnetic turbulence, in particular, at kinetic scales.

In this work we are applying a method known as *MultiFractal Detrended Fluctuation Analysis* (MFDFA) (Kantelhardt *et al.* 2002) to study the magnetic fluctuations of solar-wind-like plasmas in thermodynamic equilibrium (represented by Maxwell velocity distribution functions), and out of equilibrium plasma represented by Tsallis distribution functions, characterized by a parameter κ. In first place we studied magnetic fluctuations through *Particle in Cell* (PIC) (Viñas *et al.* 2014) simulations of a magnetized plasma compound by ions and electrons, where we calculated the multifractality in time series, extracted in the simulations, to establish relations between the multifractality of the system and the kappa parameter.

The MFDFA method allows us to study and extract valuable information of any time series, that is the reason it has been applied in different research areas, for example, there are studies in the biology field associated to DNA sequences (Peng *et al.* 1994) and also heartbeat time series (Peng *et al.* 1995); furthermore, there are studies in seismic complex networks (Pastén *et al.* 2018) and different applications in economics and finance (Grech 2016). Those applications and results show that this method is an interesting tool to explore in different areas of science.

This article organizes as follows: section 2 describes the computer simulation where the magnetic fluctuation data were obtained, associated to each velocity distribution functions; section 3 describes the MFDFA method and the steps we have to carry out; in section 4 we show the preliminary results when we applied the MFDFA method to the magnetic fluctuation data. Finally, in section 5 we summarize our main conclusions.

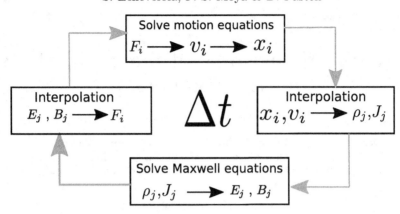

Figure 1. The figure shows how the electromagnetic fields are calculated in each cell. Starting with the position and the velocity of the particles it is possible to determine the density and the current generated in each cell. Then, Maxwell equations are solved to determine the fields in each cell.

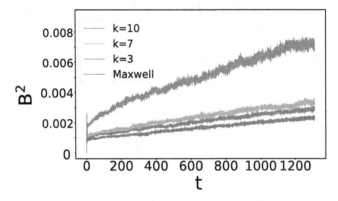

Figure 2. Magnetic fluctuations for different velocity distribution functions. We can observe the results associate with each value of κ. Fields and time are dimensionless. Time is normalized to the electron gyrofrequency.

2. Particle In Cell Simulation (PIC)

We have analyzed magnetic fluctuations obtained in a Particle In Cell (PIC) simulation (Viñas *et al.* 2014). We studied a turbulent plasma compound by ions and electron that are treated kinetically and periodic boundary conditions are imposed. Time is normalized in units of the electron cyclotron frequency and particle positions are normalized to the electron inertial lengths. Fig. 1 shows the general idea of the method.

Then, the resultants magnetic fluctuation associated to each velocity distribution functions: Maxwellian and Tsallis (for $\kappa = 3, 7, 10$) are shown in Fig. 2. We can observe that while the κ parameter increases, magnetic fluctuations converge to the Maxwellian equilibrium as is expected.

3. MultiFractal Detrended Fluctuation Analysis (MFDFA)

In order to determinate the fractality of times series, we have to carry out a series of steps (Kantelhardt *et al.* 2002). For this, let us consider that B_k is a time series of length N; first, we have to determine the profile:

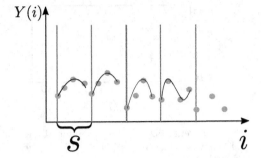

Figure 3. When the profile is determinated, polynomial adjust are calculated for each window. Note that we have $2N_s$ segments because of the data that may remain in the profile.

$$Y(i) = \sum_{k}^{i} B_k - <B>,$$

where $i = 1, ..., N$. Then, divide the profile into $N_s = \text{int}\,(N/s)$ nonoverlapping segments of equal length s. Since the length N is not often a multiple of the length s a short part of the profile may remain. To consider this part, the same analysis will be carry out starting from the end of the series. The next step is to determine the local trend for each segments using a polynomial adjust:

$$F^2\,(v,\,s) = \frac{1}{s} \sum_{i=1}^{s} (Y\,[(v-1)\,s+i] - y_v\,(i))^2\;,$$

where y_v is the polynomial adjust in each window. Finally, overage over all segments to obtain the qth order fluctuation function:

$$F_q(s) = \left(\frac{1}{2N_s} \sum_{v=1}^{2N_s} \left(F^2(v,s)\right)^{q/2} \right)^{1/q}, \tag{3.1}$$

where q and $F^2(v, s)$ are the generalize dimension index and the variance in each window, respectively.

The general idea of the method is shown in Fig. 3.

The objective is to determine how Eq. (3.1) is related with s for different values of q. With this function we can determine the generalized Hurst exponent $h(q)$, further more, it increases for large values of s as a power-law:

$$F_q(s) \sim s^{h(q)}.$$

4. Results

We applied the MFDFA method to magnetic fluctuations given by Fig. 2. Results are shown in Fig. 4. Here we compute Eq. (3.1) for different values of q ($q = \pm 20, \pm 15, \pm 10, \pm 5, 0$), then, in log-log plots we obtained values for the generalized Hurst exponent associated to each q.

5. Conclusions

In this article we studied the magnetic fluctuation of a collisionless plasma with different velocity distribution functions using the MFDFA method. First, we explained the simulation where the data were obtained; then, we describe the general ideas and steps of the MFDFA method to apply them in the simulation data. Finally, we computed the relations between the κ parameter and the generalize Hurst exponent.

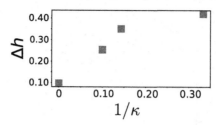

Figure 4. Relations between the generalize Hurst exponent of the time series and the velocity distribution functions. Here we took $\Delta h_\kappa = h_\kappa(q=10) - h_\kappa(q=-10)$ The limit $1/\kappa = 0$ represents the Maxwellian distribution.

The results obtained in Fig. 4 shows the multifractal spectrum of the time series analyzed. These results suggest a mono fractal behavior for the four time series, this is due that for each distribution Δh is lower than 0.5. It is interesting to notice that, while the κ parameter increases, the monofractality in the time series decreases. This result also suggests that, for lower values of κ there is no long-range correlations or, this correlations tend to zero quickly in these time series. Nevertheless, the results are still preliminary and there are different ways to corroborate and analyze them. In a future work we pretend to build the singularity spectra of fractal dimensions and analyze more time series to characterize this complex behavior.

The motivation of this research is to characterize turbulent plasma through time series analysis using the MFDFA method, with the objective of describe the dependencies associated to the different velocity distribution functions. We expect our analysis to be useful for the characterization of the electromagnetic turbulence in a collisionless space plasma, such as the solar wind. Furthermore, this tool may be useful to extract valuable information about the plasma when high resolution particle detectors are not available.

Acknowledgements

We are grateful for the support of CONICyT, Chile through FONDECyT grants 1191351 (P.S.M.), and 11160452 (D.P). S.E.V. also thanks the Travel grant handed by the IAU 354 Symposium organizing committees.

References

Bale, S. D., Kasper, J. C., Howes, G. G., Quataert, E., Salem, C., & Sundkvist, D. 2009, *Phys. Rev.* 103, 211101

Grech D. 2016, *Chaos, Solitons & Fractals* 88, 183

Kantelhardt, J. W., Zschiegner, S. A., Koscielny-Bunde, E., Havlin, S., Bunde, A. & Stanley, H. 2002, *Physica A: Statistical Mechanics and its Applications*, 315, 87

Kasper, J., Lazarus, A. & Gary, P. 2002, *Geophys. Res.* 29, 1839

Pastén, D., Czechowski, Z., & Toledo, B. 2018, *Chaos* , 28, 083128

Peng, C.-K., Havlin, S., Stanley, H., & Goldberger, A. L. 1995, *Chaos*, 5, 82

Peng, C.-K., Buldyrev, S. V., Havlin, S., Simons, M., Stanley, H., & Goldberger, A. L. 1994, *Phys. Rev*, 49, 1685

Telesca L., Czechowski, Z., & Lovallo, M. 2015, *Chaos*, 25, 063113

Viñas, A. F., Moya, P. S., Navarro, R., & Araneda, J. 2014, *Phys. Plasmas*, 21, 012902

Chapter 8. Mechanisms of flaring and CME activity on the Sun and stars

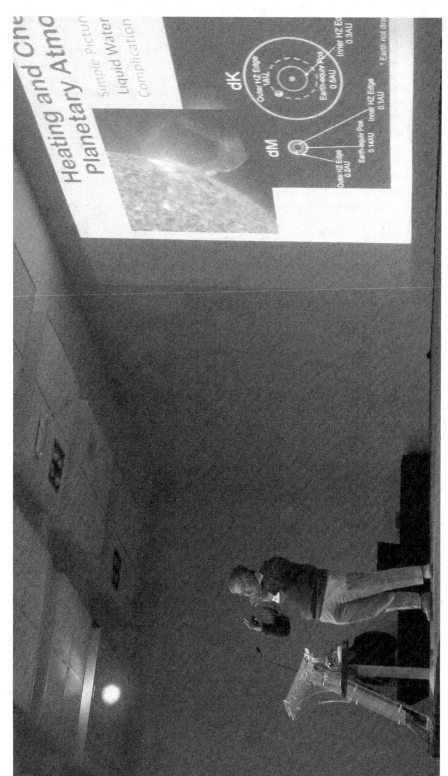

Alexander Brown

Solar and Stellar Magnetic Fields: Origins and Manifestations
Proceedings IAU Symposium No. 354, 2019
A. Kosovichev, K. Strassmeier & M. Jardine, ed.
doi:10.1017/S1743921320000162

The UV/X-ray radiation fields and particle (CME) flows of M dwarf exoplanet host stars

Alexander Brown🆔

Center for Astrophysics and Space Astronomy, 389 UCB,
University of Colorado, Boulder, CO 80309-0389, USA
email: `alexander.brown@colorado.edu`

Abstract. The high energy X-ray and UV radiation fields of host stars play a crucial role in determining the atmospheric conditions and habitability of potentially-habitable exoplanets. This paper focuses on the major surveys of the UV/X-ray emissions of M- and K-type exoplanet hosts that have been undertaken by the MUSCLES and MegaMUSCLES Hubble Space Telescope (HST) Treasury programs and associated contemporaneous X-ray and ground-based observations. The quiescent and flaring radiation (both photons and implied particles) were observed from this extensive sample of relatively old, low mass, exoplanet host stars and show that, from the viewpoint of a habitable-zone exoplanet, there is no such thing as an "inactive" M dwarf star. The resulting implications are significant for planetary habitability. Extensive monitoring of the X-ray/UV emission from a representative younger M dwarf is also presented and the direct stellar effects that influence exoplanets during the earlier phases of their formation and evolution discussed.

Keywords. stars: late-type, stars: activity, stars: flare, planetary systems, X-rays: stars, ultraviolet: stars

---

1. Introduction to M Dwarfs and their Exoplanets

Low mass M and late-K dwarf stars are currently the focus of major research efforts to discover and study their exoplanet systems. Discovering planets orbiting such low mass stars is far easier than for solar-like stars using both transit and radial velocity techniques, because the effects of the planets on the stellar signal is far larger. Additionally, studies have shown the earth-like or super-earth planets are commoner around M dwarfs and large gas-giant planets are rarer than for higher mass stars. M dwarfs show very strong surface magnetic fields (see e.g. Shulyak *et al.* (2019)). These complex 3-6 kG magnetic fields fill and control the outer atmospheres of all M dwarfs (Afram & Berdyugina (2019)) and lead to bright, variable coronal X-ray and chromospheric/transition-region ultraviolet emission.

Habitable zones.
The concept of habitable zones is fundamental to discussions of whether life, particularly as we know it on Earth, would be possible on exoplanets. The zeroth order definition of habitability is to consider whether an exoplanet orbits its host star at a distance that provides an equilibrium temperature compatible with the presence of liquid water on its surface. Beyond this initial starting point, there are many complicating factors including the properties of the planet's atmosphere and magnetic field. An excellent discussion of habitability is provided by Shields, Ballard & Johnson (2016).

M dwarfs are far less luminous than the Sun and consequently their liquid water habitable zones must be much closer to the star. While the habitable zone for an early-G dwarf lies between 0.8 and 2 AU from the star, the same region for an M dwarf is at

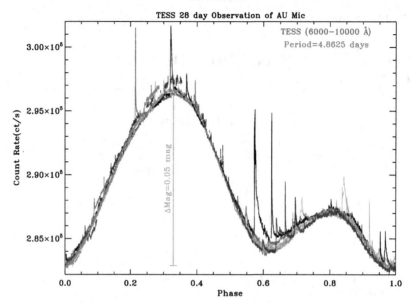

Figure 1. Broadband optical TESS light-curve of AU Mic phase-folded using a 4.8625 day rotational period. Clear starspot rotational modulation with a 5% full amplitude is the dominant repetitive variability (See Wisniewski *et al.* (2019)).

radii between 0.1 and 0.2 AU. Flares on M dwarfs are at least as strong as solar flares, so they can have a far greater effect on their habitable zones planets, than solar flares on the Earth.

Role of different spectral regions.

Stellar radiation in different spectral regions affects an exoplanet in different ways. The optical/IR radiation is the majority of the radiated energy and controls the atmospheric and surface heating of the planet. X-ray and EUV ($\leq$ 912 Å) radiation plays a major role in thermospheric heating and atmospheric erosion. The intermediate FUV/NUV radiation controls the atmospheric chemistry via molecular formation and photolysis. The FUV radiation is dominated by the 1215.67 Å H I Lyman-α emission line (Youngblood *et al.* (2016)). The EUV region contains a mixture of coronal and transition region emission lines that must be reconstructed using spectral information recorded in the X-ray and FUV regions.

2. Young Active M Dwarfs: Conditions Encountered by Newly-formed Exoplanets

Example: The young star AU Mic. The intense stellar activity shown by young M dwarfs is well illustrated by recent observations of the dMe star AU Mic. AU Mic is a relatively massive M0 V star with a well established age of 24±3 Myr, based on its membership in the Beta Pictoris Moving Group (Bell, Mamajek & Naylor (2015)). It has an edge-on dust debris disk indicative of the presence of a protoplanetary system (Kalas *et al.* (2004), Wisniewski *et al.* (2019)), which is known to contain at least two exoplanets based on transit measurements.

The presence of optical flaring on AU Mic is obvious in a 28 day TESS observation obtained between 2018 Jul 25 - Aug 22. Even though the sensitivity of TESS lies in the red part of the optical spectral region (6000-10000 Å) where flaring is far harder to detect than in the blue, many flare enhancements are clearly detected in the rotational-phase-folded light-curve (Fig. 1) .

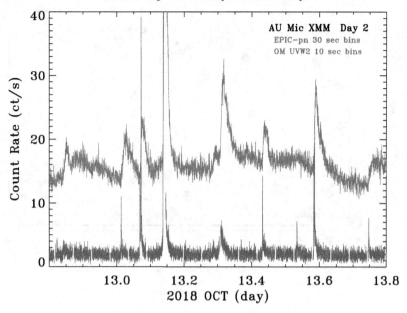

Figure 2. One day of XMM-Newton flaring variability from a 550 kilosecond observation of AU Mic. The upper curve (blue) shows the EPIC-pn soft-X-ray variability and the lower curve (red) shows the OM UVW2 NUV variability. The X-ray flares show exponential rise and decay time-scales roughly 10 times longer than those seen for the NUV flares. The NUV emission is produced during the impulsive phase of the flare when high energy particles are being accelerated into the stellar surface, while the soft X-rays are emitted during the gradual phase when the heated coronal plasma is cooling. Clearly, the planets orbiting AU Mic are impacted by multiple flare events on a daily basis.

The major flares produced by magnetic energy release are revealed in even more detail by a 7 day XMM-Newton Large Project (PI Adam Kowalski). Soft X-ray and near-ultraviolet emission were monitored for a total time of 550 kiloseconds. In Fig. 2 the EPIC-pn (soft X-ray) and OM UVW2 (NUV) variability observed during a 1 day time interval are presented. These data were obtained 11 stellar rotations after the TESS data shown in Fig. 1. On average $\sim$4 flares were detected per day with 19 flares detected in both X-rays and NUV, and a further 8 X-ray flares seen when the OM was not working. The largest flare had an X-ray energy of 10^{33} ergs and 21 flares had X-ray energies $\geq 10^{32}$ ergs.

AU Mic is capable of producing even more extreme flares than those seen in our XMM observation. In 1992 during the commissioning phase of the EUVE satellite an almost 4 day observation of AU Mic contained a very large flare outburst that lasted 2 days (Cully *et al.* (1993)). The EUVE Deep Survey (DS) detector sampled coronal emission in the 65-190 Å spectral region and the DS light-curve peaked at a luminosity of 10^{30} ergs s^{-1}. The total flare energy in the DS bandpass was 3×10^{34} ergs. Based on the temperature evolution derived for this flare by Monsignori Fossi *et al.* (1996)) and Katsova, Drake, & Livshits (1999)), the corresponding soft X-ray (0.3-10 keV) energy would have been $\sim 2 \times 10^{35}$ ergs and thus a factor of 100 larger than the largest flare seen in the 2018 XMM-Newton observation. Such large flare events are likely to cause severe atmospheric erosion from close-in exoplanets and may well be responsible for large-scale structures that are observed moving outwards in the AU Mic disk (Boccaletti *et al.* (2018), Wisniewski *et al.* (2019)).

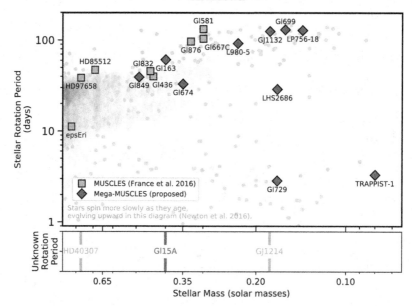

Figure 3. MUSCLES/MegaMUSCLES sample of K/M dwarfs as a function of stellar mass and rotation period. These stars are all exoplanet hosts, typically with super-Earth planets within or near to their habitable zones. These stars provide an as-yet unique sampling of stellar magnetic activity on low-activity, low mass stars. (Observed targets GJ176, GJ649, and GJ676A are not shown on this plot.)

3. MUSCLES/MegaMUSCLES: Exploring Conditions Around Mature M Dwarfs

While considerable efforts have been devoted to studying young active M dwarfs in the ultraviolet and X-ray regions, until recently comparatively little was known about the activity levels of older "inactive" M dwarfs.

M dwarf sample. The MUSCLES (125 orbits) and MegaMUSCLES (157 orbits) HST Treasury programs have conducted an in-depth study of the UV and X-ray spectral energy distributions of K and M dwarf exoplanet host stars with a range of rotation periods and activity levels. A sample of 24 stars with spectral types from K1 to M8 have been studied with HST UV observations and supporting X-ray observations from Chandra and XMM-Newton (see Fig. 3). The stars range from still fast-rotating (few day period) stars to older stars with ~100 day rotation periods. Specific aims include characterizing the energetic radiation environment in the stars' habitable zones, measure the flare distributions on these less active stars, and providing basic observational inputs to modeling the atmospheric photochemistry and the production of molecular tracers.

Panchromatic spectral energy distributions (SEDs). A fundamental product of the MUSCLES/MegaMUSCLES programs are SEDs that cover the complete wavelength range from the infrared through the optical and ultraviolet to the X-ray region (Loyd *et al.* (2016)). These SEDs provide vital input to the modeling of exoplanets and their atmospheres. All the results from the MUSCLES project and growing datasets from MegaMUSCLES are available at https://archive.stsci.edu/prepds/muscles/.

FUV/NUV balance and atmospheric chemistry. The balance between the FUV (912-1700 Å) and NUV (1700-3200 Å) radiation has a strong influence on oxygen chemistry of Earth-like habitable zone exoplanets and the abundances of molecular oxygen and ozone. The FUV/NUV ratio increases from 10^{-3} for a Sun-like star to 10^{-2} by early-K spectral type and reaches 0.2-0.7 for M dwarfs. This increase in FUV dominance can lead

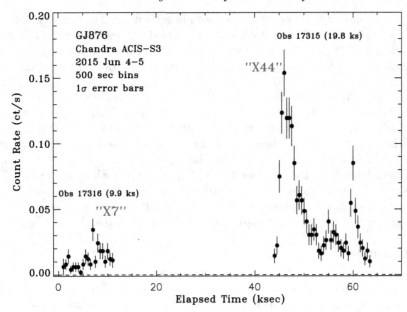

Figure 4. Chandra ACIS-S soft X-ray light-curve of the slowly rotating M5 dwarf GJ876. The flare in the 1st observation is equivalent to an X7 solar flare in the star's habitable zone, while the large flare in the 2nd observation is equivalent to an X44 flare. For comparison, on the Sun an X10 flare occurs once every few years, while the largest known solar flare, the Carrington Flare of 1859, is thought to have been an X45 event.

to the production of significant O_2 and O_3 without the involvement of biotic processes and the false interpretation of the presence of such molecules as biosignatures.

Flares: UV and X-ray variability. Perhaps the most important result from the MUSCLES and MegaMUSCLES surveys is the discovery that strong UV/X-ray flare activity is present on all M dwarfs even down to the oldest, slowly rotating and lowest mass stars sampled. (Wilson *et al.* (2019)). FUV and X-ray flares were observed that reached peaks at 10-100 times the quiescent emission on timescales of 10^2-10^3 seconds. These observations were enabled by the temporal variability studies possible using the Cosmic Origins Spectrograph (COS) on HST (Green *et al.* (2012)) and the photon counting X-ray imagers on Chandra and XMM-Newton. COS provides both flare-related flux and spectral line profile monitoring. Loyd *et al.* (2018) studied the FUV variability seen in the COS data for the MUSCLES sample and found that most of the nominally "inactive" stars showed significant flaring in transition region and chromospheric emission lines. While the FUV emission declines as M dwarfs age and rotate more slowly, the level of flaring emission relative to quiescent emission remains constant.

Almost all the stars observed show X-ray variability, often in the form of large flare outbursts and at other times as clear changes in quiescent flux between observations at different times. Several examples are described by Youngblood *et al.* (2017), including the multiple flares seen by Chandra from the slowly-rotating ($P_{rot} = 97$ days) M4 dwarf GJ876 (see Fig. 4). These flares when viewed from the habitable zone are remarkable in comparison to present day solar flares that impact the Earth.

Flares: CMEs and energetic particles. Exoplanets are vulnerable not only to the high energy radiation generated by stellar flares but also any energetic particles within coronal mass ejections (CMEs) released as part of the flare magnetic reconnection process. However, it is difficult to estimate the particle flux associated with a particular flare. This is particularly difficult because the observed flare properties indicate flare processes and conditions that appear to be beyond those seen in solar flares. For example, COS spectra of a flare on the intermediate rotation (P_{rot} = 33 days) M2.5 star GJ674 showed an unexpected blue FUV continuum below 1200 Å with a color temperature of ~40,000 K, which requires a far denser emitting flare footpoint than provided by existing radiative hydrodynamic flare models (Froning *et al.* (2019)). One approach is to extrapolate from solar flares and CMEs to try to estimate the potential particle flux. Youngblood *et al.* (2017) used correlations between solar flare soft X-rays and CME fluxes and stellar X-ray and He II 1640 Å fluxes to derive rough estimates of M dwarf CME proton fluxes. These estimates indicate that severe atmospheric changes would result. However, the magnetic fields of M dwarfs are far stronger than on the Sun and it will likely be far harder for a CME to break loose from the star (see Gome (2020)). Even with a cutoff of CME release for flares with lower energies, there are still very many large flares occurring on both younger and older M dwarfs that, even if only a small number have associated CMEs, dramatic effects on exoplanet atmospheres are almost inevitable.

4. Conclusions

High energy X-ray and UV radiation has a significant influence on exoplanets orbiting M dwarf stars. This radiation is extremely intense during the early phases of protoplanetary system evolution, but, while it declines with stellar age and is weaker with decreasing stellar mass, it continues to be important for all M dwarfs throughout their lives. The X-ray/UV radiation is inherently variable and produces exoplanetary illumination that requires time-dependent modeling to adequately study exoplanet atmospheres. While still a matter of debate, any coronal mass ejections associated with the frequent large flares will have severe effects on the exoplanet atmospheres. Despite the significant high energy radiation from M dwarf host stars, it is still unclear how its effects would influence the presence of life on an exoplanet. Obvious mitigating factors include the role of surface water in absorbing high energy photons and how the presence of strong planetary magnetic fields might shield the surface from high energy particles.

Acknowledgements

I thank Drs. Kevin France, Cynthia Froning, and Adam Kowalski (all PIs of projects described in this paper) for the opportunity to collaborate on such exciting investigations. The work presented here from the MUSCLES and MegaMUSCLES programs has been supported by STScI grants HST-GO-12464.01, HST-GO-13650.01, and HST-GO-15071.02 to the University of Colorado at Boulder. The corresponding X-ray work was supported by Chandra grants GO4-15014X, GO5-16155X, and GO8-19017X from the Smithsonian Astrophysical Observatory and NASA XMM grant NNX16AC09G to the University of Colorado at Boulder.

References

Afram, N. & Berdyugina, S. V. 2019, *A&A*, 629, A83
Bell, C. P. M., Mamajek, E. E., & Naylor, T. 2015, *MNRAS*, 454, 593
Boccaletti, A., Sezestre, E., Lagrange, A.-M., Thébault, Gratton, R., Langlois, M., Thalmann, C., Janson, M., *et al.* 2018, *A&A*, 614, A52
Cully, S. L., Siegmund, O. H. W., Vedder, P. W., & Vallerga, J. V. 1993, *ApJ*(Letters), 414, 49
France, K., Loyd, R. O. P., Youngblood, A., Brown, A., Schneider, P. C., Hawley, S.l., Froning, C. S., Linsky, J. L., *et al.* 2016, *ApJ*, 820, 89
Froning, C. S., Kowalski, A., France, K., Loyd, R. O. P., Schneider, P. C., Youngblood, A., Wilson, D., Brown, A., *et al.* 2019, *ApJ*(Letters), 871, 26
Gomez, J. A. 2020, this proceedings.

Green, J. C., Froning, C. S., Osterman, S., Ebbets, D., Heap, S. H., Leitherer, C., Linsky, J. L., Savage, B. D., *et al.* 2012, *ApJ*, 744, 60

Kalas, P., Liu, M. C., & Matthews, B. C. 2004, *Science*, 303, 1990

Katsova, M. M., Drake, J. J., & Livshits, M. A. 1999, *ApJ*, 510, 986

Loyd, R. O. P., France, K., Youngblood, A., Schneider, C., Brown, A., Hu, R., Linsky, J., Froning, C. S., *et al.* 2016, *ApJ*, 824, 102

Loyd, R. O. P., France, K., Youngblood, A., Schneider, C., Brown, A., Hu, R., Segura, A., Linsky, J., *et al.* 2018, *ApJ*, 867, 71

Monsignori Fossi, B. C., Landini, M., Del Zanna, G., & Bowyer, S. 1996, *ApJ*, 466, 427

Shields, A. L., Ballard, S., & Johnson, J. A. 2016, *Physics Reports*, 663, 1

Shulyak, D., Reiners, A., Nagel, E., Tal-Or, L., Caballero, J. A., Zechmeister, M., Béjar, V. J. S., Cortés-Contreras, M. *et al.* 2019, *A&A*, 626, A86

Wheatley, P. J., Louden, T., Bourrier, V., Ehrenreich, D. & Gillon, M. 2017, *MNRAS*, 465, L74

Wilson, D., Froning, C. S., France, K., Youngblood, A., Schneider, P. C., Berta-Thompson, Z., Brown, A., Buccino, A. P., *et al.* 2019, *ApJ*, in prep.

Wisniewski, J. P., Kowalski, A. F., Davenport, J. R. A., Schneider, G., Grady, C. A., Hebb, L., Lawson, K. D., Augereau, J-C., *et al.* 2019, *ApJ(Letters)*, 883, L8

Youngblood, A., France, K., Loyd, R. O. P., Brown, A., Mason, J. P., Schneider, P. C., Tilley, M. A., Berta-Thompson, Z. K., *et al.* 2017, *ApJ*, 843, 31

Youngblood, A., France, K., Loyd, R. O. P., Linsky, J. L., Redfield, S., Schneider, P. C., Wood, B. E., Brown, A., *et al.* 2016, *ApJ*, 824, 101

Discussion

LUHMANN: How well do we know the interplanetary conditions around M dwarf stars? Are they similar to those of the Solar System?

BROWN: We know far less about the interplanetary conditions around M dwarfs than around the Sun. Detailed modeling of the magnetic field structure has been performed for a variety of young M dwarfs, which typically shows stronger and more complex field geometry than seen on the Sun. Far less is known about the conditions around older, less active, M dwarfs.

LUHMANN: Extreme SEP events in our solar system do not necessarily scale with indicators, such as flare X-rays, because they modify the shocks that are producing them. Thus, care needs to be taken with scaling relationships.

BROWN: Indeed, it is unclear how to extrapolate solar relationships to the far more active M dwarf situation. However, provided that large M dwarf flares are related to plasma release above some critical energy, the CME-related particle release will almost inevitably lead to important effects in the atmospheres of habitable-zone exoplanets.

STRASSMEIER: Could the interplanetary material flip the FUV/NUV ratio again when measured at a distance similar to Jupiter's?

BROWN: Detailed studies of the circumstellar environments of M dwarfs show almost no gas and only a little dust present, even for the evolutionary stage represented by stars like AU Mic, so the FUV/NUV ratio should not be changed from the exoplanet's perspective.

Solar and Stellar Magnetic Fields: Origins and Manifestations
Proceedings IAU Symposium No. 354, 2019
A. Kosovichev, K. Strassmeier & M. Jardine, ed.
doi:10.1017/S1743921319009980

Exploring Flaring Behaviour on Low Mass Stars, Solar-type Stars and the Sun

L. Doyle[1,2] ⓘ, G. Ramsay[1], J. G. Doyle[1], P. F. Wyper[3], E. Scullion[2], K. Wu[4] and J. A. McLaughlin[2]

[1]Armagh Observatory, College Hill, Armagh, BT61 9DG, UK

[2]Mathematics, Physics and Electrical Engineering, Northumbria University,
Newcastle upon Tyne, NE1 8ST, UK

[3]Department of Mathematical Sciences, Durham University, Durham, DH1 3LE, UK

[4]Mullard Space Science Laboratory, University College London,
Holmbury St Mary, Surrey RH5 6NT

Abstract. We report on our project to study the activity in both the Sun and low mass stars. Utilising high cadence, Hα observations of a filament eruption made using the CRISP spectropolarimeter mounted on the Swedish Solar Telescope has allowed us to determine 3D velocity maps of the event. To gain insight into the physical mechanism which drives the event we have qualitatively compared our observation to a 3D MHD reconnection model. Solar-type and low mass stars can be highly active producing flares with energies exceeding 10^{33} erg. Using K2 and TESS data we find no correlation between the number of flares and the rotation phase which is surprising. Our solar flare model can be used to aid our understanding of the origin of flares in other stars. By scaling up our solar model to replicate observed stellar flare energies, we investigate the conditions needed for such high energy flares.

Keywords. Sun: chromosphere, Sun: flares, stars: low-mass, stars: flare

1. Introduction

Solar flares represent a sudden increase in radiation which results from a rapid reconfiguration of the coronal magnetic field. These events are extremely powerful and are observed across the entire electromagnetic spectrum, possessing energy outputs up to 10^{32} erg (Fletcher *et al.* 2011). Magnetic energy released from solar flares can be observed as multiple phenomena including flare ribbons and post-flare arcades, but also in filament eruptions (e.g. Schmieder *et al.* 2013), coronal mass ejections (CME's) (Karpen *et al.* 2012) and blow-out jets (e.g. Young & Muglach 2014). Overall, the pre-flare magnetic topology is responsible for determining which of these phenomena will manifest to produce a solar flare.

In addition to solar flares, stellar flares have been observed on stars similar to the the Sun and less massive stars over many decades with energies exceeding 10^{33} erg (e.g. Schaefer *et al.* 2000). Known as 'superflares' these large outbursts can have severe consequences for any orbiting planets atmosphere, therefore, understanding their frequency and origin is vital for the existence of life. Solar-type stars have a similar interior structure to our Sun, yet they possess stronger magnetic fields producing higher levels of activity and stronger flares (Maehara *et al.* 2012). In low-mass stars with spectral types later than M4, their interiors are thought to be fully convective (Hawley *et al.* 2014) so they posses no tachocline (the boundary between the radiative and convective zones) and must generate their magnetic fields through a different dynamo mechanism. However, despite

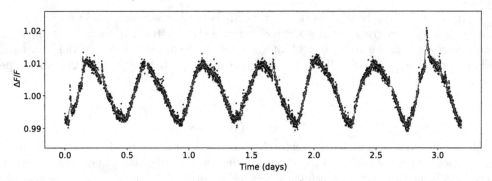

Figure 1. A section of the K2 lightcurve for the known flare star GJ 3225 (EPIC 210758829) which covers ~ 3 days. This star has a spectral type of M4.5 and rotation period, P_{rot}, of 0.45 days. The black points represent the K2 data points which have a cadence of 1 min and the red line is the Savitzky-Golay filtered, smoothed data.

this, these stars can also show increased levels of flaring activity with flares reaching energies much greater than our Sun.

As the Sun is our nearest star we are able to collect detailed spatial observations of its many phenomena from large scale flares and CME's to the smaller scale granulation and spicules. In addition, there are historical data including spot observations and number, and since the launch of the Solar Dynamics Observatory in 2010 the Sun is observed 24/7. Along with all of these observations comes a deep knowledge and understanding of the mechanisms which are at play on our nearest star and how they can affect the Earth and Solar System. In stellar physics, although the number of stars now observed by missions such as Kepler, TESS and Gaia is nearing 2 billion, the lack of detailed and long-term observations remains an issue. The capabilities of our technology and the vast distances between us and our neighbouring stars restricts our ability to produce observations which show details of the magnetic activity. Therefore, we should be looking to use the knowledge gained from detailed solar observations to illuminate our understanding of stellar flares. In this paper, we look at the solar-stellar flare connection through detailed observations of a confined solar flare event and use the results to provide insights into large scale flare events observed on other stars.

2. Stellar Flare Studies

The brightness of many stars show periodic changes as the they rotate. This is widely thought to be the result of a large dominant starspot which is cooler than its surroundings moving in and out of view as the star rotates (see Figure 1). From observations of the Sun, we know flares typically originate near sunspots so it is natural to expect flares to originate from starspots in other stars. If the analogy between solar and stellar flares holds, then we would expect to see a correlation between the timing of stellar flares and flare numbers.

The relationship between sunspots and flares in solar physics has been studied for decades with both of these phenomena being closely linked. For example, Guo *et al.* (2014) used flares from solar cycles 22 and 23 to compute a statistical study on the dependence of flares with sunspots and rotational phase. Overall, they found X-class flares were in phase with the solar cycle, suggesting flares follow the same 11 year cycle as sunspots. In a sample of solar-type stars Maehara *et al.* (2017) looked at the correlation between starspots and superflares using Kepler data, finding the superflares tend to originate from regions which hosted larger spots.

Despite the comprehensive work on stellar flares over the years, one area which has not been studied in great detail is the correlation between stellar flares and starspots. In these next sections we go on to look at the work we have carried out on the rotational phase dependency of stellar flares in multiple samples of both low mass and solar-type stars.

2.1. *Low Mass Stars*

Over the course of nine years, Kepler/K2 has provided a wealth of photometric observations for over half a million stars revolutionising the field of stellar physics. In our initial study Doyle *et al.* (2018) (henceforth Paper 1), we utilised K2 short cadence 1-min data from Fields 1 – 9 to conduct a statistical analysis of the flares on 34 M dwarfs. Our stellar sample ranged in spectral type from M0 – L1 and mass from $0.58M_\odot$ – $0.08M_\odot$. Each target was observed for $\sim$ 70 – 80 days producing a near continuous lightcurve over this period, additionally the short cadence (1-min) data is of great importance as it allows for the detection of flares with durations within a few minutes, providing a more comprehensive view of the stellar activity.

We derive rotation periods for each of the stars in our sample using the rotational modulation within the lightcurve. We utilise a Lomb-Scargle (LS) periodogram to determine the rotation period, P_{rot} and define phase zero, ϕ_0, as the minimum of the rotational modulation. Once complete, we looked to identifying and cataloging all flares in each of our targets. To do this we used an IDL suite of programs called FBEYE (see Davenport (2014) for more details). This program produces a comprehensive list of all flares and their properties in each star including start, peak and end times, flux peak, and equivalent duration. All of the flares were manually checked by eye to validate their nature. Next, the energies of the flares are calculated within the Kepler bandpass as the quiescent lumionosity, L_*, multiplied by the equivalent duration (area under flare lightcurve). We used PanStarrs magnitudes in the g, r, i and z bands to construct a template spectrum of each star and convolved it with the Kepler bandpass, deriving the quiescent flux of the stars. The quiescent luminosity is then calculated by multiplying the flux by $4\pi d^2$, where the distance, d has been determined from Gaia DR2 parallaxes.

We can now go on to investigate the rotational phase of the flares and determine whether there is a preference for certain rotational phases. For this analysis we can only use stars which possess rotation periods shorter than the observation length and as a result 8 of our targets were omitted from further analysis. In order to test whether the phase distribution of the flares is random within our M dwarf sample we utilise a simple χ_ν^2 statistical test. Flares were split into low and high energy with a cut-off determined by the median energy of and rotational phase was split into 10 bins. The χ_ν^2 was determined for each star in the all, low and high energy flare categories, where the degrees of freedom, ν, is 9. These results showed there was no preference for any rotational phase within all of the stars individually suggesting the flares do not originate from the large starspot producing the rotational modulation. This result comes as a surprise as it goes against where we believe flares should originate from.

Kepler's successor, the Transiting Exoplanet Survey Satellite (TESS: Ricker *et al.* 2015) was launched in April 2018 and has since been making observations of the northern hemisphere as part of its 2-year prime mission. Unlike Kepler/K2, TESS makes month long observations at 2-min cadence of stars brighter than $\sim$13 mag. In our second study, Doyle *et al.* (2019) (henceforth Paper 2), we use TESS short cadence 2-min photometric data from a sample of 149 M dwarfs made in Sectors 1 – 3. In Paper 1 we calculate the rotation periods, quiescent luminosity of the stars and identify all flares determining the energies for each. For Paper 2 using TESS data, we use the same methods as Paper 1, however,

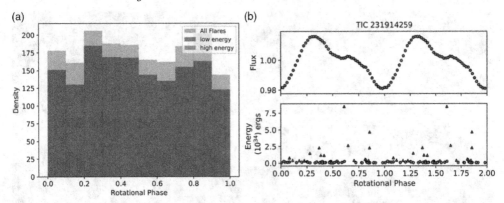

Figure 2. Panel (a) is the rotational phase distribution for all 1776 flares from the sample of 149 M dwarfs observed with TESS. Panel (b) shows the binned, folded lightcurve with $P_{rot} = 1.43$ days (upper) and the rotational phase distribution of the flares on 2MASS J0030-6236 (lower). The coverage of $\phi = 1.0 - 2.0$ is a repeat of $\phi = 0.0 - 1.0$ and the triangles represent higher energy flares and the circles represent lower energy flares with a cut off of 3.16×10^{33} erg.

instead of PanStarrs magnitudes we use SkyMapper (Wolf *et al.* 2018) magnitudes as Sectors $1-3$ are in the southern hemisphere. We also utilised the same χ^2_ν statistical test to determine whether the distribution of flares within our sample of 149 M dwarfs was random or whether there was a preference for certain rotational phases. Firstly, there was no evidence for a correlation between rotational phase of the flares in any of the 149 stars in our sample. Our χ^2_ν test indicates the flares are randomly distributed and we show the histogram of that distribution in Figure 2(a) where there is a uniform spread in low, high and all energy flares. Similarly, in individual stars, see Figure 2(b), the same χ^2_ν test was also applied and again showed there was no preference for rotational phase within the stars individually.

We highlight 4 scenarios to explain the lack of a correlation between flares and the large starspot on the surface on these stars. Firstly, there is the potential for star-planet interactions (SPI's) causing flaring activity at all rotational phases. Similar to this, there is also the potential for star-star interactions which, like SPI's, would depend on the binary orbital period. A third scenario is the presence of polar spots (Strassmeier 1996) on the surface of these stars interacting with emerging active regions and quiet sun regions as the star rotates. Polar spots are not present on our Sun due to its dynamo mechanism. However, can be present in low-mass stars where the inclination of these stars could be such that these polar spots would be visible at all rotational phases. Lastly, there is the potential for multiple spots locations across the disk of the star. A lightcurve with a one or two spot model does not produce the sinusoidal pattern observed in many low mass stars and if one spot were present on the disk we would always observe flat-top lightcurves. Therefore, the sinusoidal pattern could be produced by multiple active regions hosting spots where, there is one possessing a larger spot/group of spots. In theory, this active region should still produce higher energy flares according to McIntosh 1990 and so, a correlation between flare number and rotational phase should still be observed.

2.2. *Solar-Type Stars*

We have extended our study of stellar flares using TESS to look at solar-type stars. In this study, we present a statistical analysis of stellar flares on solar-type stars from F7 $-$ K2 spectral type, using photometric data in 2-min cadence from TESS of the whole Northern hemisphere (Sectors $1-13$). Overall, we have 210 solar-type stars in our sample

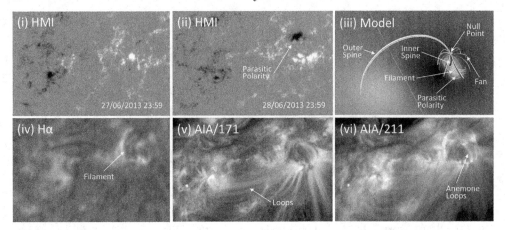

Figure 3. This selection of images shows the active region in multi-wavelengths from GONG, AIA and HMI. Panel (i) and (ii) show the HMI magnetogram before and after the parasitic polarity appears, (iii) shows the pre-flare magnetic field within the simulation and (iv) – (vi) shows the region after the parasitic polarity appears in Hα, AIA 171Å and AIA 211Å.

and have identified 1974 flares with energies between $10^{31} - 10^{36}$ erg. We are not only looking at the rotational phase distribution of the flares but also flare occurrence and year-long observations of a handful of stars, monitoring the variability of stellar activity. So far, the preliminary results of this study show no correlation between the rotational phase of stellar flares and flare number. Again, these stars all show rotational modulation as a result of starspots present on the disk so, this result comes as a surprise. As a benchmark, we are using historic GOES data of solar flares to identify the solar rotation and activity cycles to detail the close relationship between solar flares and sunspots. This will in turn help us to understand the relationship between stellar flares and starspots on other stars.

3. The Solar Study

3.1. *Observation*

Throughout this study we use multi-wavelength observations from the Swedish Solar Telescope (SST: Scharmer *et al.* 2008), Solar Dynamics Observatory (SDO: Pesnell *et al.* 2012) and the Global Oscillations Network Group (GONG: Harvey *et al.* 2011). On the 30th June 2013 in AR 11778 SST/CRISP observed a filament eruption associated with a C1.5 class solar flare. These observations consist of a series of images scanning the Hα spectral line in the range of ±1.38Å resulting in 33 spectral line positions. Overall, this active region was observed for one hour with a temporal resolution of 7.27 seconds where the eruption and flare occurred within the first 5 minutes. Full-Disk-H-alpha (FDHA) images were acquired from GONG and used as context to identify and monitor the filament as it evolves and erupts. Images from SDO/AIA and HMI were also used to provide a larger FOV in comparison to SST and to monitor the development of the active region. In particular we used images from AIA wavelengths 131Å 171Å 211Å and 304Å which represents the chromosphere, transition region, corona and flaring regions.

Figure 3 shows the pre-flare structure of the active region. In panel (i) we see the HMI magnetogram before the emergence of a parasitic polarity (a patch of negative field within a positive region) and (ii) shows the parasitic polarity. Panels (iv) – (vi) then show the FDHA and AIA images of the same time frame as (ii) detailing the filament structure which lies over the parasitic polarity and loop structure within the active region. Panel

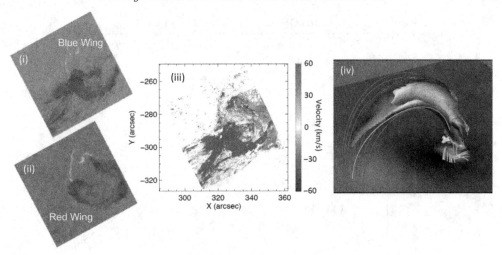

Figure 4. Panels (i) and (ii) shows the SST/CRISP images from both the red and blue wings of the Hα absorption profile, panel (iii) is the amalgamated velocity map of the blue and red wings created from Doppler velocities of the SST/CRISP Hα profiles and panel (iv) is the model of the eruption with an isosurface of velocity to show the jet.

(iii) then shows the corresponding pre-flare magnetic field structure of the 3D MHD simulation showing similar structures to those observed within the multi-wavelength observations. After this time the parasitic polarity begins to break down and disperse which is followed by the eruption of the filament. During the eruption brightnenings are observed in AIA 171Å adjacent to the filament in loops and underneath the filament as it erupts, all of which are signatures of magnetic reconnection. Once the filament erupts it propagates eastwards as a helical jet along the coronal loops (see Figure 4) towards the second footpoint and is completely confined within the active region.

In order to understand more about the plasma outflows and the development of the rotation we carried out a line fitting process on the SST/CRISP Hα line profiles. Every pixel was fitted with a single, double and triple Gaussian where the χ^2 statistic was minimised to achieve the best fit to the profile. The output was then a structure containing the Gaussian parameters such as centroid wavelength, amplitude and FWHM for each time. For full details on the fitting method please refer to the paper Doyle *et al.* (2019) which describes the methods used and the resulting output in much greater detail. From the output, the centroid wavelengths were used to calculate corresponding Doppler velocities of the line profiles representing upflows and downflows. An example of the velocity map created from these Doppler velocities can be seen in Figure 4(iii) where the map represents an amalgamation of the blue (panel (i)) and red (panel (ii)) wing components detailing the strongest flow movements of the filament/jet plasma. These maps detail the helical nature of the plasma flows where the red represents the plasma falling back to the surface under gravity and the blue the top of the helicity which is being ejected upwards.

3.2. *Model*

To investigate the details of our event further we utilised a 3D MHD simulation for qualitative comparison with the observed filament eruption and jet. The simulated filament channel progresses in the same way as previous coronal hole jet simulations reported in Wyper *et al.* (2017, 2018). An example of what the model shows can be seen in Figure 4(iv) where the isosurface shows the helical motion of the jet as it propagates away from

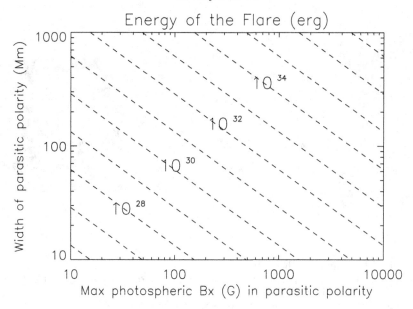

Figure 5. This plot represents a scale up of the solar 3D MHD simulation. The x-axis represent the field strength of the parasitic polarity and the y-axis represent the size of the parasitic polarity. The dashed lines then represent the varying output energies of the flare.

the surface along the coronal loops. Overall, both the observation and the simulation are in good qualitative agreement with each other. For full details of the simulation including the setup conditions and the evolution of the eruption, please refer to Doyle *et al.* (2019).

4. Discussion & Conclusion

In summary, our studies of stellar flares reported in Papers 1 & 2 using one and two-minute photometric data from K2 and TESS we conducted a statistical study on flares in samples of 34 and 149 M dwarfs respectively. Utilising a simple statistical test, we investigated whether the distribution of the flares was random and concluded that none of the stars in the sample showed any preference for certain phase distributions. This was a big surprise, as it indicates other stars do not behave like the Sun where the relationship between solar flares and sunspots is well established.

In a study involving a confined solar flare we used ground-and-space-based observations from SST and SDO and GONG where the event was compared to a 3D MHD simulation. These observations provide the evidence to validate the simulation eruption model, showing it can be applied to not only jets and Coronal Mass Ejections but also confined eruptions and flares. Overall, this study explores the finer details of solar flares and their associated eruptive phenomena, providing a unique perspective when applying this knowledge to stellar flare scenarios.

To compare the observations of stellar flares and solar flares the key lies within the 3D MHD simulation. This simulation can be scaled up to see how it would produce flares of greater energies like the ones observed in both low mass and solar-type stars. An example of this scale up can be seen in Figure 5 which shows the conditions needed in both the magnetic strength and size of the parasitic polarity in order to produce flares of greater energies. Overall, our Sun can show flares with energy outputs ranging from 10^{24} - 10^{32} erg. However, studies, like Paper 1 & Paper 2, of low mass and solar-type stars have revealed flares with energies exceeding 10^{32} erg, with 'superflares' having energies up to 10^{38} erg (Schaefer *et al.* 2000).

From Figure 5, to produce a flare of energy 10^{34} erg would require a parasitic polarity of size 200Mm and field strength 2kG. For an M dwarf with spectral type M3/M4 this means the active region hosting the parasitic polarity, which may or may not host spots, would be half the size of the visible stellar disk. In terms of the Sun the field strength of 2kG would be possible as sunspots tend to be in the region of 1kG – 4kG, however a sunspot which is a third of the stellar disk is extremely unlikely. In a study by Aulanier *et al.* (2013), they scaled up their 3D MHD simulation for eruptive flares calculating the parameters needed for larger energy flares. In their highly sheared bipole model, a flare of energy 10^{34} erg would require a bipole the size of 100Mm and field strength of 4kG. Overall, they conclude that solar-type stars which produce superflares with energies $> 10^{33}$ erg would require a much stronger dynamo than the Sun.

References

Aulanier G., Démoulin P., Schrijver C. J., *et al.* 2013, *A&A*, 549, A66

Davenport J. R. A. 2014, *ApJ*, 797, 122

Doyle L., Ramsay G., Doyle J. G., Wu K., Scullion E. 2018, *MNRAS*, 480, 2153

Doyle L., Ramsay G., Doyle J. G., Wu K. 2019, *MNRAS*, 489, 437

Doyle L., Wyper, P. F., Scullion E., *et al.* 2019, *ApJ*

Fletcher L., Dennis B. R., Hudson H. S., *et al.* 2011, *Space Science Reviews*, 159, 19

Guo J., Lin J., & Deng Y. 2014, *MNRAS*, 441, 2208

Harvey J., *et al.* 2011, *Bulletin of American Astronomical Scociety*, Vol. 43

Hawley S. L., Davenport J. R., Kowalski A. F., Wisniewski J. P., *et al.* 2014, *ApJ*, 797, 121

Karpen J. T., Antiochos S. K., *et al.* 2012, *ApJ*, 760, 81

Maehara H., *et al.* 2012, *Nature*, 485, 478

Maehara H., Notsu Y., Notsu S., *et al.* 2017, *PASJ*, 69

McIntosh P. S. 1990, *Sol. Phys*, 125, 251

Pesnell W. D., Thompson B. J. & Chamberlin P. C. 2012, *Sol. Phys*, 275, 3

Ricker G. R. *et al.* 2015, *JATIS*, 1, 014003

Schaefer B. E., King J. R. Deliyannis C. P. 2000, *ApJ*, 529, 1026

Scharmer G. B., Narayan G., Hillberg T., *et al.* 2008, *ApJ Letters*, 689, L69

Schmieder B., Demoulin P., & Aulanier G. 2013, *Advances in Space Research*, 51, 1967

Strassmeier K. G. 1996, *IAU Symposium 176*, p289

Wolf C. *et al.* 2018, *PASA*, 35, 010

Wyper P. F., Antiochos S. K. & DeVore C. R. 2017, *Nature*, 544, 452

Wyper P. F., DeVore C. R. & Antiochos S. K. 2018, *ApJ*, 852, 98

Young P., & Muglach K. 2014, *Sol Phys*, 289, 3313

Discussion

KOSOVICHEV: Did you find any stars with a clear solar-like behaviour?

DOYLE: No we did not, in fact all of the stars we have looked at both low mass and solar-type are very un-solar like. Overall they produce much higher energy flares more frequently and all rotate faster than the Sun's 27 day rotation period. It would be interesting to look at stars which have rotation periods similar to the Sun, however, this is a limitation with TESS which has an observation length of ~ 27 days.

MEDINA: Does the rotation period effect whether you see flares at all rotational phases?

DOYLE: No it does not. When we split up our sample in terms of rotation period to check for a presence of rotational phase it makes no difference and we still find the flares are randomly distributed and present at all rotational phases.

Solar and Stellar Magnetic Fields: Origins and Manifestations
Proceedings IAU Symposium No. 354, 2019
A. Kosovichev, K. Strassmeier & M. Jardine, ed.
doi:10.1017/S1743921319009943

Trigger mechanisms of the major solar flares

Shuhong Yang[1,2]🆔

[1]CAS Key Laboratory of Solar Activity, National Astronomical Observatories,
Chinese Academy of Sciences, Beijing 100101, China
email: shuhongyang@nao.cas.cn

[2]School of Astronomy and Space Science, University of Chinese Academy of Sciences,
Beijing 100049, China

Abstract. Solar flares, suddenly releasing a large amount of magnetic energy, are one of the most energetic phenomena on the Sun. For the major flares (M- and X-class flares), there exist strong-gradient polarity-inversion lines in the pre-flare photospheric magnetograms. Some parameters (e.g., electric current, shear angle, free energy) are used to measure the magnetic non-potentiality of active regions, and the kernels of major flares coincide with the highly non-potential regions. Magnetic flux emergence and cancellation, shearing motion, and sunspot rotation observed in the photosphere are deemed to play an important role in the energy buildup and flare trigger. Solar active region 12673 produced many major flares, among which the X9.3 flare is the largest one in solar cycle 24. According to the newly proposed block-induced eruption model, the block-induced complex structures built the flare-productive active region and the X9.3 flare was triggered by an erupting filament due to the kink instability.

Keywords. Sun: atmosphere, Sun: flares, Sun: magnetic fields, Sun: photosphere, sunspots

---

1. Introduction

Solar flares are one of the most energetic phenomena on the Sun (Priest and Forbes 2002; Benz 2017). During a solar flare, a great amount of energy is released. Magnetic reconnection is deemed to be an efficient way for the sudden release of free energy to drive solar flares and stellar flares (Parker 1957; Rosner *et al.* 1985; Haisch *et al.* 1991; Yang *et al.* 2015). In some flares, a bulk of plasma and magnetic structure can be ejected into the interplanetary space, thus forming a coronal mass ejection (CME; Chen 2011; Schmieder *et al.* 2015; Kilpua *et al.* 2017). CMEs may interact with the Earth and consequently impact on the terrestrial environment and the human activities (Schwenn 2006; Pulkkinen 2007).

Solar flares were first independently discovered in the white light as sudden enhancements of emission in the visible continuum by Carrington (1859) and Hodgson (1859). Actually, solar flares can be observed as conspicuous brightenings in different lines, e.g., Hα (see Fig. 1(a)), Ultraviolet (UV), EUV, X-ray, and radio. They have been frequently studied with the space-based instruments (e.g., the Solar and Heliospheric observatory (SOHO; Domingo *et al.* 1995), the Reuven Ramaty High-Energy Solar Spectroscopic Imager (RHESSI; Lin *et al.* 2002), the Hinode (Kosugi *et al.* 2007), the Solar Dynamics Observatory (SDO; Pesnell *et al.* 2012), and the Interface Region Imaging Spectrograph (IRIS; De Pontieu *et al.* 2014)), and the ground-based ones (e.g. the Goode Solar Telescope (GST; former the New Solar Telescope; Cao *et al.* 2010), the Optical and Near-infrared Solar Eruption Tracer (ONSET; Fang *et al.* 2013), the New Vacuum Solar Telescope (NVST; Liu *et al.* 2014), and the MingantU SpEctral Radioheliograph (MUSER; Yan *et al.* 2009; Yan *et al.* 2016; Chen *et al.* 2019)). Although solar flares can

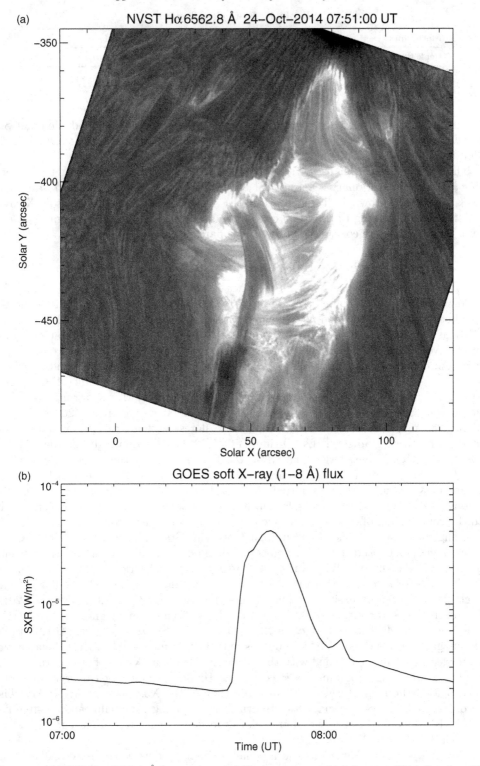

Figure 1. NVST Hα 6562.8 Å image (panel (a)) showing an M4.0 flare on 2014 October 24, and the corresponding GOES soft X-ray (1-8 Å) flux variation (panel (b)).

Table 1. Flare classification.

Flare class	A	B	C	M	X
Peak flux[1] $(\mathbf{W/m^2})$	$<10^{-7}$	$10^{-7} - 10^{-6}$	$10^{-6} - 10^{-5}$	$10^{-5} - 10^{-4}$	$>10^{-4}$

Notes:[1] According to the GOES soft X-ray (1-8 Å) flux.

occur almost everywhere on the Sun, including active regions (ARs) and the quiet Sun, large flares tend to take place in ARs with a complex geometry (Benz & Krucker 1998; Berghmans *et al.* 1998; Régnier & Canfield 2006).

The flare classification uses the letters A, B, C, M and X, according to the peak flux of the Geostationary Operational Environmental Satellite (GOES) soft X-ray (1-8 Å) flux, as shown in Table 1. Fig. 1(b) shows the variation of the soft X-ray 1-8 Å flux corresponding to the M4.0 class flare displayed in Fig. 1(a). Generally, M-class and X-class flares are considered to be major flares.

2. Magnetic properties before major flares

Solar flares tend to occur within ARs with strong magnetic fields and strong field gradients (Jing *et al.* 2006; Wang & Liu 2015; Toriumi & Wang 2019). The separation line between the positive and negative magnetic fields is termed "polarity-inversion line" (PIL; as shown in Fig. 2(a)). Zirin & Wang (1993) investigated the strength and direction of transverse magnetic fields in 6 delta-spots with the spectroscopic measurements. The magnetic fields were found to be parallel to the PIL and the field strengths were as strong as 3980 G. For AR 11035 in 2009 December, Jaeggli (2016) studied the polarized Stokes spectra, and found that the magnetic field near the PIL was strong and nearly horizontal. The largest field strengths were 3500-3800 G. Schrijver (2007) analyzed about 289 major flares, and found that these flares, without exception, were associated with strong-gradient PILs. Toriumi *et al.* (2017) systematically studied 51 flare events with GOES levels larger than M5-class. They found that there were only two flares without strong-gradient PILs. Recently, in a PIL between two opposite-polarity umbrae, Okamoto & Sakurai (2018) reported clear evidence of magnetic field of 6250 G. The strong field was parallel to the solar surface, which was suggested to be generated due to the compression of one umbra pushed by the horizontal flow from the other umbra.

Using the data from the Huairou Solar Observing Station (HSOS) in Beijing, China, Wang *et al.* (1996) studied the relationship between flare occurrence and electric currents. They found that flare activity was closely associated with vertical electric currents. AR 11158 which produced several major eruptions (including the first X-class flare of solar cycle 24) has been extensively studied (e.g., Vemareddy *et al.* 2012; Song *et al.* 2013). Based on the vector magnetograms from the Helioseismic and Magnetic Imager (HMI; Scherrer *et al.* 2012; Schou *et al.* 2012) on board the SDO, Sun *et al.* (2012) studied the magnetic fields of AR 11158 in details and their results revealed that the area with large electric current coincided with the initiation site of the X-class flare. With the high quality vector magnetograms observed by the Hinode satellite, Schrijver *et al.* (2008) investigated the magnetic fields around the time of the X3.4 flare in AR 10930. They reconstructed the coronal magnetic structures by applying the nonlinear force-free field (NLFFF) modeling, and found that, before the flare, there were large currents at the initiation site of the X flare.

Magnetic shear is defined as the angular difference between the potential field and the observed field. Hagyard *et al.* (1984) quantitatively studied the magnetic shear along the PIL in an AR. Their results revealed that the shear angle was non-uniform along the PIL, and the maxima were at the locations of repeated flare onsets. They also suggested

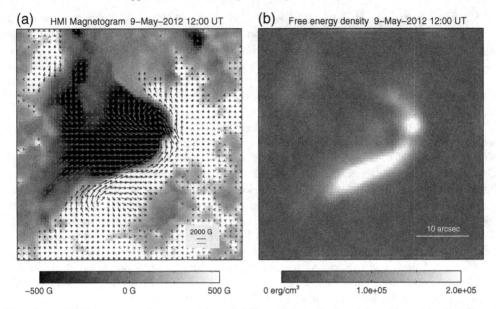

Figure 2. HMI vector magnetogram (panel (a)) and the calculated free energy density (panel (b)) before an M4.7 flare on 2012 May 09.

that continued magnetic field evolution caused the maximum shear to exceed a critical value, which resulted in a flare occurring around the site of maximum shear. AR 10486 is a super AR in solar cycle 23. It produced several major flares larger than X10 (i.e., X28, X17.2, X10) flares, in 2003. Chen & Wang (2012) quantified the characteristics of this AR using the vector magnetograms taken by the Solar Magnetic Field Telescope at HSOS. It is clear that, at the PIL, the free energy was quite high and the shear angles were very large. At some areas along the PIL, the shear angles were larger than 80 degrees. Fig. 2(a) displays the vector magnetic fields before an M4.7 flare on 2012 May 09 in AR 11476. We can see that the horizontal magnetic fields around the PIL are almost parallel to the PIL, and the highly sheared fields indicate that the magnetic fields are non-potential. Based on the vector field observation, the magnetic free energy density in the photosphere is calculated. The distribution of the free energy is displayed in Fig. 2(b). It reveals that the high free energy area (i.e., the bright region) is located along the PIL between the opposite polarities and corresponds to the initiation site of the M4.7 flare.

3. Photospheric dynamics

What have happened before the major flares? According to the previous studies, the emergence and cancellation of magnetic flux, the photospheric shearing motion, and the sunspot rotation are observed frequently.

Magnetic flux emergence and cancellation are thought to play an important role in triggering major flares and CMEs (e.g., Zirin & Wang 1993; Schmieder *et al.* 1997; Choudhary *et al.* 1998; Zhang *et al.* 2001; Burtseva, & Petrie 2013). Wang & Shi (1993) examined the associations of flares to flux emergence and cancellation. They found that the flux emergence and its driven flux cancellation with the pre-existing flux are one inseparable and elementary process, which is favorable for the occurrence of solar flares. Zhang *et al.* (2001) investigated the X5.7 flare on 2000 July 14 in AR 9077. After the detailed examination of the magnetic evolution, they found that the only obvious change was flux cancellation. The results indicated that the magnetic reconnection manifested as flux cancellation led to the global instability and thus resulted in the major flare. Burtseva, & Petrie (2013) studied 77 X-class and M-class flares with the help of 1-minute cadence Global Oscillation Network Group (GONG) full-disk magnetograms, and the importance of flux cancellation in triggering major flares was proved. Muhamad *et al.* (2017) conducted 3-dimensional magnetohydrodynamic simulations, and suggested that the data-constrained simulation involving both the large-scale magnetic structure and small-scale disturbance, e.g., emerging flux, is efficient in discovering a flare-producing AR.

The rapid shearing flows in the photosphere are crucial in the buildup of free energy which can be released to power major solar flares (Harvey & Harvey 1976; Meunier & Kosovichev, (2003); Yang *et al.* 2004). AR 10486 produced 8 X-class flares from 2003 October 23 to November 6. Yang *et al.* (2004) analyzed the high spatial resolution white-light observations with a 1 min cadence prior to an X10 flare in AR 10486, and found strong shearing flows along the PIL. These shearing flows were as high as 1.6 km s^{-1}, and they were well correlated with white-light flare kernels. With the observations from the Hinode, Shimizu *et al.* (2014) studied an X5.4 flare on 2012 March 7 and reported on a remarkable high-speed horizontal material flow along the PIL between two flare ribbons. The material flow was considered to contribute to increase the magnetic shear and to develop magnetic structures favorable for the flare initiation. In AR 11158, two emerging bipoles P1-N1 and P2-N2 collided against and sheared with each other and produced a highly sheared PIL, where the major flare kernel was located (Toriumi *et al.* 2014). Park *et al.* (2018) determined the photospheric shearing flows with a large data set of 2548 pairs of AR vector magnetograms, and investigated the shearing flows along strong magnetic PILs. They studied the relationship between the shearing flow parameters and the waiting time until the next major flare (M1.0 or above). Their results revealed that large ARs with widespread and/or strong shearing flows along PILs tend to produce major flares within 24 hr. Furthermore, Chintzoglou *et al.* (2019) demonstrated that the opposite polarities belonging to different bipolar magnetic regions collide, resulting in shearing and cancellation of magnetic flux, and named this kind of motion "collisional shearing". Fig. 3 shows a series of HMI intensity maps displaying the movements of two sunspots within AR 11476. The positive sunspot (marked with "P") significantly sheared with the negative one (marked with "N"). During this process, several M-class flares occurred.

Besides shearing motion, sunspot rotation is also very important in the storage and release of free energy (e.g., Stenflo 1969). Brown *et al.* (2003) showed that some sunspots rotated up to 200 degrees around their center, and the corresponding coronal loops were twisted and finally erupted as flares. AR 10930 has been extensively studied (e.g., Zhang *et al.* 2007; Abramenko *et al.* 2008; Min & Chae 2009; Yan *et al.* 2009; Inoue *et al.* 2012; Bamba *et al.* 2013; Gopasyuk 2015). Zhang *et al.* (2007) examined the magnetic field and sunspot evolution in AR 10930. Around the PIL region, the interaction between the fast rotating sunspot and the ephemeral regions triggered a series of brightenings and eventually the major flare occurred. They also found that the major event took place after the sunspot rotated up to 200 degrees, and the sunspot rotated at least 240 degrees about its center. Yan *et al.* (2008) statistically studied the relationship between rotating sunspots and flare productivity, and found that the sunspots with the rotating direction opposite to the global differential rotation were in favor of producing strong flares. Min & Chae (2009) studied the pattern and behavior of a rotating sunspot in AR 10930 with the high-resolution G-band images from the Solar Optical Telescope onboard the Hinode, and examined the corresponding coronal structures using the Hinode/X-Ray Telescope images. They found that the small sunspot rotated about its center by 540 degrees during five days, and the coronal loops connecting two sunspots became sigmoidal in shape. In the simulation of Amari *et al.* (2014), due to the sunspot rotation and shearing motion, the field lines were twisted, forming a flux rope gradually.

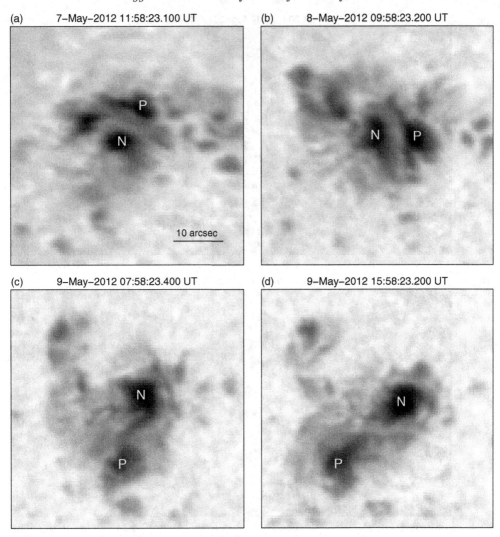

Figure 3. Sequence of HMI intensity maps showing the shearing motion of two sunspots with positive (marked with "P") and negative (marked with "N") polarities.

Then the flux rope erupted, triggering the X-class flare in AR 10930. Vemareddy *et al.* (2016) studied the major events in AR 12158 with the HMI vector magnetic field measurements and the AIA coronal EUV observations. It is shown that the time evolution of many non-potential parameters corresponded well with the sunspot rotation, and when the sunspots was rotating, two major eruptions occurred.

4. Flares and flux ropes

For the occurrence of solar flares, the CSHKP flare model has been well known for several decades (Carmichael 1964; Sturrock 1966; Hirayama 1974; Kopp & Pneuman 1976). In the following years, solar flares have been observed and investigated in details with the development of the observational instruments. According to the popular flare model, a rising filament (flux rope) stretches the overlying loops, and a current sheet is created between the anti-directed field lines beneath the flux rope (Shibata *et al.* 1995;

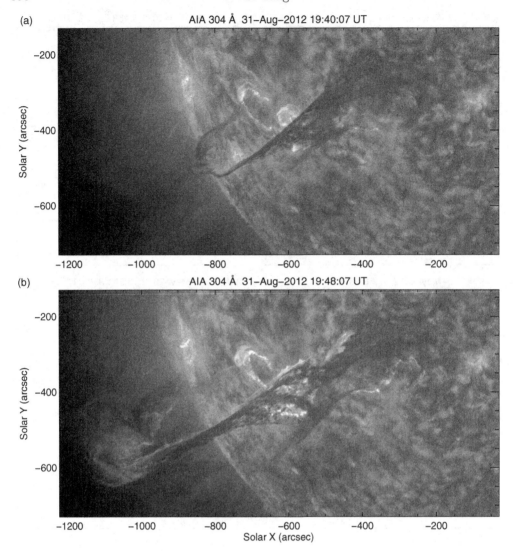

Figure 4. AIA 304 Å images showing a filament eruption accompanied by a two-ribbon flare on 2012 August 31.

Lin & Forbes 2000). Then magnetic reconnection takes place, and a solar flare occurs (Masuda *et al.* 1994). At the same time, the hot cusp-shaped coronal arcades are formed, and two ribbons at the feet of the coronal loops appear and separate. Fig. 4 shows a filament eruption observed in AIA 304 Å on 2012 August 31, which resulted in a two-ribbon flare. The eruption of the flux rope or the filament is often associated with a CME. The bright core of a CME always corresponds to the flux rope or the filament (Isenberg *et al.* 1993; Lin *et al.* 1998; Hudson *et al.* 2006).

Recently, an X8.2 flare event observed by SDO on 2017 September 10 is very consistent with the popular flare model (Fig. 5). In this event, a flux rope (denoted by the arrows in panels (a)-(b)) began to rise rapidly, behind which a long current sheet (denoted by the arrow in panel (c)) was formed. The width of the current sheet was estimated to be about 3000 km (Yan *et al.* 2018). Meanwhile, a cusp-shaped structure was formed due to the magnetic reconnection during the flare (see panel (d)). This event was also

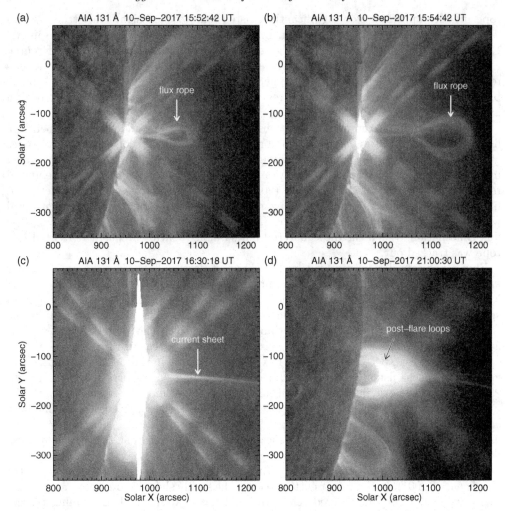

Figure 5. Sequence of AIA 131 Å images showing the occurrence of an X8.2 flare on 2017 September 10.

accompanied by a CME (Cheng *et al.* 2018). For this typical flare event, more details can be found in many papers (e.g., Hou *et al.* 2018; Li *et al.* 2018; Long *et al.* 2018; Seaton & Darnel 2018).

In ARs, the S shaped or inverse S shaped sigmoid structures in the corona are considered to be flux ropes (Rust & Kumar 1996; Canfield *et al.* 1999). If a flux rope is filled with dark material, it will be observed as a filament (Amari *et al.* 2000; Kleint *et al.* 2015). In a study of an M5.7 flare on 2012 May 10, Yang & Zhang (2018) reconstructed the coronal magnetic structures at the pre-flare stage using the NLFFF modeling (Wheatland *et al.* (2000); Wiegelmann (2004)). The results revealed that there was a flux rope above the PIL, which exactly corresponded to the Hα filament observed with the ONSET.

In the images observed with the Atmospheric Imaging Assembly (AIA; Lemen *et al.* 2012) on board the SDO, some flux ropes only can be observed in high temperature line. In the study of Cheng *et al.* (2011) about an eruptive event on 2010 November 3 observed with AIA, a flux rope rapidly moving upward was seen as a bright blob of hot plasma in 131 Å passband. Zhang *et al.* (2012) studied a flux rope which was observed as a hot channel before and during a solar eruption with the AIA observations. The flux rope initially appeared as a significantly twisted and writhed sigmoidal structure and its temperature was as high as 10 Mk. In a study about an erupting flux rope in AR 12733, Yang *et al.* (2019) determined the temperature of a flux rope with the Differential Emission Measure (DEM) method (Cheung *et al.* 2015; Su *et al.* 2018). The results revealed that when the temperature range is as high as 10-40 MK, the flux rope is much brighter than the surrounding structures, which means that this flux rope is indeed a high-temperature structure.

However, some flux ropes cannot be observed in both lower and higher temperature lines. For example, Li & Zhang (2013) presented SDO observations of two flux ropes which were tracked out by surge and filament material. When the bright mass was added into the flux rope body, the flux ropes were detected. With the high spatial and temporal resolution NVST Hα data, Yang *et al.* (2014) detected a flux rope tracked by activated filament material flow. Initially, the flux rope was invisible, and the filament material was located at one end of the flux rope. Then the filament was activated by magnetic flux cancellation. When the dark material flowed along helical threads, the twisted flux rope was tracked out.

Flux ropes can also be revealed by the shape of flare ribbons. For example, Janvier *et al.* (2014) paid attention to the double J-shaped flare ribbons during an eruptive X-class flare on 2011 February 15. They calculated the electric currents in the photosphere using the HMI vector magnetic field observations. The electric current in one ribbon was positive and in the other ribbon was negative. The shape of flare ribbons and the electric currents revealed how twisted the flux rope was in three dimensions.

When a flux rope becomes unstable, it will erupt. For the initiation of eruptions, there are several mechanisms. One possible mechanism is flux emergence (Chen & Shibata 2000). When magnetic flux emerges within the filament channel, it reconnects with the pre-existing magnetic field lines below the flux rope or on the outer edge of the filament channel, leading to the loss of equilibrium. Then the flux rope rises, a current sheet below it is formed. The fast reconnection in the current sheet induces the fast ejection of the flux rope.

Another mechanism is the tether cutting based on a single bipolar field geometry (Moore *et al.* 2001). In the tether cutting model, the highly sheared core fields are overlaid by magnetic arcades. The sheared fields slowly reconnect above the PIL, forming a large-scale twisted flux rope and some small shrinking flaring loops. The reconnection beneath the flux rope cuts off the anchoring of field lines, and allows the flux rope to rise and erupt. Chen *et al.* (2014) investigated an X4.9 flare in AR 11990 and reported the observation of tether cutting reconnection between pre-existing loops. Prior to the X4.9 flare, some pre-existing loops interacted with each other, producing a brightening region beneath the filament. Below the interaction region, a small flaring loop appeared. Meanwhile, some large-scale new helical field lines connecting two far ends of the loop structures were formed and added into the former twisted flux rope. Then due to the imbalance between the magnetic pressure and magnetic tension, the newly formed flux rope together with the filament erupted outward. This process coincides well with the tether cutting model.

A similar model is the magnetic breakout model in a multi-polar magnetic configuration (Antiochos 1998; Antiochos *et al.* (1999)). It can be regarded as the external tether cutting. In the breakout model, the reconnection occurs high in the corona above a flux rope. Since the reconnection takes between the central flux rope and the overlying field, the confinement from the overlying magnetic field is removed, like an onion-peeling process. Consequently, the flux rope begins to rise and erupt outward. Chen *et al.* (2016) reported critical observational evidence of breakout reconnection leading to an X-class flare and a CME. The observations clearly showed the presence of pairs of heated cusp-shaped loops around an X-type null point. In addition, there also existed signatures of reconnection inflows.

For a flux rope, there is a critical twist, above which the flux rope is unstable. This kind of instability is called kink instability (Hood & Priest 1979; Török & Kliem 2003, Török & Kliem 2005). The typical threshold value of the twist needed for kink instability under coronal conditions is about 3.5π, equivalent to 1.75 turns. Kumar *et al.* (2012) presented multi-wavelength AIA observations of an M3.5 limb flare associated with a CME triggered by the helical kink instability on 2011 February 24 in AR 11163. The event in their study is in agreement with the standard flare model (CSHKP). The twist of the flux rope is estimated to be 6π-8π, which is sufficient to generate the kink instability.

If the background field above a flux rope decays fast enough, the flux rope is unstable. This kind of instability is called torus instability (Kliem & Török 2006). The critical value for the torus instability is given by

$$n = -\frac{d(\log B)}{d(\log R)} > 1.5, \qquad (4.1)$$

where B is the strength of the background field at a geometrical height R above the eruption site. If the decay index (n) approaches the threshold, it will result in torus instability or partial torus instability (Aulanier *et al.* 2010; Démoulin & Aulanier 2010; Olmedo & Zhang 2010).

5. The largest flare in solar cycle 24

In 2017 September, AR 12673 produced a series of flares, including 31 major flares (4 X-class and 27 M-class flares), from September 4 to 10. After the first publication (Yang *et al.* 2017) about this super AR, a lot of more studies have been carried out (e.g., Sun & Norton 2017; Wang *et al.* 2018, Wang *et al.* 2018; Chertok 2018; Yan *et al.* 2018; Sharykin & Kosovichev 2018; Hou *et al.* 2018; Inoue *et al.* 2018; Jiang *et al.* 2018; Zou *et al.* 2019; Li *et al.* 2019; Romano *et al.* 2019; Vemareddy 2019; Getling 2019; Moraitis *et al.* 2019; Price *et al.* 2019; Anfinogentov *et al.* 2019).

Among the numerous flares in AR 12673, the X9.3 flare (Fig. 6) is the largest one in solar cycle 24. Yang *et al.* (2017) mainly focused on two questions: (1) Why was this AR so flare-productive? (2) How did the largest flare occur? In the HMI intensity maps, there was only one sunspot in the initial several days (see Fig. 1 in Yang *et al.* 2017). Then two bipoles "A" and "B" emerged nearby it successively. Due to the standing of the pre-existing sunspot, the movement of the bipoles was blocked. Thus, the bipolar patches were greatly distorted. The opposite polarities formed two semi-circular shaped structures. Then two new bipoles "C" and "D" emerged within the semi-circular zone. The newly emerging bipolar patches separated along the curved channel, and interacted with the previous fields, forming a complex system. During this process, numerous flares occurred. As noted by Sun & Norton (2017), this AR has one of the fastest magnetic flux emergence ever observed. They calculated the magnetic flux emergence rate in 6-hr chunks, and found the instantaneous magnetic flux emergence rate around 21:00 UT on September 3 was as high as $1.12^{+0.15}_{-0.05} \times 10^{21}$ Mx hr^{-1}, which occurred during the early emerging stage.

At the PIL, the magnetic fields were highly sheared, and a great deal of free magnetic energy was stored. Based on the observed photospheric vector magnetograms, Yang *et al.* (2017) extrapolated the coronal magnetic fields using the NLFFF modeling. Moreover, using the code developed by Liu *et al.* (2016), Yang *et al.* (2017) calculated the twist number T_w (Berger & Prior 2006) and squashing factor Q (Demoulin *et al.* 1996;

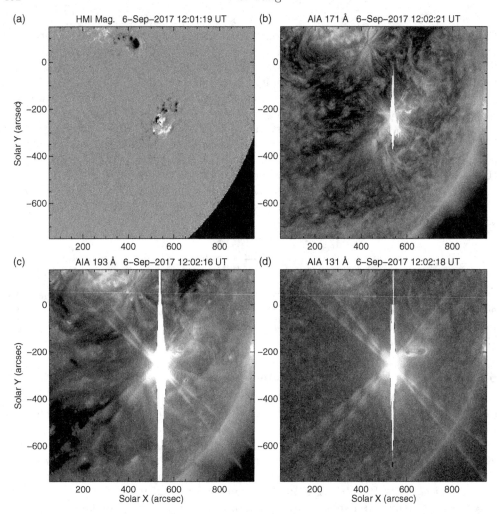

Figure 6. HMI magnetogram (panel (a)), AIA 171 Å (panel (b)), 193 Å (panel (c)), and 131 Å (panel (d)) images showing the appearance of an X9.3 flare on 2017 September 6 in AR 12673.

Titov *et al.* (2002)) of the reconstructed coronal magnetic field. The reconstructed magnetic structures revealed that there was a flux rope above the PIL, as shown by the red structure (Fig. 7). About 2 hr before the X9.3 flare, the average T_w of the inner part of the flux rope was about -1.5.

To examine the magnetic gradient across the main PIL at the AR core, Mitra *et al.* (2018) considered a slit across the PIL. Along the slit, magnetic field strength and gradient were calculated. They found that the magnetic field gradient was very sharp. The magnetic strength changes from -1000 G to 1000 G over a distance of about 1 arcsec, with the peak gradient of about 2.4×10^3 G Mm^{-1} on the PIL. With the 0.1 arcsec spatial resolution observations obtained by the GST at BBSO, Wang *et al.* (2018) found that the light bridge within this AR has usual behaviors, i.e., the strong magnetic fields and apparent photospheric twist.

At the core of the AR, the positive and negative fields sheared with each other continually. In addition, the sunspot with negative polarity rotated anticlockwise. Therefore, the twist number T_w of the flux rope continuously increased, which ultimately reached

Figure 7. Side-view of the 3-dimensional coronal structures reconstructed from the HMI photospheric vector magnetogram before the X9.3 flare on 2017 September 6.

or even exceeded the threshold of the kink instability. AIA images clearly show that a filament erupted and two ribbons appeared. The observed filament in AIA 304 Å images corresponds to the reconstructed flux rope using the NLFFF method. During the filament eruption, a kink structure appeared (Yang *et al.* 2017). It is a signature of the kink instability, which triggered the largest flare.

Based on the observations, Yang *et al.* (2017) proposed for the first time the block-induced eruption model to answer the two main questions. In this model, there was a standing sunspot, which blocked the movement of newly emerging bipoles. The block-induced complex structures built the flare-productive AR, and the X9.3 flare was triggered by an erupting flux rope due to the kink instability. When the flux rope erupted, it interacted with two nearby flux ropes (Hou *et al.* 2018; Mitra *et al.* 2018). Eventually, the multi-flux-rope system erupted outward, forming a CME.

Acknowledgements

The data are used courtesy of NVST, HMI, AIA, and GOES science teams. This work is supported by the National Natural Science Foundations of China (11673035, 11790304, 11533008, 11790300), Key Programs of the Chinese Academy of Sciences (QYZDJ-SSW-SLH050), and the Youth Innovation Promotion Association of CAS.

References

Abramenko, V., Yurchyshyn, V., & Wang, H. 2008, *Astrophys. J.*, 681, 1669
Amari, T., Luciani, J. F., Mikic, Z., & Linker, J. 2000, *Astrophys. J. Lett.*, 529, L49
Amari, T., Canou, A., & Aly, J.-J. 2014, *Nature*, 514, 465
Anfinogentov, S. A., Stupishin, A. G., Myshyakov, I. I., & Fleishman, G. D. 2019, *Astrophys. J. Lett.*, 880, L29
Antiochos, S. K. 1998, *Astrophys. J. Lett.*, 502, L181
Antiochos, S. K., DeVore, C. R., & Klimchuk, J. A. 1999, *Astrophys. J.*, 510, 485
Aulanier, G., Török, T., Démoulin, P., & DeLuca, E. E. 2010, *Astrophys. J.*, 708, 314
Bamba, Y., Kusano, K., Yamamoto, T. T., *et al.* 2013, *Astrophys. J.*, 778, 48
Benz, A. O., & Krucker, S. 1998, *Sol. Phys.*, 182, 349

Benz, A. O. 2017, *Living Rev. Sol. Phys.*, 14, 2

Berger, M. A., & Prior, C. 2006, *J. Phys. A*, 39, 8321

Berghmans, D., Clette, F., & Moses, D. 1998, *Astron. Astrophys.*, 336, 1039

Brown, D. S., Nightingale, R. W., Alexander, D., *et al.* 2003, *Sol. Phys.*, 216, 79

Burtseva, O., & Petrie, G. 2013, *Sol. Phys.*, 283, 429

Canfield, R. C., Hudson, H. S., & McKenzie, D. E. 1999, *Geophys. Res. Lett.*, 26, 627

Cao, W., Gorceix, N., Coulter, R., *et al.* 2010, *Astro. Nachr.*, 331, 636

Carmichael, H. 1964, *NASA Special Publication*, 50, 451

Carrington, R. C. 1859, *Mon. Not. Roy. Astron. Soc.*, 20, 13

Chen, A. Q., & Wang, J. X. 2012, *Astron. Astrophys.*, 543, A49

Chen, H., Zhang, J., Cheng, X., *et al.* 2014, *Astrophys. J. Lett.*, 797, L15

Chen, P. F. 2011, *Living Rev. Sol. Phys.*, 8, 1

Chen, P. F., & Shibata, K. 2000, *Astrophys. J.*, 545, 524

Chen, X., Yan, Y., Tan, B., *et al.* 2019, *Astrophys. J.*, 878, 78

Chen, Y., Du, G., Zhao, D., *et al.* 2016, *Astrophys. J. Lett.*, 820, L37

Cheng, X., Li, Y., Wan, L. F., *et al.* 2018, *Astrophys. J.*, 866, 64

Cheng, X., Zhang, J., Liu, Y., *et al.* 2011, *Astrophys. J. Lett.*, 732, L25

Chertok, I. M. 2018, *Res. Notes AAS*, 2, 20

Cheung, M. C. M., Boerner, P., Schrijver, C. J., *et al.* 2015, *Astrophys. J.*, 807, 143

Chintzoglou, G., Zhang, J., Cheung, M. C. M., *et al.* 2019, *Astrophys. J.*, 871, 67

Choudhary, D. P., Ambastha, A., & Ai, G. 1998, *Sol. Phys.*, 179, 133

De Pontieu, B., Title, A. M., Lemen, J. R., *et al.* 2014, *Sol. Phys.*, 289, 2733

Demoulin, P., Henoux, J. C., Priest, E. R., & Mandrini, C. H. 1996, *Astron. Astrophys.*, 308, 643

Démoulin, P., & Aulanier, G. 2010, *Astrophys. J.*, 718, 1388

Domingo, V., Fleck, B., & Poland, A. I. 1995, *Sol. Phys.*, 162, 1

Fang, C., Chen, P.-F., Li, Z., *et al.* 2013, *Res. Astron. Astrophys.*, 13, 1509–1517

Getling, A. V. 2019, *Astrophys. J.*, 878, 127

Gopasyuk, O. S. 2015, *Adv. Space Res.*, 55, 937

Hagyard, M. J., Smith, J. B., Jr., Teuber, D., & West, E. A. 1984, *Sol. Phys.*, 91, 115

Haisch, B., Strong, K. T., & Rodono, M. 1991, *Annu. Rev. Astron. Astrophys.*, 29, 275

Harvey, K. L., & Harvey, J. W. 1976, *Sol. Phys.*, 47, 233

Hirayama, T. 1974, *Sol. Phys.*, 34, 323

Hodgson, R. 1859, *Mon. Not. Roy. Astron. Soc.*, 20, 15

Hood, A. W., & Priest, E. R. 1979, *Sol. Phys.*, 64, 303

Hou, Y. J., Zhang, J., Li, T., *et al.* 2018, *Astron. Astrophys.*, 619, A100

Hudson, H. S., Bougeret, J.-L., & Burkepile, J. 2006, *Space Sci. Rev.*, 123, 13

Inoue, S., Shiota, D., Yamamoto, T. T., *et al.* 2012, *Astrophys. J.*, 760, 17

Inoue, S., Shiota, D., Bamba, Y., & Park, S.-H. 2018, *Astrophys. J.*, 867, 83

Isenberg, P. A., Forbes, T. G., & Demoulin, P. 1993, *Astrophys. J.*, 417, 368

Jaeggli, S. A. 2016, *Astrophys. J.*, 818, 81

Janvier, M., Aulanier, G., Bommier, V., *et al.* 2014, *Astrophys. J.*, 788, 60

Jiang, C., Zou, P., Feng, X., *et al.* 2018, *Astrophys. J.*, 869, 13

Jing, J., Song, H., Abramenko, V., Tan, C., & Wang, H. 2006, *Astrophys. J.*, 644, 1273

Kilpua, E., Koskinen, H. E. J., & Pulkkinen, T. I. 2017, *Living Rev. Sol. Phys.*, 14, 5

Kleint, L., Battaglia, M., Reardon, K., *et al.* 2015, *Astrophys. J.*, 806, 9

Kliem, B., & Török, T. 2006, *Phys. Rev. Lett.*, 96, 255002

Kopp, R. A., & Pneuman, G. W. 1976, *Sol. Phys.*, 50, 85

Kosugi, T., Matsuzaki, K., Sakao, T., *et al.* 2007, *Sol. Phys.*, 243, 3

Kumar, P., Cho, K.-S., Bong, S.-C., Park, S.-H., & Kim, Y. H. 2012, *Astrophys. J.*, 746, 67

Lemen, J. R., Title, A. M., Akin, D. J., *et al.* 2012, *Sol. Phys.*, 275, 17

Li, T., & Zhang, J. 2013, *Astrophys. J. Lett.*, 770, L25

Li, X., Zhang, J., Yang, S., & Hou, Y. 2019, *Publ. Astron. Soc. Jpn.*, 71, 14

Li, Y., Xue, J. C., Ding, M. D., *et al.* 2018, *Astrophys. J. Lett.*, 853, L15

Lin, J., & Forbes, T. G. 2000, *J. Geophys. Res-Space Phys.*, 105, 2375

Lin, J., Forbes, T. G., Isenberg, P. A., & Démoulin, P. 1998, *Astrophys. J.*, 504, 1006

Lin, R. P., Dennis, B. R., Hurford, G. J., *et al.* 2002, *Sol. Phys.*, 210, 3

Liu, R., Kliem, B., Titov, V. S., *et al.* 2016, *Astrophys. J.*, 818, 148

Liu, Z., Xu, J., Gu, B.-Z., *et al.* 2014, *Res. Astron. Astrophys.*, 14, 705–718

Long, D. M., Harra, L. K., Matthews, S. A., *et al.* 2018, *Astrophys. J.*, 855, 74

Masuda, S., Kosugi, T., Hara, H., Tsuneta, S., & Ogawara, Y. 1994, *Nature*, 371, 495

Meunier, N., & Kosovichev, A. 2003, *Astron. Astrophys.*, 412, 541

Min, S., & Chae, J. 2009, *Sol. Phys.*, 258, 203

Mitra, P. K., Joshi, B., Prasad, A., *et al.* 2018, *Astrophys. J.*, 869, 69

Moore, R. L., Sterling, A. C., Hudson, H. S., *et al.* 2001, *Astrophys. J.*, 552, 833

Moraitis, K., Sun, X., Pariat, É., & Linan, L. 2019, *Astron. Astrophys.*, 628, A50

Muhamad, J., Kusano, K., Inoue, S., & Shiota, D. 2017, *Astrophys. J.*, 842, 86

Olmedo, O., & Zhang, J. 2010, *Astrophys. J.*, 718, 433

Okamoto, T. J., & Sakurai, T. 2018, *Astrophys. J. Lett.*, 852, L16

Park, S.-H., Guerra, J. A., Gallagher, P. T., Georgoulis, M. K., & Bloomfield, D. S. 2018, *Sol. Phys.*, 293, 114

Parker, E. N. 1957, *J. Geophys. Res-Space Phys.*, 62, 509

Pesnell, W. D., Thompson, B. J., & Chamberlin, P. C. 2012, *Sol. Phys.*, 275, 3

Price, D. J., Pomoell, J., Lumme, E., & Kilpua, E. K. J. 2019, *Astron. Astrophys.*, 628, A114

Priest, E. R., & Forbes, T. G. 2002, *Astron. Astrophys. Rev.*, 10, 313

Pulkkinen, T. 2007, *Living Rev. Sol. Phys.*, 4, 1

Régnier, S., & Canfield, R. C. 2006, *Astron. Astrophys.*, 451, 319

Romano, P., Elmhamdi, A., & Kordi, A. S. 2019, *Sol. Phys.*, 294, 4

Rosner, R., Golub, L., & Vaiana, G. S. 1985, *Annu. Rev. Astron. Astrophys.*, 23, 413

Rust, D. M., & Kumar, A. 1996, *Astrophys. J. Lett.*, 464, L199

Scherrer, P. H., Schou, J., Bush, R. I., *et al.* 2012, *Sol. Phys.*, 275, 207

Schmieder, B., Aulanier, G., Demoulin, P., *et al.* 1997, *Astron. Astrophys.*, 325, 1213

Schmieder, B., Aulanier, G., & Vršnak, B. 2015, *Sol. Phys.*, 290, 3457

Schou, J., Scherrer, P. H., Bush, R. I., *et al.* 2012, *Sol. Phys.*, 275, 229

Schrijver, C. J. 2007, *Astrophys. J. Lett.*, 655, L117

Schrijver, C. J., DeRosa, M. L., Metcalf, T., *et al.* 2008, *Astrophys. J.*, 675, 1637

Schwenn, R. 2006, *Living Rev. Sol. Phys.*, 3, 2

Seaton, D. B., & Darnel, J. M. 2018, *Astrophys. J. Lett.*, 852, L9

Sharykin, I. N., & Kosovichev, A. G. 2018, *Astrophys. J.*, 864, 86

Shibata, K., Masuda, S., Shimojo, M., *et al.* 1995, *Astrophys. J. Lett.*, 451, L83

Shimizu, T., Lites, B. W., & Bamba, Y. 2014, *Publ. Astron. Soc. Jpn.*, 66, S14

Song, Q., Zhang, J., Yang, S.-H., *et al.* 2013, *Res. Astron. Astrophys.*, 13, 226

Stenflo, J. O. 1969, *Sol. Phys.*, 8, 115

Sturrock, P. A. 1966, *Nature*, 211, 695

Su, Y., Veronig, A. M., Hannah, I. G., *et al.* 2018, *Astrophys. J. Lett.*, 856, L17

Sun, X., Hoeksema, J. T., Liu, Y., *et al.* 2012, *Astrophys. J.*, 748, 77

Sun, X., & Norton, A. A. 2017, *Res. Notes AAS*, 1, 24

Titov, V. S., Hornig, G., & Démoulin, P. 2002, *J. Geophys. Res-Space Phys.*, 107, 1164

Toriumi, S., Iida, Y., Kusano, K., *et al.* 2014, *Sol. Phys.*, 289, 3351

Toriumi, S., Schrijver, C. J., Harra, L. K., *et al.* 2017, *Astrophys. J.*, 834, 56

Toriumi, S., & Wang, H. 2019, *Living Rev. Sol. Phys.*, 16, 3

Török, T., & Kliem, B. 2003, *Astron. Astrophys.*, 406, 1043

Török, T., & Kliem, B. 2005, *Astrophys. J. Lett.*, 630, L97

Vemareddy, P. 2019, *Astrophys. J.*, 872, 182

Vemareddy, P., Ambastha, A., & Maurya, R. A. 2012, *Astrophys. J.*, 761, 60

Vemareddy, P., Cheng, X., & Ravindra, B. 2016, *Astrophys. J.*, 829, 24

Wang, H., & Liu, C. 2015, *Res. Astron. Astrophys.*, 15, 145–174

Wang, H., Yurchyshyn, V., Liu, C., *et al.* 2018a, *Res. Notes AAS*, 2, 8

Wang, J., & Shi, Z. 1993, *Sol. Phys.*, 143, 119

Wang, J., Shi, Z., Wang, H., *et al.* 1996, *Astrophys. J.*, 456, 861

Wang, R., Liu, Y. D., Hoeksema, J. T., Zimovets, I. V., & Liu, Y. 2018b, *Astrophys. J.*, 869, 90

Wheatland, M. S., Sturrock, P. A., & Roumeliotis, G. 2000, *Astrophys. J.*, 540, 1150

Wiegelmann, T. 2004, *Sol. Phys.*, 219, 87

Yan, X.-L., Qu, Z.-Q., & Kong, D.-F. 2008, *Mon. Not. Roy. Astron. Soc.*, 391, 1887

Yan, X.-L., Qu, Z.-Q., Xu, C.-L., *et al.* 2009, *Res. Astron. Astrophys.*, 9, 596

Yan, X. L., Wang, J. C., Pan, G. M., *et al.* 2018, *Astrophys. J.*, 856, 79

Yan, X. L., Yang, L. H., Xue, Z. K., *et al.* 2018, *Astrophys. J. Lett.*, 853, L18

Yan, Y., Zhang, J., Wang, W., *et al.* 2009, *Earth Moon Planets*, 104, 97

Yan, Y., Chen, L., & Yu, S. 2016, IAU Symp. 320 in Solar and Stellar Flares and their Effects on Planets, ed. A. G. Kosovichev, S. L. Hawley, & P. Heinzel (Cambridge: Cambridge Univ. Press), 427

Yang, G., Xu, Y., Cao, W., *et al.* 2004, *Astrophys. J. Lett.*, 617, L151

Yang, S., & Zhang, J. 2018, *Astrophys. J. Lett.*, 860, L25

Yang, S., Zhang, J., Liu, Z., & Xiang, Y. 2014b, *Astrophys. J. Lett.*, 784, L36

Yang, S., Zhang, J., Song, Q., Bi, Y., & Li, T. 2019, *Astrophys. J.*, 878, 38

Yang, S., Zhang, J., & Xiang, Y. 2014a, *Astrophys. J. Lett.*, 793, L28

Yang, S., Zhang, J., & Xiang, Y. 2015, *Astrophys. J. Lett.*, 798, L11

Yang, S., Zhang, J., Zhu, X., & Song, Q. 2017, *Astrophys. J. Lett.*, 849, L21

Zhang, J., Wang, J., Deng, Y., & Wu, D. 2001, *Astrophys. J. Lett.*, 548, L99

Zhang, J., Li, L., & Song, Q. 2007, *Astrophys. J. Lett.*, 662, L35

Zhang, J., Cheng, X., & Ding, M.-D. 2012, *Nat. Commun.*, 3, 747

Zirin, H., & Wang, H. 1993, *Nature*, 363, 426

Zirin, H., & Wang, H. 1993, *Sol. Phys.*, 144, 37

Zou, P., Jiang, C., Feng, X., *et al.* 2019, *Astrophys. J.*, 870, 97

Discussion

K. STRASSMEIER: Do you see a relation to radio-II bursts along with X-class flares?

S. YANG: For these X-class flares, we have not paid attention to the radio-II bursts. It is indeed worth studying.

J. LUHMANN: Given the obvious complexity of the initiation of the flare you describe, what advice would you have for the flare forecasting community?

S. YANG: Magnetic field is the most important factor for the flare initiation, and now only the magnetic field in the photosphere can be accurately measured. If we want to forecast the flare occurrence, we should focus on the structure and evolution of the photospheric magnetic field. We can examine the complexity of the vector magnetograms of active regions, e.g., the polarity-inversion lines with strong shear angles. In addition, we can examine the photospheric evolution, especially the significant flux emergence and sunspot rotation.

Y. YAN: You showed the largest flare processes in solar cycle 24. There was also a largest sunspot produced many X-class flares and there were not many CMEs accompanied. Do you compare any difference between these two event?

S. YANG: The largest sunspot group in solar cycle 24 is within active region 12192 in 2014 October. It produced 6 X-class flares and none of them was accompanied by CME. This is mainly due to the strong confinement from the overlying field above the flaring core region of active region 12192.

Solar and Stellar Magnetic Fields: Origins and Manifestations
Proceedings IAU Symposium No. 354, 2019
A. Kosovichev, K. Strassmeier & M. Jardine, ed.
doi:10.1017/S1743921320001465

(Simulating) Coronal Mass Ejections in Active Stars

Julián D. Alvarado-Gómez[1,2]† ⓘ, Jeremy J. Drake[2], Cecilia Garraffo[3],
Sofia P. Moschou[2], Ofer Cohen[4], Rakesh K. Yadav[3] and
Federico Fraschetti[2,5]

[1]Leibniz Institute for Astrophysics Potsdam
An der Sternwarte 16, 14482 Potsdam, Germany
email: julian.alvarado-gomez@aip.de | 🐦 @AstroRaikoh

[2]Center for Astrophysics | Harvard & Smithsonian
60 Garden Street, Cambridge, MA 02138, USA

[3]Institute for Applied Computational Science, Harvard University
33 Oxford Street, Cambridge, MA 02138, USA

[4]University of Massachusetts at Lowell, Department of Physics & Applied Physics
600 Suffolk Street, Lowell, MA 01854, USA

[5]Department of Planetary Sciences-Lunar and Planetary Laboratory
University of Arizona, Tucson, AZ 85721, USA

Abstract. The stellar magnetic field completely dominates the environment around late-type stars. It is responsible for driving the coronal high-energy radiation (e.g. EUV/X-rays), the development of stellar winds, and the generation transient events such as flares and coronal mass ejections (CMEs). While progress has been made for the first two processes, our understanding of the eruptive behavior in late-type stars is still very limited. One example of this is the fact that despite the frequent and highly energetic flaring observed in active stars, direct evidence for stellar CMEs is almost non-existent. Here we discuss realistic 3D simulations of stellar CMEs, analyzing their resulting properties in contrast with solar eruptions, and use them to provide a common framework to interpret the available stellar observations. Additionally, we present results from the first 3D CME simulations in M-dwarf stars, with emphasis on possible observable signatures imprinted in the stellar corona.

Keywords. Magnetohydrodynamics (MHD), stars: activity, stars: coronae, stars: flare, stars: winds, outflows, Sun: coronal mass ejections (CMEs), Sun: flares

---

1. Introduction

Flares and Coronal Mass Ejections (CMEs) are spectacular manifestations of magnetic energy release in the Sun and cool stars. Flares correspond to a temporal increase in the electromagnetic radiation (across the entire spectrum) up to several orders of magnitude, in rare cases briefly exceeding a star's quiescent state bolometric luminosity (in particular wavelengths). A CME is characterized by the release of relatively dense, magnetized material to the outer corona and the stellar wind (Webb & Howard 2012, Benz 2017). Large flares on the Sun (X-class, or $\geq 10^{31}$ ergs in 1–8 Å soft X-rays) are almost always associated with CMEs (e.g, Yashiro & Gopalswamy 2009). The most energetic events pose threats to life on Earth, with recent examples including the X15 and X4 class flares (nearly 10^{33} erg of X-ray energy) of March 6 1989 that triggered a CME that caused the collapse of Quebec's electricity grid; and the 1859 "Carrington Event" that ignited

† Karl Schwarzschild Fellow

telegraph lines and spread aurorae as far south as Hawaii, Cuba, and even Colombia (Carrington 1859, Moreno Cárdenas *et al.* 2016). An extraordinary event in AD 774–775, that was discovered in $^{14}C/^{12}C$ data from Japanese cedar tree rings, was also likely caused by a huge solar flare that would have destroyed 20% of the ozone layer (Melott & Thomas 2012).

Despite their importance, the energetics of flares and CMEs is still poorly constrained. Flares originate with the sudden conversion of magnetic energy into plasma heating and acceleration of electrons and protons within a magnetic loop. Observations have revealed that the radiated energy in white light completely dominates the soft X-ray emission in solar flares—by factors of up to 100 (Kretzschmar 2011, Emslie *et al.* 2012). In turn, the kinetic energy of the CMEs associated with large flares are typically larger than the bolometric luminosity by factors of 3 or so. Moreover, multi-spacecraft observations have shown that only up to 20% of the kinetic energy of CMEs is channelled into generation of energetic particles (Emslie *et al.* 2012). The solar data, then, indicate that soft X-rays constitute only a few percent of the total dissipated energy, and that CMEs carry more energy than flares.

The situation in the stellar regime may be radically different. Large flares on active M-dwarfs and T Tauri stars can reach total soft X-ray fluences of $10^{34} - 10^{36}$ erg—between three and five orders of magnitude larger than in X-class solar flares (e.g. Güdel *et al.* 2004, Guarcello *et al.* 2019). Furthermore, the coronae of very active stars appear to be continuously flaring (e.g. Güdel 1997, Drake *et al.* 2000, Huenemoerder *et al.* 2010), and their light curves can be well-modeled using a superposition of flares (e.g. Kashyap *et al.* 2002, Caramazza *et al.* 2007). This presents a serious energy problem. The most active stars are observed to emit about 1/1000th of their bolometric luminosity in soft X-rays—the so-called coronal saturation limit (e.g. Wright *et al.* 2011, Wright *et al.* 2018). If these X-rays originate from flares, and only about 1% of the flare energy is in the form of soft X-rays like in solar flares, then the implication is that magnetic energy dissipation—mainly in the form of CMEs—amounts to 100 times the X-ray flux or about 10% of a star's total energy output (Drake *et al.* 2013). Apart from placing an implausibly high energy requirement, such elevated stellar CME activity largely disagrees with observations (or the lack of, see Leitzinger *et al.* 2014, Crosley *et al.* 2016, Villadsen 2017, Crosley & Osten 2018), where only one event has been confirmed so far (Argiroffi *et al.* 2019). It appears then that strong flares and CMEs on very active stars have a quite different energy partition compared to their solar counterparts.

Briefly discussed by Drake *et al.* (2016), this re-distribution of energy could be related with a suppression mechanism of CMEs, in which the stellar large-scale magnetic field would entrap the plasma ejecta (up to a certain energy), allowing only the radiation and a fraction of particles accelerated at the flare site to escape (see also Odert *et al.* 2017, Fraschetti *et al.* 2019). We investigate this possibility through 3D magnetohydrodynamic (MHD) simulations, applying realistic models currently used in space weather studies of the solar system (e.g. Jin *et al.* 2017a). The CMEs evolve in the corona and stellar wind conditions imposed by surface field configurations—in terms of field strength and topology—compatible with observations of young Sun-like stars (e.g. Donati & Landstreet 2009, Alvarado-Gómez *et al.* 2015) and a state-of-the-art dynamo simulation of the fully-convective M-dwarf Proxima Centauri (see Yadav *et al.* 2016). We provide here a summary of the results and refer the reader to Alvarado-Gómez et al. (2018, 2019) for additional details.

2. Numerical Models

Three different models are considered in our investigation. The first one is the Alfvén Wave Solar Model (AWSoM, Sokolov *et al.* 2013, van der Holst *et al.* 2014), which is used to compute the quiescent conditions of the stellar wind and corona (steady-state solution), driven by the magnetic field configuration at the stellar surface. Coupled to AWSoM, the flux-rope models of Gibson & Low (1998, GL), and Titov & Démoulin (1999, TD), serve to drive the CME simulations. These models are part of the Space Weather Modeling Framework (Gombosi *et al.* 2018, SWMF), commonly used for solar system research and forecast (e.g. Manchester *et al.* 2008, Jin *et al.* 2013, Jin *et al.* 2017b, Oran *et al.* 2017).

3. Results and Discussion

3.1. *Suppression of CMEs by a large-scale magnetic field*

We begin by considering a surface field configuration suitable for testing the large-scale magnetic confinement of stellar CMEs. For this, we use the well-studied synoptic magnetogram of the solar Carrington Rotation (CR) 2107 (e.g. Sokolov *et al.* 2013, Jin et al. 2017a, 2017b), and enhance the dipole component (aligned with the stellar rotation axis) to 75 G‡. While much stronger large-scale fields are reported for very active stars (up to kG levels, see Donati 2011), this assumption is commensurable with observed surface magnetic fields in $\sim 0.4 - 0.8$ Gyr old F-G-K main sequence stars (e.g. Morgenthaler *et al.* 2012, Jeffers *et al.* 2014, Hussain *et al.* 2016). Nominal solar values for mass, radius, and rotation period are assumed in our simulations.

In addition, we take advantage of the solar CME numerical calibration study (based on the GL flux-rope model) performed by Jin *et al.* (2017a), and employ the same eruption parameters in our modified stellar simulations. Figure 1 shows the same flux-rope eruption taking place under the modified CR 2107 + 75 G large-scale dipole field (left) and the fiducial solar CR 2107 magnetic configuration (right). While this particular eruption, with an associated poloidal flux $\Phi_p \simeq 2.0 \times 10^{22}$ Mx (or an equivalent X5.0 GOES class; see Alvarado-Gómez *et al.* 2018), produces a relatively strong CME

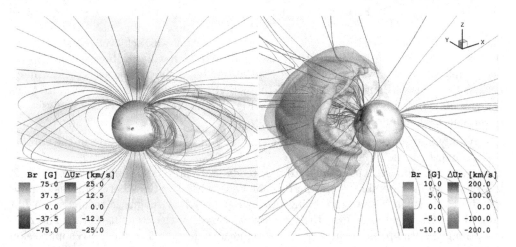

Figure 1. Results for a GL flux-rope eruption taking place in AWSoM simulations driven by the CR 2107 + 75 G large-scale dipole case (left), and the nominal CR 2107 (right). The central sphere corresponds to the stellar surface, colored by the magnetic field driving the model. The secondary color scale denotes the coronal Doppler shift velocity (ΔU_r) with respect to the pre-CME conditions. The identified eruption is visualized as a translucent yellow iso-surface. Selected magnetic field lines surrounding the eruptive active region (magenta) and associated with the large-scale field (gray) are included. The field of view in both panels is 10 $R_\star$.

‡ Restricted by computational limitations.

in the solar case ($M^{\mathrm{CME}} \sim 10^{17}$ g, $E_{\mathrm{K}}^{\mathrm{CME}} \sim 10^{32}$ erg), is totally confined in the stellar simulation. The perturbed material (yellow iso-surface in the visualization) follows the overlying field, remaining bound to the lower regions of the corona. As discussed in Alvarado-Gómez *et al.* (2018), we found that eruptive events with equivalent flare energies up to $\sim$X20 in the GOES classification, would be mitigated by this particular configuration of the large-scale magnetic field.

We also considered sufficiently strong flux-rope eruptions so they would escape the large-scale field confinement. Our analysis revealed that the overlying field significantly reduced the final CME speeds (and therefore the associated kinetic energies) in contrast with extrapolations from solar data. On the other hand, the total mass perturbed in our simulated events (confined and escaping) roughly followed the solar flare-CME relation extended to the stellar regime (Aarnio *et al.* 2012, Drake *et al.* 2013). Interestingly, these two predictions are consistent with the observed properties of the best stellar CME candidates observed so far (Moschou et al. 2017, 2019, Vida *et al.* 2019), as well as with the only direct detection currently available (Argiroffi *et al.* 2019).

3.2. *Magnetically-suppressed CME events in M-dwarf stars: Coronal response*

Similar to the Sun-like models, our M-dwarf CME simulations consist of a steady-state description of the stellar wind and the corona, which is then used as initial condition for a time-dependent flux-rope eruption model (the TD model was considered in this case). We drive AWSoM using the surface field configuration predicted by a self-consistent dynamo simulation of a fully-convective star (Yadav *et al.* 2016). Fine-tuned to the archetypical flare star Proxima Centauri—in terms of mass, radius, and rotation period—this dynamo model yields a long-term variability time-scale for the stellar magnetic field compatible with the observed activity cycle in this star ($P_{\mathrm{cyc}} \sim 7$ yr, Suárez Mascareño *et al.* 2016, Wargelin *et al.* 2017).

As discussed in Alvarado-Gómez *et al.* (2019), we focused on the coronal response during a CME, using eruption parameters expected on the stellar regime ($M^{\mathrm{FR}} = 4 \times 10^{14}$ g, $E_{\mathrm{B,free}}^{\mathrm{FR}} \simeq 6.5 \times 10^{34}$ erg), and under different large-scale confinement conditions (weak, moderate, strong). The latter was achieved by scaling the strength of the dynamo-generated surface field to match values observed in low- to moderately-active M-dwarfs (Reiners 2014). We analyzed five cases, with surface magnetic field strengths ranging from 600 G to 1400 G in 200 G increments. Visualizations of the $\pm$ 800 G scaling case are presented in Fig. 2. The main results of our analysis can be summarized as follows:

• Weakly and partially confined CMEs generated a flare-like signature in the corona, predominantly in the integrated X-ray emission ($0.2 - 2.5$ keV), with increments of up to one order of magnitude with respect to the steady-state pre-CME conditions (e.g. Fig. 2, top-right). While resembling a normal stellar flare, with durations between tens of minutes up to an hour, these brightenings are not powered by magnetic reconnection but instead by strong compression of the coronal material by the escaping CME. The relative strength of the peak in these transient signatures decreases as the CME suppression increases.

• The resulting flare-like events show a characteristic hot-to-cool, red-to-blue evolution in their Doppler shift profiles, transitioning from hotter ($\log(T) \gtrsim 6.8$) to cooler ($\log(T) \lesssim 6.0$) coronal lines, and with velocities within $\pm$ 200 km s^{-1} (for the CME and stellar parameters here considered). As with the general behavior of stellar CMEs, stronger confinement leads to lower relative velocities (<100 km s^{-1}).

• A gradual brightening of the soft X-ray corona (by factors of $\sim 2 - 3$) is observed in fully suppressed CME events. Extending over the course of several hours, the associated

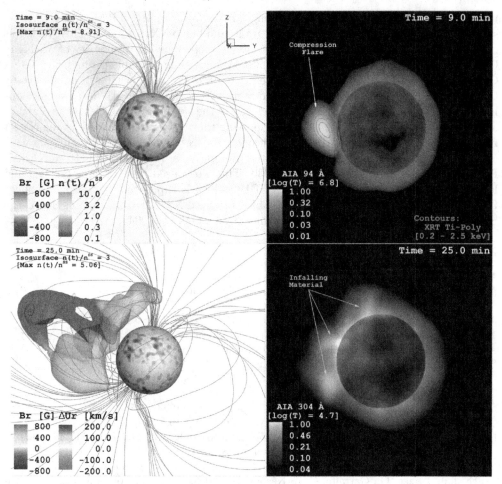

Figure 2. Snapshots at different times of a CME simulation in the flare star Proxima Centauri. The magnetic geometry driving the AWSoM corona/stellar wind solution is provided by the fully-convective dynamo simulation of Yadav *et al.* (2016). The case shown corresponds to a scaling in the surface field between $\pm$ 800 G. *Left:* Three-dimensional visualizations of the density contrast $(n(t)/n^{SS}$, top), and the coronal Doppler shift velocity (ΔU_r, bottom). The listed isosurface value is used to identify the perturbation. Selected large-scale magnetic field lines are shown in gray. *Right:* Simulated line-of-sight images of the stellar corona (in arbitrary normalized units) synthesized in two AIA/SDO filters (94 Å, top; 304 Å, bottom). Magenta contours in the top panel localize the X-ray emission (Ti-poly filter of XRT/Hinode; $0.2 - 2.5$ keV) at the peak of the CME-induced compression flare. Arrows in the bottom panel indicate bight kernels of infalling material due to the partial confinement of the CME by the large-scale field.

emission is redshifted (< -50 km s^{-1}), indicative of infalling material, which we designate as a *coronal rain cloud*. A similar infalling process, with corresponding brightenings in the lower layers of the corona, is obtained for partially confined events where CME fragmentation takes place (see Fig. 2, bottom-right).

Acknowledgments

J.D.A.G. would like to thank the organizers of the IAU Symposium 354 for the invitation to present this work and for the financial support received to attend the conference. J.D.A.G. was also supported by Chandra GO5-16021X and HST GO-15326 grants. J.J.D. was funded by NASA contract NAS8-03060 to the Chandra X-ray Center and thanks

the director, Belinda Wilkes, for continuing advice and support. S.P.M. and O.C. were
supported by NASA Living with a Star grant number NNX16AC11G. This work was
carried out using the SWMF/BATSRUS tools developed at The University of Michigan
Center for Space Environment Modeling (CSEM) and made available through the NASA
Community Coordinated Modeling Center (CCMC). We acknowledge the support by
the DFG Cluster of Excellence "Origin and Structure of the Universe". Some simulations
have been carried out on the computing facilities of the Computational Center for Particle
and Astrophysics (C2PAP). Resources supporting this work were provided by the NASA
High-End Computing (HEC) Program through the NASA Advanced Supercomputing
(NAS) Division at Ames Research Center. Simulations were performed on NASA's
Pleiades cluster under award SMD-17-1330. This work used the Extreme Science and
Engineering Discovery Environment (XSEDE), which is supported by National Science
Foundation grant number ACI-1548562. This work used XSEDE Comet at the San Diego
Supercomputer Center (SDSC) through allocation TG-AST170044.

References

Aarnio, A. N., Matt, S. P., & Stassun, K. G. 2012, *ApJ*, 760, 9

Alvarado-Gómez, J. D., Drake, J. J., Cohen, O., Moschou, S. P., & Garraffo, C. 2018, *ApJ*, 862, 93

Alvarado-Gómez, J. D., Drake, J. J., Moschou, S. P., *et al.* 2019, *ApJ Letters*, 884, L13

Alvarado-Gómez, J. D., Hussain, G. A. J., Grunhut, J., *et al.* 2015, *A&A*, 582, A38

Argiroffi, C., Reale, F., Drake, J. J., *et al.* 2019, Nature Astronomy, 3, 742

Benz, A. O. 2017, Living Reviews in Solar Physics, 14, 2

Caramazza, M., Flaccomio, E., Micela, G., *et al.* 2007, *A&A*, 471, 645

Carrington, R. C. 1859, *MNRAS*, 20, 13

Crosley, M. K., & Osten, R. A. 2018, *ApJ*, 856, doi:10.3847/1538-4357/aaaec2

Crosley, M. K., Osten, R. A., Broderick, J. W., *et al.* 2016, *ApJ*, 830, 24

Donati, J.-F. 2011, in IAU Symposium, Vol. 271, Astrophysical Dynamics: From Stars to Galaxies, ed. N. H. Brummell, A. S. Brun, M. S. Miesch, & Y. Ponty, 23–31

Donati, J.-F. & Landstreet, J. D. 2009, *Annual Review of Astronomy and Astrophysics*, 47, 333

Drake, J. J., Cohen, O., Garraffo, C., & Kashyap, V. 2016, in IAU Symposium, Vol. 320, Solar and Stellar Flares and their Effects on Planets, ed. A. G. Kosovichev, S. L. Hawley, & P. Heinzel, 196–201

Drake, J. J., Cohen, O., Yashiro, S., & Gopalswamy, N. 2013, *ApJ*, 764, 170

Drake, J. J., Peres, G., Orlando, S., Laming, J. M., & Maggio, A. 2000, *ApJ*, 545, 1074

Emslie, A. G., Dennis, B. R., Shih, A. Y., *et al.* 2012, *ApJ*, 759, 71

Fraschetti, F., Drake, J. J., Alvarado-Gómez, J. D., *et al.* 2019, *ApJ*, 874, 21

Gibson, S. E. & Low, B. C. 1998, *ApJ*, 493, 460

Gombosi, T. I., van der Holst, B., Manchester, W. B., & Sokolov, I. V. 2018, Living Reviews in Solar Physics, 15, 4

Guarcello, M. G., Micela, G., Sciortino, S., *et al.* 2019, *A&A*, 622, A210

Güdel, M. 1997, *ApJ Letters*, 480, L121

Güdel, M., Audard, M., Reale, F., Skinner, S. L., & Linsky, J. L. 2004, *A&A*, 416, 713

Huenemoerder, D. P., Schulz, N. S., Testa, P., *et al.* 2010, *ApJ*, 723, 1558

Hussain, G. A. J., Alvarado-Gómez, J. D., Grunhut, J., *et al.* 2016, *A&A*, 585, A77

Jeffers, S. V., Petit, P., Marsden, S. C., *et al.* 2014, *A&A*, 569, A79

Jin, M., Manchester, W. B., van der Holst, B., *et al.* 2017a, *ApJ*, 834, 172

—. 2013, *ApJ*, 773, 50

—. 2017b, *ApJ*, 834, 173

Kashyap, V. L., Drake, J. J., Güdel, M., & Audard, M. 2002, *ApJ*, 580, 1118

Kretzschmar, M. 2011, *A&A*, 530, A84

Leitzinger, M., Odert, P., Greimel, R., *et al.* 2014, *MNRAS*, 443, 898

Manchester, IV, W. B., Vourlidas, A., Tóth, G., *et al.* 2008, *ApJ*, 684, 1448

Melott, A. L. & Thomas, B. C. 2012, *Nature*, 491, E1

Moreno Cárdenas, F., Cristancho Sánchez, S., & Vargas Domínguez, S. 2016, Advances in Space Research, 57, 257

Morgenthaler, A., Petit, P., Saar, S., *et al.* 2012, *A&A*, 540, A138

Moschou, S.-P., Drake, J. J., Cohen, O., Alvarado-Gomez, J. D., & Garraffo, C. 2017, *ApJ*, 850, 191

Moschou, S.-P., Drake, J. J., Cohen, O., *et al.* 2019, *ApJ*, 877, 105

Odert, P., Leitzinger, M., Hanslmeier, A., & Lammer, H. 2017, *MNRAS*, 472, 876

Oran, R., Landi, E., van der Holst, B., Sokolov, I. V., & Gombosi, T. I. 2017, *ApJ*, 845, 98

Reiners, A. 2014, in IAU Symposium, Vol. 302, Magnetic Fields throughout Stellar Evolution, ed. P. Petit, M. Jardine, & H. C. Spruit, 156–163

Sokolov, I. V., van der Holst, B., Oran, R., *et al.* 2013, *ApJ*, 764, 23

Suárez Mascareño, A., Rebolo, R., & González Hernández, J. I. 2016, *A&A*, 595, A12

Titov, V. S. & Démoulin, P. 1999, *A&A*, 351, 707

van der Holst, B., Sokolov, I. V., Meng, X., *et al.* 2014, *ApJ*, 782, 81

Vida, K., Leitzinger, M., Kriskovics, L., *et al.* 2019, *A&A*, 623, A49

Villadsen, J. R. 2017, PhD thesis, California Institute of Technology

Wargelin, B. J., Saar, S. H., Pojmański, G., Drake, J. J., & Kashyap, V. L. 2017, *MNRAS*, 464, 3281

Webb, D. F. & Howard, T. A. 2012, Living Reviews in Solar Physics, 9, 3

Wright, N. J., Drake, J. J., Mamajek, E. E., & Henry, G. W. 2011, *ApJ*, 743, 48

Wright, N. J., Newton, E. R., Williams, P. K. G., Drake, J. J., & Yadav, R. K. 2018, *MNRAS*, 479, 2351

Yadav, R. K., Christensen, U. R., Wolk, S. J., & Poppenhaeger, K. 2016, *ApJ Letters*, 833, L28

Yashiro, S. & Gopalswamy, N. 2009, in IAU Symposium, Vol. 257, Universal Heliophysical Processes, ed. N. Gopalswamy & D. F. Webb, 233–243

Discussion

LUHMANN: You have assumed very solar-like CME settings. Have you considered that the magnetic fields of the stars may differ, e.g. rapid global field reconfigurations that may not be so confining?

ALVARADO-GÓMEZ: We have indeed assumed solar CME models for our simulations. While other eruption mechanisms may work in the stellar case, using solar validated models permit a better understanding of the model results with respect to the observations (solar and stellar). This is critical as there are almost no constraints on these type of events in the stellar regime that could inform the models. Regarding the rapid global field configuration, it is important to remember that the typical flare/CME time-scale is much shorter compared to the observed (through spectropolarimetric data) and expected (via dynamo models) evolution of the large-scale field. Therefore, the large-scale confining conditions are expected to be relatively stable, with noticeable changes on activity/magnetic cycle time-scales.

Solar and Stellar Magnetic Fields: Origins and Manifestations
Proceedings IAU Symposium No. 354, 2019
A. Kosovichev, K. Strassmeier & M. Jardine, ed.
doi:10.1017/S1743921320001039

Diagnostics of non-thermal-distributions from solar flare EUV line spectra

Elena Dzifčáková, Alena Zemanová, Jaroslav Dudík, and Juraj Lörinčík

Astronomical Institute of the Czech Academy of Sciences,
Fričova 298, 251 65 Ondřejov, Czech Republic
email: elena.dzifcakova@asu.cas.cz

Abstract. Spectral line intensities observed by the Extreme Ultraviolet Variability Experiment (EVE) on board the Solar Dynamics Observatory (SDO) during 2012 March 9 M6.3 flare were used to diagnose a presence of a non-thermal electron distribution represented by a κ-distribution. The diagnosed electron densities ($\approx 2 \times 10^{11}$ cm^{-3}) are affected only a little by the presence of the non-thermal distribution, and are within the uncertainties of observation. On the other hand, the temperature diagnostics based on the line ratios involving different ionization degrees is strongly affected by the type of the electron distribution. The distribution functions diagnosed from relative Fe line intensities demonstrate the presence of strongly non-thermal distributions during the impulsive phase of the flare and later their gradual thermalization.

Keywords. Sun: flares, Sun: UV radiation, techniques: spectroscopic

1. Introduction

Solar flares are the most energetic manifestations of the solar magnetic activity. The magnetic reconnection transforms the energy of magnetic field mainly into the heating, waves, motions, and particle acceleration (e.g. Priest & Forbes 2000). Typically, a supra-thermal component (high-energy tail of electron distribution) is observed in flares (e.g. Brown 1971, Fletcher *et al.* 2011) and solar wind (Maksimovic *et al.* 1997). Recent theoretical papers (Bian *et al.* 2014) and RHESSI observations of coronal X-ray sources (Kašparová & Karlický 2009, Oka *et al.* 2013, 2015) suggested that the electron distribution function could have a form of κ-distributions. These distributions have a power-law high-energy tail and they are able to approximate an effect of the power-law electron beam on line spectra. The presence of the high-energy tail of the electron distribution during solar flares affects the ionization and excitation state of flaring plasma, and thus the spectrum emitted (e.g. Dzifčáková & Dudík 2013, Dzifčáková & Kulinová 2010). Dzifčáková *et al.* (2018) investigated diagnostics of the κ-distributions from the flare spectra observed by the EVE instrument onboard the Solar Dynamics Observatory (Woods *et al.* 2012) and indicated the presence of the electron distribution with a high-energy tail in the X-class solar flare from 2012 March 7. In this paper, similar diagnostics were applied on EVE spectra of the 2012 March 9, M6.3 flare (03:22 - 03:50 - 04:18 UT).

2. Kappa-distribution and Spectrum Calculation

The expression for the κ-distribution is (e.g. Olbert 1968, Dudík *et al.* 2017):

$$f_\kappa(E)dE = A_\kappa \frac{2}{\pi^{1/2}(kT)^{3/2}} \left(1 + \frac{E}{(\kappa - 1.5)kT}\right)^{-(\kappa+1)} E^{1/2}dE \qquad (2.1)$$

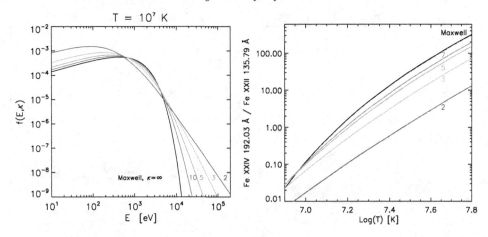

Figure 1. *Left:* Comparison of the Maxwellian distribution (black line) with the κ-distributions for $\kappa = 2$ (red), 3 (orange), 5 (green), and 10 (blue). The mean energies of distributions are the same. *Right:* Ratio of Fe XXIV 192.02 Å to Fe XXII 135.79 Å for the Maxwellian distribution (black line) and κ-distributions with $\kappa = 2$ (red), 3 (orange), 5 (green), and 7 (blue).

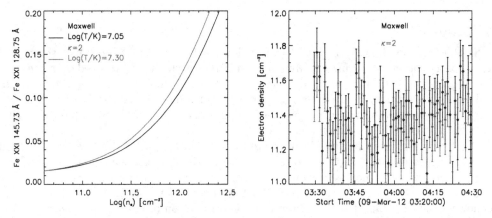

Figure 2. Density sensitive ratio of Fe XXI 145.73 Å to Fe XXI 128.75 Å for the Maxwellian distribution (black line) and the κ-distributions with $\kappa = 2$ (red line). Temperatures correspond to the maxima of the line intensities for these two distributions (*left*). Electron densities and their error-bars diagnosed under the assumption of the Maxwellian distribution are shown in black and for the κ-distributions with $\kappa = 2$ are red (*right*).

where κ is a free patameter, T is temperature, A_κ is normalization constant, m is particle mass and E epresents the particle kinetic energy (Figure 1, *left*). In comparison with the Maxwellian distribution, the κ-distributions have only one extra parameter κ, which models its shape. Strongly non-thermal distribution has $\kappa \to 1.5$ and the Maxwellian distribution corresponds to $\kappa \to \infty$. The mean energy $\langle E \rangle = 3kT/2$ of a κ-distribution is the same as for the Maxwellian distribution.

Synthetic spectra in the SDO/EVE spectral range $60 - 600$ Å were calculated for $\log (T/K) = 6.5 - 8.0$, electron densities $10^{10} - 10^{13}$ cm^{-3}, and $\kappa = 2$, 3, 5, 7, 10, and Maxwellian distribution. To do so, we used the KAPPA package (Dzifčáková *et al.* 2015, http://kappa.su.cas.cz) based on the CHIANTI database 7.1. The ionization equilibria for the κ-distributions were taken from Dzifčáková & Dudík (2013). Ratios of synthetic line intensities were used to diagnose physical parameters of the flaring plasma (Figure 1, *right*, Figure 2, *left*).

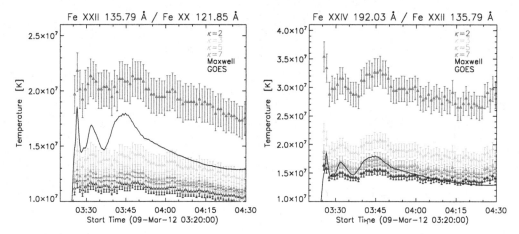

Figure 3. Behaviour of the flare temperature for the Maxwellian distribution (black points with their error bars) and different κ-distributions (collor points with their error bars) derived from Fe XXII 135.79 Å /Fe XX 121.85 Å (*left*) and Fe XXIV 192.03 Å /Fe XXII 135.79 Å (*right*). GOES temperature is shown for a comparison (black line).

3. Results

SDO/EVE provides spectra with relatively low resolution ($\approx$1 Å) and time resolution of 10 s. Spectra observed during the M6.3 flare of 2012 March 9 were averaged over 1 minute for the density and temperature diagnostics, and over 2 minutes for diagnostics of κ to reduce the uncertainty in the determination of the line intensity ratios. After subtraction of preflare spectrum, the spectra observed during 03:27–04:30 UT were fitted by XCFIT (SolarSoft) to obtain the line intensities.

Electron densities in flaring plasma were found to be about 2×10^{11} cm^{-3} with local maxima up to 6×10^{11} cm^{-3} (Figure 2, *right*). During the gradual phase of the flare, the electron density increased slowly. The densities for κ-distribution with $\kappa = 2$ are approximately 0.1 deg lower than those for the Maxwellian distribution. This difference is however smaller than the uncertainties in determination of the electron density.

Diagnosed temperature depends on the line ratio used and as well as on the assumed electron distribution (Figure 3). Local maxima of temperature correspond approximately to the local maxima of the GOES temperature and their timing corresponds also to the maxima of the diagnosed electron density. Different temperatures obtained from different line ratios indicate the presence of the multi-thermal or non-equilibrium plasma. The shift in the diagnosed temperature due to the assumption of the different distributions can be up to a factor of 2 due to the presence of the high-energy tail. This effect shows how important is the diagnostics of electron distribution function for determination of the temperatures and flare energetics.

For diagnostics of the distribution function, the method proposed by Dzifčáková *et al.* (2018) were used. Figure 4 *left* shows the observed line ratios for the whole time interval 03:27–04:30 UT. The right panel displays only data from the time interval 03:50–03:55 UT, when the line Fe XX 424.3 Å was visible. The colour coding of points with their error bars corresponds to time, it increases from black, ought blue, green, and orange to red. Observed line ratios from the impulsive phase or from flare maximum (blue points) indicate strongly non-thermal distribution. They are far away from line ratios corresponding to the non-thermal κ-distributions with $\kappa = 2$. This is a signature of strongly non-thermal plasmas, which could also be out of ionization equilibrium. During decay phase, flare emission shows signatures of successive thermalisation with time.

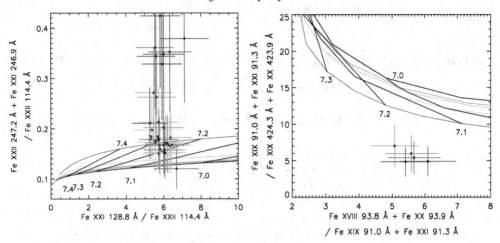

Figure 4. Diagnostics of the distribution function. The theoretical ratio-ratio diagrams are based on the Fe XXI, Fe XXII lines (*left*) and Fe XVIII, Fe XIX, Fe XX lines (*right*). Maxwellian distribution is represented by black lines, while $\kappa = 10$ by blue lines, 5 (green), 3 (orange), and 2 (red). Observed line ratios are shown with their error bars. Thin black lines connect points with the same temperature in different distributions.

4. Acknowledgements

This work was supported by the Grant No. 18-09072S and 17-16447S of the Grant Agency of the Czech Republic, as well as institutional support RVO:67985815 from the Czech Academy of Sciences. This work was also supported by the Charles University, project GA UK 1130218.

References

Bian, N. H., Emslie, A. G., Stackhouse, D. J., & Kontar, E. P. 2014, *ApJ*, 796, 142
Brown, J. C. 1971, *Solar Phys.*, 18, 489
Dudík, J., Dzifčáková, E., Meyer-Vernet, N., *et al.* 2017, *Solar Phys.*, 292, 100
Dzifčáková, E., & Kulinová, A. 2010, *Solar Phys.*, 263, 25
Dzifčáková, E., & Dudík, J. 2013, *ApJS*, 206, 6
Dzifčáková, E., Dudík, J., Kotrč, P., Fárník, F., & Zemanová, A. 2015, *ApJS*, 217, 14
Dzifčáková, E., Zemanová, A., Dudík, J., Lörinčík, J. 2018, *ApJ*, 853, 158
Fletcher, L., Dennis, B. R., Hudson, H. S., *et al.* 2011, *SSRv*, 159, 19
Kašparová, J. & Karlický, M. 2009, *Astron. Astrophys.*, 497, L13
Maksimovic M., Pierrard V., & Lemaire J. F. 1997, *Astron. Astrophys.*, 324, 725
Oka, M., Ishikawa, S., Saint-Hilaire, P., Krucker, S., & Lin, R. P. 2013, *ApJ*, 764, 6
Oka, M., Krucker, S., Hudson, H. S., & Saint-Hilaire, P. 2015, *ApJ*, 799, 129
Olbert, S. 1968, in R. D. L. Carovillano & J. F. McClay (eds.), *in Physics of the Magnetosphere* (Dordrecht: Reidel), p. 641
Priest, E. & Forbes, T. 2000, *Magnetic Reconnection* (Cambridge: Cambridge Univ. Press)
Woods, T. N., Eparvier, F. G., Hock, R., *et al.* 2012, *Solar Phys.*, 275, 115

Solar and Stellar Magnetic Fields: Origins and Manifestations
Proceedings IAU Symposium No. 354, 2019
A. Kosovichev, K. Strassmeier & M. Jardine, ed.
doi:10.1017/S1743921320000423

Linking radio flares with spots on the active binary UX Arietis

Christian A. Hummel[1] and Anthony Beasley[2]

[1]European Southern Observatory,
Karl-Schwarzschild-Str. 2, 85748 Garching, Germany
email: `chummel@eso.org`

[2]National Radio Astronomy Observatory, Charlottesville, VA 22903, USA
email: `tbeasley@nrao.edu`

Abstract. Signs of stellar activity such as large surface spots and radio flares are often related to binarity. UX Arietis is one of the most active members of the RS CVn class of binaries in which spin-up of a sub-giant/giant star by a close companion leads to the creation of magnetic fields. UX Arietis exhibits these signs of activity, originating mostly on the K0 sub-giant primary component. We measured the orbit with the CHARA interferometer and made images of a single large spot rotating in and out of view over a month in 2012. The rotation of the stars is synchronous with the orbit, and long-term photometric observations show that the spot or spots do not move much during intervals of a year. Our aim is to relate the positions of the stars and the spots on the primary to astrometry of the radio components observed during outbursts.

Keywords. techniques: interferometric, stars:activity, stars:flare, stars: magnetic fields

1. Introduction

Radio emission of active stars is caused by the gyro-synchrotron process in plasma contained by the large scale magnetic fields (Franciosini & Chiuderi Drago 1995) in the coronae of these stars. The geometry and extent of magnetospheres in magnetically active stars controls stellar winds and therefore the angular momentum evolution in binary systems. Detailed modeling of the radio sources for past and future epochs requires an unambiguous understanding of the stellar alignment in the system and absolute sub-milliarcsecond astrometry. With orbital periods between a few days and a few weeks, RS CVn stars are resolved only by optical interferometry (or in the radio by VLBI).

Radio emission of UX Arietis may be modeled by a single radio component related to the active sub-giant (Peterson *et al.* 2011), or can be modeled by two distinct sources separated by angles commensurate with the optical orbit size (Mutel *et al.* 1985; Ros & Massi 2007). In the former case, Peterson *et al.* (2011) were able to measure the absolute motion of the single radio component around the common center of mass of the close binary and determine a preliminary orbit. Hummel *et al.* (2017) determined the final orbital elements using near-IR interferometry with CHARA (ten Brummelaar *et al.* 2005). They also observed a co-rotating large spot on the primary component.

To explain the double-peaked radio emission of HR 1099, Ransom *et al.* (2002) considered two models where a magnetic loop structure is attached to the poles or equatorial regions of the cooler K star, or is part of a joint magnetosphere of both stars in the binary. Observations of the radio emission of Algol by Peterson *et al.* (2010) detected a large coronal loop straddling a radio-bright KIV subgiant. This observation clearly is only consistent with the predictions of the polar loop model.

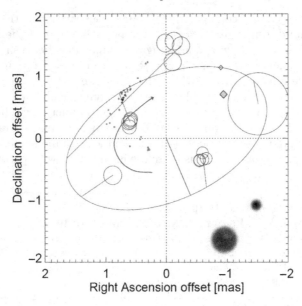

Figure 1. Retrograde orbit of the G0V secondary around the spotted K0IV primary at the center. Locations of the secondary star are connected with a red line to the position of its radio counterpart at epochs when two radio components were observed. The big circle to the upper right is the measurement by Mutel *et al.* (1985), with the size of the circle indicating the synthesized beam size. The single small circle to the lower left is due to Beasley et al. (in prep.). Three clusters of small circles correspond to the daily averages of observations by Ros & Massi (2007), with the size of the circles indicating the RMS of the scatter during the observations. While the average radio position is near (but inside) the position of secondary on Sep 23 and 26, the radio position on Sep 25 is nowhere near it. The small dots correspond to hourly averages during a major flare on the last day moving outwards from the K0IV primary (lower right inset, with spot appearing) in a clock-wise fashion (blue curved arrow) towards the secondary's position marked with a small diamond. The image at the lower right shows the positions of the stars during the flare and a spot on the sub-giant (from modelling light-curves). The color scale of the image is *inverted* heat.

In the following, we reconstruct the relative positions of the optical components in UX Arietis during radio observations which were modelled with two emission components. The earliest radio map produced by VLBI observations of UX Arietis during an outburst was presented by Mutel *et al.* (1985) and shows two components, the eastern one associated with the K0IV primary and a western "halo" component associated with the secondary component (or the joint magnetosphere of the system), the alignment having been described as "conjectural" by the authors, but now confirmed by our orbit. Observations with the VLBA of the quiet and flaring radio emission at four epochs within less than a week was obtained by Ros & Massi (2007), who also used a model of two Gaussian components. The emission during the last epoch of observations was dominated by a flare, causing the radio components to move significantly with respect to each other in a 6 hour period. The relative astrometry also showed significant dispersion during the second, while the first and third epoch observations showed very stable relative positions.

As shown in Fig. 1, where we adopt an identification of one of the two radio component as potentially associated with the optical secondary, during four observations the secondary radio component indeed was located close to the optical secondary. However, there are exceptions, both during an observation of the quiescent radio emission, and during a flare. Intriguingly, the location of the radio secondary appeared to move during the flare in an arc-shaped path from a location near the optical primary towards the location of the optical secondary. Hence, our reconstruction of the alignment of the optical components during radio observation appears to support the model that radio emission in UX Arietis may flow along magnetic flux-tubes between the stars. In this case, the two components would be directly related to the two stars in a scenario where they are magnetically coupled, i.e., where the main site of the energy release is on the more active subgiant close to the mid-latitude spot groups, while the magnetic field of the dwarf star acts as a passive magnetic foot print, bright in microwaves because the field is enhanced there. The magnetic loop model of Franciosini & Chiuderi Drago (1995), on the other hand, would predict separations between the radio components much smaller than between the binary components and is therefore inconsistent with pattern seen in Fig. 1.

References

Franciosini, E. & Chiuderi Drago, F. 1995, *A&A*, 297, 535

Mutel, R. L., Lestrade, J. F., Preston, R. A., & Phillips, R. B. 1985, *ApJ*, 289, 262

Peterson, W. M., Mutel, R. L., Güdel, M., & Goss, W. M. 2010, *Nature*, 463, 207

Peterson, W. M., Mutel, R. L., Lestrade, J.-F., Güdel, M., & Goss, W. M. 2011, *ApJ*, 737, 104

Ransom, R. R., Bartel, N., Bietenholz, M. F., *et al.* 2002, *ApJ*, 572, 487

Ros, E. & Massi, M. 2007, *Mem. S.A.It.*, 78, 298

Hummel, C. A., Monnier, J. D., Roettenbacher, R. M., *et al.* 2017, *ApJ*, 844, 115

ten Brummelaar, T. A., McAlister, H. A., Ridgway, S. T., *et al.* 2005, *ApJ*, 628, 453

Solar and Stellar Magnetic Fields: Origins and Manifestations
Proceedings IAU Symposium No. 354, 2019
A. Kosovichev, K. Strassmeier & M. Jardine, ed.
doi:10.1017/S1743921320000150

CME deflections due to magnetic forces from the Sun and Kepler-63

F. Menezes[1], Y. Netto[1,2], C. Kay[3], M. Opher[4] and A. Valio[1]

[1]Universidade Presbiteriana Mackenzie, CRAAM, São Paulo, Brazil
email: menezes.astroph@gmail.com

[2]Osservatorio Astrofisico Di Catania, Catania, Italy

[3]NASA, Goddard Space Flight Center, Greenbelt, USA

[4]Boston University, Department Of Astronomy, Boston, USA

Abstract. The stellar magnetic field is the driver of activity in the star and can trigger energetic flares, CMEs and ionized wind. These phenomena, specially CMEs, may have an important impact on the magnetosphere and atmosphere of the orbiting planets. To predict whether a CME will impact a planet, the effects of the background on the CME's trajectory must be taken into account. We used the MHD code ForeCAT – a model for CME deflection due to magnetic forces – to perform numerical simulations of CMEs being launched from both the Sun and Kepler-63, which is a young, solar-like star with high activity. Comparing results from Kepler-63 and the Sun gives us a panorama of the distinct activity level and star-planet interactions of these systems due to the difference of stellar ages and star-planet distances.

Keywords. Sun: coronal mass ejections (CMEs), stars: activity, stars: magnetic fields, stars: spots, MHD

1. Introduction

The average surface magnetic field of the Sun is about 1 G, however in some regions the field may reach values up to several hundred gauss in photospheric features such as sunspots and faculae. These features are observable proxies of solar magnetic activity, providing a window to the unseen internal dynamo and acting as tracers of magnetic topology. Sunspots and faculae are the surface emanations of internal magnetic fields caught in turbulent flows that eventually become areas of amplified magnetic fields when erupting through the photosphere. The occurrence of these features has an 11-year modulation. This is known as the solar activity cycle, in which the magnetic field of the Sun reverses its polarity forming a magnetic cycle of 22 years. During solar maxima, energetic events like coronal mass ejections (CME), increased solar wind, and flares are more common.

The stellar activity is related to the age of the star. The Sun for instance with an age of 4.6 billion years is considered just a mildly active star, while Kepler-96 – an active solar-type star of 2.3 billion years – is a very active star presenting super flares of 1.8^{35} ergs in its light curves (Estrela & Valio 2018). Stars interact with their orbiting planets through their magnetic field and energetic events, specially CMEs that may impact on the magnetosphere and atmosphere of these planets.

Despite being at 1 AU away from the Sun, on Earth CMEs can cause geomagnetic storms and affect the communication systems, power lines, satellites orbits, etc. Observations from Mars have shown that CMEs can have a significant impact on the atmosphere and the long-term atmospheric evolution of the planet, suggesting that space weather can affect the habitability of a planet through atmospheric losses (Jakosky 2015). Close-in planets – such as hot Jupiters around solar-like stars and planets in the habitable zone of active M dwarfs – may experience more extreme space weather than on Earth, including frequent CME impacts leading to atmospheric erosion and leaving the surface exposed to extreme radiation from flare activity (Kay *et al.* 2016).

It is very important to be able to predict CMEs trajectories since that is what determines if a CME will hit a planet and cause important impact in space weather. Our goals are to simulate CMEs trajectories and deflections which are subject to the magnetic forces, being launched from the Sun and Kepler-63 – a young, highly active, solar-like star with ∼210 Myr, 0.984 $M_\odot$, 0.901 $R_\odot$ and a 5.4-day rotation period (Sanchis-Ojeda *et al.* 2013) – and then compare results from both stars in order to have a panorama of the distinct activity level and star-planet interactions of these systems due to the difference of stellar ages and star-planet distances.

2. Numerical simulations

To run the simulations we used the ForeCAT (Forecasting a CME's Altered Trajectory, Kay *et al.* 2013), an MHD code created to simulate CME deflections and rotations due to magnetic forces which are the magnetic pressure gradients and the magnetic tension. It divides the propagation of the CME in three parts: gradual rise, impulsive acceleration, and propagation. Also, it includes CME expansion and the effects of drag on the CME's deflection. The CME simulations are initiated by setting the launch site on the stellar surface (latitude and longitude), tilt (measured clockwise with respect to the equatorial plane), shape parameters, and other properties of the CME such as mass, final propagation speed, initial radius, and initial magnetic strength (Kay *et al.* 2015). The shape parameters A, B and C describe an elliptical toroidal CME. A is defined as the axis in the direction of the nose of the CME (the point on the surface of the torus with greatest radial distance) and C is the initial distance of the nose from the star surface. The radius B describes the cross section of the torus.

This code uses the Potential Field Source Surface (PFSS) model to calculate the stellar background magnetic field, which assumes that the magnetic field can be described as current-free and becomes entirely radial above the source surface (Kay *et al.* 2015). The source surface is typically taken to be 2.5 $R_\star$ for PFSS models of the Sun. The smaller the radius, the greater the influence of the active regions and the smaller the influence of the coronal holes on the source surface magnetogram, and vice versa. Therefore, we performed simulations with source surfaces of 1.05 and 2.5 $R_\star$ to explore the influence of the magnetic density flux in the spots of both stars.

To calculate the stellar background magnetic field, the input parameter is a photospheric magnetogram, which in the case of the Sun a magnetogram from 1 solar rotation (CR2203), however such data is not available for Kepler-63. The only information on the photosphere of the star from transits light curves from Kepler Space Telescope, were the starspots mapped by Netto & Valio (2019). These starspots were mapped using the model from Silva (2003), which allows determining the physical characteristics of the spots, such as size, temperature, and location. Also, Kepler-63 was chosen because it harbors a hot Jupiter (Kepler-63 b)orbiting in an almost polar orbit, which allows mapping starspots at different latitudes. Figure 1 shows all mapped spots.

Then, to obtain a magnetogram we used the relation between spots intensity and magnetic field determined from 32,000 sunspots (Valio *et al.* 2019):

$$B = (4848 \pm 15) - (4008 \pm 20) \times I, \qquad (2.1)$$

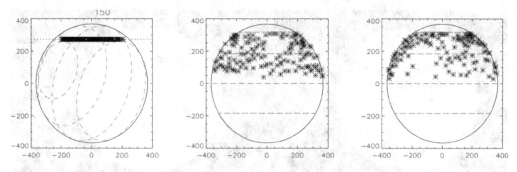

Figure 1. Kepler-63's spots. *Left panel:* spots over-plotted in a referential frame as seen from Earth. *Right panel:* all mapped spots in a frame rotating with the star

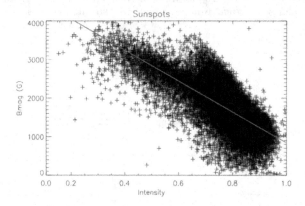

Figure 2. Sunspot intensity vs sunspot magnetic density flux. The red line is the linear equation fitted (Eq. 2.1)

where B is the maximum magnetic density flux and I is the spot intensity (normalized to the star intensity at the center of the disk). To obtain magnetic field intensities of spots on Kepler-63, we used the values of spot intensities from the fit of the light curves vaiations observed on the 88th transit of the planet. The intensities of the spots – 0.398, 0.416 and 0.486 (from South to North in the left panel in Figure 3) – yield respectively magnetic density flux values of 3258, 3186 and 2906 G (or Mx/cm^2). The magnetogram was created by inserting the magnetic flux in a dipole spread within the starspot area (right panel in Figure 3). Also, for the background magnetic field, we used the CR2203 magnetogram without the large active regions, and multiplied by three, since the active regions in Kepler-63 were three time stronger than the Sun's.

For both CME simulations we set initial position at $1.1R_\star$, so the shape parameter C was $0.1R_\star$ and the ratios A/C and B/C were 1.0 and 0.2, respectively. The final position was set at $10R_\star$, initial latitude at 16.5°, initial longitude at 316.8°, initial tilt of -35° and masses equal to 10^{15} g.

3. Results and conclusions

We performed a total of 4 simulations – 2 for the Sun and 2 for Kepler-63 – with the source surface taken at 2 different values – 1.05 and 2.5 $R_\star$. Figure 4 shows the CME simulations output parameters – tilt, latitude, longitude and radial velocity over the distance from the star. The tilt of the deflection is larger in the case of the CME launched from Kepler-63 (57°) than that of the Sun (45°). Simulations from Kepler-63 with the source surface at 1.05 $R_\star$ presented all output parameters, except radial velocity,

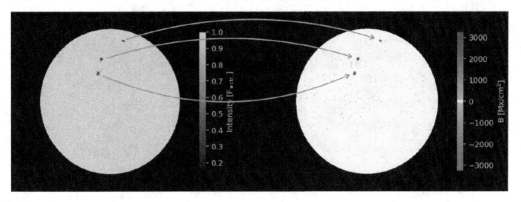

Figure 3. Kepler-63's mapped starspots from the transit no. 88. *Left panel:* relative intensity color scale *Right panel:* magnetic density flux color scale.

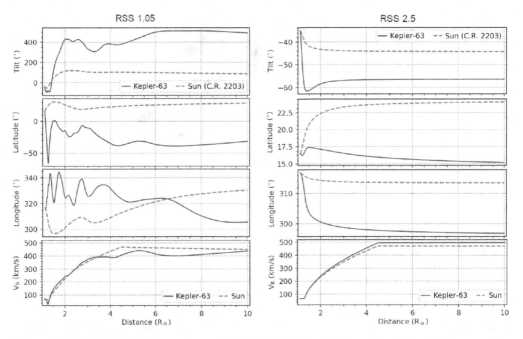

Figure 4. CME simulations outputs. Tilt, latitude, longitude and radial velocity over the distance from the star. The blue solid lines represents the Kepler-63 and the red dashed lines represents the Sun.

to have significant variations over the distance up to $\sim 3R_\star$, and for both source surface values, the output parameters from Kepler-63 – tilt, latitude and longitude – are more intense than the simulations from the Sun. Despite the initial discrepancy, the velocities of the sun and Kepler-63 CME approximate each other at 10 radii.

With the source surface taken at lower heights, such as 1.05 $R_\star$, the influence of the active regions are stronger and the background magnetic field presents an intricated configuration, causing the trajectory and rotation of the CME to present more variation. The background magnetic field tends to approximate to the heliospheric current sheet with the source surface taken at 2.5 $R_\star$, but due to a 3-time stronger magnetic field, the CMEs from Kepler-63 undergo higher deflections and rotation.

References

Estrela, R. & Valio, A. 2018, Astrobiology, 18, 1414

Jakosky, B. M. 2015, Science, 350, 643

Kay, C., Opher, M., & Evans, R. M. 2013, *Astrophys. J.*, 775, 5

Kay, C., Opher, M., & Evans, R. M. 2015, *Astrophys. J.*, 805, 168

Kay, C., Opher, M., & Kornbleuth, M. 2016, *Astrophys. J.*, 826, 195

Netto, Y. & Valio, A. 2019, *A&A* (accepted)

Sanchis-Ojeda, R., Winn, J. N., Marcy, G. W., *et al.* 2013, *Astrophys. J.*, 775, 54

Silva, A. V. R. 2003, *Astrophys. J. Lett.*, 585, L147

Valio, A., Spagiari, E., Marengoni, M., & Selhorst, C. 2019, Manuscript in preparation

Solar and Stellar Magnetic Fields: Origins and Manifestations
Proceedings IAU Symposium No. 354, 2019
A. Kosovichev, K. Strassmeier & M. Jardine, ed.
doi:10.1017/S1743921320000575

Coronal dimming as a proxy for stellar coronal mass ejections

M. Jin[1,2]⊙, M. C. M. Cheung[2,3], M. L. DeRosa[2], N. V. Nitta[2], C. J. Schrijver[2]⊙, K. France[4], A. Kowalski[4], J. P. Mason[4] and R. Osten[5]

[1]SETI Institute, 189 N Bernardo Ave suite 200, Mountain View, CA 94043, USA
email: jinmeng@lmsal.com

[2]Lockheed Martin Solar and Astrophysics Lab (LMSAL), 3251 Hanover St.,
Bldg. 252, Palo Alto, CA 94304, USA

[3]Hansen Experimental Physics Laboratory, Stanford University,
452 Lomita Mall, Stanford, CA 94305, USA

[4]Laboratory for Atmospheric and Space Physics, University of Colorado at Boulder,
1234 Innovation Dr, Boulder, CO 80303, USA

[5]Space Telescope Science Institute, 3700 San Martin Dr, Baltimore, MD 21218, USA

Abstract. Solar coronal dimmings have been observed extensively in the past two decades and are believed to have close association with coronal mass ejections (CMEs). Recent study found that coronal dimming is the only signature that could differentiate powerful flares that have CMEs from those that do not. Therefore, dimming might be one of the best candidates to observe the stellar CMEs on distant Sun-like stars. In this study, we investigate the possibility of using coronal dimming as a proxy to diagnose stellar CMEs. By simulating a realistic solar CME event and corresponding coronal dimming using a global magnetohydrodynamics model (AWSoM: Alfvén-wave Solar Model), we first demonstrate the capability of the model to reproduce solar observations. We then extend the model for simulating stellar CMEs by modifying the input magnetic flux density as well as the initial magnetic energy of the CME flux rope. Our result suggests that with improved instrument sensitivity, it is possible to detect the coronal dimming signals induced by the stellar CMEs.

Keywords. magnetohydrodynamics (MHD) – methods: numerical – solar wind – Sun: corona – Sun: coronal mass ejections (CMEs)

---

1. Introduction

"Coronal dimming" refers as the reduction in intensity on or near the solar disk across a large area during solar eruptive events. It was first observed in white light corona and described as a "depletion" (Hansen *et al.* 1974) and later was found in solar X-ray observations as "transient coronal holes" (Rust & Hildner 1976). The studies about coronal dimming dramatically increased with Solar and Heliospheric Observatory (SOHO)/Extreme-ultraviolet Imaging (EIT) observations (Delaboudinière *et al.* 1995), with which the coronal dimming was first observed in multiple EUV channels with different emission temperatures (Thompson *et al.* 1998). The coronal dimming was also found usually associated with coronal EUV waves (also called "EIT waves", Thompson *et al.* 1999). Recently, with EUV observations of unprecedented high temporal (∼12 s) and spatial resolution (∼0.6 arcsec) in seven channels from Solar Dynamics Observatory (SDO; Pesnell *et al.* 2012)/Atmospheric Imaging Assembly (AIA; Lemen *et al.* 2012) as well as the high spectral resolution data from SDO/Extreme Ultraviolet Variability

Experiment (EVE; Woods *et al.* 2012), it provides us an unique opportunity for in-depth studies about coronal dimming.

Two decades of solar observations suggest that all coronal dimmings are associated with coronal mass ejections (CMEs; e.g., Sterling & Hudson 1997; Reinard & Biesecker 2008). Furthermore, since observations show simultaneous and co-spatial dimming in multiple coronal lines (e.g., Zarro *et al.* 1999; Sterling *et al.* 2000) and the spectroscopic observations show that the dimming region has up-flowing expanding plasma (e.g., Harra & Sterling 2001; Harra *et al.* 2007; Imada *et al.* 2007; Jin *et al.* 2009; Attrill *et al.* 2010; Tian *et al.* 2012), it is widely accepted that the coronal dimming is due to the plasma evacuation during the CME and the dimming area is believed to be the footpoints of the erupting flux rope. Recent magnetohydrodynamics (MHD) modeling results of coronal dimming (e.g., Cohen *et al.* 2009; Downs *et al.* 2012) also suggest that the dimming is mainly caused by the CME-induced plasma evacuation, and the spatial location is well correlated with the footpoints of the erupting magnetic flux system (Downs *et al.* 2015). Although, there are other known mechanisms that could cause coronal dimming in observations (Mason *et al.* 2014).

Due to its close association with CMEs, coronal dimming encodes important information about CME's mass, speed, energy etc. (e.g., Hudson *et al.* 1996; Sterling & Hudson 1997; Harrison *et al.* 2003; Zhukov & Auchère 2004; Aschwanden *et al.* 2009; Cheng & Qiu 2016; Krista & Reinard 2017; Dissauer *et al.* 2018a,b), therefore provide critical estimations for space weather forecast. For example, Krista & Reinard (2013) found correlations between the magnitudes of dimmings/flares and the CME mass by studying variation between recurring eruptions and dimmings. Using SDO/EVE observations, Mason *et al.* (2016) found that the CME velocity/mass can be related to coronal dimming properties (e.g., dimming depth and dimming slope), which could be used to estimate CME mass and speed in the space weather forecast operations. On the other hand, by exploring the characteristics of 42 X-class solar flares, Harra *et al.* (2016) found that coronal dimming is the only signature that could differentiate powerful flares that have CMEs from those that do not. Therefore, dimming might be one of the best candidates to observe the stellar CMEs on distant Sun-like stars.

In this study, we first model the coronal dimming associated with a realistic CME event on 2011 February 15 (Schrijver *et al.* 2011; Jin *et al.* 2016) to demonstrate the capability of the model to reproduce solar observations. We then extend the model for simulating stellar CMEs by changing the input magnetic flux density as well as the initial energy of the CME flux rope. In §2 we briefly introduce the global MHD model used in this study, followed by the results and discussion in § 2.

2. Global Corona & CME Models

In this study, the model for reconstructing the global corona environment is the Alfvén Wave Solar Model (van der Holst *et al.* 2014) within the Space Weather Modeling Framework (SWMF; Tóth *et al.* 2012). AWSoM is a data-driven global MHD model with inner boundary specified by observed magnetic maps and simulation domain extending from the upper chromosphere to the corona and heliosphere. Physical processes included in the model are multi-species thermodynamics, electron heat conduction (both collisional and collisionless formulations), optically thin radiative cooling, and Alfvén-wave turbulence that accelerates and heats the solar wind. The Alfvén-wave description includes non-Wentzel-Kramers-Brillouin (WKB) reflection and physics-based apportioning of turbulent dissipative heating to both electrons and protons. AWSoM has demonstrated the capability to reproduce high-fidelity solar corona conditions (Sokolov *et al.* 2013; van der Holst *et al.* 2014; Oran *et al.* 2013, 2015; Jin *et al.* 2016; Jin *et al.* 2017a).

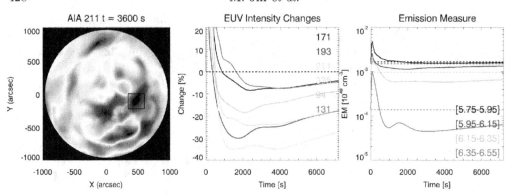

Figure 1. Coronal dimming evolution in the simulation of 2011 February 15 event. Left panel: synthesized AIA 211 Å base difference image at t = 1 hour. The black box shows the sub-region where the EUV intensity and Emission Measure (EM) are derived. Middle panel: the EUV intensity changes in 6 synthesized AIA channels. Right panel: EM evolution in the simulation. The dashed lines show the pre-event EM value.

Based on the steady-state global corona solution, we initiate the CME by using an analytical Gibson-Low (GL) flux rope model (Gibson & Low 1998). This flux rope model has been successfully used in numerous modeling studies of CMEs (e.g., Manchester *et al.* 2004a,b; Lugaz *et al.* 2005; Manchester *et al.* 2014; Jin *et al.* 2016; Jin *et al.* 2017a). Jin *et al.* (2017b) developed a module (EEGGL) to calculate the GL flux rope parameters based on near-Sun observations so that this first-principles-based MHD model could be utilized as a forecasting tool. Analytical profiles of the GL flux rope are obtained by finding a solution to the magnetohydrostatic equation $(\nabla \times \mathbf{B}) \times \mathbf{B} - \nabla p - \rho \mathbf{g} = 0$ and the solenoidal condition $\nabla \cdot \mathbf{B} = 0$. After inserting the flux rope into the steady-state solar corona solution: i.e. $\rho = \rho_0 + \rho_{\mathrm{GL}}$, $\mathbf{B} = \mathbf{B_0} + \mathbf{B_{GL}}$, $p = p_0 + p_{\mathrm{GL}}$, the combined background-flux-rope system is in a state of force imbalance and thus erupts immediately when the numerical model is advanced forward in time. The simulation setup in this study is similar to that in our previous work (Jin *et al.* 2016).

3. Results & Discussion

To demonstrate the capability of the model for reproducing the solar coronal dimming, we synthesize the EUV emissions of 6 AIA wavebands for two hours after the CME onset in the 2011 February 15 event (Schrijver *et al.* 2011; Jin *et al.* 2016). In addition, we calculate the Emission Measure (EM) in 4 temperature bins ($5.75 < lgT < 6.55$). In Figure 1, we show the core dimming (near the source region) evolution in the simulation. The left panel in Figure 1 shows a base-difference image (by substracting the pre-event intensity) in AIA 211 Å at $t = 1$ hour. The black box shows the sub-region where the EUV intensities and EMs are derived. In the middle panel of Figure 1, the EUV intensity evolution (represented by relative changes in percentage comparing with the pre-event values) in 2 hours for all 6 AIA wavebands is shown. We can see that the intensities in all 6 wavebands drop, which suggests the plasma depletion included by the CME. This is also evident in the EM calculation (right panel of Figure 1). By fitting the intensity curves, we can estimate the dimming recovery time is about ∼9 to 16 hours.

With higher mean surface magnetic flux density $\langle |fB| \rangle$ on the M dwarf or young Sun-like stars, their coronas are believed to be hotter than the solar case. By varying the $\langle |fB| \rangle$ in the model by 5, 10, 20, 30 times, we demonstrate this effect quantitatively in Figure 2. With the increasing magnetic field strength, the peak EM temperature shifts from ∼1 MK for the solar case to ∼5 MK for the stellar case with magnetic field 30 times

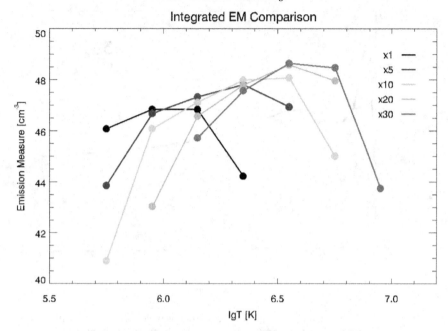

Figure 2. Integrated EM for different input magnetic flux densities. "x1" represents the solar case, and "x30" means the magnetic flux density is 30 times the solar case.

stronger. Note that the peak EM temperature for the solar case is consistent with the dominant dimming lines in the solar observation (i.e., 171 Å and 193 Å, with emission temperatures around 1 MK). Therefore, even without simulating the CMEs, the steady-state corona solution is useful to estimate the dominant coronal dimming lines in the stellar cases with different magnetic flux densities.

We then initiate CMEs with varying initial energies and into different steady-state coronal conditions. The simulation is switched to time-accurate mode to capture the CME eruption, and the MHD equations are solved in conservative form to guarantee the energy conservation during the eruption process. Here, we show two representative cases: one confined eruption and one explosive eruption. The confined eruption is due to the strong overlaying global coronal magnetic field that prevents the CME to escape (Alvarado-Gómez *et al.* 2018). In this case, since there is no plasma depletion involved, the EUV lines show no dimming features. Instead, due to the redistribution of the magnetic energy in the corona, it leads to a second peak in some EUV intensity profiles as shown in Figure 3. The second peak is more evident in the lines with higher emission temperature (e.g., 335 Å and 284 Å), which suggests the coronal plasma is dominant in these temperatures. This phenomenon is also frequently observed on the Sun and referred as EUV late-phase (e.g., Woods *et al.* 2011).

The second case is an explosive eruption in which the coronal mass is released into the interstellar space and leads to EUV coronal dimmings. In this case, the input $\langle |fB| \rangle$ is 5 times higher than the solar case. And the initial CME flux rope energy is about 10^{33} ergs. The resulting CME speed is ~3000 km s^{-1}, which is only slightly higher than the fast solar CMEs due to the stronger coronal field confinement. We simulate the coronal dynamical evolution for 4 hours after the CME eruption and synthesize the EUV line emissions. To further investigate the detectability of the dimming signals, we apply the instrument performance estimates from the Extreme-ultraviolet Stellar Characterization for Atmospheric Physics and Evolution (ESCAPE) mission concept, which provides extreme- and far-ultraviolet spectroscopy (70 - 1800 Å) to characterize the high energy radiation

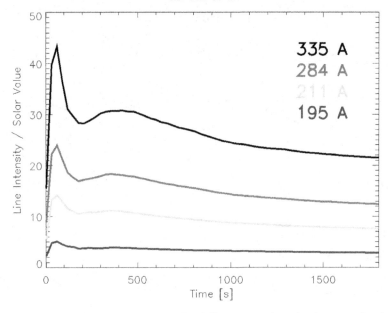

Figure 3. The EUV intensity evolution of 4 different wavelengths for a confined eruption case. The intensities are scaled with the steady-state solar intensities.

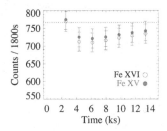

Figure 4. Synthesized Fe XV 284 Å and Fe XVI 335 Å line intensity evolution after CME onset from ESCAPE instrument performance estimates. The detector counts are based on 30 minutes exposure time.

environment in the habitable zones around nearby stars (France *et al.* 2019). The EUV detector of the ESCAPE mission will be $\sim$100 times more sensitive than the previous EUVE mission (Craig *et al.* 1997). Figure 4 shows the simulated Fe XV 284 Å and Fe XVI 335 Å line emissions by assuming the star at 6 pc and using 30 minutes exposure time. The ISM absorption effect has been taken into account assuming $N(\mathrm{H\ I}) = 10^{18}$ cm^{-2} (France *et al.* 2018). This preliminary result suggests that with better instrumentation, the stellar coronal dimmings associated with CMEs can be observed.

We summarize the main results as follows:

• The coronal dimmings encode important information about CME energetics, CME-driven shock properties, and magnetic configuration of erupting flux ropes.

• With higher magnetic flux density, the stellar dimming occurs in the higher temperature range than the solar case. With better instrumentation, the coronal dimming could be detected from the distant stars.

• Our results show a proof-of-concept that the MHD model can be used for quantitative studies of CME-dimming relationships and applied to stellar cases. To get useful CME information from the future stellar observations, more detailed modeling studies are needed.

Acknowledgments

M. Jin was supported by NASA's SDO/AIA contract (NNG04EA00C) to LMSAL. We are thankful for the use of the NASA Supercomputer Pleiades at Ames and its helpful staff for making it possible to perform the simulations presented in this study. SDO is the first mission of NASA's Living With a Star (LWS) Program.

References

Alvarado-Gómez, J. D., Drake, J. J., Cohen, O., *et al.* 2018, *Astrophys. J.*, 862, 93

Attrill, G. D. R., Harra, L. K., van Driel-Gesztelyi, L., & Wills-Davey, M. J. 2010, *Solar Phys.*, 264, 119

Aschwanden, M. J., Nitta, N. V., Wuelser, J.-P., *et al.* 2009, *Astrophys. J.*, 706, 376

Cheng, J. X. & Qiu, J. 2016, *Astrophys. J.*, 825, 37

Cohen, O., Attrill, G. D. R., Manchester, W. B., IV, & Wills-Davey, M. J. 2009, *Astrophys. J.*, 705, 587-602

Craig, N., Abbott, M., Finley, D., *et al.* 1997, *Astrophys. J. Supp. Series*, 113, 131

Delaboudinière, J.-P., Artzner, G. E., Brunaud, J., *et al.* 1995, *Solar Phys.*, 162, 291

Dissauer, K., Veronig, A. M., Temmer, M., *et al.* 2018, *Astrophys. J.*, 855, 137

Dissauer, K., Veronig, A. M., Temmer, M., *et al.* 2018, *Astrophys. J.*, 863, 169

Downs, C., Roussev, I. I., van der Holst, B., Lugaz, N., & Sokolov, I. V. 2012, *Astrophys. J.*, 750, 134

Downs, C., Török, T., Titov, V., *et al.* 2015, AAS/AGU Triennial Earth-Sun Summit, 1, 304.01

Gibson, S. E. & Low, B. C. 1998, *Astrophys. J.*, 493, 460

Hansen, R. T., Garcia, C. J., Hansen, S. F., & Yasukawa, E. 1974, *Pub. Astron. Soc. Pac.*, 86, 500

Harra, L. K. & Sterling, A. C. 2001, *Astrophys. J. Lett.*, 561, L215

Harra, L. K., Hara, H., Imada, S., *et al.* 2007, *Pub. Astron. Soc. Japan*, 59, S801

Harra, L. K., Schrijver, C. J., Janvier, M., *et al.* 2016, *Solar Phys.*, 291, 1761

Harrison, R. A., Bryans, P., Simnett, G. M., & Lyons, M. 2003, *Astron. Astrophys.*, 400, 1071

Hudson, H. S., Acton, L. W., & Freeland, S. L. 1996, *Astrophys. J.*, 470, 629

Imada, S., Hara, H., Watanabe, T., *et al.* 2007, *Pub. Astron. Soc. Japan*, 59, S793

Jin, M., Ding, M. D., Chen, P. F., Fang, C., & Imada, S. 2009, *Astrophys. J.*, 702, 27

Jin, M., Schrijver, C. J., Cheung, M. C. M., *et al.* 2016, *Astrophys. J.*, 820, 16

Jin, M., Manchester, W. B., van der Holst, B., *et al.* 2017a, *Astrophys. J.*, 834, 172

Jin, M., Manchester, W. B., van der Holst, B., *et al.* 2017b, *Astrophys. J.*, 834, 173

France, K., Arulanantham, N., Fossati, L., *et al.* 2018, *Astrophys. J. Supp. Series*, 239, 16

France, K., Fleming, B. T., Drake, J. J., *et al.* 2019, *Proc. SPIE* 11118, UV, X-Ray, and Gamma-Ray Space Instrumentation for Astronomy XXI, 1111808

Krista, L. D. & Reinard, A. 2013, *Astrophys. J.*, 762, 91

Krista, L. D. & Reinard, A. A. 2017, *Astrophys. J.*, 839, 50

Lemen, J. R., Title, A. M., Akin, D. J., *et al.* 2012, *Solar Phys.*, 275, 17

Lugaz, N., Manchester, IV, W. B., & Gombosi, T. I. 2005, *Astrophys. J.*, 634, 651

Manchester, W. B., Gombosi, T. I., Roussev, I., de Zeeuw, D. L., Sokolov, I. V., Powell, K. G., Tóth, G., & Opher, M. 2004a, Journal of Geophysical Research (Space Physics), 109, 1102

Manchester, W. B., Gombosi, T. I., Roussev, I., Ridley, A., de Zeeuw, D. L., Sokolov, I. V., Powell, K. G., & Tóth, G. 2004b, Journal of Geophysical Research (Space Physics), 109, 2107

Manchester, IV, W. B., van der Holst, B., & Lavraud, B. 2014, Plasma Physics and Controlled Fusion, 56, 064006

Mason, J. P., Woods, T. N., Caspi, A., Thompson, B. J., & Hock, R. A. 2014, *Astrophys. J.*, 789, 61

Mason, J. P., Woods, T. N., Webb, D. F., *et al.* 2016, *Astrophys. J.*, 830, 20

Oran, R., van der Holst, B., Landi, E., *et al.* 2013, *Astrophys. J.*, 778, 176

Oran, R., Landi, E., van der Holst, B., *et al.* 2015, *Astrophys. J.*, 806, 55

Pesnell, W. D., Thompson, B. J., & Chamberlin, P. C. 2012, *Solar Phys.*, 275, 3

Reinard, A. A. & Biesecker, D. A. 2008, *Astrophys. J.*, 674, 576-585

Rust, D. M. & Hildner, E. 1976, *Solar Phys.*, 48, 381

Schrijver, C. J., Aulanier, G., Title, A. M., Pariat, E., & Delannée, C. 2011, *Astrophys. J.*, 738, 167

Sokolov, I. V., van der Holst, B., Oran, R., *et al.* 2013, *Astrophys. J.*, 764, 23

Sterling, A. C. & Hudson, H. S. 1997, *Astrophys. J. Lett.*, 491, L55

Sterling, A. C., Hudson, H. S., Thompson, B. J., & Zarro, D. M. 2000, *Astrophys. J.*, 532, 628

Tian, H., McIntosh, S. W., Xia, L., He, J., & Wang, X. 2012, *Astrophys. J.*, 748, 106

Thompson, B. J., Plunkett, S. P., Gurman, J. B., *et al.* 1998, *Geophys. Res. Lett.*, 25, 2465

Thompson, B. J., Gurman, J. B., Neupert, W. M., *et al.* 1999, *Astrophys. J. Lett.*, 517, L151

Thompson, B. J., Cliver, E. W., Nitta, N., Delannée, C., & Delaboudinière, J.-P. 2000, *Geophys. Res. Lett.*, 27, 1431

Tóth, G., *et al.* 2012, Journal of Computational Physics, 231, 870

van der Holst, B., Sokolov, I. V., Meng, X., *et al.* 2014, *Astrophys. J.*, 782, 81

Woods, T. N., Hock, R., Eparvier, F., *et al.* 2011, *Astrophys. J.*, 739, 59

Woods, T. N., Eparvier, F. G., Hock, R., *et al.* 2012, *Solar Phys.*, 275, 115

Zarro, D. M., Sterling, A. C., Thompson, B. J., Hudson, H. S., & Nitta, N. 1999, *Astrophys. J. Lett.*, 520, L139

Zhukov, A. N. & Auchère, F. 2004, *Astron. Astrophys.*, 427, 705

Discussion

CHRISTOFFER KAROFF: How would this look if you looked in the optical and not the UV?

MENG JIN: Because the mass loss of the CME is mainly from the corona, the coronal dimming is not seen in the optical bands that dominate by photospheric emissions.

Chapter 9. Surface magnetic fields of the Sun and stars

Classifications of magnetised star-planet interactions

Bow-shocks

comet-like tails

inspiraling accretion str

Matsakos+1

Villarreal D'Angelo+18

Daley-Yates & Stevens 19

Aline Vidotto

Solar and Stellar Magnetic Fields: Origins and Manifestations
Proceedings IAU Symposium No. 354, 2019
A. Kosovichev, K. Strassmeier & M. Jardine, ed.
doi:10.1017/S1743921319009840

On the properties of the magnetic Chemically Peculiar B, A, and F-type stars

Kutluay Yüce[1] , Saul J. Adelman[2], Diane M. Pyper[3] and Robert J. Dukes[4]

[1]Dept. of Astronomy and Space Sciences, Faculty of Science, University of Ankara,
TR-06100 Tandoğan, Ankara, Turkey
email: Kutluay.Yuce@ankara.edu.tr

[2]Dept. of Physics, The Citadel, 171 Moultrie Street, Charleston, U.S.A.

[3]Dept. of Physics/Astronomy, University of Nevada, Las Vegas, U.S.A.

[4]Dept. of Physics and Astronomy, The College of Charleston, Charleston, U.S.A.

Abstract. We present a preliminary analysis of the Strömgren *uvby* photometry of the magnetic CP stars obtained using the Four College Automated Photometric Telescope for its 21.5 years of operation ending in Fall 2012. We summarize the photometry for all the FCAPT mCP stars that have been published to date. We do not find any significant correlation between the amplitudes of variation in the *uvby* filters and the periods. A small number of stars show anomalous behaviour of the *v* filter which will be discussed in a future study.

Keywords. stars: magnetic chemically peculiar, photometry

1. Introduction

If magnetic fields cause elemental abundances to change in the stellar photosphere, then all magnetic Chemically Peculiar (mCP) stars should exhibit signatures of this effect in both their spectra and flux distributions. Given that all stars rotate, these stars should be magnetic, spectrum, and photometric variables, albeit sometimes of low amplitude. The generally accepted explanation for the light variations of magnetic stars is inhomogenous brightness distribution on the surface of the rotating star. This explanation was given earlier by the discoveries of this effect (Guthnik & Prager 1918). The magnetic fields and light curves vary periodically with a range from about one-half day to decades (e.g., Preston 1971, Adelman & Woodrow 2007, Mathys 2017). Adelman (2002) found some evidence for the expectation that as mCP stars move away from the ZAMS their rotational velocities decrease. He noticed that many, but not all, of the most rapidly rotating mCP stars are close to the ZAMS and some of the least rapidly rotatings are the furthest from the ZAMS. Recently, Hümmerich *et al.* (2018) detect observed periods ranging from 0.84 d to 9.6 d, and effective amplitudes ranging from 0.6 mmag to 90.5 mmag based on Kepler data.

The differential Strömgren *uvby* photometric studies from the Four College Automated Photometric Telescope (FCAPT, Fairborn Observatory in Southern Arizona) for the magnetic CP stars have both improved periods and better defined the shapes of their light curves (e.g. Pyper & Adelman 2017, Dukes & Adelman 2018). In the literature a series of papers entitled "The FCAPT *uvby* Photometry of the mCP Stars" by Adelman & others (e.g. see Dukes & Adelman 2018) discuss mCP stars that have spectral types between B2 and A2 on or near the Main Sequence of the HR diagram. The long-term FCAPT observation program was completed in 2012, after 21.5 years of operation. In

Table 1. Descriptive statistics in Periods and in *uvby* of FCAPT magnetic CP stars.

	0.5<P<7850 d	0.5<P<18 d	*u*	*v*	*b*	*y*
Mean	166.339	5.140	0.050	0.037	0.033	0.029
Stand.dev.	931.487	5.071	0.035	0.033	0.024	0.026
Median	3.873	3.093	0.045	0.030	0.028	0.024
Sample Variance	867668.651	25.719	0.00124	0.00106	0.00057	0.00070
Range	7849.481	17.969	0.180	0.155	0.120	0.140
Min.	0.519	0.519	0.000	0.000	0.000	0.000
Max.	7850	18.488	0.180	0.155	0.120	0.140
Count	76	63 (83 %)	81	81	81	81

this work we present a preliminary study of the photometric properties of 81 FCAPT mCP stars, that have been published so far, focusing on their amplitudes of variation as a function of period for each of the *uvby* filters.

2. Findings

Observationally, the shortest known period for our mCP stars was found to be about 1/2 day for CU Vir (Adelman *et al.* 1999, Pyper *et al.* 2013) while HD 15980 (Ap Si) appears to be minimally variable, with a period of at least 5 yr (Adelman & Woodrow 2007) and HD 9996 (B9pCrEu) has a period of 21.5 yr (Pyper & Adelman 2017) in the FCAPT program. Since the majority of these stars of Pyper & Adelman (2017) have unusual light curves that cannot be easily explained by the oblique rotator theory, they will be discussed in more detail in a future study.

Most of the stars (83 %) have periods between 0.51899 and 18.4877 days. Because of that the distribution of 76 mCP stars as a function of period were investigated for two different groups: 0.51899 < P < 18.4877 d and 0.51899 < P < 7850 d. Descriptive statistics are given in Table 1. The mean and median values are measures of central tendency. The median is a robust to outstanding observations and it indicates that the period is about 3.873 days.

We did not observe a significant relationship between photometric periods and amplitudes for our mCP stars. Correlation coefficients between period and *u*,*v*, *b*, *y* bandpasses of 75 stars are -0.118, 0.395, 0.161, and 0.461, respectively.

For the *uvby* bands of mCP stars the descriptive statistics are as in Table 1. The mean and median for each band are nearly equal. The amplitudes from the highest to lowest value of our data are from the *u* to the *y* band. The means of these distribution are 0.050 ± 0.035 mag for *u*, 0.037 ± 0.033 mag for *v*, 0.033 ± 0.024 mag for *b*, and 0.029 ± 0.026 mag for the *y* band.

The mean amplitudes and their standard deviations decreases with increasing wavelength. These values are nearly equal for *v* and *b*. They are slightly smaller than those of Adelman & Woodrow (2007).

Figure 1 shows the dependence of the light amplitudes on the wavelength. The observed amplitudes of our mCP stars in the four bandpasses tend to change the same manner. However, the structure of the amplitude-wavelength relation for the various stars is not simple. For example, we note for a few stars (the SrCrEu stars HD 86592, HR 7575, HD 81009), larger amplitudes have been observed in the *v* band compared to the other bands. HD 32966 has the largest amplitudes in each band in the group of 81 magnetic CP stars.

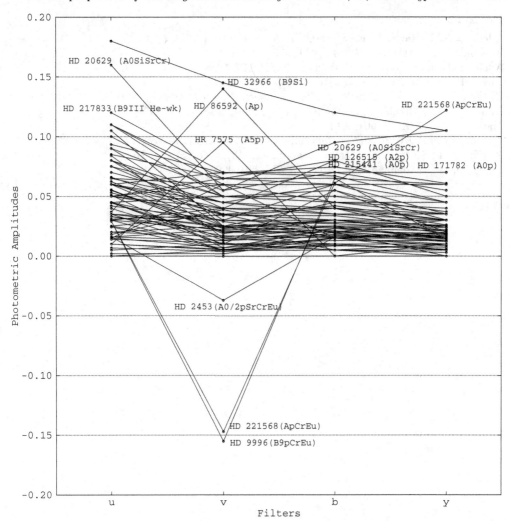

Figure 1. Amplitudes of light variations in stellar magnitudes vs. wavelength relation.

3. Conclusion

We investigated the distribution of periods and amplitudes of magnetic chemically peculiar stars in *uvby*. We did not observe a significant relationship between photometric periods and amplitudes for these stars. The amplitudes have a Gaussian distribution, but each band has a slight skewness to large amplitudes.

Acknowledgments

K. Yüce thanks Prof. Dr. Fikri Öztürk. This research has made use of the SIMBAD database, operated at CDS, Strasbourg, France.

References

Adelman, S. J., Rayle, K. E., & Pi, C. -L. M. 1999, *A&AS*, 136, 379
Adelman, S. J. 2002, *Balt. Astron.*, 111, 475
Adelman, S. J. & Woodrow, S. L. 2007, *PASP*, 119, 1256
Dukes, R. J. & Adelman, S. J. 2018, *PASP*, 130:044202
Guthnik, P. & Prager, R. 1918, *Veröff. Sternwarte Berlin-Babelsberg II*, H. 3
Hümmerich, S., Mikulasek, Z., Paunzen, E., & et al. 2018, *A&A*, 619, 98
Mathys, G. 2017, *A&A*, 601, A14
Preston, G. W. 1970, *in Stellar Rotation, ed. A. Sletteback (Dordrecht: D. Reidel)*, 254
Pyper, D. M., Stevens, R. I., & Adelman, S. J. 2013, *MNRAS*, 431, 2106
Pyper, D. M. & Adelman, S. J. 2017, *PASP*, 129:104203

Solar and Stellar Magnetic Fields: Origins and Manifestations
Proceedings IAU Symposium No. 354, 2019
A. Kosovichev, K. Strassmeier & M. Jardine, ed.
doi:10.1017/S1743921319009724

Impact of small-scale emerging flux from the photosphere to the corona: a case study from IRIS

Salvo L. Guglielmino[1]📶, Peter R. Young[2,3,4], Francesca Zuccarello[1], Paolo Romano[5] and Mariarita Murabito[6]

[1]Dipartimento di Fisica e Astronomia "Ettore Majorana" – Sezione Astrofisica,
Università degli Studi di Catania,
Via S. Sofia 78, 95123, Catania, Italy
email: `salvatore.guglielmino@inaf.it`

[2]Code 671, NASA Goddard Space Flight Center,
Greenbelt, MD 20771, USA

[3]College of Science, George Mason University,
Fairfax, VA 22030, USA

[4]Northumbria University,
Newcastle upon Tyne, NE1 8ST, United Kingdom

[5]INAF – Osservatorio Astrofisico di Catania,
Via S. Sofia 78, 95123 Catania, Italy

[6]INAF – Osservatorio Astronomico di Roma,
via Frascati 33, I-00078 Monte Porzio Catone, Italy

Abstract. We report on multi-wavelength ultraviolet (UV) high-resolution observations taken with the *IRIS* satellite during the emergence phase of an emerging flux region embedded in the unipolar plage of active region NOAA 12529. These data are complemented by measurements taken with the spectropolarimeter aboard the *Hinode* satellite and by observations from *SDO*. In the photosphere, we observe the appearance of opposite emerging polarities, separating from each other, and cancellation with a pre-existing flux concentration of the plage.

In the upper atmospheric layers, recurrent brightenings resembling UV bursts, with counterparts in all UV/EUV filtergrams, are identified in the EFR site. In addition, plasma ejections are observed at chromospheric level. Most important, we unravel a signature of plasma heated up to 1 MK detecting Fe XII emission in the core of the brightening sites.

Comparing these findings with previous observations and numerical models, we suggest evidence of several long-lasting, small-scale magnetic reconnection episodes between the new bipolar EFR and the ambient field.

Keywords. Sun: magnetic fields, Sun: photosphere, Sun: chromosphere, Sun: transition region, Sun: corona, Sun: UV radiation

1. Introduction

A decade of observations of small-scale emerging flux regions (EFRs) in the solar atmosphere, carried out with increasing spatial resolution, has reinforced the idea that magnetic reconnection is likely to occur when the newly emerging magnetic flux interacts with the pre-existing ambient fields (e.g., Guglielmino *et al.* 2010, Ortiz *et al.* 2014, 2016, Toriumi *et al.* 2017). This process results in energy release, which is able to heat the upper atmospheric layers and to drive high-temperature plasma flows, according to early models

(Shibata *et al.* 1989, Yokoyama & Shibata 1995) and more recent numerical simulations (e.g., MacTaggart *et al.* 2015, Archontis & Syntelis 2019, Isliker *et al.* 2019).

In this context, observations performed by the Interface Region Imaging Spectrograph (*IRIS*, De Pontieu *et al.* 2014) satellite have revealed the presence of intense, small-scale ($\approx 500 - 1000$ km), short-lived (~ 5 minutes) brightenings seen in ultraviolet (UV) images, called *IRIS* bombs (Peter *et al.* 2014) or UV bursts (Young *et al.* 2018). They are associated with opposite-polarity magnetic flux patches in the photosphere and seem to be caused by small-scale magnetic reconnection occurring in the low atmosphere.

Here, we describe a long-lived UV burst, which was observed in active region NOAA 12529 and showed a significant impact at coronal heights (Guglielmino *et al.* 2018, 2019).

2. Observations

Active region NOAA 12529 was observed in April 2016 (Fig. 1). It passed at the central solar meridian between April 13 and 14, being located at heliocentric angle $\mu \approx 0.96$. At that time, an EFR was emerging in the plage of the trailing polarity of the active region (see the solid box in Fig. 1).

During the EFR evolution, the *IRIS* satellite acquired an observing sequence between 22:34 UT on April 13 and 01:55 UT on April 14. This sequence consists of six large dense 64-step raster scans, covering a field of view of $22'' \times 128''$ (see the dashed box in Fig. 1). UV observations included spectra of the C II 1334.5 Å and 1335.7 Å, Si IV 1394 and 1402 Å, Mg II k 2796.3 and h 2803.5 Å lines, as well as of other faint lines around the chromospheric O I 1355.6 Å line, comprising the coronal forbidden Fe XII 1349.4 Å line. The duration of each scan was about 33 minutes, with a 31.5 s step cadence and 0.″33 step size. Simultaneous slit-jaw (SJ) images were acquired in the 1400 (Si IV 1402 Å) and 2796 Å (Mg II k) passbands. These SJ images have a cadence of 63 s for consecutive frames in each passband, covering a field of view of $144'' \times 128''$. Further information about this *IRIS* data set can be found in Guglielmino *et al.* (2018).

We used full-disk continuum filtergrams and line-of-sight magnetograms, taken along the Fe I 6173 Å line by the Helioseismic and Magnetic Imager (HMI) on board the *SDO* satellite, to study the photospheric configuration of the EFR. Coronal images acquired by the Atmospheric Imaging Assembly (AIA) in the UV/EUV channels were also considered in this analysis. Moreover, observations acquired by the *Hinode* spectropolarimeter were used to determine the fine structure of the EFR in the photosphere.

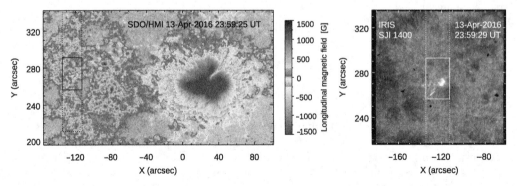

Figure 1. *Left panel*: Active region NOAA 12529 as seen in the *SDO*/HMI magnetogram. *Right panel*: Simultaneous *IRIS* SJ image in the 1400 Å passband. The solid-line box frames the EFR analyzed in the text. The dashed-line box indicates the area scanned by the *IRIS* slit.

3. Results

The analysis of the photospheric evolution of the EFR from *SDO*/HMI indicates that it emerged in a unipolar plage, where pre-existing flux concentrations were harbored, corresponding to pores in continuum images. In particular, the negative flux patch of the EFR approached a positive polarity pore that became smaller and finally disappeared, while the emerging magnetic flux formed a new pore with negative polarity.

At the flux peak, *Hinode* measurements show a mixed polarity pattern in the emergence zone, suggesting a serpentine magnetic field structure. In this region, elongated granules in the continuum as well as enhanced linear polarization signals were also observed.

During the three hours of the *IRIS* observing sequence, recurrent UV brightenings were observed in the EFR site. Comparing *SDO*/HMI magnetograms to *IRIS* radiance maps (see Fig. 2), we found that the brightenings occurred near the contact region between the positive pre-existing polarity field and the new emerging flux patch with negative polarity. Surge-like ejections, with a length of about 10″, were repeatedly observed close to the UV brightenings, in particular at chromospheric heights (O I, Mg II and C II lines in Fig. 2). These were also seen in transition region lines (e.g., Si IV line in Fig. 2), showing an asymmetric appearance between the blue and red wings of the lines, reminiscent of the dynamics of the ejection (see Fig. 2). All of the *SDO*/AIA channels exhibited a long-lasting counterpart of the event, as seen for instance in the the 193 Å channel (Fig. 2).

UV spectra in the brightening core revealed the presence of plasma components with different velocity, mostly blueshifted. Some spectral features, like the absence of the O IV lines and the Mg II triplet emission, suggest that the burst occurred at low atmospheric heights. Notably, we also detected Fe XII emission (log T [K]= 6.2) in the brightening core. This indicates that plasma is locally heated up to 1 MK and that the enhancements seen in the *SDO*/AIA channels have a genuine coronal origin.

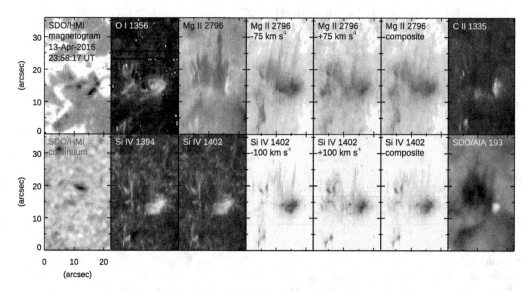

Figure 2. Synoptic view of the EFR at different atmospheric heights, during the third *IRIS* raster scan. The *SDO*/HMI magnetogram and continuum filtergram relevant to the half time of the *IRIS* scan are shown as a reference. Top panels display the morphology of the EFR in the chromosphere at increasing height. Bottom panels show the appearance of the EFR in the transition region, including a *SDO*/AIA filtergram in the 193 Å passband at the half time of the *IRIS* scan. For the chromospheric Mg II 2796 Å line and for the transition region Si IV 1402 Å line, we also show the morphology of the EFR in the blue and red wings of the lines.

4. Implications

The analysis of this event strongly suggests that it is a result of magnetic reconnection between the emerging and the pre-existing field. Indeed, we find a general agreement of the evolution of the UV brightening with radiative magnetohydrodynamic numerical simulations concerning surges and UV bursts observed in flux emergence experiments (Nóbrega-Siverio *et al.* 2016, 2018). However, reconnection appears to occur at higher levels with respect to UV bursts, explaining the observed coronal counterpart.

We remark that the interaction between the pre-existing field and the new emerging flux is also important for the interpretation of larger events and greater amount of released energy (e.g., Romano *et al.* 2014), which might result in severe Space Weather phenomena (Zuccarello *et al.* 2013).

Acknowledgments

The research has received funding from the European Union's Horizon 2020 research and innovation programme under grant agreement nos. 739500 (PRE-EST project) and 824135 (SOLARNET project). This work was also supported by the Università degli Studi di Catania through the program: Piano per la Ricerca 2016-2018 - Linea di intervento 2 "Dotazione Ordinaria" and by the Space Weather Italian COmmunity (SWICO) Research Program. Support from INAF (National Institute for Astrophysics) – Catania Astrophysical Observatory is gratefully acknowledged. P.R.Y. acknowledges funding from NASA grant NNX15AF48G, and he thanks ISSI Bern for supporting the International Team Meeting "Solar UV Bursts – a New Insight to Magnetic Reconnection."

References

Archontis, V. & Syntelis, P. 2019, *Philosophical Transactions of the Royal Society A*, 377, 20180387

De Pontieu, B., Title, A .M., Lemen, J. R., *et al.* 2014, *Solar Physics*, 289, 2733

Guglielmino, S. L., Bellot Rubio, L. R., Zuccarello, F., *et al.* 2010, *ApJ*, 724, 1083

Guglielmino, S. L., Zuccarello, F., Young, P. R., Murabito, M., & Romano, P. 2018, *ApJ*, 856, 127

Guglielmino, S. L., Young, P. R., & Zuccarello, F. 2019, *ApJ*, 871, 82

Isliker, H., Archontis, V., & Vlahos, L. 2019, *ApJ*, 882, 57

MacTaggart, D., Guglielmino, S. L., Haynes, A. L., Simitev, R., & Zuccarello, F. 2015, *A&A*, 576, A4

Nóbrega-Siverio, D., Moreno-Insertis, F., & Martínez-Sykora, J. 2016, *ApJ*, 822, 18

Nóbrega-Siverio, D., Moreno-Insertis, F., & Martínez-Sykora, J. 2018, *ApJ*, 858, 8

Ortiz, A., Bellot Rubio, L. R., Hansteen, V. H., de la Cruz Rodríguez, J., & Rouppe van der Voort, L. 2014, *ApJ*, 781, 126

Ortiz, A., Hansteen, V. H., Bellot Rubio, L. R., *et al.* 2016, *ApJ*, 825, 93

Peter, H., Tian, H., Curdt, W. *et al.* 2014, *Science*, 346, 1255726

Romano, P., Zuccarello, F. P., Guglielmino, S. L., & Zuccarello, F. 2014, *ApJ*, 794, 118

Rouppe van der Voort, L., De Pontieu, B., Scharmer, G. B., *et al.* 2017, *ApJ Lett.*, 851, L6

Shibata, K., Tajima, T., Steinolfson, R. S., & Matsumoto, R. 1989, *ApJ*, 345, 584

Toriumi, S., Katsukawa, Y., & Cheung, M. C. M. 2017, *ApJ*, 836, 63

Yokoyama, T. & Shibata, K. 1995, *Nature*, 375, 42

Young, P. R., Tian, H., Peter, H., *et al.* 2018, *Space Science Reviews*, 214, 120

Zuccarello, F., Balmaceda, L., Cessateur, G., *et al.* 2013, *Journal of Space Weather and Space Climate*, 3, A18

Solar and Stellar Magnetic Fields: Origins and Manifestations
Proceedings IAU Symposium No. 354, 2019
A. Kosovichev, K. Strassmeier & M. Jardine, ed.
doi:10.1017/S1743921319009967

Multi-flux-rope system in solar active regions

Yijun Hou[1,2]🄳, Jun Zhang[1,2], Ting Li[1,2] and Shuhong Yang[1,2]🄳

[1]CAS Key Laboratory of Solar Activity, National Astronomical Observatories
Chinese Academy of Sciences, Beijing 100101, China

[2]University of Chinese Academy of Sciences, Beijing 100049, China
email: yijunhou@nao.cas.cn

Abstract. Magnetic flux rope (MFR) is closely connected with solar eruptions, such as flares and coronal mass ejections. The classical scenario assumes a single MFR for each eruption, but it is reasonable to expect multiple MFRs in a complex active region (AR). Statistically investigating AR 11897, we verify the existence of multiple MFR proxies during the AR evolution. Recently, AR 12673 in 2017 September produced the two largest flares in Solar Cycle 24. The evolutions of the AR magnetic fields and the two large flares reveal that significant flux emergence and successive interactions between different emerging dipoles resulted in the formations of multiple MFRs and twisted loop bundles, which successively erupted like a chain reaction within several minutes before the peaks of the two flares. We propose that the eruptions of a multi-flux-rope system can rapidly release enormous magnetic energy and result in large flares in solar AR.

Keywords. Sun: activity, Sun: atmosphere, Sun: filaments, Sun: flares, Sun: magnetic fields

---

1. Introduction

Solar flares and coronal mass ejections (CMEs) are explosive phenomena in the solar atmosphere and release dramatic free magnetic energy into the interplanetary space, which can severely affect the space environment around the earth. The magnetic flux rope (MFR) is a set of magnetic field lines winding around a central axis and is widely believed to play a key role in triggering the solar eruptive events (Priest and Forbes 2002; Schmieder *et al.* 2015). Thus, a complete research on the MFR is necessary to obtain a clear understanding of solar flares and CMEs, which will undoubtedly result in accurate forecasts of eruptive activities and associated space weather. With high-resolution observations, the existence of MFR in the solar atmosphere has been unambiguously evidenced (Guo *et al.* 2010; Cheng *et al.* 2011; Li and Zhang 2013; Chintzoglou *et al.* 2015; Hou *et al.* 2019). Moreover, some recent works have implied that MFRs may be ubiquitous on the Sun and could gather in the solar active regions (Zhang *et al.* 2015; Awasthi *et al.* 2018; Jiang *et al.* 2018).

2. Observations and Results

Based on observations from Atmospheric Imaging Assembly (AIA; Lemen *et al.* 2012) and Helioseismic and Magnetic Imager (HMI; Schou *et al.* 2012) of the *Solar Dynamics Observatory* (*SDO*; Pesnell *et al.* 2012), we statistically study MFR proxies in active region (AR) 11897. Then we investigate the X9.3 flare on 2017 September 06 occurring in AR 12673, which produced 4 X-class flares from September 04 to September 10. In the X9.3 flare, multiple MFRs were detected to successively erupt within five minutes before the flare peak. The results of nonlinear force-free field (NLFFF) modeling (Wiegelmann *et al.* 2012) also confirm the existence of a multi-flux-rope system in AR 12673. The similar phenomenon was also observed during the X8.2 flare on September 10.

Table 1. Distribution of the detected MFR proxies in AR 11897.

	Nov. 14	Nov. 15	Nov. 16	Nov. 17	Nov. 18	Nov. 19
Site1	3					
Site2	8	1				
Site3			2	5	1	1
Site4		3	4	1	1	

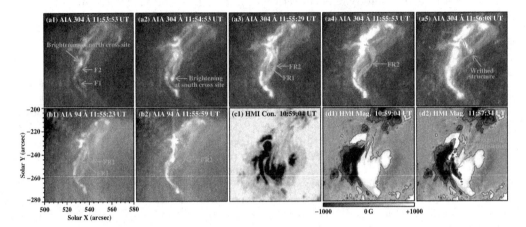

Figure 1. AIA 304 Å and 94 Å images showing eruption of the double-decker MFR configuration and corresponding HMI continuum intensitygram and LOS magnetograms displaying the magnetic fields of the AR core region before the flare peak at 12:02 UT.

During the evolution of AR 11897 from 2013 November 14 to 19, we identify MFR proxies for 30 times in 4 different sites, that is, 5 times per day on average. The daily distribution of these MFR proxies is shown in Table 1. Here we notice that some MFR proxies appeared in one location for several times. It is possible that 7 MFRs were detected in 4 different sites and repeatedly illuminated for 30 times in total (see more details in Hou *et al.* 2016).

On 2017 September 6, an X9.3 flare took place in AR 12673, which is the largest flare in Solar Cycle 24 (Yang *et al.* 2017; Yan *et al.* 2018; Liu *et al.* 2018). Here we investigate the evolutions of this large flare and the associated complex magnetic system (see more details in Hou *et al.* 2018). By examining the AIA 304 Å observations, we detected two sets of filament threads located in the AR core region before the occurrence of the X9.3 flare (see F1 and F2 in Fig. 1), which forms a double-decker MFR configuration (Liu *et al.* 2012). Around 11:53:53 UT and 11:54:53 UT, brightenings appeared at the north and south cross sites of these two filament threads (also two MFRs, FR1 and FR2), implying the interaction between rising FR1 and FR2 (see green arrows in panels (a1)-(a2)). The two MFRs then were tracked completely by the brightening material, and FR2 began to moved upward as well (see panel (a3)). In panels (a4) and (a5), FR2 showed obvious twisted threads and writhed structure, implying the occurrence of kink instability (Kliem *et al.* 2004; Török *et al.* 2011). In 94 Å channel, the two MFRs were also observed clearly (panels (b1)-(b2)). Panels (d1)-(d2) show that the north ends of the two MFRs were rooted in a negative-polarity patch. Before the onset of the X9.3 flare, this negative magnetic patch kept moving northwestward along the semicircular PIL and successively sheared with the adjacent positive fields. Meanwhile, the HMI continuum intensitygrams reveal that this negative patch exhibited a counterclockwise rotation motion (panel (c1)).

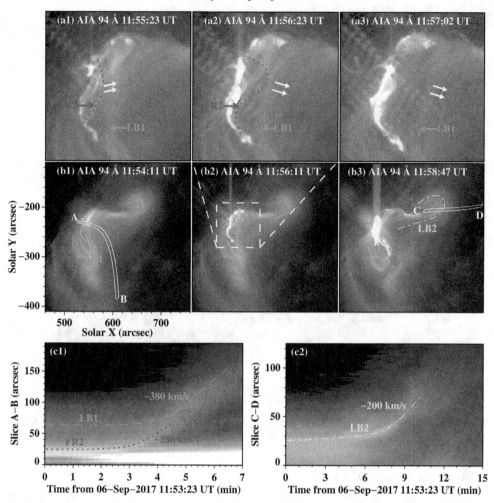

Figure 2. Dynamic evolutions of the complex system consisting of multiple flux ropes and twisted loop bundles during the X9.3 flare.

During the X9.3 flare, a total of two MFRs (FR1 and FR2) and two twisted loop bundles (LB1 and LB2) are identified in the flaring region (see Fig. 2). AIA 94 Å images of panels (a1)-(a3) show the interaction between the kink-unstable FR2 and the nearby loop bundles (LB1). Around 11:56:23 UT, FR2 and LB1 interacted with each other in their middle parts. Then LB1 began to rise up rapidly and disturbed another set of loop bundles (LB2) with a larger scale. Along the two white arc-sector domains "A-B" and "C-D", we make two time-space plots and show them in panels (c1) and (c2), where the eruptions of FR2, LB1, and LB2 are clearly visible. After the successive eruptions of multiple MFRs and twisted LBs, the X9.3 flare reached its peak at 12:02 UT. In order to verify these structures illuminated in EUV channels and study their magnetic topologies, we reconstruct 3D magnetic field above the AR and show the results in Figure 3. It is clear that before the onset of X9.3 flare, two MFRs (FR1 and FR2) with the twist number $(T_w) \leq -1.0$ are located above the PIL in the AR core region. Two sets of twisted LBs (LB1 and LB2) are also tracked nearby.

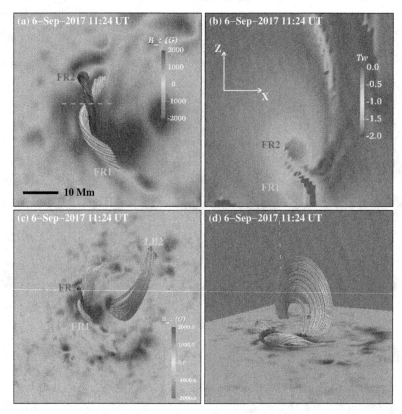

Figure 3. Extrapolated 3D NLFFF structures corresponding to FR1, FR2, LB1, and LB2 at 11:24 UT on 2017 September 6.

3. Summary and Discussion

Employing the *SDO* observations, for the first time, we detect multiple MFR proxies for 30 times in AR 11897 at four different locations during six days. These new observations imply that multiple MFRs can exist in an AR and that the complexity of AR magnetic configurations is far beyond our imagination. Furthermore, we investigate the X9.3 flare on 2017 September 06 occurring in AR 12673, which is the largest flare in Solar Cycle 24. Aided by the NLFFF modeling, we identify a double-decker MFR configuration above the PIL in the AR core region. The north ends of these two MFRs were rooted in a negative-polarity magnetic patch, which began to move along the PIL and rotate anticlockwise before onset of the X9.3 flare. The strong shearing motion and rotation contributed to the destabilization of the two MFRs, of which the upper one eventually erupted upward due to the kink-instability. Then another two sets of twisted loop bundles beside these MFRs were disturbed and successively erupted within five minutes like a chain reaction.

MFRs have been thought to be closely connected with CMEs and solar flares. The classical scenario assumes a single MFR for each eruption, but it is natural to imagine the existence of multiple MFRs if the AR is complex and has extended curved PIL (Liu *et al.* 2012; Shen *et al.* 2013; Awasthi *et al.* 2018). Török *et al.* (2011) presented a 3D MHD simulation to investigate three consecutive filament eruptions. In the observational domain, Shen *et al.* (2012) reported the simultaneous occurrence of a partial and a full filament eruption in two neighboring source regions. Therefore, based on the statistical and case studies mentioned above, we propose that the eruption of a multi-flux-rope system in solar AR could rapidly release enormous magnetic energy and trigger large flares, such as the largest flare in Solar Cycle 24: the X9.3 flare on 2017 September 6.

Acknowledgments

The data are used courtesy of the *SDO* science team. The authors are supported by the National Natural Science Foundations of China (11903050, 11790304, 11773039, 11533008, 11673035, 11673034, 11873059, and 11790300), the NAOC Nebula Talents Program, the Youth Innovation Promotion Association of CAS (2017078 and 2014043), Young Elite Scientists Sponsorship Program by CAST (2018QNRC001), and Key Programs of the Chinese Academy of Sciences (QYZDJ-SSW-SLH050).

References

Awasthi, A. K., Liu, R., Wang, H., Wang, Y., & Shen, C. 2018, *ApJ*, 857, 124
Cheng, X., Zhang, J., Liu, Y., & Ding, M. D. 2011, *ApJ* (Letters), 732, L25
Chintzoglou, G., Patsourakos, S., & Vourlidas, A. 2015, *ApJ*, 809, 34
Guo, Y., Schmieder, B., Démoulin, P., *et al.* 2010, *ApJ*, 714, 343
Hou, Y. J., Li, T., & Zhang, J. 2016, *A&A*, 592, A138
Hou, Y. J., Zhang, J., Li, T., Yang, S. H., & Li, X. H. 2018, *A&A*, 619, A100
Hou, Y., Li, T., Yang, S., & Zhang, J. 2019, *ApJ*, 871, 4
Jiang, C., Zou, P., Feng, X., *et al.* 2018, *ApJ*, 869, 13
Kliem, B., Titov, V. S., & Török, T. 2004, *A&A*, 413, L23
Lemen, J. R., Title, A. M., Akin, D. J., *et al.* 2012, *Solar Phys.*, 275, 17
Li, T. & Zhang, J. 2013, *ApJ* (Letters), 778, L29
Liu, L., Cheng, X., Wang, Y., Zhou, Z., Guo, Y., Cui, J. 2018, *ApJ* (Letters), 867, L5.
Liu, R., Kliem, B., Török, T., *et al.* 2012, *ApJ*, 756, 59
Pesnell, W. D., Thompson, B. J., & Chamberlin, P. C. 2012, *Solar Phys.*, 275, 3
Priest, E. R. & Forbes, T. G. 2002, *A&AR*, 10, 313
Schmieder, B., Aulanier, G., & Vršnak, B. 2015, *Solar Phys.*, 290, 3457
Schou, J., Scherrer, P. H., Bush, R. I., *et al.* 2012, *Solar Phys.*, 275, 229
Shen, C., Li, G., Kong, X., *et al.* 2013, *ApJ*, 763, 114
Shen, Y., Liu, Y., & Su, J. 2012, *ApJ*, 750, 12
Török, T., Panasenco, O., Titov, V. S., *et al.* 2011, *ApJ* (Letters), 739, L63
Wiegelmann, T., Thalmann, J. K., Inhester, B., *et al.* 2012, *Solar Phys.*, 281, 37
Yan, X. L., Wang, J. C., Pan, G. M., *et al.* 2018, *ApJ*, 856, 79
Yang, S., Zhang, J., Zhu, X., & Song, Q. 2017, *ApJ* (Letters), 849, L21
Zhang, J., Yang, S. H., & Li, T. 2015, *A&A*, 580, A2

Solar and Stellar Magnetic Fields: Origins and Manifestations
Proceedings IAU Symposium No. 354, 2019
A. Kosovichev, K. Strassmeier & M. Jardine, ed.
doi:10.1017/S1743921320000071

The 3D structure of the penumbra at high resolution from the bottom of the photosphere to the middle chromosphere

Mariarita Murabito[1][ID]**, Ilaria Ermolli[1], Fabrizio Giorgi[1],**
Marco Stangalini[1], Salvo L. Guglielmino[2], Shahin Jafarzadeh[3,4],
Hector Socas-Navarro[5,6], Paolo Romano[7] and Francesca Zuccarello[2]

[1]INAF–Osservatorio Astronomico di Roma,
Via Frascati 33, I-00078, Monte Porzio Catone, Italy
email: `mariarita.murabito@inaf.it`

[2]Dipartimento di Fisica e Astonomia–Sezione Astrofisica
Università degli studi di Catania, Via S. Sofia 78, I-95123 Catania, Italy

[3]Rosseland Centre for Solar Physics, University of Oslo,
P.O. BOX 1029 Blindern, NO-0315 Oslo, Norway

[4]Institute of Theoretical Astrophysics, University of Oslo,
P.O. BOX 1029 Blindern, NO-0315 Oslo, Norway

[5]Instituto de Astrofisica de Canarias,
C/Via Lactea s/n, E-38205 La Laguna, Tenerife, Spain

[6]Departamento de Astrofisica, Universidad de La Laguna,
E-38205 La Laguna, Tenerife, Spain

[7]INAF–Osservatorio Astrofisico di Catania,
Via S. Sofia 78,I-95123, Catania, Italy

Abstract. Sunspots are the most prominent feature of the solar magnetism in the photosphere. Although they have been widely investigated in the past, their structure remains poorly understood. Indeed, due to limitations in observations and the complexity of the magnetic field estimation at chromospheric heights, the magnetic field structure of sunspot above the photosphere is still uncertain. Improving the present knowledge of sunspot is important in solar and stellar physics, since spot generation is seen not only on the Sun, but also on other solar-type stars. In this regard, we studied a large, isolated sunspot with spectro-polarimeteric measurements that were acquired at the Fe I 6173 nm and Ca II 8542 nm lines by the spectropolarimeter IBIS/DST under excellent seeing conditions lasting more than three hours. Using the Non-LTE inversion code NICOLE, we inverted both line measurements simultaneously, to retrieve the three-dimensional magnetic and thermal structure of the penumbral region from the bottom of the photosphere to the middle chromosphere. Our analysis of data acquired at spectral ranges unexplored in previous studies shows clear spine and intra-spine structure of the penumbral magnetic field at chromopheric heights. Our investigation of the magnetic field gradient in the penumbra along the vertical and azimuthal directions confirms results reported in the literature from analysis of data taken at the spectral region of the He I 1083 nm triplet.

Keywords. Sun: magnetic fields, Sun: chromosphere, Sun: photosphere

1. Introduction

Space- and ground-based observations have been used to study the sunspot penumbra, from its formation (Murabito *et al.* 2016, 2017, 2018) to the three dimensional (3D) magnetic structure (Joshi *et al.* 2016, 2017). Balthasar 2018 reported about the unsolved problem of the estimation of the magnetic field gradient in sunspots and how this gradient changes due to the different techinque employed. All previous studies were on photospheric observations taken through several lines, but only one chromospheric data acquired at the He I triplet at 1083.0 nm. Recently, a different chromospheric diagnostic, the Ca II 854.2 nm line, was used by Joshi & de la Cruz Rodríguez 2018 to study the variation of atmospheric parameters associated with umbral flashes, reporting a decrease of the magnetic field of -0.5 G km^{-1} when moving from the photosphere to the chromosphere. In this contribution, we summarize the results obtained from our study of the vertical gradient of the magnetic field in a sunspot penumbra, based on photospheric Fe I 617.3 nm and chromospheric Ca II 854.2 nm lines.

2. Observations

We used data acquired on 2016 May 20 at the Dunn Solar Telescope of the National Solar Observatory. The data consists of spectropolarimetric observations of a large sunspot located near the disk center in photosphere and chromosphere taken at the Fe I 617.3 nm and Ca II 854.2 nm lines. These data were obtained from the Interferometric BIdimensional Spectrometer (IBIS, Cavallini 2006) at 21 spectral points for both lines, with a spectral sampling of 20 mÅ and 60 mÅ for the Fe I and Ca II lines, respectively. The data were inverted with the Non-LTE inversion COde (NICOLE Socas-Navarro *et al.* 2015) to infer the magnetic field at the atmospheric heigths sampled by the data. For further details about the observations, data reduction and data inversion, see Stangalini *et al.* 2018 and Murabito *et al.* 2019.

Based on the results from the response functions (RFs) computation, we focused our analysis to the plasma conditions at three atmospheric heigths in the photosphere (log $\tau \approx -0.5$, -1.0, and -1.5) and one in the chromosphere (log$\tau \approx -4.6$).

3. Results

Fig. 1 shows circular and linear polarization maps derived from the measurements taken at the Fe I and Ca II lines. The photospheric spine/intraspine structure of the penumbra is clearly displayed in panels a and b, while the same is attenuated in chromosphere as obtained from the circular and linear polarization signals estimated from the Ca II data. We derived the vertical gradient of the magnetic field strength following the formula:

$$\left(\frac{\Delta B}{\Delta log\tau}\right)_{a,b} = \frac{(\Delta B)_{a,b}}{(\Delta log\tau)_{a,b}} = \frac{B(b) - B(a)}{b - a} \tag{3.1}$$

used by Joshi *et al.* 2017, where a and b denote the lower and upper log τ, respectively. Fig. 2 (top panels) shows the maps of $(\Delta B/\Delta log\tau)$ for the subfov A marked in Fig. 1. In particular, the panel a displays the photospheric vertical gradient of the magnetic field by considering as [a,b] = [log $\tau = -0.5$, log $\tau = -1.5$], while the rigth panel shows the gradient between the photosphere and the chromosphere when [a,b] = [log $\tau = -1$, log $\tau = -4.6$]. The photospheric vertical gradient exhibits a ring-like structure, where the gradient has negative values, meaning that the magnetic field decreases with optical depth. The outer penumbra, instead, is charactherized by postive values. The radial dependence of the vertical gradient was calculated considering the average field values along 80 isocontours in the Fe I line continuum. This radial dependence is shown in Fig. 2 panels c and d. In particular, we can note that in the inner penumbra at r/R$_{spot} = 0.5$ the gradient has values of about -100 G/log τ. Moving onward to r/R$_{spot} = 0.6$, it sligthly increase reaching average values of about 300 G/ log τ.

On the other hand, both the map and the plot of the gradient between photosphere and chromosphere (panel b and d) show that in the inner penumbra, at r/R$_{spot} = 0.5$, the values are about 200 G/log τ. Moving away from the inner to the outer penumbra,

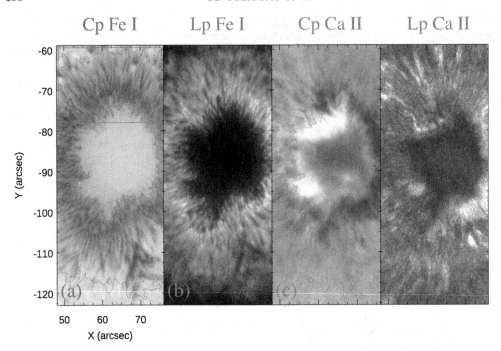

Figure 1. Circular and linear polarization maps derived from the IBIS Fe I (panels a and b) and Ca II (panels c and d) data. The red contours represent the umbra-penumbra (UP) boundary and the outer penumbra contour. The red box labelled with A indicates the subfov used for the analysis shown in Fig. 2.

the gradient slowly decreases, with an average value of 170 G/log τ up to r/R_{spot} = 0.6, then it has a constant value of 100 G/log τ.

4. Implications

We presented the analysis of photospheric and chromospheric data taken at different spectral range and inverted with different methods than those reported in the literature. We estimated the vertical gradient of the magnetic field strength in the photosphere and chromosphere. The ring-like structure visible in the maps of the photospheric vertical gradient derived from our study is similar to the one reported by Joshi *et al.* 2017 even if our average gradient is higher. We estimated the gradient beetween the photosphere and the chromosphere at the atmospheric heigths more sensitive to field pertubations in our data. The field gradient in the chromosphere assumes an average value of 100 G/logτ, corresponding to ≈ 0.3 G km^{-1}. This value is within the range of values reported by Joshi *et al.* 2017 from analysis of data taken at the He I 1083.0 nm.

We conclude that in order to produce a complete 3D magnetic picture of the penumbra at higher atmospheric heigths an analysis of multiple spectral diagnostics of the photosphere and chromosphere is needed.

Acknowledgments

The research leading to these results has received funding from the European Research Council under the European Union's Horizon 2020 Framework Programme for Research and Innovation, grant agreements H2020 PRE-EST (no. 739500) and H2020 SOLARNET

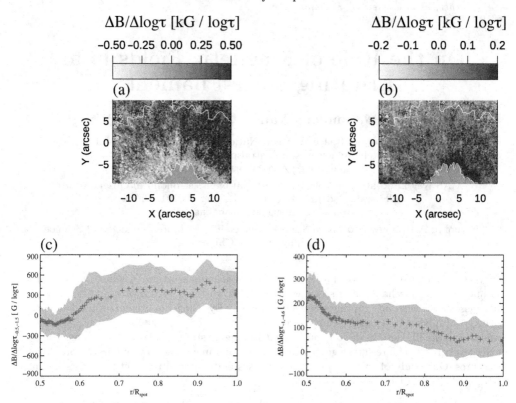

Figure 2. Top panels: Maps of the vertical gradient of the magnetic field strength considering two atmospheric heights in the photosphere (left panel) and two heights representative of the photosphere and chromosphere (rigth panel). Bottom panels: Variation of the vertical gradient as a function of r/R_{spot} for the subarray A displayed in the top panels.

(no. 824135). This work was also supported by INAF Istituto Nazionale di Astrofisica (PRIN-INAF-2014).

S.J. acknowledges support from the European Research Council under the European Union's Horizon 2020 research and innovation programme (grant agreement no. 682462) and from the Research Council of Norway through its Centres of Excellence scheme (project no. 262622).

References

Balthasar H. 2018, *Solar Physics*, 293, 120

Stangalini M., Jafarzadeh S., Ermolli I., *et al.* 2018, *ApJ*, 869, 110

Murabito M., Romano P., Guglielmino S. L., *et al.* 2016, *ApJ*, 825, 75

Murabito M., Zuccarello F., Guglielmino S. L., *et al.* 2018, *ApJ*, 885, 58

Murabito M., Romano P., Guglielmino S. L., *et al.* 2017, *ApJ*, 834, 76

Joshi J., Lagg J., Solanki S. K., *et al.* 2016, *A&A*, 596, A8

Joshi J., Lagg J., Hirzberger J., *et al.* 2017, *A&A*, 604, A98

Joshi J., Lagg J., Hirzberger J., *et al.* 2017, *A&A*, 599, A35

Cavallini F. 2006, *Solar Physics*, 236, 415

Socas-Navarro H., de la Cruz Cruz Rodríguez J., Asensio Ramos A., *et al.* 2015, *A&A*, 577, A7

Murabito M., Ermolli I., Giorgi F., *et al.* 2019, *Apj*, 873, 126

Tiwari S., K., van Noort, M., Lagg, A., *et al.* 2013, *A&A*, 557, A25

Joshi, J. & de la Cruz Rodríguez, J. 2018, *A&A*, 619, A63

Solar and Stellar Magnetic Fields: Origins and Manifestations
Proceedings IAU Symposium No. 354, 2019
A. Kosovichev, K. Strassmeier & M. Jardine, ed.
doi:10.1017/S1743921320000149

On the Role of Magnetic Fields in an Erupting Solar Filament

Qiao Song[1] ⓘ, Shuhong Yang[2,3] ⓘ and Jing-Song Wang[1]

[1] Key Laboratory of Space Weather, National Center for Space Weather,
China Meteorological Administration, Beijing 100081, China
email: songq@cma.cn, wangjs@cma.cn

[2] CAS Key Laboratory of Solar Activity, National Astronomical Observatories,
Chinese Academy of Sciences, Beijing 100101, China
email: shuhongyang@nao.cas.cn

[3] School of Astronomy and Space Science, University of Chinese Academy of Sciences,
Beijing 100049, China

Abstract. A filament eruption may lead to a coronal mass ejection (CME), which is one of the main driving mechanisms of space weather. This work analyses a slow and flareless CME event associated with an erupting quiescent filament. By using the extreme ultraviolet images of the Atmospheric Imaging Assembly onboard the Solar Dynamics Observatory, we trace the evolution of the filament in detail, and present the manifestations of the role of magnetic fields in the low corona. The results suggest the existence of a magnetic flux rope in the pre-eruption structures. Our study of this complex magnetic system may lead to a better understanding of CMEs and their impact on the space weather.

Keywords. Sun: filaments, Sun: magnetic fields, Sun: coronal mass ejections (CMEs)

---

1. Introduction

A coronal mass ejection (CME) is a significant release of magnetized plasma from the solar corona into the interplanetary space. It is one of the main driving mechanisms of space weather events that may lead to major geomagnetic storms. Solar quiescent filaments are dark curves-like structures that sometimes appear in the quiet Sun. When the dark filaments are above the limb of the Sun, they will turn to bright and significant, so they are also called solar prominences. The magnetic field is the controlling force in the solar corona and it dominates the formation, evolution and eruption of filaments. The magnetic structure of filaments can provide gravitational support and thermal isolation (see Gibson 2018 and references therein). The flux rope, the sheared-arcade and other magnetic structural models provide a framework for understanding the evolution of the filament in the solar corona.

2. Data

This work uses the extreme ultraviolet (EUV) images of the Atmospheric Imaging Assembly (AIA) onboard the Solar Dynamics Observatory (SDO), because of its excellent temporal coverage and spatial resolution (Lemen *et al.* 2012). The data set is processed by using the standard routines in the Solar SoftWare.

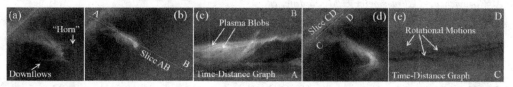

Figure 1. Dynamic features of an evolving filament in SDO/AIA 171 Å channel.

3. Results

A quiescent filament appeared on the western limb of the Sun around June 13, 2018. Although the overall evolution of the filament was slow, it was full of highly dynamic features in small-scale. Inside a coronal cavity, the filament slowly extends outward, and becomes bright on its top in the EUV images of the SDO/AIA. There was a horn-like structure on the top of the filament (Fig. 1), and it may relate to a magnetic flux rope. The horn-like structure appeared bifurcation and brightening which may suggest that it had a magnetic reconnection and lead to downflows back to the solar surface. As we can see from the time-distance graph of slice AB, some magnetized plasma blobs were moving upward and they may accumulate magnetic fields for the final eruption of the filament. We also analyses rotational motions in the filament. A slice (CD) is made at the location near the feet of the filament, and the dark materials in the middle of the slice appear a wave-like pattern in the time-distance graph. It may indicate that these materials have rotational motions which may be involved to the final eruption.

After a long period of evolution, the filament finally erupted and caused a slow CME. The erupting structure moved to the sides and outward. In the SDO/AIA 211 Å waveband image, we can see a long bright line that are nearly perpendicular to the solar surface, while there is no such bright line in the AIA 171 Å waveband images. The different observations in these different wavebands may indicate that this long line structure has a high temperature and it may be a current sheet produced by the magnetic reconnection of the eruption. The eruption was without significant X-ray enhancement, but had growing post-eruption arcades which are similar to the post-flare loops (Song *et al.* 2016) and can be explained by the standard flare model. A CME can be observed from the image of SOHO/LASCO and it may cause disturbance to the interplanetary space.

4. Summary

This work investigated the evolution of a solar quiescent filament and its final eruption. Coronal cavity, horn-like structure, downflows, plasma blobs and other fine structures are observed during its evolution. It suggests that the magnetic flux rope plays an important role for the erupting filament, the CME and the space weather.

Acknowledgments

This work is supported by the National Natural Science Foundation of China (41404136, 41774195 and 11673035) and the Youth Innovation Promotion Association of CAS.

References

Gibson, S. E. 2018, *Living Rev Sol Phys*, 15, 7
Lemen, J. R., Title, A. M., Akin, D. J., *et al.* 2012, *Sol Phys*, 275, 1–2, 17
Song, Q., Wang, J. S., Feng, X. S., & Zhang, X. X. 2016, *ApJ*, 821, 2, 83

Solar and Stellar Magnetic Fields: Origins and Manifestations
Proceedings IAU Symposium No. 354, 2019
A. Kosovichev, K. Strassmeier & M. Jardine, ed.
doi:10.1017/S174392131900992X

Fast downflows in a chromospheric filament

K. Sowmya[1], A. Lagg[1], S. K. Solanki[1,2] and J. S. Castellanos Durán[1]

[1]Max-Planck-Institut für Sonnensystemforschung
Justus-von-Liebig-Weg 3, 37077 Göttingen, Germany
email: **krishnamurthy@mps.mpg.de**

[2]School of Space Research, Kyung Hee University, YongIn, Gyeonggi 446–701, Korea

Abstract. An active region filament in the upper chromosphere is studied using spectropolarimetric data in He I 10830 Å from the GREGOR telescope. A Milne-Eddingon based inversion of the Unno-Rachkovsky equations is used to retrieve the velocity and the magnetic field vector of the region. The plasma velocity reaches supersonic values closer to the feet of the filament barbs and coexist with a slow velocity component. Such supersonic velocities result from the acceleration of the plasma as it drains from the filament spine through the barbs. The line-of-sight magnetic fields have strengths below 200 G in the filament spine and in the filament barbs where fast downflows are located, their strengths range between 100 - 700 G.

Keywords. Sun: filaments, Sun: chromosphere, Sun: magnetic fields

---

1. Introduction

Solar filaments are dense threads suspended in the chromosphere and held in place by strong magnetic fields in the solar atmosphere. A filament appears dark when observed on the disk with the chromospheric lines such as the He I 10830 Å triplet, as the gas inside it is cooler than the photospheric plasma below. Its structure resembles that of a centipede with the elongated body called 'spine' and the many legs called 'barbs' which are rooted in the solar surface.

High speed downflows have been observed in the filament barbs (see e.g., Joshi *et al.* 2013; Sasso *et al.* 2011, 2014) where the magnetic field is more vertical than in the filament itself. Recently, Díaz Baso *et al.* (2019a,b) explored the dynamic and magnetic properties of an active region (AR) filament observed with the ground based 1.5 m GREGOR telescope (Schmidt *et al.* 2012), with an emphasis on understanding its magnetic topology. An overview about chromospheric filaments can be found in these papers and the references therein. We study the same AR filament that was studied by these authors but focus on the line-of-sight (LOS) velocity distribution and magnetic field strength at selected locations in the barbs.

2. Observations

The spectropolarimetric raster scan of the filament associated with the AR 12087 was recorded on 17 June 2014 using the GREGOR Infrared Spectrograph (GRIS; Collados *et al.* 2012). The scan (ID: 17jun14.005) lasted approximately 15 minutes (09:13 UT – 09:27 UT), covering about 19″ in the scan direction (x-axis) and 63″ along the slit (y-axis), with a step size of 0.126″ and a pixel size of 0.135″ along the slit. The coordinates of the center of the field-of-view (FOV) correspond to a heliocentric angle of 23° ($\mu = \cos\theta = 0.92$, $x = +180″$ and $y = -320″$). The data was reduced with the standard GRIS data reduction software. The image resolution determined by averaging the power spectrum along the slit direction is 0.5″. The observed spectral window of about 18 Å includes the

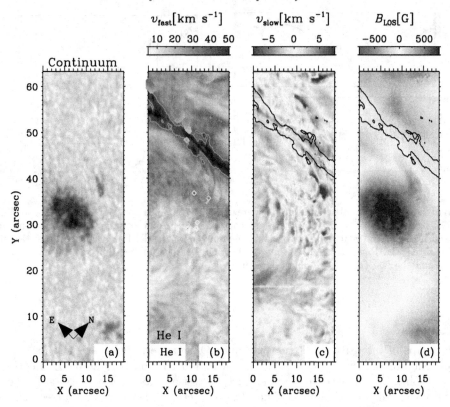

Figure 1. Panel (a): continuum intensity image with the arrows showing the directions of solar north and east. Panel (b): normalized intensity in the He I line at 10830.34 Å with the overplotted colored symbols representing the LOS velocity of the fast component. Panel (c): LOS velocity map for the slow component. Panel (d): LOS magnetic field strength map for the slow component saturated at ± 800 G. The green and black contours at $0.6\,I_c$ mark the filament spine.

chromospheric He I 10830 Å triplet, photospheric Si I 10827 Å and Ca I 10839 Å lines and a telluric blend at 10832 Å. The average noise in the Stokes parameters is $1.2 \times 10^{-3} I_c$, where I_c is the continuum intensity. Panels (a) and (b) in Fig. 1 show the intensity images in the continuum redward of the Si line and at He I 10830.34 Å. In the observed FOV, a part of the filament spine and a few barbs are visible in He I.

3. Method

We invert the spectropolarimetric data using the HeLIx$^+$ inversion code assuming Milne-Eddington atmosphere and taking into account the Zeeman effect in the incomplete Paschen–Back regime (Lagg *et al.* 2004, 2009). The Stokes profiles at some locations along the filament barbs show the presence of a second velocity component which is strongly redshifted. A single component inversion does not yield a good fit to such profiles (see panels (a) and (b) in Fig. 2). Therefore, we invert the He I triplet by using two atmospheric components which are mixed through a filling factor. We call the two components 'slow' (± 10 km s^{-1}) and 'fast' (downflows up to 50 km s^{-1}). From such inversions, we retrieve the LOS velocity and the magnetic field vector for the two atmospheric components.

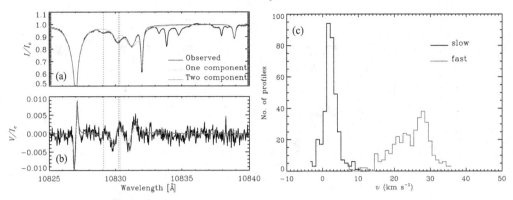

Figure 2. Panel (a): observed intensity profile (black) at a location with the coordinates $x = 8.7''$ and $y = 52.5''$ (see Fig. 1), a fit to the observed profile obtained from one component inversion (green) and a fit from two component inversion (red). Panel (b): same as panel (a) but for Stokes V. Panel (c): histograms of the LOS velocity at locations where both the slow (black) and the fast (red) components are present.

4. Results

The maps of intensity, LOS velocity and LOS magnetic field strength for the slow component are shown in Fig. 1. The green and black contours at $0.6\,I_c$ identify the filament spine. The slow component is nearly at rest and its LOS velocity (see panel (c)) follows a Maxwellian distribution. The LOS magnetic field strength map (see panel (d)) is saturated at $\pm 800\,$G. Chrompsheric magnetic fields stronger than $800\,$G are clearly present in the sunspot. The filament is located above the polarity inversion line where the magnetic field is changing sign. Inside the filament spine, the LOS component of the magnetic field is weaker than $200\,$G, in agreement with the previous reports (see e.g., Kuckein *et al.* 2009).

The locations where a fast component with a filling factor greater than 20 % and a signal in any Stokes polarization parameter higher than three times the noise (within $\pm 0.35\,$Å from it's wavelength position) is present are marked by red symbols in the He I intensity image in Fig. 1. The colorbar indicates the velocity of the fast component. We found no location where more than two atmospheric components are present. It is apparent that the fast component lies along the filament barbs. The velocity seems to be increasing down the barb reaching supersonic values (higher than the local sound speed of $\sim 10\,$km s^{-1}) closer to the footpoints. This increase can be explained by the gravitational acceleration of the plasma from the spine as it drains along the barbs.

Panels (a) and (b) in Fig. 2 show Stokes I and V profiles at one of the locations corresponding to $x = 8.7''$ and $y = 52.5''$, where the fast component has a velocity of $\sim 30\,$km s^{-1}. Stokes Q and U signals are close to the noise level and hence we do not show them here. The observed profiles are shown in black, the fits from one component and two component inversions are shown in green and red, respectively. Clearly, two components are needed to properly model such anomalous profiles. We identified 360 locations where a fast component is present simultaneously with the slow component. Fig. 2(c) shows the velocity distribution for the fast (red) and slow (black) components at those locations. Most of the downflow velocities range between $15 - 30$ km s^{-1} and in some extreme cases they reach 35 km s^{-1} or more. Along the filament barbs, the fast downflow component has LOS magnetic field strength ranging between $100 - 700\,$G.

5. Conclusions

We studied an active region filament observed by the ground based GREGOR telescope. Using two component Milne-Eddington inversions, we determined the velocity and the magnetic field vector at chromospheric heights where He I forms. We identified supersonic downflows, as high as $35\,\mathrm{km\,s^{-1}}$, coexisting with a slow flow component. The high donwlfow velocities can be understood with the acceleration of the plasma as it drains along the filament barbs. The slow component could be originating from He I layer at a different height or could be due to straylight. We find that the LOS component of the magnetic field is weaker than $200\,\mathrm{G}$ within the filament spine. In filament barbs where strong downflows are found, the LOS magnetic field is as strong as $700\,\mathrm{G}$.

Acknowledgments

Marie Skłodowska-Curie grant agreement No. 797715. The 1.5-meter GREGOR solar telescope was built by a German consortium under the leadership of the Leibniz Institut für Sonnenphysik in Freiburg with the Leibniz Institut für Astrophysik Potsdam, the Institut für Astrophysik Göttingen, and the Max-Planck Institut für Sonnensystemforschung in Göttingen as partners, and with contributions by the Instituto de Astrofísica de Canarias and the Astronomical Institute of the Academy of Sciences of the Czech Republic. The GRIS instrument was developed thanks to the support by the Spanish Ministry of Economy and Competitiveness through the project AYA2010-18029 (Solar Magnetism and Astrophysical Spectropolarimetry). This study has made use of SAO/NASA Astrophysics Data System's bibliographic service

References

Collados, M., López, R., Páez, E., Hernández, E., Reyes, M., *et al.* 2012, *AN*, 333, 872

Díaz Baso, C. J., Martínez González, M. J., & Asensio Ramos, A. 2019a, *A&A*, 625, A128

Díaz Baso, C. J., Martínez González, M. J., & Asensio Ramos, A. 2019b, *A&A*, 625, A129

Joshi, A. D., Srivastava, N., Mathew, S. K., & Martin, S. F. 2013, *SoPh*, 288, 191

Kuckein, C., Centeno, R., Martínez Pillet, V., Casini, R., Manso Sainz, R., & Shimizu, T. 2009, *A&A*, 501, 1113

Lagg, A., Woch, J., Krupp, N., & Solanki, S. K. 2004, *A&A*, 414, 1109

Lagg, A., Ishikawa, R., Merenda, L., Wiegelmann, T., Tsuneta, S., & Solanki, S. K. 2009, *ASPC*, 415, 327

Sasso, C., Lagg, A., & Solanki S. K. 2011, *A&A*, 526, A42

Sasso, C., Lagg, A., & Solanki S. K. 2014, *A&A*, 561, A98

Schmidt, W., von der Lühe, O., Volkmer, R., Denker, C., Solanki, S. K., *et al.* 2012, *Astronomische Nachrichten*, 333, 796

Chapter 10. Observations of solar eclipses and exoplanetary transits

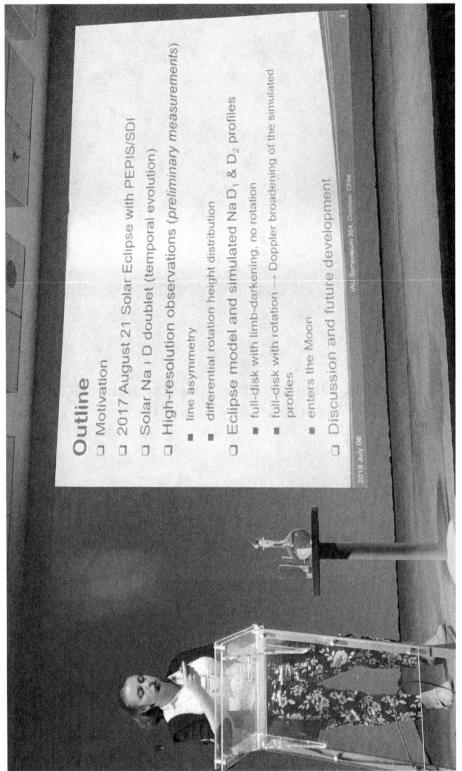

Ekaterina Dineva

Solar and Stellar Magnetic Fields: Origins and Manifestations
Proceedings IAU Symposium No. 354, 2019
A. Kosovichev, K. Strassmeier & M. Jardine, ed.
doi:10.1017/S1743921320000174

Characterization of stellar activity using transits and its impact on habitability

Raissa Estrela[1,2]⬤, Adriana Valio[2]⬤ and Sourav Palit[3]

[1] Jet Propulsion Laboratory, California Institute of Technology, 4800 Oak Grove Dr, Pasadena, CA 91109
email: `rlf.estrela@gmail.com`

[2] Center for Radioastronomy and Astrophysics Mackenzie, Mackenzie Presbyterian University Rua da Consolacao 896, Sao Paulo, SP 01302-907, Brazil
email: `adrivalio@gmail.com`

[3] Department of Physics, Indian Institute of Technology, Bombay, Powai, Mumbai 400076

Abstract. Stellar magnetic field is the driver of activity in stars and can trigger spots, energetic flares, coronal plasma ejections and ionized winds. These phenomena play a crucial role in understanding the internal mechanisms of the star, but can also have potential effects in orbiting planets. During the transit of a planet, spots can be occulted producing features imprinted in the transit light curve. Here, we modelled these features to characterize the physical properties of the spots (radius, intensity, and location). In addition, we monitor spots signatures on multiple transits to estimate magnetic cycles length of Kepler stars. Flares have also been observed during transits in active stars. We derive the properties of the flares and analyse their UV impact on possible living organisms in planets orbiting in the habitable zone.

Keywords. stars: flare, stars: activity, astrobiology, ultraviolet: stars

---

1. Introduction

In the Sun, the sunspot number is used as an estimator of its 11 year solar cycle. Similarly, the number of spots that appears at the surface of a star varies in accordance with the stellar magnetic cycle. Estrela & Valio (2016) uses the passage of a planet in front of its host star to find evidence of starspots. During the transit, the planet can occult spots present in the stellar disk producing small "bumps" as signatures (increase in stellar luminosity) in the planetary transit lightcurves. Hence, the variation of the number of spots per transit can give an estimate of the magnetic cycle of the star.

Also, during the maxima of a cycle there is a higher frequency of flares, coronal mass ejections (CMEs) and protons events (McIntosh *et al.* (2015)). These space weather pheonomena can drive the emission of ultraviolet and X-ray radiation as well as energetic particles causing potential effects on the planetary atmosphere. For example, Carrington-class solar flares are usually associated with fast and dense CME events that can erode and compress Earth's magnetosphere and ionosphere (Airapetian *et al.* (2017)). Other stars can be more active and show flares that are 1000 times more energetic than solar flares. Studies show that these flares when associated with energetic particles can indirectly destroy the ozone in Earth-like planets due to the formation of nitrogen oxides (Segura *et al.* (2010); Venot *et al.* (2016)).

The solar-type star analysed here, Kepler-96, has several flares in its lightcurve, and also during the transits of planet Kepler-96 b that orbits this star every 16.23 days. The strongest flare, visible in the middle of the 48th transit (966.70 BJD − 2.454.833

days), released a total of 1.8×10^{35} ergs, within the range of superflares (Maehara *et al.* (2015)). A study by Vida *et al.* (2017) using the data from the Kepler spacecraft in the K2 program found frequent flaring activity during the 80 days of observations of TRAPPIST-1. A total of 42 flares were detected and the strongest eruption emitted energy of roughly 10^{33} ergs in white light, which is more energetic than the largest flare ever recorded from the Sun. These energetic events could threaten the habitability of the planets in the system as they orbit much closer to their host star (0.029, 0.037, and 0.0451 AU, respectively) than Earth.

2. Stellar magnetic cycles

By monitoring the number of spots during the approximate 4 years of observation of the Kepler stars it is possible to estimate stellar cycles. Estrela & Valio (2016) applied two new methods to investigate the existence of a magnetic cycle: spot modelling and transit residuals excess. They found agreement between the results of the two methods for the solar-type stars Kepler-63 and Kepler-17. With the first method, they obtained $P_{cycle} = 1.12 \pm 0.16$ yr (Kepler-17) and $P_{cycle} = 1.27 \pm 0.16$ yr (Kepler-63), and for the second approach: $P_{cycle} = 1.35 \pm 0.27$ yr (Kepler-17) and $P_{cycle} = 1.27$. The second method is much faster to determine magnetic cycles because it only requires to integrate the area of the residuals due to the activity (spots) in the transit light curve. We used this method to estimate the magnetic cycle of two more active stars observed by Kepler: HAT-P-11 (Kepler-3) and Kepler-96. For the former, we obtained a cycle of 0.83 ± 0.16 yr and for the latter, 1.50 ± 0.35 yr.

3. UV radiation from flares on Kepler-96 and Trappist-1 systems

During a flare, the solar flux in XUV (20–300Å) band increases by a factor of 1000 and in the EUV (300–1215Å) band by up to a factor of 20 (Woods & Rottman (2005); Airapetian *et al.* (2017)). Longer UV wavelenghts also have an increase, Woods *et al.* (2004) reported that one of the most intense solar flare detected, a X17 GOES class observed in 2003, increased by 12% the Mg II h and k emissions (279.58–279.70 nm), which is within the MUV range.

Here we compute the UV contribution due to the energetic flares observed in Kepler-96 and Trappist-1. We focus on the MUV region of the spectra (200–300) because at short wavelengths (0.1–200 nm) the UV is attenuated in the top of the atmosphere, considering an Earth-like atmosphere with strong absorbers like N_2, CO_2 or O_2. Radiation at wavelengths between 200–320 nm can partially reach the surface of the planet depending if the planet has or not an ozone layer, and are very harmful to life.

On Earth, the ozone layer is responsible for absorbing most of the solar ultraviolet radiation arriving at our planet. In particular, the radiation that is the most threatening for life, like UVC (100–280 nm), is completely absorbed by the ozone layer, while UVB (280–315 nm) has an absorption of 95% and UVA (315–400 nm) can reach the Earth's surface. Therefore, the ozone layer acts as a shield that protects lifeforms living on the surface of our planet from the harmful ultraviolet radiation.

Estrela & Valio (2018) estimated the UV flux contribution of Kepler-96 flares using the MUV flux measured from the most intense solar flares, that was observed in 2003. For Trappist-1, along with the flux values of the flare observed by Vida *et al.* (2017), we use spectral and lightcurve information of a similar observed superflare of the M dwarf star AD Leo presented by Hawley & Pettersen (1991) to find the approximate spectra of the Trappist-1 flare, obtained by systematic extrapolation following the methods of Segura *et al.* (2010). The estimated UV (180–400 nm) spectra of the Trappist-1 flare are shown in Fig. 1.

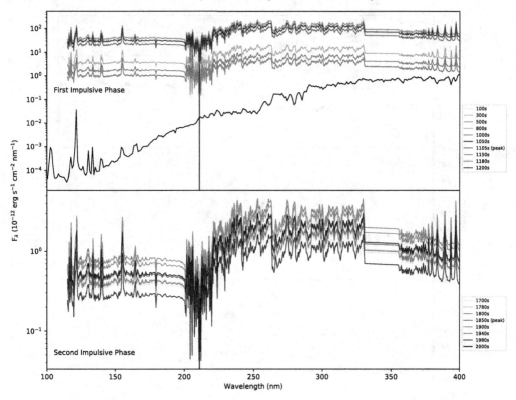

Figure 1. Estimated UV spectra of the strongest flare of Trappist-1 from Vida *et al.* (2017) for every 5 minutes during the evolution of the flare.

To compute the UV radiation (180–300 nm) through the atmospheres of the Trappist-1 planets, we use a two stream radiative transfer code from Ranjan *et al.* (2017). This code computes the UV fluxes and intensities at the top of the atmosphere and at the surface of the planets orbiting M-dwarfs, under a specified atmosphere and surface conditions. As input, the code requires the stellar spectrum, and the temperature, pressure and composition (gas molar concentration) as a function of altitude. We partitioned the atmosphere into 55 layers, each having thickness of 1km, and two atmospheric scenarios were used as input to the code: (i) a 1 bar CO_2 dominated atmosphere (0.9 bar N_2, 0.1 bar CO_2), similar to the Archean Earth at 3.9 Gyr and (ii) a modern atmosphere with ozone. For the former, we adopted a pre-biotic model already provided by the code. While for the latter, we consider an atmosphere composed of N_2, O_2, CO_2, H_2O, CH_4, O_3, and SO_2.

3.1. *Results*

The net UV flux at the top of the atmosphere (dashed lines) and on the surface (solid lines) of the three TRAPPIST-1's HZ planets are shown in Fig 2. The fluxes reach the highest values during the first impulsive phase ($\sim 10^2$ Wm^{-2}nm^{-1}), but has a considerable increase during the second peak of the flare. These values are in accordance with the UV surface flux from Segura *et al.* (2010) for an Earth-like planet during the energetic flare (10^{33} erg) of the M dwarf star AD Leo. In the presence of ozone, the UV radiation shortwards of 280 nm, which are the most dangerous to life, is absorbed. Just a small amount of UVB radiation arrives at the surface. However, the UVB and

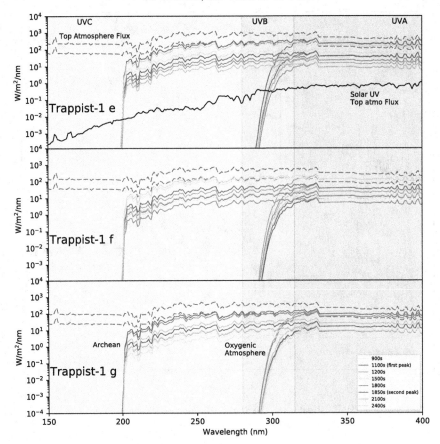

Figure 2. Ultraviolet flux at the top of the atmosphere (dashed line) and at the surface (solid line) of planets Trappist-1 e (top), Trappist-1 g (middle) and Trappist-1 f (bottom) during the evolution of the flare. The UV flux arriving at the surface is transmitted by two atmosphere models: 1 bar CO_2 dominated atmosphere (Archean) and a present day Earth-like atmosphere with ozone. The quiescent solar UV flux received at the top of the atmosphere of Earth is shown in the top panel for comparison (black solid line).

UVA flux during the impulsive phase of the flare can still be $\sim$100 times higher than the flux received by Earth (solid black line in Fig. 2). For an Archean atmosphere, only UV wavelengths smaller than 200 nm are absorbed, which means that the planetary surfaces receive UVB and UVC fluxes that are $\sim$5 times more biologically harmful than those on the present-day Earth.

4. Biological Impact

4.1. *On the surface*

To determine the survival of two bacteria on the surface of the Trappist-1's HZ planets we calculate the overall effective UV flux (E_{eff}) that falls in a biological body, considering the strongest observed flare of Trappist-1 and Kepler-96. Details of this calculation are presented in Estrela & Valio (2018). The threshold for the E_{eff} was chosen using the maximum UV flux for 10% survival of these bacteria.

Table 1 and 2 summarizes the results found for the E_{eff} considering the UV increase during the two impulsive phases of the flares in Trappist-1 and for a hypothetical planet at 1AU of Kepler-96. Thus, it is possible to observe that for a sudden increase of 5400%

Table 1. Biological effective irradiance, E_{eff} (J/m^2), due to the strongest superflare on Kepler-96.

Planet	Bacteria	Archean	Present day
Kepler-96 (1 AU)	*E. coli*	1.4×10^4	21.5
	D. radiodurans	8.0×10^3	7.5

Table 2. Biological effective irradiance, E_{eff} (J/m^2), due to the two impulsive phases of the superflare. To obtain the values in Joules, we multiplied the values in Watts by the total duration of the each peak (146s and 158s, respectively).

First Impulsive Phase			
Planet	Bacteria	Archean	Present day
TRAPPIST-1e	*E. coli*	1.3×10^6	59.4
	D. radiodurans	6.6×10^6	19
TRAPPIST-1f	*E. coli*	8.2×10^5	14
	D. radiodurans	4×10^5	4.4
TRAPPIST-1g	*E. coli*	5.5×10^5	12
	D. radiodurans	2.7×10^5	3.8
Second Impulsive Phase			
Planet	Bacteria	Archean	Present day
TRAPPIST-1e	*E. coli*	3.8×10^5	17
	D. radiodurans	2×10^5	5.4
TRAPPIST-1f	*E. coli*	2.3×10^5	4
	D. radiodurans	1.2×10^5	1.3
TRAPPIST-1g	*E. coli*	1.6×10^5	3.5
	D. radiodurans	7.8×10^4	1.1

from the strongest flare in Kepler-96, both micro-organisms can live in the surface if the planet has an atmosphere like the one we find on the present-day Earth with ozone. The same is true for the Trappist-1 planets. The UV flux received by the bacteria in a planet with a primitive atmosphere is very high ($\sim 10^5$ J/m^2) compared to the one with ozone. For a planet with a present-day atmosphere (with ozone), *E. Coli* could not survive in Trappist-1 e only during the first impulsive phase of the flare. However, both bacteria could survive in the other HZ planets under the presence of ozone.

4.2. On the ocean

The exposure to the high UV irradiation of flares imposes difficulties to the survival of microorganisms on the surface of the planet. A deep ocean could provide a safe refuge for micro-organisms against this extreme environment, by attenuating the effects of this radiation. Therefore, depending on the absorption of the UV radiation by the water, which varies with the ocean depth, the aquatic environment is more likely to host life. The UV spectral irradiance as a function of ocean depth can be calculated using the following equation:

$$I(\lambda, z) = I_0(\lambda)e^{-K(\lambda)z} \tag{4.1}$$

where $I(\lambda, z)$ is the UV spectral irradiance at depth z, $I_0(\lambda)$ is the UV spectral irradiance with the superflare contribution passing through a Primitive/Present-day atmosphere, and reaching the water surface and $K(\lambda)$ is the diffuse attenuation coefficient for water given by the sum of the absorption coefficient of water and the scattering coefficient.

Then, to quantify the effects on a micro-organism living in the ocean, we compute the biological effective irradiance. For that, we convolve the UV irradiation at a certain ocean depth z with the action spectrum of the bacteria to estimate at which depth the micro-organisms would receive an UV radiation dosage that they can tolerate.

Estrela & Valio (2018) assumed that an hypothetical Earth at 1 AU orbiting the star Kepler-96 has a calm and flat Archean ocean and found that *D. Radiodurans* and *E. Coli* would need to live at a depth of 12m and 28m, respectively. In the case of the Trappist-1 planets, during the first impulsive phase, E. Coli and D. Radiodurans could survive approximately at 40 and 20m, respectively, below the ocean surface of the three HZ planets. While for the second impulsive phase, they could survive at lower ocean depths: ∼38 and 21m, respectively.

5. Conclusions

In this work we estimated short magnetic cycles using planetary transits for the stars HAT-P-11 and Kepler-96, and we obtained cycles of 0.83 ± 0.16 yr and 1.50 ± 0.35 yr, respectively. Moreover, we estimated the impact that the UV increase due to high energetic flares from Kepler-96 would have if there was a hypothetical planet orbiting this star in the habitable zone (1 AU). We also applied similar analysis for the Trappist-1 system. We find that to survive the impacts from the strongest flare observed in Kepler-96 and in Trappist-1 microorganisms would need the protection from the ozone. An ocean in these planets could also provide a safe refuge for the lifeforms under the high UV irradiation of the flares. For the bacteria analysed in this work, *E. Coli* and *D. Radiodurans* they could escape from the hazardous UV effects of Kepler-96 flare at a depth of 28m and 12m below the ocean surface,respectively. In the three Trappist-1 HZ planets, they could survive at approximately 40m and 20m, respectively, below the ocean surface during the first impulsive phase of the superflare.

Acknowledgements

Raissa Estrela acknowledges a FAPESP fellowship (2016/25901-9 and 2018/09984-7). This research was carried out at Center for Radioastronomy and Astrophysics Mackenzie and at Jet Propulsion Laboratory, California Institute of Technology, under a contract with the National Aeronautics and Space Administration.

References

Airapetian, V. S., Glocer, A., Khazanov, G. V., *et al.* 2017, *ApjL*, 836, L3
Estrela, R. & Valio, A. 2016, *ApJ*, 831, 57
Estrela, R. & Valio, A. 2018, *Astrobiology*, 18, 1414–1424
Maehara, H., Shibayama, T., Notsu, Y., *et al.* 2015, *Earth, Planets, and Space*, 67, 59
McIntosh, S. W., Leamon, R. J., Krista, L. D., *et al.* 2015, *Nature Communications*, 6, 6491
Hawley, S. L. & Pettersen, B. R. 1991, *ApJ*, 378, 725
Ranjan, S., Wordsworth, R., & Sasselov, D. D. 2017, *ApJ*, 843, 110
Segura, A., Walkowicz, L.M., Meadows, V., Kasting, J., & Hawley, S. 2010, *Astrobiology*, 10, 751–771
Tilley, M. A., Segura, A., Meadows, V., *et al.* 2019, *Astrobiology*, 19, 64
Vida, K, Kovari, Zs., Pal, A., K. Olah, K, & Kriskovics, L. 2017, *ApJ*, 841, 124–129
Venot, O., Rocchetto, M., Carl, S., *et al.* 2016, *ApJ*, 830, 77
Woods, T. N. & Rottman, G. 2005, *Solar Physics*, 230, 375
Woods, T. N., Eparvier, F. G., Fontenla, J. *et al.* 2004, *Geophysical Research Letters*, 31, L10802

Solar and Stellar Magnetic Fields: Origins and Manifestations
Proceedings IAU Symposium No. 354, 2019
A. Kosovichev, K. Strassmeier & M. Jardine, ed.
doi:10.1017/S1743921320000198

Discovering the atmospheres of hot Jupiters

P. Wilson Cauley[iD]

LASP, CU Boulder 1234 Innovation Dr., Boulder, CO 80305
email: `paca7401@lasp.colorado.edu`

Abstract. Hot Jupiters are an extraordinary class of exoplanets, orbiting their host stars with periods of hours to a few days. Some of these objects have day-side temperatures approaching photospheric temperatures of late K-type stars. I will give an overview of how we characterize the atmospheres of these fascinating objects and some the more recent exciting results to come from ground and space-based telescopes, as well as what the future holds for detailed characterization of short-period exoplanet atmospheres.

Keywords. planetary systems, techniques: spectroscopic

1. Introduction

Some of the first dedicated searches for planets outside of our solar system, or exoplanets, were performed in the 1970's and 1980's using radio telescopes to attempt detection of the radio emission from Jupiter-mass exoplanets (Yantis *et al.* (1977); Winglee *et al.* (1986)). Although unsuccessful, these searches, along with similarly unsuccessful astrometric searches (Lippincott (1977)), jump-started the serious scientific effort to find worlds around other stars.

That effort culminated in 1995 when the first planet around a main-sequence star other than our own was detected. However, the nature of this planet was entirely unexpected: a Jupiter-mass planet on a 4.2 day orbit (Mayor & Queloz (1995)). These planets, subsequently named hot Jupiters due to their very high equilibrium temperatures, are now understood to be fairly common and are present around ≈1% of main-sequence FGK stars (Winn & Fabrycky (2015)). Hot Jupiters are relatively easy to detect due to the large RV amplitudes of their host stars and they have a high transit probability, making them the first planets to have their transit light curves measured (Charbonneau *et al.* (2000)). Finally, it is now understood that hot Jupiter radii are inflated relative to their cooler counterparts and that the high levels of radiation received from their parent stars is responsible (Thorngren & Fortney (2018)).

2. Transmission spectroscopy

Hot Jupiters have large scale heights due to their high atmospheric temperatures and thus are amenable to atmospheric characterization. The primary tool used to measure the structure and composition of these atmospheres is transmission spectroscopy. These measurements consist of comparing in-transit stellar spectra with out-of-transit spectra and looking for small differences that can be attributed to absorption in the planet's atmosphere. The first detection of an exoplanet atmosphere was made using the *Hubble Space Telescope* by Charbonneau *et al.* (2002), who measured Na I absorption around the hot Jupiter HD 209458 b. Ground-based detections would soon follow (Snellen *et al.* (2008); Redfield *et al.* (2008)), demonstrating that hot Jupiter atmospheres can be characterized using fairly standard spectrographs on large telescopes.

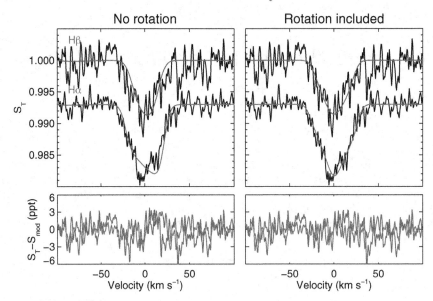

Figure 1. Balmer line transmission spectra and atmospheric models from Cauley *et al.* (2019) for the ultra-hot Jupiter KELT-9 b. The Balmer lines are useful tracers of the upper thermosphere in hot Jupiters.

In general, transmission spectroscopy observations can be split into two categories: low-resolution, which probes broad molecular features and the pressure-broadened wings of atomic lines, and high-resolution, which provides velocity-resolved detail on the cores of atomic lines and individual molecular transitions. The bulk of the important low-resolution observations have been performed by *HST* and have revealed a plethora of atmospheres around hot Jupiters, including everything from featureless spectra to objects with strong H_2O, Na I, K, and Rayleigh scattering signatures (Sing *et al.* (2016)).

High-resolution observations are more amenable to ground-based observing since echelle spectra are most often not flux-calibrated. This allows absorption to be measured against the local continuum rather than in absolute counts or flux. There are two methods of extracting absorption signatures from high-resolution spectra. The first involves a straightforward search for absorption lines in the transmission spectrum, which can often be accomplished manually since the only lines present in a majority of hot Jupiter atmospheres are strong resonance lines, such as Na I D or the Balmer lines (Figure 1). Automated searches that are agnostic to the atomic parameters are also useful if one suspects many different absorbing species. The second method requires a model template of the planet's atmosphere which can be cross-correlated with the transmission spectrum in order to find peaks in the cross-correlation function. This method is especially useful when many weak features are present from the same atom since the signal in the cross-correlation function is increased with each absorption feature. Cross-correlation is also uniquely suited to identifying molecular features in high-resolution near-infrared transmission spectra (Brogi *et al.* (2016)).

3. High-resolution absorption features in hot Jupiter atmospheres

The first feature ever measured in an exoplanet atmosphere was atomic sodium (Charbonneau *et al.* (2002)). Since that pioneering measurement, high-resolution transmission spectrsocopy has revealed a wide variety of atomic and molecular absorption signals in a diverse set of hot Jupiters. Sodium continues to be an important diagnostic

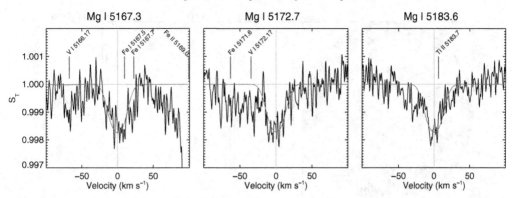

Figure 2. The first detection of the neutral magnesium triplet in an exoplanet atmosphere from Cauley *et al.* (2019) for KELT-9 b. Magnesium is an important coolant in hot Jupiter atmospheres and is a good mass loss tracer.

of the lower thermosphere (Wyttenbach *et al.* (2015); Khalafinejad *et al.* (2017); Seidel *et al.* (2019)). Potassium at 7699 and 7665 Å has also been detected in a number of planets (Sing *et al.* (2016); Keles *et al.* (2019)) and has similar formation conditions as the Na I D lines.

The hydrogen Balmer lines, and most prominently Hα, have become widely used tracers of the extended thermospheres of hot Jupiters (Cauley *et al.* (2015); Jensen *et al.* (2018); Yan & Henning (2018); Casasayas-Barris *et al.* (2019)). Balmer line absorption was first detected around HD 189733 b by Jensen *et al.* (2012) and subsequently confirmed by Cauley *et al.* (2015) and Cauley *et al.* (2016). Balmer line absorption is especially prominent around hot Jupiters orbiting A-type stars, such as KELT-9 b and KELT-20 b, due to the intense UV flux incident on the planet, which produces high rates of ionization and recombination to the $n = 2$ electronic state (Yan & Henning (2018); Cauley *et al.* (2019); Casasayas-Barris *et al.* (2019)).

Recently, absorption in the meta-stable neutral helium line at 10830 Å has been measured around a number of hot Jupiters. This line is a wonderful tracer of the thermosphere due to its ability to absorb photons out to large distances beyond the planet's optical radius (Oklopčić & Hirata (2018)). The first He I 10830 Å detection was made by Spake *et al.* (2018) for the hot gas giant WASP-107 b. This absorption was later characterized from the ground, revealing a cloud of helium extending to twice the planet's optical radius (Allart *et al.* (2019)). A number of He I 10830 Å immediately followed, including for HAT-P-11 b (Allart *et al.* (2018)), HD 189733 b (Salz *et al.* (2018)), WASP-69 b (Nortmann *et al.* (2018)), and HD 209458 b (Alonso-Floriano *et al.* (2019)).

Most spectacularly, the ultra-hot Jupiter KELT-9 b, which has an equilibrium temperature of $\approx 4000K$ (Gaudi *et al.* (2017)), displays absorption in a number of neutral and singly ionized metals, including Mg I (Figure 2), Fe I and Fe II, Ti II, Cr II, Sc II, and Y II (Hoeijmakers *et al.* (2018); Cauley *et al.* (2019); Hoeijmakers *et al.* (2019)). KELT-9 b also shows prominent Balmer line absorption and is estimated to be losing mass at a rate of $\approx 10^{12}$ g s^{-1} (Yan & Henning (2018); Cauley *et al.* (2019)).

4. Stellar activity and transmission spectra

When looking for atmospheric absorption by strong atomic lines, it is natural to ask whether or not active regions on the stellar surface could be producing the signals rather than the planet's atmosphere. This is possible due to the nature of transmission spectroscopy: during transit the planet occults different portions of the stellar disk at various times during the transit. If the planet occults an active region that is brighter or emits

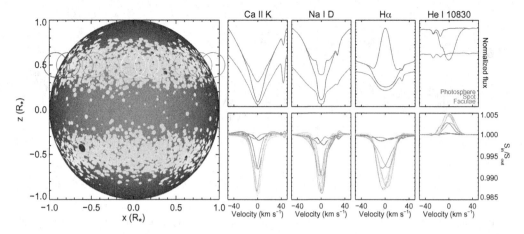

Figure 3. Simulated hot Jupiter transit of a very active stellar surface (left) and the resulting in-transit spectra for Ca II H and K, Na I D, Hα, and He I 10830 Å(bottom right). Transiting active regions can produce signals that mimic absorption in the planet's atmosphere, although the planet must continuously transit strongly emitting active regions to produce a consistent signal.

a different spectrum than the rest of the stellar disk, the difference will manifest in the transmission spectrum as an absorption or emission feature with strength roughly comparable to $(R_p/R*)^2$.

This question was investigated by Cauley *et al.* (2018) specifically for hot Jupiters transiting active stars (see Rackham *et al.* (2018) for a similar exercise involving rocky planets transiting M-dwarfs). They found that in order to produce the strength of the observed transmission spectra for systems like HD 189733 that the planet needs to transit almost directly across an active latitude and that the strength of the chromospheric emission features needs to be similar to the intensity produced by a moderate solar flare. Figure 3 shows an example of a hot Jupiter transiting a very active stellar latitude. Notably, the observed He I 10830 Å absorption signatures cannot be caused by active region transits since He I 10830 Å is in *absorption* in normal stellar chromospheres and thus would produce an emission feature when occulted. Thus stellar activity likely contributes to Balmer, Na I D, Ca II, and He I signals but the specific geometries required make it unlikely that the absorption is due entirely to occulted active regions.

5. Atmospheric dynamics from high-resolution spectra

One of the main advantages high-resolution observations have over low-resolution observations is the ability to resolve velocity features in the transmission spectrum. Hot Jupiters are expected to have extremely powerful winds and jets in their upper atmospheres with velocities on the order of a few kilometers per second Rauscher & Kempton (2014). Even though they are likely tidally liked, the equatorial rotational velocities of hot Jupiters are also on the order of a few kilometers per second. The combination of these strong winds and rotation can shift and broaden line profiles, which can then be extracted using relatively simple models of the atmospheric dynamics.

The first measurement of a hot Jupiter's wind speed was reported by Snellen *et al.* (2010) who showed that the CO signal in the planet's atmosphere was blue-shifted by ≈ 2 km s^{-1}, indicative of day-to-night side winds in the lower thermosphere. Since that pioneering study, only a handful of dynamics signatures have been published in the literature. Among them are jet and wind speeds for HD 189733 b (Louden & Wheatley (2015); Brogi *et al.* (2016); Cauley *et al.* (2017a)), rotation for KELT-9 b (Cauley *et al.* (2019)), and wind speeds and atmospheric expansion for HAT-P-11 b (Allart *et al.* (2018)).

One of the limiting factors in obtaining a reliable measurement of velocity shifts and line broadening in a transmission spectrum is the signal-to-noise of the transmission spectrum itself. In-transit variability, such as active region transits, can also confuse the velocity centroids. Confident velocity measurements from individual in-transit spectra can only be arrived at using exposures from 10-meter class telescopes (see Cauley *et al.* (2017a) and Cauley *et al.* (2019) for examples). The upcoming era of extremely large telescopes and ultra-stable spectrographs will usher in a new age of atmospheric dynamics measurements for exoplanets by greatly expanding the number of systems accessible to such experiments and increasing the precision of the in-transit observations.

6. Asymmetries due to circumplanetary material

One of the defining features of hot Jupiters is their potentially large mass loss rates. Although not high enough to evaporate a significant fraction of the atmosphere on evolutionary timescales, most hot Jupiters lose enough material to produce observable exospheres. These highly extended atmospheres can produce fantastic transit signals that show striking asymmetric absorption, i.e., absorption strength that changes as a function of time. Unfortunately, the most useful atomic line to measure the exosphere is Lyman-α at 1216 Å and must be observed using a space-based spectrograph. The only instruments currently available to perform these experiments are STIS and COS on the *Hubble Space Telescope*. The limited amount of available *HST* time and the low signal-to-noise at Lyman-α has prevented the detection of a large number of hot Jupiter exospheres.

The most famous of the detected exosphere is that of the hot Neptune GJ 436 b. Ehrenreich *et al.* (2015) reported an enormous transit depth of $\approx 40\%$ in the wings of the Lyman-α line. The transit in Lyman-α began about three hours before the optical transit and continued out to $\approx$30 hours after the optical transit. This incredible feature is nicely explained by a cometary tail of hydrogen atoms that have escaped from the planet and are being swept away from the star by radiation pressure and interactions with the stellar wind. Although not as spectacular, hydrogen exospheres have also been measured around GJ 3470 b (Bourrier *et al.* (2018)), HD 209458 b (Vidal-Madjar *et al.* (2003)), and HD 189733 b (Lecavelier des Etangs *et al.* (2010)).

Asymmetric transit features have also been noted for HD 189733 b using the Balmer lines (Cauley *et al.* (2015, 2016)) and WASP-12 b in the near-UV Mg II doublet (Fossati *et al.* (2010)). Both systems showed an early ingress, suggestive of material orbiting ahead of the planet and crossing the stellar disk before the optical transit. Although the Balmer line signal for HD 189733 b may be due to variability in the stellar activity level, there is evidence that such changes only occur near a planetary transit (Cauley *et al.* (2017b)). Thus the pre-transit signals are likely due to planetary material or some manifestation of a star-planet interaction. WASP-12 b similarly shows variability in the pre-transit absorption (Nichols *et al.* (2015)). Further investigation is needed to understand the nature of these interesting signals.

7. Summary and future prospects

The race has now begun to detect the imprint of biologically generated gasses in the atmospheres of rocky planets orbiting M-dwarf stars. This is one of the primary exoplanet goals of the *James Webb Space Telescope*. It is important to keep in mind, however, that hot Jupiters still have much to teach us concerning exoplanet atmospheres and, unlike small rocky planets, we have no solar system analog to examine for clues. Such a wonderful class of planets deserves continuing study.

The next generation of extremely large ground-based telescopes will greatly expand the number of transiting hot Jupiter systems for which transmission spectra can be obtained. This will evenutally allow population-level statistics to be obtained on things such as hot

Jupiter rotation, wind speeds, atmospheric structure, and chemistry. Although we are still in the discovery phase of exoplanet atmospheric characterization, the future is bright and all roads lead to a deeper statistical understanding of these amazing worlds.

References

Allart, R. *et al.* 2018, *Science*, 362, 1384
Allart, R. *et al.* 2019, *A&A*, 623, 58
Alonso-Floriano, F. J. *et al.* 2019, *A&A*, 629, 7
Bourrier, V. *et al.* 2018, *A&A*, 620, 147
Brogi, M. *et al.* 2016, *ApJ*, 817, 106
Casasayas-Barris, N. *et al.* 2019, *A&A*, 628, 9
Cauley, P. W. *et al.* 2015, *ApJ*, 810, 13
Cauley, P. W. *et al.* 2016, *AJ*, 152, 20
Cauley, P. W. *et al.* 2017, *AJ*, 153, 217
Cauley, P. W. *et al.* 2017, *AJ*, 153, 185
Cauley, P. W. *et al.* 2018, *AJ*, 156, 189
Cauley, P. W. *et al.* 2019, *AJ*, 157, 69
Charbonneau, D. *et al.* 2000, *ApJ*, 529, 45
Charbonneau, D. *et al.* 2002, *ApJ*, 568, 377
Ehrenreich, D. *et al.* 2015, *Nature*, 522, 459
Fossati, L. *et al.* 2010, *ApJ*, 714, 222
Gaudi, B. S. *et al.* 2017, *Nature*, 546, 514
Hoeijmakers, H. J. *et al.* 2018, *Nature*, 560, 453
Hoeijmakers, H. J. *et al.* 2019, *A&A*, 627, 165
Jensen, A. G. *et al.* 2012, *ApJ*, 751, 86
Jensen, A. G. *et al.* 2018, *AJ*, 156, 154
Keles, E. *et al.* 2019, *MNRAS*, 489, 37
Khalafinejad, S. *et al.* 2017, *A&A*, 598, 131
Lecavelier des Etangs, A. *et al.* 2010, *A&A*, 514, 72
Lippincott, S. L. 1977, *AJ*, 82, 925
Louden, T. & Wheatley, P. J. 2015, *ApJ*, 814, 24
Mayor, M. & Queloz, D.
Nichols, J. D. *et al.* 2015, *ApJ*, 803, 9
Nortmann, L. *et al.* 2018, *Science*, 362, 1388
Oklopčić, A. & Hirata, C. M. 2018, *ApJ*, 855, 11
Rackham, B. V. *et al.* 2018, *ApJ*, 853, 122
Rauscher, E. & Kempton, E. M. R. 2014, *ApJ*, 790, 79
Redfield, S. *et al.* 2008, *ApJ*, 673, 87
Salz, M. *et al.* 2018, *A&A*, 620, 97
Seidel, J. V. *et al.* 2019, *A&A*, 623, A166
Sing, D. K. *et al.* 2016, *Nature*, 529, 59
Snellen, I. A. G. *et al.* 2008, *A&A*, 487, 357
Snellen, I. A. G. *et al.* 2010, *Nature*, 456, 1049
Spake, J. J. *et al.* 2018, *Nature*, 557, 68
Thorngren, D. P. & Fortney, J. J. 2018, *AJ*, 155, 214
Vidal-Madjar, A. *et al.* 2003, *Nature*, 422, 143
Winglee, R. M., Dulk, G. A., & Bastian, T. S. 1986, *ApJL*, 309, L59
Winn, J. N. & Fabrycky, D. C. 2015, *ARA&A*, 53, 409
Wyttenbach, A. *et al.* 2015, *A&A*, 577, 62
Yan, F. & Henning, T. 2018, *Nature Astronomy*, 2, 714
Yantis, W. F., Sullivan, W. T., III, & Erickson, W. C. 1977, *Bulletin of the American Astronomical Society*, 9, 453

Solar and Stellar Magnetic Fields: Origins and Manifestations
Proceedings IAU Symposium No. 354, 2019
A. Kosovichev, K. Strassmeier & M. Jardine, ed.
doi:10.1017/S1743921319010019

Sun-as-a-star observations of the 2017 August 21 solar eclipse

Ekaterina Dineva[1,2], Carsten Denker[1], Meetu Verma[1], Klaus G. Strassmeier[1,2], Ilya Ilyin[1] and Ivan Milic[3]

[1]Leibniz-Institut für Astrophysik Potsdam (AIP), An der Sternwarte 16,
14482 Potsdam, Germany
email: **edineva@aip.de**

[2]Universität Potsdam, Institut für Physik und Astronomie, Karl-Liebknecht-Str. 24/25,
14476 Potsdam, Germany

[3]University of Colorado Boulder, Laboratory for Atmospheric and Space Physics,
1234 Innovation Drive, Boulder, CO 80303-7814, U.S.A.

Abstract. The Potsdam Echelle Polarimetric and Spectroscopic Instrument (PEPSI) is a state-of-the-art, thermally stabilized, fiber-fed, high-resolution spectrograph for the Large Binocular Telescope (LBT) at Mt. Graham, Arizona. During daytime the instrument is fed with sunlight from the 10-millimeter aperture, fully automated, binocular Solar Disk-Integrated (SDI) telescope. The observed Sun-as-a-star spectra contain a multitude of photospheric and chromospheric spectral lines in the wavelength ranges 4200–4800 Å and 5300–6300 Å. One of the advantages of PEPSI is that solar spectra are recorded in the exactly same manner as nighttime targets. Thus, solar and stellar spectra can be directly compared. PEPSI/SDI recorded 116 Sun-as-a-star spectra during the 2017 August 21 solar eclipse. The observed maximum obscuration was 61.6%. The spectra were taken with a spectral resolution of $\mathcal{R} \approx 250\,000$ and an exposure time of 0.3 s. The high-spectral resolution facilitates detecting subtle changes in the spectra while the Moon passes the solar disk. Sun-as-a-star spectra are affected by changing contributions due to limb darkening and solar differential rotation, and to a lesser extend by supergranular velocity pattern and the presence of active regions on the solar surface. The goal of this study is to investigate the temporal evolution of the chromospheric Na D doublet during the eclipse and to compare observations with synthetic line profiles computed with the state-of-the-art Bifrost code.

Keywords. Sun: chromosphere, instrumentation: spectrographs, techniques: spectroscopic, methods: data analysis

1. Introduction

Data collected during solar eclipse observations provide valuable input for studies of the outer atmosphere of the Sun including the chromosphere (Athay 1976). Its name originates from the red flash of Hα emission, which appears within the totality phase. The chromosphere is an intermediate layer between the relatively cool photosphere and the million-degree corona. The physical boundaries of this intermediate layer are in the range 500–5000 km above the solar surface (Vernazza *et al.* 1981; Rutten 2007) but the chromosphere is highly structured, both horizontally and vertically, e.g., spicules, mottles, rosettes, and various kinds of filaments, so that an exact definition of its borders is difficult and maybe not even appropriate. Aimanova & Gulyaev (1976) derived the formation height among other height-dependent properties of the Na I D doublet, using slitless spectrograms of the 1976 July 10 total solar eclipse over Chukotka, Russia.

The obscuration of the solar disk by the Moon presents a unique setting for researchers to address a broad range of topics (Pasachoff 2009). Major parts of eclipse research concern the corona (e.g., Saito & Tandberg-Hanssen 1973; Habbal *et al.* 2011; Koutchmy *et al.* 2019). However, observations in the visible spectrum, as well as in the adjacent infrared (IR), ultraviolet (UV), and extreme ultraviolet (EUV) bands, are popular in solar physics (e.g., Dunn *et al.* 1968; Bazin & Koutchmy 2013), solar-stellar connection studies (Schleicher *et al.* 2004), and exoplanet research (e.g., Takeda *et al.* 2015; Reiners *et al.* 2016). Albeit rarely in the focus of eclipse observations, manifestations of magnetic activity are interesting because they affect the radial velocity measurements of active stars (Haywood *et al.* 2016; Cauley *et al.* 2018, e.g.,). Although the Sun is not an extremely active star, only the solar case delivers detailed knowledge about the evolution and properties of the active regions (e.g., Verma *et al.* 2016; Verma 2018). Supported by models of stellar atmospheres (Beeck *et al.* 2012), the above studies have two significant outcomes, i.e., an improved understanding of the underlying physical processes and additional input information to increase the precision of models.

The chromosphere is highly dynamics and feature-rich. Observing the Sun in various strong chromospheric absorption lines, e.g., Hα, Ca II K & H, and the infrared Ca II triplet, allows us to investigate the fine structure as well as the global behaviour of the solar chromosphere. The solar alkali spectrum, in particular Sodium Na I D λ5890 Å and Potassium K I λ7699 Å, gain significant popularity in solar and stellar research (e.g., Nikolov *et al.* 2016; Lendl *et al.* 2017). For example, conveniently used in helioseismology (Bruls *et al.* 1992), the Na I D doublet offers diagnostics tools linking photospheric and lower chromosphere dynamics (Leenaarts *et al.* 2010).

During the 2012 May 21 solar eclipse, Takeda *et al.* (2015) obtained 184 disk-integrated spectra of with the High Dispersion Echelle Spectrograph (HIDES, Izumiura 1999)) at the Okayama Astrophysical Observatory, Japan. The high-resolution spectra ($\mathcal{R} \approx 520\,000$) covered the wavelength range 4400–7500 Å. An exposure time of 10 s made it possible to trace the eclipse with very good temporal resolution. Deriving the latitude dependence of the solar differential rotation was the primary goal of this study. Thereby, the Sun served as a template that can be utilized in the analysis of stellar radial velocity curves, deriving sensitive information about stellar systems, star-planet interactions, and eclipsing binaries and their rotational parameters.

In the spirit of the aforementioned study, Reiners *et al.* (2016) investigated the the properties of disk-integrated photospheric absorption line spectra observed during the 2015 March 20 solar eclipse. Employing a Fourier Transform Spectrometer (FTS, Davis *et al.* 2001) at the 50-cm Vacuum Vertical Telescope (VVT) of the Institut für Astrophysik, Göttingen (IAG), they obtained 159 disk-integrated spectra of the eclipse in two spectral ranges, i.e., 5000–6200 Å and 6500–6700 Å. Similar to the observations by Takeda *et al.* (2015), the Göttingen eclipse was not on the path of totality, which is however irrelevant in the context of exoplanet research. Reiners *et al.* (2016) analyzed the velocity curves computed from the eclipse spectra and quantified all solar contributions. Thus, the crucial impact of the convective blueshift was revealed and the need for more realistic models was demonstrated.

The studies by Takeda *et al.* (2015) and Reiners *et al.* (2016) motivated our investigation of the Na I D doublet during the 2017 August 21 solar eclipse and the modeling of these strong chromospheric absorption lines based on magneto-hydrodynamics (MHD) simulations. Additional impetus arises from the strong interest in the behavior of the Fraunhofer lines within the stellar and exoplanet research communities (Czesla *et al.* 2015), in particular provoked by the Na I D abundance in the atmospheres of hot Jupiter planets (Yan *et al.* 2015).

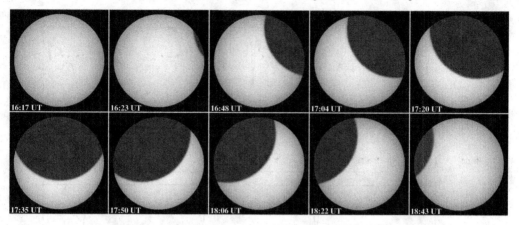

Figure 1. Full-disk broad-band images observed with the SDI guiding telescope during the 2017 August 21 solar eclipse. Active regions NOAA 12671 and 12672 are visible near disk center and at the eastern limb, respectively.

2. Observations and data reduction

The 2017 August 21 total solar eclipse was only visible as a partial eclipse at Mt. Graham Observatory, Arizona, U.S.A. The partial eclipse started at 16:17 UT and ended at 19:06 UT, while the maximum obscuration (61.6%) took place at 17:38 UT, i.e., the elevation of the Sun rose from 42.6° to 68.8° during the eclipse, which corresponds to an air mass of 1.48 at the beginning of the eclipse and about unity at its conclusion. More details of solar eclipses can be obtained from the Solar Eclipse Computer (SEC) provided by the U.S. Naval Observatory (Bartlett *et al.* 2019). Due to technical problems with the telescope drive between 16:23 and 16:49 UT, a chunk of data is missing. Increasing overcast towards the end of the eclipse terminated observations after 18:43 UT. Overall, however, the weather and seeing conditions were favorable during the eclipse.

The Potsdam Echelle Polarimetric and Spectroscopic Instrument (PEPSI, Strassmeier *et al.* 2015, 2018) was designed to record stellar spectra with a spectral resolution of up to $\mathcal{R} \approx 250\,000$, exploiting the light gathering capability of the 2×8.4-meter diameter Large Binocular Telescope (LBT, Hill *et al.* 2006). The 10-millimeter diameter Solar-Disk Integrated (SDI) telescope is located on the LBT's kitchen balcony. It observes the Sun as a star, feeding a pair of 300 μm-core fibers that guide the light to the blue and red arm of the spectrograph, while a small full-disk telescope provides accurate guiding. Selected full-disk images tracing the evolution of the eclipse are shown in Figure 1. The time-series contains more than 800 full-disk images at a cadence of 10 s. Two small active regions (NOAA 12671 and 12672) were present on the disk, one at disk center and one near the eastern limb, which were successively concealed by the passing Moon.

In a dedicated observing campaign during the solar eclipse, 116 high-resolution spectra were obtained with PEPSI/SDI, covering the two spectral spectral windows 4200–4800 Å and 5300–6300 Å with an average dispersion of 2 mÅ. An exposure time of 0.3 s followed by a CCD readout time of 40–60 s, yields a cadence of approximately 60 s between consecutively acquired observation. The average signal-to-noise (S/N) ratio is approximately 700:1 for a single exposure. Stacking spectra to improve the S/N ratio is not an option because a good temporal resolution is needed to track the evolution of the eclipse.

In the present study, we investigate the temporal evolution of the solar Na I D doublet during the eclipse. Facilitated by the eclipse geometry, the 116 high-resolution spectra provide snapshots with different contributions from solar differential rotation and limb darkening. In the top panel of Figure 2, we display a "quiet-Sun" Na I D spectrum $I_{\mathrm{qS}}(\lambda)$,

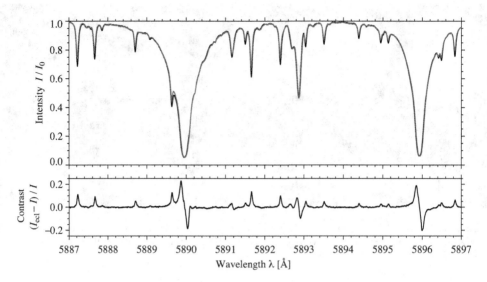

Figure 2. Sun-as-a-star intensity profiles of the chromospheric Na I D doublet (*top*). This spectral range was extracted from a PEPSI/SDI spectrum covering the spectral window 5300–6300 Å. A spectrum $I(\lambda)$ before the eclipse (*blue*) is compared with a spectrum $I_{\text{ecl}}(\lambda)$ recorded at 18:06 UT (*red*), i.e., close to the culmination of the eclipse. Both spectra were normalized so that the continuum intensity I_0 corresponds to unity. The contrast profile was computed according to Equation 2.1 and exhibits signals of up to 20% of the local intensity.

which was recorded just before the onset of the eclipse. An eclipse Na I D spectrum $I_{\text{ecl}}(\lambda)$ is shown for comparison. The differences are small and hard to detect by visual inspection. Some telluric lines exhibit a variation of line depth during the eclipse, which can be attributed to the decreasing air mass. The sample eclipse spectrum was selected 27 min after the culmination of the eclipse because the eclipse geometry will produce a spectrum at maximum obscuration that is almost indistinguishable from the quiet-Sun profile, besides the lower intensity. Computing contrast profiles (see bottom panel of Figure 2) will enhance and quantify the eclipse signal according to

$$C(\lambda,\,t) = \frac{I_{\text{ecl}}(\lambda,\,t) - I_{\text{qS}}(\lambda)}{I_{\text{qS}}(\lambda)}. \tag{2.1}$$

After the culmination of the eclipse, the shape of contrast profile resembles that of the derivative of a Gaussian, with the positive lobe at shorter wavelengths. The peak-to-valley (PTV) contrast difference is 0.41. Before the eclipse both lobes change sign. An inclination of the Moon's path with respect to solar North-South will lead to an asymmetry of minima and maxima of the contrast profiles and also affect the wings.

Despite the minimum of Solar Cycle 24, two small active regions were present, which will however not significantly affect the spectral profiles. The Moon started to cover the spots of active region NOAA 12671 at 16:45 UT, completely covered them at 16:48 UT, and uncovered them at 18:15 UT. Due the location of active region NOAA 12672 near the eastern limb, the first spot was covered at 17:41 UT. Observing its reemergence was obstructed by clouds. The presence of sunspot groups facilitated the proper alignment of the images taken by SDI, which in turn confirmed the eclipse geometry. The impact of active regions was neglected in the later comparison with the contrast profiles derived from Bifrost simulations (Section 3). In principal, the time-dependent supergranular velocity pattern will also affect the spectral profile but in the comparison, we

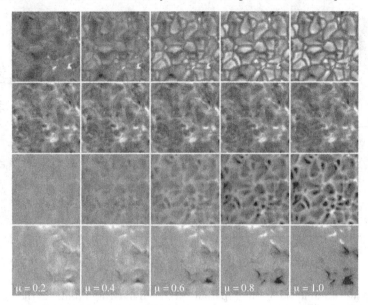

Figure 3. Center-to-limb variation (*left to right*) of a network element in the quiet Sun based on a Bifrost simulation. Displayed are emergent continuum intensity, line-core intensity, center-of-gravity velocity, and maximum degree of circular polarization (*top to bottom*), which were derived from the strong chromospheric absorption line $Na\,D_2$. The intensities are scaled for clarity, and the line-core intensity is shown on a logarithmic scale. The velocity is scaled between ± 2 km s^{-1} and the degree of polarization between $\pm 20\%$.

limited the velocity distribution across the disk to solar differential rotation including the accurate B_0-angle (heliographic latitude of the central point of the solar disk).

3. Contrast profiles derived from Bifrost simulations

Bifrost is a state-of-the-art parallel numerical code for MHD simulations of stellar atmospheres from the convection zone to the corona (Gudiksen *et al.* 2011). The version employed in this study builds on many generations of numerical codes, e.g., the Oslo Stagger code. Five spatio-spectral data cubes of the $Na\,I\,D$ doublet describe the CLV of a quiet-Sun region containing a strong magnetic network element. The cosine of the heliocentric angle corresponds to $\mu = 0.2$, 0.4, 0.6, 0.8, and 1.0. The full Stokes vector is available for each of the 200×200 pixels. The Stokes profiles are sampled at 1 mÅ covering a 10 Å-wide spectral window in the range 5887–5897Å. The simulated spectra contain the $Na\,I\,D$ doublet as well as the photospheric $Fe\,I$ $\lambda 5892.7$ Å and $Ni\,I$ $\lambda 5892.9$Å lines. In Figure 3, we present the CLV of four physical parameters describing the $Na\,I\,D_2$ line, i.e., emergent continuum intensity, line-core intensity, line-of-sight (LOS) velocity, and degree of circular polarization. The field-of-view (FOV) contains a strong magnetic network element in the lower right corner embedded in a quit-Sun background. Ideally, such simulations would blanket the entire solar surface. This, however, will require immense computational efforts and does not take into account the three-dimensional nature of an atmosphere in a spherical shell encompassing the entire Sun. Thus, some simplifications are needed.

The goal is to model the PEPSI/SDI Sun-as-a-star spectra of the eclipse with those derived from Bifrost simulations. In a first step, the 200×200 Stokes-I spectra are averaged for each of the five spatio-spectral data cubes. Our model Sun has a diameter of 1024 pixels, and for each pixel on the disk a spectrum is needed, which is appropriate

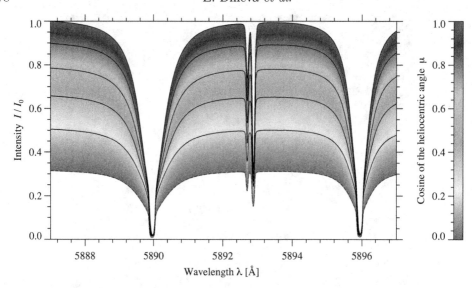

Figure 4. Center-to-limb variation of spatially averaged intensity profiles of the Na I D doublet, which is based on the Bifrost simulation shown in Figure 3. The initial five profiles (*solid lines*) correspond to a different value of the cosine of the heliocentric angle $\mu = 0.2$, 0.4, 0.6, 0.8, and 1.0. The $\mu = 0.0$ profile (*dashed line*) is a scaled version of the $\mu = 0.2$ profile adjusted for the CLV of the continuum intensity. The six intensity profiles of the Na I D doublet are the basis for an interpolation to an equidistant grid in μ with a resolution of $\delta\mu = 0.001$. The intensity profiles of the Na I D doublet are plotted with increasing distance from disk center (*violet to red*).

for a given μ and the Doppler shift due to solar differential rotation at the given location. Active regions and the imprint of the supergranular velocity pattern are currently neglected. Thus, in a second step, the five average Na I D profiles were interpolated to an equidistant grid in μ with a grid spacing of $\delta\mu = 0.001$. Since no Na I D profiles are available for $\mu < 0.2$, we just scaled the profile for $\mu = 0.2$ with an appropriate CLV value of the continuum intensity. The results of this interpolation step are summarized in Figure 4.

Finally, we computed disk-integrated spectra for the eclipse with a time step of 1 min using the exact eclipse geometry at the Mt. Graham Observatory. As mentioned in Section 2, the differences are better seen in contrast profiles. A disk-integrated quiet-Sun profile without any obscuration served as the reference profile. Each disk-integrated eclipse spectrum was normalized such that the continuum intensity is unity. In principle, model light curves can be computed for continuum and line-core intensity. Signals in the contrast profiles are detected at the 10^{-5} level already one minute after the start of the eclipse, which is comparable to the signal from a Mercury transit. The results of this modeling effort are plotted in Figure 5 for every second computed contrast profile. The red-blue color coded spectra vary smoothly from the start to the end of the eclipse. The maximum of the PTV contrast difference of 0.63 occurs at 18:24 UT, which is about 50% higher and occurs 18 min later than in the observations. The shape of the contrast profiles resembles that of a derivative of a Gaussian, switching sign after the culmination of the eclipse. The asymmetries in the lobes and the wings of the contrast profiles are a direct consequence of the eclipse geometry, breaking symmetries related to the CLV and solar differential rotation. In summary, we demonstrated successfully that forward modeling using the Bifrost simulations produces reasonable disk-integrated contrast profiles of the 2017 August 21 eclipse but a more detailed comparison will be carried out in a forthcoming study.

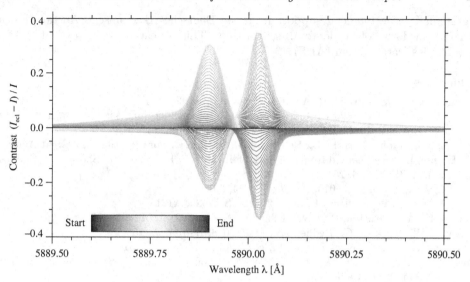

Figure 5. Series of contrast profiles covering the 2017 August 21 solar eclipse. The color scale indicate the progress of the partial solar eclipse from start to end as observed at the Mt. Graham Observatory. The contrast profiles were derived from a Bifrost simulation of the CLV of the $Na \, I \, D_2$ intensity profile.

4. Conclusion

The Sun is the only star, where we can obtain high-resolution spectra and spatially resolved images. This allows us to study closely and to quantify with formidable precision phenomena such as convection, magnetic fields, solar activity, and differential rotation. Their detailed examination provides a validation environment connecting solar and stellar physics. Recent advances were made possible by state-of-the-art instruments, facilities, and experiments such as the Synoptic Optical Long-term Investigations of the Sun (SOLIS, Keller *et al.* 2003) program, High Accuracy Radial velocity Planet Searcher (HARPS, Pepe *et al.* 2002), the Calar Alto high-Resolution search for M-dwarfs with Exoearths with Near-infrared and optical Echelle Spectrographs (CARMENES, Quirrenbach *et al.* 2018), and last but not least PEPSI (Strassmeier *et al.* 2015) The last three instruments exploit during nighttime a spectral resolution that rivals high-resolution solar spectroscopy, thus, facilitating close spectral comparisons between the Sun and the stars. Synoptic full-disk data form space missions such as the Solar Dynamic Observatory (SDO, Pesnell *et al.* 2012) enable us to relate spectral signatures from Sun-as-a-star observations to activity and dynamics on the solar disk (e.g., Denker & Verma 2019). In this context, the present study of the 2017 August 21 eclipse serves as an example demonstrating the power of high-resolution Sun-as-a-star spectroscopy.

Acknowledgments

PEPSI and SDI were made possible by longtime support of the Bundesministerium für Bildung und Forschung (BMBF) for the collaborative research projects 05AL2BA1/3 and 05A08BAC. This study was supported by grant DE 787/5-1 of the Deutsche Forschungsgemeinschaft (DFG). The support by the European Commission's Horizon 2020 Program under grant agreements 824064 (ESCAPE – European Science Cluster of Astronomy & Particle physics ESFRI research infrastructures) and 824135 (SOLARNET – Integrating High Resolution Solar Physics) is highly appreciated. ED is grateful for the financial support from the Deutsche Akademische Austauschdienst (DAAD) in form of a doctoral scholarship. This research made use of NASA's Astrophysics Data System (ADS).

References

Aimanova, G. K. & Gulyaev, R. A. 1976, *Soviet Astron.*, 20, 201

Athay, R. G. 1976, The Solar Chromosphere and Corona: Quiet Sun, *Astrophys. Space Sci. Lib.*, 53, D. Reidel Publishing Company, Dordrecht, Holland

Bartlett, J. L., Frouard, M. R. C., Bell, S., *et al.* 2019, in Celebrating the 2017 Great American Eclipse: Lessons Learned from the Path of Totality, S. R. Buxner, L. Shore, & J. B. Jensen (eds.), *ASP-CS*, 516, 251

Bazin, C. & Koutchmy, S. 2013, *J. Adv. Res.*, 4, 307

Beeck, B., Collet, R., Steffen, M., *et al.* 2012, *A&A*, 539, A121

Bruls, J. H. M. J., Rutten, R. J., & Shchukina, N. G. 1992, *A&A*, 265, 237

Cauley, P. W., Kuckein, C., Redfield, S., *et al.* 2018, *Astron. J.*, 156, 189

Czesla, S., Klocová, T., Khalafinejad, S., *et al.* 2015, *A&A*, 582, A51

Davis, S. P., Abrams, M. C., & Brault, J. W. 2001, Fourier Transform Spectrometry, Academic Press, San Diego, California

Denker, C. & Verma, M. 2019, *SoPh*, 294, 71

Dunn, R. B., Evans, J. W., Jefferies, J. T., *et al.* 1968, *ApJSS*, 15, 275

Gudiksen, B. V., Carlsson, M., Hansteen, V. H., *et al.* 2011, *A&A*, 531, A154

Habbal, S. R., Druckmüller, M., Morgan, H., *et al.* 2011, *ApJ*, 734, 120

Haywood, R. D., Collier Cameron, A., Unruh, Y. C., *et al.* 2016, *MNRAS*, 457, 3637

Hill, J. M., Green, R. F., & Slagle, J. H. 2006, in Ground-based and Airborne Telescopes, L. M. Stepp (ed.), *Proc. SPIE*, 6267, 62670Y

Izumiura, H. 1999, in Observational Astrophysics in Asia and its Future, P. S. Chen (ed.), 4, 77

Keller, C. U., Harvey, J. W., & Giampapa, M. S. 2003, in Innovative Telescopes and Instrumentation for Solar Astrophysics, S. L. Keil & S. V. Avakyan (eds.), *Proc. SPIE*, 4853, 194

Koutchmy, S., Baudin, F., Abdi, S., *et al.* 2019, *A&A*, in press

Leenaarts, J., Rutten, R. J., Reardon, K., *et al.* 2010, *ApJ*, 709, 1362

Lendl, M., Cubillos, P. E., Hagelberg, J., *et al.* 2017, *A&A*, 606, A18

Nikolov, N., Sing, D. K., Gibson, N. P., *et al.* 2016, *ApJ*, 832, 191

Pasachoff, J. M. 2009, *Nature*, 459, 789

Pepe, F., Mayor, M., Rupprecht, G., *et al.* 2002, *Messenger*, 110, 9

Pesnell, W. D., Thompson, B. J., & Chamberlin, P. C. 2012, *SoPh*, 275, 3

Quirrenbach, A., Amado, P. J., Ribas, I., *et al.* 2018, in Ground-based and Airborne Instrumentation for Astronomy VII, C. J. Evans, L. Simard, & H. Takami (eds.), *Proc. SPIE*, 10702, 107020W

Reiners, A., Lemke, U., Bauer, F., *et al.* 2016, *A&A*, 595, A26

Rutten, R. J. 2007, in The Physics of Chromospheric Plasmas, P. Heinzel, I. Dorotovič, & R. J. Rutten (eds.), *ASP-CS*, 368, 27

Saito, K. & Tandberg-Hanssen, E. 1973, *SoPh*, 31, 105

Schleicher, H., Wiedemann, G., Wöhl, H., *et al.* 2004, *A&A*, 425, 1119

Strassmeier, K. G., Ilyin, I., Järvinen, A., *et al.* 2015, *AN*, 336, 324

Strassmeier, K. G., Ilyin, I., & Steffen, M. 2018, *A&A*, 612, A44

Takeda, Y., Ohshima, O., Kambe, E., *et al.* 2015, *PASJ*, 67, 10

Verma, M. 2018, *A&A*, 612, A101

Verma, M., Denker, C., Balthasar, H., *et al.* 2016, *A&A*, 596, A3

Vernazza, J. E., Avrett, E. H., & Loeser, R. 1981, *ApJSS*, 45, 635

Yan, F., Fosbury, R. A. E., Petr-Gotzens, M. G., *et al.* 2015, *A&A*, 574, A94

Solar and Stellar Magnetic Fields: Origins and Manifestations
Proceedings IAU Symposium No. 354, 2019
A. Kosovichev, K. Strassmeier & M. Jardine, ed.
doi:10.1017/S1743921320004068

Solar astrometry with planetary transits

Marcelo Emilio[1,2,3]🄳, **Rock Bush**[4], **Jeff Kuhn**[3] and **Isabelle Scholl**[3]

[1]Universidade Estadual de Ponta Grossa, 84030-900 Ponta Grossa, PR, Brazil

[2]Observatório Nacional, MCTIC, 20921-400 Rio de Janeiro, RJ, Brazil
email: `memilio@uepg.br`

[3]Institute for Astronomy, University of Hawaii, Maui, HI 96768, USA
emails: `kuhn@ifa.hawaii.edu`, `ifscholl@hawaii.edu`

[4]Stanford University, Stanford, CA, 94305, USA
email: `rock@sun.stanford.edu`

Abstract. Planetary transits are used to measure the solar radius since the beginning of the 18th century and are the most accurate direct method to measure potentially long-term variation in the solar size. Historical measures present a range of values dominated by systematic errors from different instruments and observers. Atmospheric seeing and black drop effect contribute as error sources for the precise timing of the planetary transit ground observations. Both Solar and Heliospheric Observatory (SOHO) and Solar Dynamics Observatory (SDO) made observations of planetary transits from space to derive the solar radius. The International Astronomical Union approved the resolution B3 in 2015, defining a nominal solar radius of precisely 695,700 km. In this work, we show that this value is off by more than 300 km, which is one order of magnitude higher than the error of the most recent solar radius observations.

Keywords. Sun: fundamental parameters, Sun: photosphere, Sun: activity

---

1. Introduction

The solar radius in theoretical models is defined as the photospheric region where optical depth is equal to the unity. In practice, helioseismic inversions determine this point using f-mode analysis, but most experiments which measure the solar radius optically use the inflection point of the Limb Darkening Function (LDF) as the definition of the solar radius. The stellar photosphere is defined as the layer from which its visible light originates, that is, where the optical depth is two-thirds in the star's continuum, since this is the average level in the atmosphere from which photons escape. Measurements of the solar radius found in literature varies from 958".54 Sánchez (1995) to 960".62 Wittmann (2003). Systematic errors among the experiments are the explanation of those differences. Planetary transits technique can only affected by second-order systematic errors and it provides an independent way to measure the plate scale. Space-based observations have the advantage of being not dependent on the Earth's Atmospheric error sources in the measurements. Emilio *et al.* (2012) measured the solar radius with the Michelson Doppler Imager (MDI) aboard the Solar and Heliospheric Observatory during the 2003 and 2006 mercury transits. The value found was 960".12 ± 0".09 (696,342 ± 65 km). In 2012 during the Venus transit the value found was 959".57 ± 0".02 (695,946 ± 15 km) with the Helioseismic and Magnetic on board the Solar Dynamics Observatory Emilio *et al.* (2015). Section 2 discusses historical and methods to measure the solar radius, including planetary transits. Section 3 compares modern measurements of the Solar Radius with International Astronomical Union Resolution B3, and finally, in section 4, we discuss why this value should be changed.

2. Historical Solar Radius Measurements

The Greeks, around 270 B.C., made the first attempts to measure the Solar radius. The value of 900" was much later compared by Auwers (1891) and Ambronn & Schur (1905) measurements, using a heliometer. We point out that the result obtained by Auwers, subtracted from "irradiation correction," was 959".63, and the standard value for more than a century. Different authors analyzed sets of measurements of the solar diameter. Among them, Gilliland (1981) studying a data set, distributed over 258 years, such as meridian observations, Mercury transit, and Solar eclipses, evidenced the existence of an 11-year modulation, in addition to a variation of 76 years, in phase, with measurements using meridian circles and Mercury transit, with amplitudes of 0".2, high even in the face of dispersion. Toulmonde (1997), analyzing measurements obtained through solar eclipses and Mercury transit, compared to astrolabes, intends that the variations found are solely due to advances in precision, and therefore due to optical effects, improved with advancement observational and instrumental techniques. This contradicts similar analyzes conducted by Ribes, Ribes & Barthalot (1987) and Gilliland (1981), which show secular variations. However, none of these authors ruled out the possibility of fluctuations on smaller time scales.

2.1. *Micrometer measurements*

Louis XIV, King of France, founded the Royal Academy of Sciences in 1666 and authorized the Paris Observatory construction, thus giving Astronomy an integral part of the programs of that academy. The Sun's study and the orbital parameters of the Earth occupied a prominent place in the scientific works of the Paris Observatory. Jean Picard, a member of the academy, dedicated an essential part of his activities to the Sun's problems. Observing the solar diameter, he determined the eccentricity of the Earth's orbit. Moreover, with the sunspot movements, Picard measured the solar rotation. After he died in 1682, his student Philippe de La Hire continued his work, having the programs observed covered the Maunder minimum period. A seasonal variation was removed from the annual average obtained at one astronomical unit for the period between 1666 and 1719. This value is 1" higher when calculated during the Maunder period. Were observed few sunspots, as expected, but Picard measurements also show a loss of speed of rotation of the Sun at the equator and a higher number of sunspots in the southern hemisphere than in the solar north.

La Hire observed, with the same instrument, for more than four solar cycles, and his observations are compared to those of Halley (1715), made during the total solar eclipse. The measurements taken from the total eclipse were more accurate than those of La Hire at the time. Thus, Ribes, Ribes & Barthalot (1987) calibrated the measures of La Hire. It was necessary to subtract approximately 3" from his measurements to correspond to the measurements made by Halley. Even so, the value found in 1683 was 962".5, which is still about 2,000 km higher than the modern values of the diameter (Ribes *et al.* (1991)).

This value corresponds to 6 times the average deviation and 20 times higher than the 11-year variation found by Laclare *et al.* (1996). Between 1680 and 1690, a decade corresponding to the end of Maunder minimum, the diameter decreases by 3", similar to the value found by Halley in 1715. Ribes *et al.* (1991) concluded that an increase in the semi-diameter occurred in the Maunder minimum period due to low solar magnetic activity. The rotation speed found was 3% slower than the current speed. Morrison, Stephenson & Parkinson (1988) affirm that careful observations of the shadow edges of the total eclipse of 1715 imply that the value of the solar radius found is essentially the same observed today. Toulmonde (1997) concluded that there is no evidence of a secular variation after revising the analysis of several measurements. The analysis by Ribes *et al.* (1989) shows an oscillation of 9.6 years. This analysis included up of 7,000 measurements made by La Hire. Other fluctuations were detected, particularly periods of the order of 2 to 3 years and 17 months. The amplitudes of the periods are in the order of two to three times the noise level. The conclusion about a possible secular variation of the solar radius from the Picard and La Hire measurements is still a matter of study in the literature.

2.2. *Meridian circles*

One of the observational programs maintained by the Royal Observatory at Greenwich was solar. Through the measurements of instants of transit of the Sun's limbs by the local meridian, and using reference stars (Cullen 1926), it was possible to obtain a reference system used at the time. This program, which included observation of planets, the moon, and small planets, was maintained on a routine basis from 1836 to 1953. Through a combination of the Sun's transit times, it was possible to obtain a series of measurements of its diameter; the first analyzes looking for periodicity were unsuccessful (Gething 1955). On the other hand, Howse (1975) obtained a variation of 0".01/year from 1890, corroborated by Eddy & Boornazian (1979) from 1836. The measurements were challenging since difficulties imposed by constant modifications introduced since 1936 in Greenwich, and the fact measures were made with a tangency in a small part of the solar limb and a reticule. It should also keep in mind that the moment of transit was interpolated by the observer between two successive beats of a pendulum, used as a time pattern, whose methodology was replaced by stopwatches from 1854. The program conducted by many observers, made the measures strongly dependent on personal equations, translated by each individual's to define the moment of tangency at the transit. There are still transit measurements made with the primary objective of measuring the size of the Sun (Auwers 1891; Gething 1955). In these cases, the primary source of error was the Earth's atmosphere.

2.3. *Mercury transits*

Observations of the time interval that the planet Mercury takes to cross the Sun resulted in one of the most accurate solar diameter measurements to detect long-term changes. Due to the particular geometry of Earth and Mercury's orbit, the passage occurs only in May or November, at a frequency of 14 times per century. The maximum duration of central transit in May is ~8h and in November ~6h. The precision for the solar diameter is of the order of 0".1. However, due to an observation difficulty in discerning Mercury's contact instant with the Sun's limb, the average deviation of observations for each transit is typically between 0".5, and 1" on ground observations. In total, there are four instants of contact between the limbs of the Sun and Mercury. There are two contacts with Mercury for each solar limb, one internally and one externally. The contact observations (t1 and t4) in which Mercury appears ultimately outside the limb of the Sun are said to be external. The internal contacts (t2, t3) are much more defined than the external ones (Parkinson, Morrison & Stephenson (1980)) and are used to measure solar diameter variations (Fig. 1). More than 2,000 contact measures in 30 transits of Mercury, distributed over the past 250 years, were collected by Morrison and Ward (Morrison & Ward 1975a,b), and analyzed by Parkinson, Morrison & Stephenson (1980). These measures, collected mainly to determine the variations in the Earth's rotation rate and the relativistic advance of Mercury's orbit's perihelion, allow a combination of the diameters of the Sun and Mercury to be obtained, through the angular separation between these objects. Looking those transit measurements from 1723 through 1973, Parkinson, found a decrease of 0".14 ± 0".08 (in agreement with Shapiro's (1980) variation of 0."15/sec) and a periodic variation of eight years with an amplitude of 0".24 ± 0".08, besides a sub-harmonic of approximately twenty years. The long term variations are consistent with Bush, Emilio & Kuhn (2010) null result and upper limit to secular variations obtained from Michelson

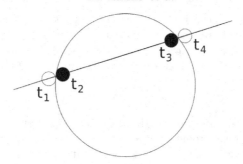

Figure 1. Planetary transit

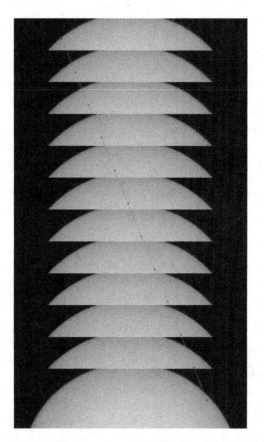

Figure 2. Composite image of Mercury transit 2003 May observed with Michelson Doppler Imager (MDI) aboard the Solar and Heliospheric Observatory with a 28-minute cadence

Doppler Imager (MDI) imagery of 0".12 per century. Sveshnikov (2002), analyzing 4500 archival contact-timings between 1631 and 1973, found that the secular decrease did not exceed 0".06 ± 0".03. Our group made the first analysis with high-quality images outside the Earth's atmosphere of Mercury transits to obtain the solar radius (Fig. 2). The value found of 960".12 ± 0."09 is consistent with earlier MDI absolute radius measurements after taking into account systematic corrections and a calibration error in the 2004 optical distortion measurements (Emilio *et al.* (2012)). Within our accuracy, no variation of the solar radius was observed over three years between the 2003 and 2006 transits.

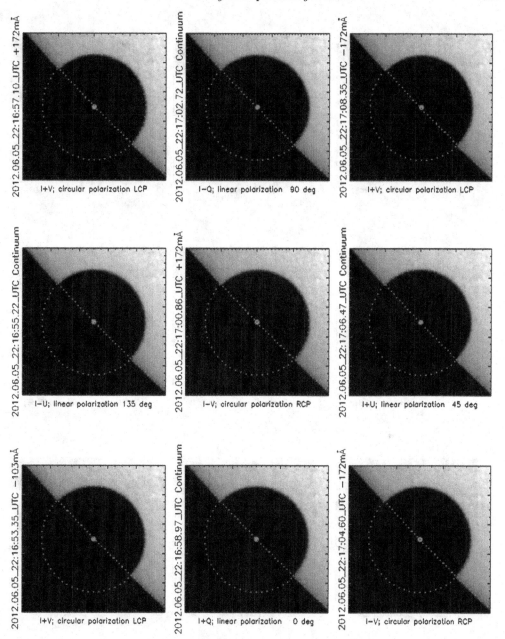

Figure 3. Venus 2012 transit ingress from HMI/SDO.

2.4. *Venus transit*

Transits of Venus occur in pairs of transits eight years apart separated by long gaps of 121.5 years and 105.5 years. They were used historically to estimate the size of the solar system. Besides Venus's apparent size being bigger than Mercury, his atmosphere brings another factor of difficulty in making precise measurements. The 2012 Venus transit was observed by the Helioseismic and Magnetic Imager (HMI) (Figs. 3 and 4) aboard the Solar Dynamics Observatory (SDO) in seven wavelengths across the Fe I absorption line at 6173 Å (Emilio *et al.* 2015). After applying a correction for the instrumental point

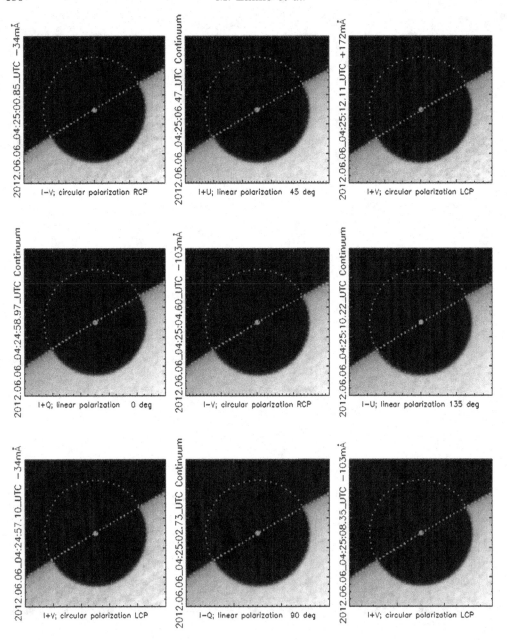

Figure 4. Venus 2012 transit egress from HMI/SDO.

spread function (PSF) of the HMI images, the value found at 1 AU was 959".57 ± 0".02 (695,946 ± 15 km). Inside the Fe *I* it was possible to measure the heights of the line formation. The difference in the solar radius determined from measurements near the line core and in the continuum wing was 0".23 (167 km).

2.5. *Solar Eclipses*

An alternative method to detect possible changes in the solar diameter comes from eclipses. The admitted accuracy of these measures, assuming the Moon's profile is known, is 0".2. In practice, the measure derives from the time interval between the Sun's light's disappearance and reappearance, seen in the Moon's irregular limb (Parkinson, Morrison & Stephenson 1980). The dataset includes the 1715 solar eclipse observed by Halley, the eclipses occurred between 1842 and 1925, and the photographs obtained in 1966. In addition to the average values of 959".63 for the Sun and 932".58 for the Moon, the corrections of the Moon's profile irregularities were taken into account (Watts 1963). The profile of the Moon, as seen from Earth, is irregular enough that it cannot be used immediately as an intermediate reference surface. It is even necessary to know the shape of the lunar limb so that we can remove from the observational residues that part due to the selenographic irregularities of the marginal zone of the Moon. For the average value of the Sun's radius, a correction of 0".22 ± 0".20 was found. Parkinson, Morrison & Stephenson (1980) still obtained, through linear regression, a secular variation of 0".08 ± 0".07. More recently Lamy *et al.* (2015) using synthetic light curves calculated from high-accuracy ephemerides and lunar-limb profiles constructed from the topographic model of the Moon provided by the Kaguya lunar space mission found the value of 959".99 ± 0".06 (696,246 ± 45 km). The value corresponds to an average of four solar eclipses between 2010 to 2015.

2.6. *Drift Scans*

In 1951, Pettit proposed an observational method for determining the solar diameter based on the monochrome photometric curves of the Sun's limb (Wittmann 1973, 1977, 1980). The experiment consists of scanning the solar disk in both directions, east-west, and north-south, using two photodiodes. The difference in signals allows the keep the telescope Zeiss, ($\phi = 5.0$ cm and $f_{eff} = 273.9$ cm) positioned with a precision of 1" while the attitude of the telescope can be changed both in straight ascension and in declination. The scan spans 208", and 6144 equidistant points characterize the measurements. Each point of intensity relative to the Sun's center is obtained photoelectrically every 63 μs. An average of every 64 readings is calculated and recorded (equivalent to about 0".06 in resolution), followed by the observation moment. The limb darkening functions are drawn for each scan, and the respective inflection points are obtained. An astrometric reduction follows to determine the diameter. Wittmann (1973) used this method and determined the diameter using a telescope at the Locarno observatory (Lat.:+46°10'41" and Lon.: −8°47'22" and altitude 2409 m). The first results of the semi-diameter found in 1972 (Wittmann (1973)) were 960".24 ± 0".16 in 5011.5 Å and 966".9± 0".4 in Hα ±5 Å. Wittmann, Alge & Bianda (1991) used yet another identical telescope using the technique described above in different locations. The telescope was a Gregory Coudé (D = 45 cm; f = 25 m). The accuracy for an isolated measurement was 1" in Izaña (latitude 46°10'40".6 N; longitude 8°47'22".9 and altitude 506 m) and 1".7 in Locarno. These values correspond to the seeing 1" of the respective sites. The average diameter found from the 1122 observations made in 1990 (472 in Izaña and 650 in Locarno) was 960".56 ± 0".04. The value is comparable to the 1773 observations made in 1981 (Wittmann, Alge & Bianda 1991) of 960.32 ± 0".02. Wittmann, Alge & Bianda (1991) found no evidence of variations in the semi-diameter greater than ± 0".3 in these observations. Wittmann, Alge & Bianda (1993) attribute a variation in phase with the solar cycle comparing the measurements made in Izaña and Locarno between 1991 and 1993 with the observations made in 1981. The variation in 10 years was 0".4. Such amplitude is twice as large as the variation found with astrolabes (Laclare *et al.* (1996); Emilio & Leister (2005)). After modifying the data acquisition system, introducing a CCD, Wittmann (1997) revised his study on the variation of the radius with the solar cycle. In other papers, Wittmann (Wittmann (1997); Wittmann & Bianda (2000)) does not attribute any variation in the solar semi-diameter higher than 0".05 with the solar cycle and considered that the variation of 0".4 was of instrumental origin to the interruptions in the observations.

2.7. *Balloon Measurements*

The Solar Disk Sextant (SDS) is an instrument developed by Sofia and collaborators to make measurements in a balloon, that allow obtaining the solar diameter, from the separation between two images of the Sun. The SDS is composed of a wedge approximately 1000" (Sofia, Heaps & Twigg (1994)), placed in front of the Cassegrain telescope objective, with a focal length of 20.5 m, producing two separate Solar images. The distance between the center of two consecutive images of the Sun produced by the instrument is given by $D = 2WF$, where W is the angle of the wedge, and F is the focal length of the telescope. Linear CCDs are placed along the solar limb, and in this way, the position of the center of the Sun's image is calculated. The solar radius is then calculated by $S = (D - d)/F$, where d is the separation between the images. The instrument is placed onboard a balloon at an altitude of 36 km. This method's advantage is that a small amount (the separation of the solar limb) is measured instead of the solar diameter. This technique increases the precision of the measurements compared to those that measure the solar diameter directly. An essential measured quantity is the separation of the images, located close to the optical axis, whose performance is optimized. The instrumental scale can be calibrated, as long as the focal distance is fixed. The distance between the two images is measured with each observation. The SDS instrument principle requires that the wedge angle remains constant. In this way, it is possible to separate the instrumental effects from variations in the solar diameter. The telescope can rotate around its axis, allowing observation at different heliographic latitudes. The deviation from the mean for any measurement is 0".2, and the instrument's sensitivity is 1 to 2 mas. Sofia, Heaps & Twigg (1994) found no significant variation in the solar radius. Therefore Sofia *et al.* (2013) found a variation of 200 ± 20 mas through 1992 to 2011, not in phase with the solar activity cycle.

2.8. *Santa Catalina Laboratories for experimental Relativity by Astrometry (SCLERA)*

SCLERA is a photometric technique that Brown and collaborators proposed using a modified meridian circle (Brown, Stebbins & Hill (1978); Brown & Christensen-Dalsgaard (1998)). From August 1981 to December 1986, an observational campaign was carried out with this instrument mounted in the mountains of Santa Carolina, north of Tucson (USA), on a site at 2609 meters above sea level. Through a filter centered close to 800 nm and with an amplitude of 10 nm, the Sun's horizontal diameter was obtained by combining the transit moments of the limbs by a series of linear CCD detectors. The scan was performed with a frequency of 32 Hz. A real-time algorithm to find the edge is used, and several other quantities are measured together with seeing and parameters to calculate refraction. The algorithm used to determine the limb was the finite Fourier transform (FFTD) described by Hill & Stebbins (1975). The process involves converting LDFs with theoretical curves. Theoretical curves are made up of a set of non-zero weighting functions only for a specific window of length a. The length of the window determines how much of the solar limb's darkening curve is involved in defining the edge (Hill & Stebbins (1975)). The edge is then defined as the center of that window in which the convolution is canceled. The FFTD has two crucial characteristics: The first is to eliminate the first-order contribution of seeing in determining the limb for a given length of the scattering point function. The position of the tip points of the limbs is highly sensitive to the variability of the seeing for each day; the second characteristic is that the FFTD depends on a free parameter called window length a.

Brown & Christensen-Dalsgaard (1998) found the solar diameter to be 1919".359 $\pm$ 0".018 at one astronomical unit using 550 measurements. The authors argue that this is not the correct diameter but an observational quantity constructed in a way independent of the vertical temperature gradient of the upper photosphere. The value obtained depends on the radiation transfer in the solar atmosphere and on the behavior of the FFTD limb definition. Physical models of the solar atmosphere are applied to obtain the diameter correction and calculate the intensity as a function of the distance from the solar center, and the brightness distribution profile identified with the edge by the FTTD. After applying two models to the solar atmosphere and correcting the diameter value, the authors obtained an average value of 958".96 ± 0".04. The authors found no significant variation in the solar radius over time.

2.9. *Solar diameter measurements in the spectral line of neutral iron 525 nm*

The Sun's diameter was also measured in the line of the neutral iron 525 nm (Ulrich & Bertello (1995)). A telescope located in Monte Wilson (USA) was used to make the measurements. The Sun's apparent radius was defined as the average distance between the image center and the point where its intensity falls at 25% of its value. A portion of the Sun's image is directed to a spectrograph in which the position of the image determined by a guide system placed close to the focal plane. A magnetogram is constructed by scanning the image over the spectrograph entry opening in alternating directions, with the scanning direction being adjusted, per day, perpendicular to the solar polar axis. The image is initially positioned randomly away from the poles, and successive scans construct each magnetogram in the main direction. Each scan line begins and ends at a fixed distance, away from the solar disk. The intensity of the disk is used to determine the scattering of light. During each scan, the acquisition system reads the intensity and the circular polarization in two spectral bands. An automatic control also keeps the opposite wings of the spectral line at 525.0 nm illuminated. The limb's position is determined during the reduction process and differs from that found in the visible. The reason is the neutral iron line formed close to the limb position where the temperature is minimal instead of the photosphere. The presence of faculae from active regions will influence the determination of the limb defined in this way. There are other corrections due to the effects of light scattering and atmospheric refraction. After these corrections, the residuals of the medium radius are obtained. Ulrich & Bertello (1995) measurements of the solar ray were made between 1982 and 1994. The residues show a direct correlation with magnetic activity with an amplitude of approximately 0".2. Since the iron 525 nm line formed in a high height of the solar atmosphere, the variation found is probably a solar atmosphere's change due to the magnetic activity.

2.10. *Helioseismic Radius*

The f modes propagate mostly on the surface, and their frequencies are independent of the stratification of the solar interior. The f modes depend mainly on global factors such as mass and radius. With precise measurements of frequencies in mode f, it is possible to determine the solar radius (Schou *et al.* (1997); Tripathy & Antia (1999)). The dispersion relation of the f modes given by (Tripathy & Antia 1999) is:

$$w^2 \sim gk = GM \frac{\sqrt{l(l+1)}}{R^3}$$

Where:
g is the acceleration of gravity on the surface;
k is horizontal wave number;
G is the gravitational constant;
l is the degree of the mode;

R is the Solar radius;

M is the Solar mass.

In practice, there are significant differences in the frequencies of this asymptotic estimate. The reason is that these modes have their maximum amplitudes in layers just below the solar surface, which corresponds to a smaller radius. Moreover, this is because the speed associated with self-functions drops exponentially with increasing depth, and density increases quickly. As a result, the density of kinetic energy increases, and the height scale of density becomes comparable to the height scale of speed. Antia (1998) estimated the solar ray through the f modes using measurements made by the GONG network at 959".34 ± 0".01. Dziembowski *et al.* (2000) found no variation in the heliossismological radius correlated with the number of spots, using data from the MDI-SOHO. However, Antia *et al.* (2000) using measurements from the GONG network between 1995 and 1998 and found changes in frequencies of mode f with solar activity. A new analysis was made by Dziembowski *et al.* (2001). This time they took into account the complete data from MDI-SOHO since 1996, whose results did not confirm the correlation with solar activity.

2.11. *Astrolabes*

The most significant disadvantage is that the star catalogs made with the astrolabe are not absolute because it is not possible to fix the equinox's position and the equator of the reference frame. Classically, the orbital parameters of the Earth contribute to obtaining the spatial orientation of the reference system, in addition to the observations of planets, those of the Sun, whose attempts until 1973 (Benevides *et al.* (1979)), had not been made due to the impossibility of knowing the instantaneous zenith distance , due to the variability of the transmission prism angle. In parallel, in 1974, Laclare (1975) modified the CERGA astrolabe with the same objective. An equilateral prism was replaced by Vitro-ceramic prisms, with low dilation, allowing observation at various zenith distances (Laclare (1983)). The solar semi-diameter was a secondary measurement. The importance of this measure grew over time until it became the main objective of solar astrolabes. One of the astrolabe advantages, compared with the meridian circle, is that the instrument allows the Sun's observation twice a day with a single prism (once in the east and once in the west). The astrolabe had a unique advantage for observing the solar diameter. Its measurements are not affected by errors in atmospheric refraction (the error is second-order). An error in determining a limb's position in a given zenith distance cancels the error of the opposite limb since the radius is found by the difference between the zenith distances of the upper and lower limb of the solar disk. The only error source is caused by a change in atmospheric properties between the two contacts. However, the measure is subject to errors due to seeing and the definition of the inflection point.

2.12. *Satellites*

Ground-based measurements limit the solar radius observations by seeing effects. Also, satellites allow observing continually with no night/day interruptions. It provides the most accurate measures of variation (if they exist) and the absolute value of the solar radius (with planetary transit observations). The first to make those observations was the Michelson Doppler Imager (MDI) instrument aboard the Solar and Heliospheric Observatory (SOHO) satellite (Emilio *et al.* (2000); Kuhn *et al.* (2014); Bush, Emilio & Kuhn (2010)). The Solar Diameter Imager and Surface Mapper (SODISM) onboard the Picard space mission was a dedicated instrument to measure the solar radius in five narrow bandpasses (Meftah *et al.* (2014)). MDI/SOHO found that fundamental changes in the solar radius synchronous with the sunspot cycle must be smaller than 23 mas peak to peak, and the average solar radius must not be changing (on average) by more than

Table 1. This table shows the difference between modern measurements of the solar radius and the IAU resolution B3 definition.

Reference	Date	Method	$R_\odot$(km)	1 σ error	Difference from IAU B3 resolution (km)
Emilio *et al.* (2012)	2003,2006	Mercury transits (MDI/SOHO)	696,345	65	645
Hauchecorne *et al.* (2014)	2012	Venus transits (SODISM)	696,149	138	449
Emilio *et al.* (2015)	2012	Venus transits (HMI/SDO)	695,946	15	246
Lamy *et al.* (2015)	2010 to 2014	Solar Eclipses	696,246	45	546

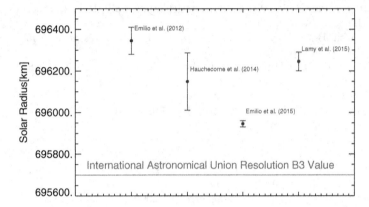

Figure 5. Modern measurements for the solar radius obtained from planetary transits and solar eclipses compared with the value adopt by the IAU B3 resolution.

1.2 mas yr^{-1} (Bush, Emilio & Kuhn (2010)). From PICARD (Meftah *et al.* (2015)), the changes in solar radius amplitudes were less than ± 20 mas (± 14.5 km) for the years 2010–2011 and not correlated with the solar cycle activity.

3. International Astronomical Union Resolution B3

Resolution B3 of the International Astronomical Union (IAU) defined the Solar radius as 695,700 km in 2015. This value is consistent with helioseismic determinations of the solar radius but not consistent with the most accurate measurements of the photospheric solar radius. Solar radius determined from helioseismic data is located below the photosphere. Table 1 shows some of the most modern measurements of the solar radius and Fig. 5 shows a plot of those values with the B3 resolution.

4. Discussion

Haberreiter, Schmutz & Kosovichev (2008) calculated the intensity profile of the limb the MDI continuum and the continuum for two atmosphere structures and compared the position of the inflection point with the radius at $\tau_{5000} = 1$ ($\tau_{Ross} = 2/3$). The difference between the seismic radius and the radius defined by the inflection point is 347 ± 6 km. This difference is consistent with some of the most recent measurements and IAU B3 definition of the solar radius found in table 1. The inflection point definition is closest to the adopt value used for evolutionary models defining stars' age and temperature where $\tau_{Ross} = 2/3$. Also, the inflection point definition is used for most of the experiments that measure the solar size. IAU B3 resolution for the solar radius must be raised by 300 km to agree with the solar photosphere's observations. Satellites measurements agreed that the upper limit of any variation of the solar radius (if any) is not bigger than 15 km with solar cycle (Bush, Emilio & Kuhn (2010); Meftah *et al.* (2015)), what is one order magnitude smaller than the 300 km IAU B3 resolution difference.

Acknowledgments

This work was partially supported by Fundação Araucária (Ed. 12/2018) and Conselho Nacional de Pesquisa (CNPq) 308871/2016-2.

References

Ambronn, L. & Schur, A. C. W. 1905, *Astronomische Mittheilungen der Koeniglichen Sternwarte zu Goettingen*, Part 7. T.: Druck der Dieterich'schen Univ.-Buchdruckerei (W. Fr., 126 p.)

Antia, H. M. 1998, *A&A*, 330, 336

Antia, H. M., Basu, B., Pintar, J., Pohl, B. *et al.* 2000, *Solar Physics*, 192, 459

Auwers, A. 1891, *Astron. Nachr.*, 128, 361

Benevides, P., Boczko, R., Clauzet, L. B. F., Leister, N. V. *et al.* 1979 *A&AS*, 36, 401

Brown, T. M., Stebbins, R. T., & Hill, H. A. 1978, *AJ*, 223, 324

Brown, T. M. & Christensen-Dalsgaard, J. 1998, *ApJ*, 500, L195

Bush, R. I., Emilio, M., & Kuhn, J. R. 2010, *ApJ*, 716 , 1381

Cullen, R. T. 1926, *MNRAS*, 86, 344

Dziembowski W. A., Goode P. R., Kosovichev A. G., Schou J. 2000, *Astrophysical Journal*, 537, 1026

Dziembowski W. A., Goode P. R., Schou J. 2001, *Astrophysical Journal*, 553, 897

Eddy J. A. & Boornazian A. A. 1979, *Bull. Am. Astr. Soc.*, 11, 437

Emilio, M., Kuhn, J. R., Bush, R. I., & Scherrer, P. 2000, *ApJ*, 543, 1007

Emilio, M. & Leister, N. V. 2005, *MNRAS*, 361, 1005

Emilio, M., Kuhn, J. R., Bush, R. I., & Scholl, I. F. 2012, *ApJ*, 750, 135

Emilio, M., Couvidat, S., Bush, R. I., Kuhn, J. R., & Scholl, I. F. 2015, *ApJ*, 798, 48

Gething, P. J. D. 1955, *MNRAS*, 115, 558

Gilliland, J. D. 1981, *Astrophys. J.*, 248, 1144

Haberreiter, M., Schmutz, W. & Kosovichev, A. G. 2008, *ApJ*, 675, L53

Hauchecorne, A., Meftah, M., Irbah, A., Couvidat, A., Bush, R., & Hochedez, J.-F. 2014, *ApJ*, 783, 127

Hill, H. A. & Stebbins, R. T. 1975, *AJ*, 200, 471

Howse, D., 1975, *Greenwich Observatory: The Royal Observatory at Greenwich and Herstmonceux 1675-1975*, Volume 3: The Buildings and Instruments, Taylor & Francis, London, 92

Kuhn, J. R.,Bush, R. I., Emilio, M. & Scherrer, P. H. 2014, *ApJ* 613, 1241

Laclare, F. 1975, *C. R. Acad. Sci. Paris*, 280, 13

Laclare, F. 1983, *A&A*, 125, 200

Laclare, F., Delmas, C., Coin, J. P., Irbah, A. *et al.* 1996, *Solar Physics*, 166, 211

Lamy, P., Prado, J.-Y., Floyd, O., Rocher, P., Faury, G. & Koutchmy, S. 2015, *Solar Physics*, 290, 2617

Meftah, M., Hochedez, J.-F., Irbah, A., Hauchecorne, A., Boumier, P., Corbard, T., Turck-Chièze, S., Abbaki, S., Assus, P., Bertran, E., Bourget, P., Buisson, F., Chaigneau, M., Damé, L., Djafer, D., Dufour, C., Etcheto, P., Ferrero, P., Hersé, M., Marcovici, J.-P. Meissonnier, M., Morand, F., Poiet, G., Prado, J.-Y., Renaud, C., Rouanet, N., Rouzé, M., Salabert, D., Vieau, A.-J., *et al.* 2014, *Solar Physics*, 289, 1043

Meftah, M., Hauchecorne, A., Irbah, A., Corbard, T., Ikhlef, R., Morand, F., Renaud, C., Riguet, F., Pradal, F., *et al.* 2015 *ApJ* 808, 4

Morrison, L. V. & Ward, C. G. 1975a, *R. Gr. Obs. Bull.*, 181, 359

Morrison, L. V. & Ward, C. G. 1975b, *MNRAS*, 173, 183

Morrison, L., Stephenson, R. & Parkinson, J. 1988, *Nature*, 331,421

Parkinson, J. H., Morrison, L. V., Stephenson, F. R. 1980, *Nature*, 288, 548

Ribes, E., Ribes, J. C., & Barthalot, R. 1987, *Nature*, 326, 52

Ribes, E., Merlin, Ph., Ribes, J. C., Barthlot R., *et al.* 1989, *Annales Geophysicae*, 7, 321

Ribes, E., Beardsley, B., Brown, T. M., Delache., Ph., Laclare, F., Leister, N. V., *et al.* 1991, "The Sun in Time", Univ. of Arizona Press, *Space Sciencie Series*, 59

Sánchez, M., Parra, F., Soler, M., & Soto, R. 1995, *A&AS*, 110, 351

Schou, J., Kosovichev, A. G., Goode, P. R., Dziembowski, W. A., *et al.* 1997, *AJ*, 489, L197

Sofia, S., Heaps, W., & Twigg, L. W. 1994, *AJ* 427, 1048

Sofia, S., Girard, T. M., Sofia, U. J., Twigg, L., Heaps, W., Thuillier, G., *et al.* 2013 *MNRAS* 436, 2151

Sveshnikov, M. L. 2002 *AstL*, 28, 115S

Tripathy, S. C. & Antia, H. M. 1999, *Solar Physics*, 186, 1

Toulmonde, M.(1997) *Astron. Astrophys.*, 325, 1177

Ulrich, R. K. & Bertello, L. 1995, *Nature*, 377, 214

Watts, C. B. 1963, *American Ephemeris*, 17, 1

Wittmann, A. D. 1973, *Solar Physics*, 29, 333

Wittmann, A. D. 1977, *A&A*, 61, 225

Wittmann, A. D. 1980, *A&A*, 83, 312

Wittmann, A. D., Alge, E., & Bianda, M. 1991, *Solar Physics*, 135, 243

Wittmann, A. D., Alge, E., & Bianda, M. 1993, *Solar Physics*, 145, 205

Wittmann A. D. 1997, *Solar Physics*, 171, 231

Wittmann, A. D. & Bianda, M. 2000, *ESASP*, 463, 113

Wittmann, A. D. 2003, *AN*, 324, 378

Discussion

ALEXANDER KOSOVICHEV: What Can you say about the shape of the Sun?

MARCELO EMILIO: The Helioseismic and Magnetic Imager (HMI) instrument on the Solar Dynamics Observatory (SDO) spacecraft has been making periodic solar shape measurements every six months since 2011. Separate the shape signal from brightness variations in the photosphere is very difficult. Our analysis shows that the Sun's oblate shape is distinctly constant and almost entirely unaffected by the solar-cycle variability. The nominal value found by our group for the solar oblateness is significantly lower than theoretical expectations. A slower differential rotation could explain this in the outer few percents of the Sun. The higher-order (hexadecapole) term is consistent with 0.

Author index

Acosta, B. – 363
Adelman, S. J. – 435
Ahuir, J. – 295
Aigrain, S. – 200
Alvarado-Gómez, J. D. – 407
Arlt, R. – 134
Astoul, A. – 195
Augustson, K. C. – 195
Azulay, R. – 189

Balthasar, H. – 38, 53, 58
Baruteau, C. – 195
Beasley, A. – 418
Benbakoura, M. – 295
Bisikalo, D. V – 268
Bolmont, E. – 195, 295
Bourrier, V. – 305
Bouvier, J. – 200
Bouy, H. – 200
Brahm, R. – 300
Brandenburg, A – 169
Broomhall, A.-M. – 94
Brown, A. – 377
Brun, A. S. – 138, 195, 215, 295
Bush, R. – 481

Campos Rozo, J. I. – 38
Cang, T. – 181
Caplan, R. M. – 3
Castellanos Durán, J. S. – 454
Cauley, P. W. – 467
Chen, L. – 17
Chen, X. – 17
Cheung, M. C. M. – 426
Chian, A. C.-L. – 351
Chou, D.-Y. – 160
Climent, J. B. – 189
Cohen, O. – 407

Davis, S. – 367
de la Cruz Rodriguez, J. – 24
Denker, C. – 38, 53, 473
DeRosa, M. L. – 426
Diercke, A. – 38
Dineva, E. – 473
Djurašević, G. – 207
Donati, J.-F. – 181
Downs, C. – 3
Doyle, J. G. – 384
Doyle, L. – 384
Drake, J. J. – 407

Dudík, J. – 414
Dukes, R. J. – 435
Dzifčáková, E. – 414

Echeverría, S. – 371
Elstner, D. – 134
Emilio, M. – 481
Ermolli, I. – 448
Espinoza, N. – 300
Esquivel, A. – 280
Estrela, R. – 461

Fares, R. – 305
Felipe, T. – 53
Folsom, C. P. – 181
Fossati, L. – 280
Fournier, A. – 138
Fournier, Y. – 134
France, K. – 426
Fraschetti, F. – 407

Gallet, F. – 195
Gallo-Méndez, I. – 367
Garcés, J. – 120, 207
Garraffo, C. – 407
Giorgi, F. – 448
Gömöry, P. – 38, 58
González Álvarez, E. – 355
González Manrique, S. J. – 38, 53, 58
Güdel, M. – 313
Guerrero, G. – 65
Guglielmino, S. L. – 439, 448
Guirado, J. C. – 189

Harvey, J. W. – 42
Hazra, S. – 138
Helling, C. – 305
Hełminiak, K. G. – 300
Hofmeister, S. – 38
Hou, Y. – 443
Huang, G.-H. – 157, 228
Hummel, C. A. – 418
Hung, C. P. – 138

Ilyin, I. – 473
Inoue, J. L. – 3
Irwin, J. – 200

Jafarzadeh, S. – 24, 448
Jardine, M. – 305
Jian, L. K. 127

Jin, C. – 123
Jin, M. – 426
Jordán, A. – 300
Jouve, L. – 138, 185

Kavanagh, R. D. – 305
Kay, C. – 421
Kiefer, R. – 94
Kilcik, A. – 232
Kitiashvili, I. N. – 86, 107, 147, 346
Kontogiannis, I. – 38, 53
Koskinen, T. – 280
Kosovichev, A. G. – 107, 232, 346
Kowalski, A. – 426
Krikova, K. – 38
Kuckein, C. – 38, 53, 58
Kuhn, J. – 481
Küker, M. – 116

Lachaume, R. – 200
Lagg, A. – 454
Lagrange, A.-M. – 286
Le Poncin-Lafitte, C. – 295
Lee, L.-C. – 157, 228
Li, T. – 443
Lin, C.-H. – 157, 160, 228
Linker, J. A. – 3
Liu, F. – 17
Llama, J. – 305
Lockwood, C. A. – 3
Lörinčík, J. – 414
Luhmann, J. G. – 127, 241

Machado Pereira, E. – 211
Maldonado, J. – 355
Mansour, N. N. – 346
Marcaide, J. M. – 189
Mason, J. P. – 426
Mathis, S. – 195, 295
McLaughlin, J. A. – 384
Meadors, E. N. – 3
Melnikov, V. – 17
Menezes, F. – 421
Mennickent, R. E. – 120, 207
Meunier, N. – 286
Milic, I. – 473
Miranda, R. A. – 351
Moraux, E. – 200
Moschou, S. P. – 407
Moutou, C. – 305
Moya, P. S. – 224, 363, 367, 371
Murabito, M. – 439, 448

Nagaraju, K. – 46
Netto, Y. – 421
Nitta, N. V. – 426

Ó Fionnagáin, D. – 305
Oláh, K. – 116
Opher, M. – 421
Osten, R. – 426
Otero, S. – 207

Palacios Hernández, J. – 38
Palit, S. – 461
Pasachoff, J. M. – 3
Pastén, D. – 363, 371
Perger, M. – 355
Perri, B. – 215
Petit, P. – 181
Pinto, V. A. – 224
Pyper, D. M. – 435

Ramsay, G. – 384
Rangarajan, K. E. – 46
Reardon, K. P. – 3
Rempel, E. L. – 351
Réville, V. – 215, 295
Reyes, P. – 224
Rocha Pinto, H. J. – 211
Rojas, G. – 207
Rojo, P. – 3
Romano, P. – 439, 448
Roquette, J. – 200
Rozelot, J. P. – 232
Rüdiger, G. – 116
Russell, C. T. – 127
Rybák, J. – 58

Sadykov, V. – 346
Sankarasubramanian, K. – 46
Scandariato, G. – 355
Schleicher, D. – 207
Schneider, G. – 3
Scholl, I. – 481
Schrijver, C. J. – 426
Schwartz, P. – 58
Scullion, E. – 384
Seaton, D. B. – 3
Sgró, M. A. – 280
Silva, S. S. A. – 351
Sittler Jr., E. C. – 333
Sittler, L. M. – 333
Sliski, A. – 3
Sliski, D. – 3
Socas-Navarro, H. – 448
Solanki, S. K. – 454
Song, Q. – 452
Sowmya, K. – 454
Stangalini, M. – 448
Sterling, A. C. – 3
Strassmeier, K. G. – 116, 473
Strugarek, A. – 138, 195, 215, 295

Suárez Mascareño, A. – 355
Szydlarski, M. – 24

Talagrand, O. – 138
Tamburrini C, A. – 367
Tan, B. – 17

Utz, D. – 38

Valio, A. – 421, 461
Verma, M. – 38, 53, 473
Vidotto, A. A. – 259, 280, 305
Villarreal D'Angelo, C. – 280
Villegas, F. – 120
Voulgaris, A. – 3

Wang, J.-S. – 452
Wang, W. – 17
Wedemeyer, S. – 24

Wheatley, P. J. – 305
Wray, A. A. – 86, 346
Wu, K. – 384
Wyper, P. F. – 384

Xu, Z. – 58

Yadav, R. K. – 407
Yan, Y. – 17
Yang, S. – 392, 443, 452
Young, P. R. – 439
Yuan, S. – 58
Yüce, K. – 435

Zaire, B. – 185
Zemanová, A. – 414
Zhang, J. – 443
Zhilkin, A. G. – 268
Zuccarello, F. – 439, 448

Printed in the United States
by Baker & Taylor Publisher Services